Concrete Petrography

Concrete Petrography

A handbook of investigative techniques

D. A. St John, A. B. Poole and I. Sims

A member of the Hodder Headline Group
LONDON • SYDNEY • AUCKLAND

Copublished in North, Central and South America
by John Wiley & Sons Inc.
New York • Toronto

First published in Great Britain 1998 by Arnold
a member of the Hodder Headline Group,
338 Euston Road, London, NW1 3BH
http:\\www.arnoldpublishers.com

Copublished North, Central and South America by John Wiley & Sons, Inc.
605 Third Avenue, New York, NY 10158

British Library Cataloguing in Publication Data
A catalogue record for this book is available from the British Library

Library of Congress Cataloging-in-Publication Data
A catalog record for this book is available from the Library of Congress

ISBN: 0 340 69266 9
ISBN: 0 470 23772 4

Printed in Great Britain by Bookcraft (Bath) Ltd, Somerset

Contents

Foreword

For many years the writer has hoped for the appearance of an authoritative and detailed, yet clear treatment of the special features associated with petrographic examination of concrete. This remarkable volume fulfils that hope to an admirable degree.

Concrete is in essence a man-made rock, with the special characteristics that stem from its hybrid character – that is, it is a conglomerate of natural rock grains (sand and coarse aggregate) encapsulated in a continuous matrix of hydrated Portland cement. Accordingly, it was to be expected that petrographers, trained in the microscopic examination of rock materials, would take a leading role elucidating the characteristics of concrete, not only the rock components but also the details of the hardened cement paste. Indeed, concrete petrographers have been extremely active in establishing the characteristics of 'healthy' concretes and in investigating the many pathologies that lead to premature deterioration and failure of concretes exposed to aggressive environments in service.

Despite the importance of concrete petrographic examination, there has been no unified, book-length exposition to date of the special petrographic procedures and the special interpretations that need to be made when investigating concrete rather than natural rock. Indeed, most concrete petrographers are rock and mineral petrographers who have moved into the field of concrete petrography without much specific training in concrete petrography *per se*. Accordingly, despite efforts to standardise concrete petrographic investigations, for example, as specified in ASTM Designation C856, 'Standard Practice for Petrographic Examination of Hardened Concrete', the results of investigations carried out by different practitioners are sometimes of uneven quality and reliability, especially with regard to the hardened cement paste features. The appearance of this book cannot help but improve this situation.

The authors of this volume have each devoted much of their professional lives to petrographic investigations of concrete. Each is an internationally recognised leader in the field with vast experience and many publications and technical presentations. The writer has been fortunate to have been a personal friend of each of them for many years, and has himself derived much benefit from contacts and consultations with them.

The volume itself takes the unskilled reader through an excellent introduction to concrete, and to petrographic examination of concretes. The importance of visual inspection and *in situ* examination, hopefully by the concrete petrographer, of the field structures from which concrete samples are derived is emphasised. This important step in the investigative process is too often neglected or ignored, sometimes leading to misleading or disastrous interpretations.

Preliminary chapters provide an intensive treatment of the types and uses of petrographic microscopes, of the place of photography in concrete microscopy, and of other methods of investigation including scanning electron microscopy and spectroscopic thermal analysis, and chemical analysis methods. They also give authoritative guidance on sampling and specimen preparation. I know of no existing source for such important, but often neglected, details.

The heart of the book is the wealth of information provided in the specific chapters on the various cement paste and aggregate components, on the textural appearance and analysis of concretes, on the details of petrographic examination of deteriorated concretes afflicted with various durability problems, and on the special characteristics of concrete products used for various non-structural purposes. The treatments are clear, comprehensive, and extremely authoritative.

The publishers are to be congratulated on the extensive and indeed almost unprecedented number of colour micrographs reproduced in this volume. These provide for the conveyance of essential details to the reader that could be accomplished in no other way, and do much to increase the effectiveness of the authors in communicating the essence of their findings in the various special areas of investigation.

It is clear from the quality of the details in the visual reproductions and figures that the authors are superb technical petrographers. It is also clear from the authoritative character of the text and the many apt citations of the literature that the authors are indeed scholars. Finally, it is clear from the quality of the prose that the authors are also excellent practitioners in the art of effective communication in the English language. It is a great pleasure to find the results of these several skills combined in one place.

I believe that this volume will stand for many years as the international standard and definitive source of information on the petrography of concrete, and congratulate the authors on their accomplishment.

Sidney Diamond
Purdue University
Indiana, USA

Preface

The authors embarked on the preparation of this book some years ago without realising the magnitude of the task they had set themselves. The original idea stemmed from the realisation that although petrography is now widely acknowledged as a tool for the investigation of hardened concrete, there are no modern texts on the subject. Older books covering the general areas of cements, clinkers, concretes and related materials may have become outdated and most are long out of print. The authors, therefore, have had to gather information from a wide range of published and unpublished sources as well as drawing on their own research material and experience.

The petrological microscope has been used for many years by geologists seeking to understand the mineralogy and genesis of rocks, and excellent textbooks are available to assist the petrographer in these studies. However, the application of microscopy to the study of concretes and related materials has a much shorter history, and it is only in recent years that the petrological microscope and the value of petrological methods in the study of hardened cementitious materials have been recognised.

The authors have attempted to provide a well-referenced, modern, comprehensive and balanced review of concrete petrography. It is hoped that the book will be of value to the technologist and engineer who may have to address materials problems with concretes or similar materials and decide on the most effective means of investigation. The book is intended primarily to provide petrographers with both the background information and some of the specialist techniques necessary for the petrographic examination of cementitious materials. Although, in general terms, cements and concretes are composed of crystalline and amorphous mineral constituents with similar optical properties to the rock-forming mineral materials familiar to all petrographers, differences arising from the method of manufacture, the chemistry, the time-dependent factors and environmental effects make the petrographic study of these materials different both in technical detail and in philosophy of approach.

In addition to reviewing the specialist techniques appropriate to the microscopical examination of cement-based materials, this book also addresses the practical problems and techniques associated with examination and sampling of concrete structures, and the special techniques appropriate to the preparation of the material for examination. It also briefly considers complementary petrological techniques which often provide invaluable additional information relating to the microstructure and compositions of these materials.

The principal area covered in this book is concerned with the use of the petrological polarising microscope for the detailed study of concretes and related materials. It is anticipated that the petrographer is already familiar with the use of the microscope and with the examination of rocks in petrographic thin-section so that the emphasis here is given to the cementitious phases and artificial materials rather than the natural aggregates and fillers that are used, except where they directly modify the chemical or physical properties of the composite material. Where appropriate, case study examples are included in the text to illustrate particular points, or to show the interrelationships between observed petrographic features, the construction history and the local environment of the structure.

It is anticipated that many petrographers will use this book as a reference source in relation to a particular problem, to identify a particular mineral or form of deterioration. Therefore, an extensive index and a glossary of minerals are included and references will direct the reader to more detailed sources of information where appropriate.

In compiling a book of this kind the authors have sought and received advice from sources too numerous to name. Many friends and colleagues around the world have given their advice and help in generous amounts, without which this book would not have been possible, and we owe them a great debt of gratitude.

Where materials such as photographs and diagrams have been made available to us for inclusion, acknowledgement has been made at the appropriate point in the text. However, the authors would like to record their particular thanks to: Eliane Wigzell, of Arnold, Ian Francis, Chris Leeding and Anna Faherty for their assistance and patience in getting this book to publication. To Norah Crammond of BRE, William French of GRS Ltd., Mike Walker of the Concrete Society, Martin Grove, Geoff Long, Kelvin Pettifer, Panos Sotiropoulos, Barry Hunt, Gunnar Idorn, Peter Fookes and Laurence Collis (now retired) for their technical help, professional advice, guidance and friendship over many years. To Kathy Abbot of Queen Mary and Westfield College and Rona Clarke of Industrial Research Ltd for their help in obtaining references, to Rebekah Chew and Gillian Tearle for help with the typing, and to many other colleagues and friends in the Geomaterials Unit at Queen Mary and Westfield College, at STATS and at the Chemistry Division DSIR, the Crown Research Institute, Industrial Research Ltd, New Zealand. Don St John also acknowledges a research grant from the New Zealand Foundation for Research, Science and Technology which enabled him to concentrate on two chapters of the book.

Finally we all owe a considerable debt of gratitude to our wives and families for their understanding and patience during the long periods of family neglect while we worked on our sections of this book.

Notes

1 Note the use of the internationally accepted cement chemists' shorthand notation: $C = CaO$, $S = SiO_2$, $A = Al_2O_3$ and $F = Fe_2O_3$, so that, for example, $C_3S = 3CaO.SiO_2$. This notation is included for individual phases in the glossary.
2 Some petrographers prefer to use the terms 'alite' and 'belite' instead of the chemical terms C_3S and C_2S respectively.

A GUIDE TO METHODS OF STUDY FOR CEMENTS AND CONCRETES

Method	Cement					Cement replacements & admixtures					Mortars & concretes			Mix proportions			Types of deterioration									
	Ordinary Portland cement	Rapid-hardening Portland cement	Sulfate-resisting cement	White Portland cement	Calcium aluminate cements	Granulated blast furnace slag	Pulverised fuel ash	Natural pozzolana	Silica fume	Chemical admixtures	Cement type (hydrated)	Aggregate types/proportions	Air voids	Aggregate/cement ratio	Coarse/fine ratio	Water/cement ratio	Chemical attack	Alkali-aggregate reaction	Carbonation	Aggressive leaching	Sulfate attack	Reinforcement corrosion	Shrinkage	Frost action	Fire damage	Quality of workmanship
Visual inspection	1	1	1	1	1						2	2					1	1		1	1	1	1	1	1	1
Binocular microscopy			1	1	1													1							1	2
Polished slice microscopy			1	1	1													1							1	
Thin-section microscopy	1	1	1	1	1	1	1	1	1	0	1	1	2	1	1	1	1	1	1	1	1	2	2	2	1	1
Reflected light microscopy	1	1	1	1	1	1	1	1	1		1															
Fluorescence microscopy	1	1	1	1	1	1																				
Etch & staining methods	1	2	1	1	2	1	1	1	1		1	2					1	1	1		2					
Scanning electron microscopy	2	2	2	2	2	1	1	1	1		1					1	1			2	1	2				
Electron probe microanalysis	2	2	2	2	2	1	1	1	1		1					1	1	1			1					
Physical methods	2	2	2	2	2	2	2	2	2	0	2			2		2	2	2	1	2	2	1	2	2	2	2
Chemical methods	2	2	2	2	2	2	2	2	2		2			2			1				2	2				
X-ray diffraction analysis																			2		2			2	2	
Infrared absorption analysis									0	0							2		2		2					
Thermal methods	2		2		2						2							2							2	

1 = Principal methods   2 = Alternative or complementary methods.   0 = Analysis difficult, but best available method.

1

Introduction

1.1 Historical background

The cementitious properties of slaked lime when exposed to the atmosphere were known to the ancient Egyptians whilst the Greeks and Romans had developed the use of pozzolana with lime sufficiently to produce the earliest examples of concrete structures, some of which still stand today (Fig. 1.1).

The importance of argillaceous impurities in limestone in the production of hydraulic lime was recognised by Smeaton when working on the construction of the new Eddystone Lighthouse commissioned in 1756. His systematic work was perhaps the first scientific study of cementitious materials. Development of these materials continued into the nineteenth century with the investigation of the calcination of wet milled mixes of chalk and clay by Vicat (1837), with J. Parker's 'Roman Cement' patented in 1796 and James Frost's patent of 1811. However, the development of an essentially modern prototype Portland cement is usually identified with the patent taken out by Joseph Aspdin in 1824. For further details, the early history of the development of Portland cement in England has been discussed by Gooding and Halstead (1952).

Although the compound optical microscope was developed early in the renaissance and demonstrated in London in 1621 by Cornellius Drebbel and the 'Micrographia' was published by Robert Hooke in 1665. The polarising microscope itself was first designed as a 'chemical microscope' for examining crystals by J. Lawrence-Smith, constructed by A. Nachet of Paris and exhibited at the Great Exhibition on 1857. A mineralogist named S. Highley realised the potential of the polarising microscope for mineralogical examination and had an instrument constructed in England to Nachet's design. He published the first description of a petrographic microscope in 1856. By 1880 several manufacturers were producing petrographic microscopes for routine mineralogical use.

The petrographic microscope was developed originally for the identification and study of rock forming minerals but early workers soon established the importance of the petrographic microscope for the investigation of artificial mineral materials, such as cement clinkers. Early studies of fine grained artificial materials were hampered by limitations of resolution, but developments in lens design during the twentieth century have led to an increasing use of the petrographic microscope and there is now a very extensive literature on the microscopic examination of materials with dimensions of 10 μm and smaller. The theoretical limit for resolution with an optical microscope is of the order of 1.0 μm, but for practical purposes 10.0 μm is a more reasonable limit, whilst the thickness of petrographic

Fig. 1.1 The Pont du Garde aqueduct, France, constructed by the Romans using masonry and lime pozzolana concrete.

thin-sections further limits the useful resolution of the petrographic microscope to between 5 and 10 μm. In the last three decades there have been rapid technological advances not only in optical design of microscopes but also in electronic means of quantifying observations. These improvements have led to the microscopic examination of clinker, cement and concrete playing an important role in the scientific study of these materials.

Le Chatelier (1882) reported the results of his studies on cement clinkers using the polarising microscope which showed that tricalcium silicate was the principal constituent. This is the first reported usage of the microscope to study Portland cement. Törnebohm (1897, 1910–11) also using a polarising microscope identified five mineral constituents of cement clinker, naming them alite, belite, celite and an apparently isotropic residual phase. The compositions of the constituents are now better understood and are explained in Chapter 4 (see page 85). The microscopic studies were confirmed by Rankin and Wright (1915) in their chemical investigation of the lime-alumina-silica phase system. However, the early microscopic studies examined cement clinker either in oil immersion or thin-section which were not suited to determining quantitative relationships. This was achieved by Tavasci (1934) who showed that the main clinker phases could be quantitatively distinguished in polished section using etching techniques. Later improvements in etching techniques and the use of polished thin-sections were described by Insley (1936) and Insley and McMurdie (1938). The history of the early work on cement clinker has been described by Mather (1952) and later work by Campbell (1986).

The microscopical study of hydrated cement and concrete proved to be more difficult for early workers. Parker and Hurst (1935) and Brown and Carlson (1936) examined thin-sections of hydrated cements with difficulty. Attempts by various workers to examine polished sections of concrete gave limited results. Johnson (1915) in his investigations of deteriorating concrete used ground and polished sections because degradation during thin-sectioning resulted in a loss of information. Insley and Fréchette (1955) in their text give little information on the examination of concrete and write of the difficulties of observation on samples obtained from field structures. A small number of investigations of concrete using thin-sections are reported but these tended to concentrate on aggregates rather than concrete as a composite material. The problem was that thin-sectioning the fabric of a concrete was not very practicable until epoxy resins became commercially available in 1946. Epoxy resin is necessary to bond together the heterogeneous components of concrete so that it can be thin-sectioned. By 1960, the methods of large-area thin-sectioning that were necessary for the examination of concrete rapidly followed as described in Section 3.7 and the petrographic examination of concrete developed rapidly. Mielenz (1962) gives an excellent review of the state of the art in the petrography of concrete up to the time when modern methods of thin-sectioning came into use.

1.2   Scope of this book

The value of the petrological microscope as a tool for the study of both natural and artificial materials is now widely recognised. It can provide information on clinker, cement, grout, mortar, concrete and related materials unobtainable by other means. When coupled with appropriate complementary techniques, such as electron microscopy and chemical analysis, it forms an effective means of studying cement raw materials, burning history and phase relationships of clinker. Microscopy is a powerful technique for the examination of the fabric of concrete and can be an efficient means of checking compliance with specification and studying deterioration and failures. Although the petrological microscope is an effective means of investigative study, its full value and usefulness remains very much dependent both on the skill, expertise and experience of the petrographer and on the understanding of the person reading the petrographic report. Adequate communication between the petrographer and the user is absolutely essential.

This book aims to provide the specialist background information and references together with the practical details of the techniques required for specimen preparation and the petrographic study of concrete. Extensive use has been made of examples which together with appropriate discussion should assist both the petrographer and the engineering technologist to understand the nature of the evidence which is basic to a petrographic analysis.

A full petrographic analysis of a premature and progressive concrete deterioration often requires additional complementary methods of study to be employed. The range of the techniques available is wide and the selection of the techniques required for the solution of a particular problem will depend on the knowledge and experience of the petrographer. Chapter 2 provides a brief summary of the additional techniques most frequently used to aid petrographic analysis. The relevant methods are introduced at appropriate points in the text, but because they are already well described in modern texts and are themselves the subject of an extensive literature, detailed descriptions are omitted but selected key references are given if further information is required.

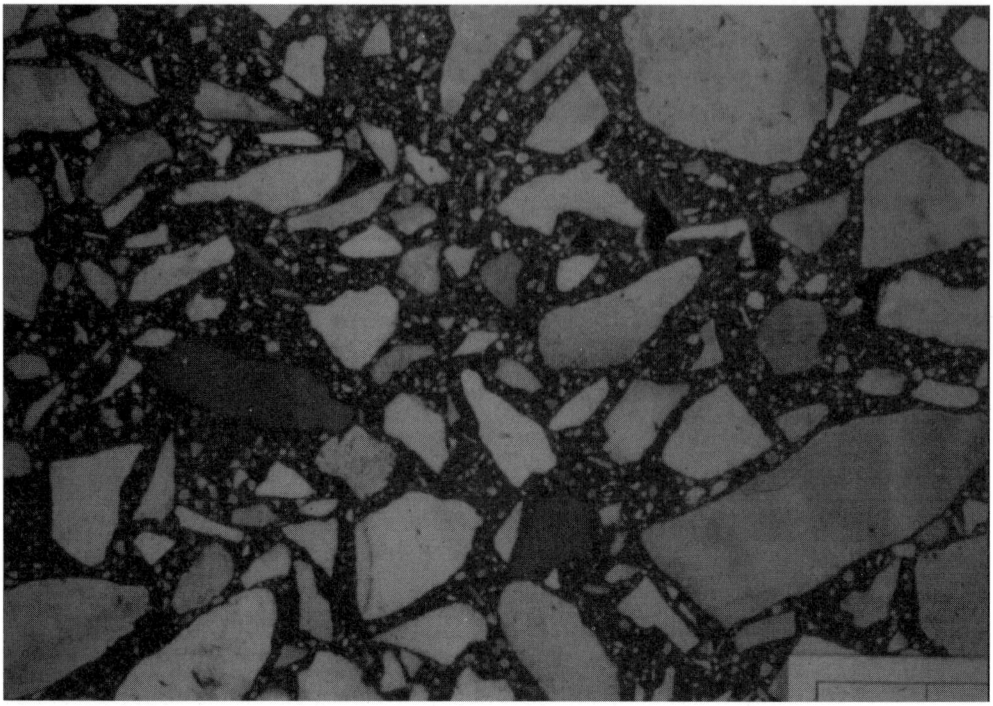

Fig. 1.2 A cut slice of concrete stained to identify the crushed rock aggregate; dolomite, white limestone pink (dark in this photograph).

Cement clinker has been the subject of numerous petrographic studies since the early work of Le Chatelier and texts in black and white by Gille *et al.* (1965) and Hofmänner (1973) and in colour by Fundal (1980) and Campbell (1986) are available so cement clinkers are dealt with only briefly in this text. In contrast, no modern texts in English relating to the petrographic study of hardened cement, grout, mortar and concrete are currently available and the information is widely dispersed through the literature. It is hoped that this text will provide an adequate and necessary summary of knowledge in this subject area. The aim is to cover all major aspects of the subject starting with the selection of sample materials and their preparation as petrographic specimens. The methods appropriate to the examination of these materials are then discussed and a review of the required features of a petrological microscope for examining concrete is given. A practical guide to setting up the microscope is outlined and a section covering photomicrography included.

Later chapters of the book discuss the chemical and mineralogical composition of concretes and related materials and include an illustrated guide to the features that may be observed and identified using a petrological microscope. There is also coverage of the defects, deterioration and failures which may occur in concretes together with the observations and petrographic evidence relating to them.

A glossary giving details of the optical properties, mineralogy, chemistry and occurrence of the more important minerals found in and associated with concrete is included, rather than giving these details in the body of the text. Although reference books covering the

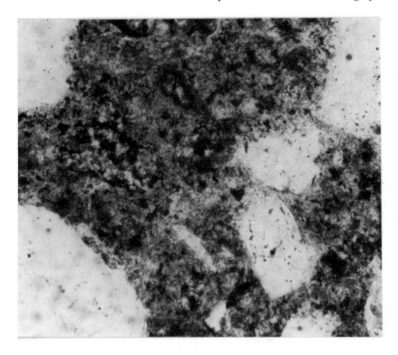

Fig. 1.3 A photomicrograph of a typical concrete in thin-section, quartz fine aggregate particles (light) set in a hydrated OPC paste. Note the small unhydrated clinker particles. Width of field 1.5 mm. A: aggregate; C: unhydrated cement clinker; P: cement paste.

optical properties of artificial materials (Winchell and Winchell 1964) are available, the occurrence of minerals associated with concrete often have unique aspects many of which are difficult to find in the literature.

1.3   Inspection and *in situ* testing of structures

The importance of visiting and carefully examining a concrete structure which is the subject of an investigation is obvious. As is described in Chapter 3, the location of areas to be sampled for detailed laboratory petrographic study is crucial to the success of the investigation.

In many cases the experienced engineer or petrographer will be able to assess the nature of the problem from visual inspection alone and be able to suggest a number of possible causes while ruling out others. This initial evaluation is of considerable value to laboratory personnel in that it will enable them to select the most effective and efficient means of studying the samples obtained. The preliminary inspection of a structure is normally limited to the external appearance of the concrete and the local environmental conditions at the site.

Consideration of the surface features noted in a careful visual inspection of a concrete structure will provide some indication of possible defects or deterioration mechanisms which may be present in a particular structure. These may range from inadequate structural design, poor mix designs, and features of poor workmanship or quality control, to defects arising from ingress of sulfates or other salts, steel reinforcement corrosion, the inclusion of pyrite or other unstable aggregate materials, carbonation or leaching of the concrete and to reactions within the material itself such as alkali-aggregate reactions.

In cases where abnormal deterioration of the structure has been identified or is suspected as a result of visual inspection, it may be necessary to consider whether a programme of *in situ* testing or monitoring would provide useful additional information.

The range of *in situ* non-destructive tests available is large and costs of implementation vary widely. A brief summary of the more commonly used techniques is included here but a thorough treatment of all the methods involved lies outside the scope of this book and the reader should refer to the appropriate specialist texts, for example Bungey (1982) and Malhotra (1976).

Non-destructive tests, although usually simple to perform, have the problem that the analysis and interpretation of the test data is difficult because of the complex nature of the concrete. Perhaps the best known non-destructive testing device is the Schmidt rebound hammer. This device provides a quick and simple means of determining the uniformity of a concrete element by providing an arbitrary measure of surface hardness from the percentage rebound of a mass impacting on the concrete surface and corrected for gravity. The variation in moisture content and surface carbonation will affect the result but it can provide a qualitative indication of the compressive strength of the concrete.

Sonic methods are also used for *in situ* testing since concrete approximates to a homogeneous elastic medium on a macroscopic scale and the theories of sound propagation can be applied to give information about voids, cracks, degree of moisture saturation and similar features. The simplest form of device produces the sound wave by striking the concrete surface with a hammer, but more sophisticated equipments utilise a pulse echo technique. Problems can arise from the presence of reinforcing steel but these methods have been found useful for assessing the integrity of concrete piles and similar concrete structural elements.

Ultrasonic pulse velocity (UPV) methods are a natural extension of the sonic method and numerous portable devices are available. Pulses between 20 and 200 kHz are generated and received by transducers placed on the concrete surface. Concrete thicknesses up to 20 m can be tested. Temperature, moisture content and reinforcing steel can all cause errors which must be corrected, but in favourable circumstances and with skilled interpretation the method can be used to assess the extent of internal cracking and the general deterioration of the concrete.

X-radiography may be used to map variation in the concrete density, position of reinforcing bars and variation in moisture content. The method is limited by the high cost and non-portability of the equipment. Gamma radiography using a radio-isotope source of gamma radiation is used to penetrate the concrete and is recorded on film detectors. The equipment is relatively cheap and portable and is limited to concrete thicknesses of less than 600 mm but it is finding increasing use in locating reinforcement and detecting voids and density variations. Three-dimensional images may be obtained by the use of stereo-pair radiographs. A recent modification to this equipment replaces the film with an electronic detection device.

The commonest method of detecting steel reinforcement is to make use of its magnetic response. Portable instruments are available and if the bar diameter is known they can provide information relating to the depth of the concrete cover.

Resistivity measurements, that is the resistance to electric current conducted through moist concrete is a well-established technique for determining the thickness of concrete pavements and for locating reinforcement. However, this method is somewhat subjective and requires careful calibration.

Measurement of the electrochemical potential of concrete by attaching one lead to the steel reinforcement and the other to a 'half cell' (usually copper in copper sulfate solution) allows the 'half cell potential' to be determined which with careful interpretation can provide information relating to the state of corrosion of the steel reinforcement.

Infrared thermography which measures small differences in surface temperatures can, with careful interpretation, provide information about delamination in slabs or other defects which cause changes in the heat radiation from surfaces.

Long-term monitoring of structures can range in scale from very simple 'tell-tale' gauges and fixed pins, to sophisticated automatic devices for recording differential movements on a continuous or regular basis. Information concerning differential movements across existing cracks or concrete sections can readily be obtained by monitoring the distance between fixed pins or studs with a variety of commonly available micrometer gauges. In the case of movements of large structures for example, the crest-line of a dam, conventional surveying of fixed reference points is perhaps most appropriate. A common device for monitoring small displacements, for example across a crack, is the Demec Gauge which will measure the distance between pairs or triangular arrays of stainless steel studs cemented to a concrete surface with a precision of 0.02 mm or better. Simple opposed and aligned pins fixed to different parts of the structure are often sufficient to provide a visual indication of differential movements as will a thin brittle glass 'tell-tale' cemented to the two sides of a crack which is subject to movement.

However, the rate of progression of differential movements or the development or extension of existing cracks can be determined only by careful repeated monitoring over a long period. Seasonal temperature fluctuations may produce movements of similar magnitudes to those produced by the deterioration process itself so these need to be excluded from the analysis.

1.4   Petrographic examination of concrete

Once the concrete samples have arrived at the laboratory, the petrographer will need to select the appropriate investigative techniques from the large number available. The choice of methods will depend on the particular objectives of the specific problem being investigated and partly on economic factors. Careful consideration of the techniques available will allow the most cost-effective selection to be made which will achieve the required objectives.

Petrographic examination at hand-specimen and petrographic thin-section levels can provide a great deal of information about the constituents of the concrete, features of deterioration (if present) and details of the mechanisms producing the deterioration. Other techniques are often required in addition, and in some cases, are more appropriate than petrographic methods. For instance, if the aim of the investigation is to identify an organic admixture in a concrete, infrared spectroscopic analysis or a chemical extraction method are perhaps more appropriate than classical petrographic techniques. Some appropriate techniques for dealing with different types of problem are illustrated in the guide to methods of study at the beginning of this chapter and described in outline in Chapter 2.

Maximum information can be obtained by careful petrographic examination of a combination of hand-specimen material, cut and polished slabs and petrographic thin-sections which should represent both the outer zones and surface of the concrete and areas distant from the surface. Clearly, if there is evidence of joints or patching between concretes, these will also require careful petrographic examination.

A variable-power binocular microscope is a very useful aid in the examination of slabs and hand-specimen material and automatic modal analysis stage movement devices will be required for the adequate statistical determination of proportions of the components present or air voids.

A full petrographic examination is able to determine the type and composition of the coarse and fine aggregates, the nature of the cement paste and the cement type. It will allow the proportions of the coarse and fine aggregates, the cement paste and entrapped and entrained air to be determined. It will also allow the original aggregate gradings to be estimated.

The disposition and size of calcium hydroxide crystals in the cement paste can in favourable circumstances provide information about the original water/cement ratio if the concrete has been placed and cured at normal temperatures. The size, state and type of remnant clinker particles can give information about cement type and age. The extent of carbonation can readily be observed and the development of secondary sulfate minerals can indicate sulfate attack. The texture and colour of the cement hydrate gels in the cement paste can also provide insight into the past history of the concrete. The development of microcracks within the concrete and the degree of cracking in terms of number of cracks traversed per metre allows comparison between samples and often provides indication of the cause of the microcrack damage. Infilling of cracks and voids can provide evidence of chemical reactions, leaching and precipitation of materials. While much of this data requires experience for its interpretation, a competent petrographic description of a normal concrete sample should be able to provide the following information.

1. General features
    (a) volume proportions of coarse aggregate, fine aggregate, cement paste and air voids
    (b) aggregate gradings and shape, whether natural gravel or crushed rock
    (c) presence or absence of artificial aggregates and cement replacements. The type, size, shape and volume proportions of such particulate additions
    (d) nature, type and proportions of air voids.

2. Specific features
    (a) rock and mineral types present in coarse and fine aggregates and their relative proportions; the grade of weathering of the aggregate particles; whether they are cracked or have interacted with the surrounding paste or have been degraded whilst in the concrete
    (b) the nature, sizes and state of remnant clinker grains in the cement, features of the cement hydrate gels and the nature, size and disposition of calcium hydroxide crystals in the paste; the nature and extent of any carbonation of the cement paste
    (c) the presence and nature of any void or crack infillings; evidence and nature of anomalous reactions within the cement paste or between aggregate particles and the paste
    (d) evidence of segregation or preferred orientation of the aggregate, steel reinforcement corrosion and its severity, original uneven distribution of water in the mix,

incomplete mixing of the constituents of the concrete; leaching of components of the hardened concrete

(e) evidence of sulfate attack, frost damage, types of chemical attack and alkali-aggregate reactions if present, and in favourable circumstances estimates of their severity.

2
Petrographic Examination Techniques

2.1 Introduction

As indicated in Chapter 1 the range of techniques used in petrography cover a very large field and it is not possible to discuss them all in detail. In the following sections on microscopy emphasis has been laid on detailing the parts of the microscope, ancillary items of equipment and their use because these are given little emphasis in many texts. Discussion of complementary petrographic techniques has been limited to detailing the type of results that can be expected from the use of these techniques. Where possible the limitations of a technique are also indicated.

2.2 Initial methods of examination

Samples which arrive at the laboratory for petrographic analysis range from small quantities of powder to large samples from concrete structures. As an example concrete specimens are often supplied as drilled cores 75 to 100 mm in diameter. They can vary from a few millimetres to many metres in length. Thus one of the problems in the initial laboratory examination of such material is handling the sheer size and weight of these specimens.

Irrespective of the examination procedures used, all specimens must be photographed, as received, for record purposes. The photographic methods required are discussed in Section 2.10. Unless special facilities are devised it is difficult to examine a long concrete core without sawing it into more manageable lengths. For most purposes this is acceptable, providing that all the pieces are marked for identification and orientation, and care is taken to retain evidence.

The purpose of the initial examination is to observe any unusual features as well as to provide a description of the general characteristics of the sample. With concrete specimens the examination method may have to cover large surface areas including minute crystals that may be present in pores and cracks. This requires examination with the unaided eye, hand lens and the stereo microscope. The combined use of both the unaided eye and the stereo microscope is vital as the unaided eye will often fail to see fine cracks and details until they are observed through the stereo microscope. On the other hand, some features such as surface discoloration clearly visible with the unaided eye tend to 'disappear' under the stereo microscope. Thus the rule is look, use the stereo microscope, and repeat the procedure until it is felt the observations are complete. An important element in such observations is to have sufficient light at a suitable orientation. Strong daylight for the unaided eye is superior to artificial light while a well adjusted bench or nose lamp will be required for the stereo microscope. Generally oblique lighting at a 45° angle to the surface is adequate but overhead illumination from a ring light may reveal additional detail.

Large samples required for thin-sectioning, determination of cement content and chemical analyses are easily removed with a diamond saw. For sampling local areas of hardened cement paste and other materials, the workshop motor with flexible shaft and handpiece is a useful tool. Shaft speeds of at least 12 000 rpm are required so that diamond tips, grinders and cutters can be used.

Sampling efflorescence, alkali silica gel and small surface deposits from pores and cracks requires manipulative skill and the use of pointed scalpels and probes. Particles are often less than 100 μm in size and in many cases the material is fragile and deposits are present only in limited quantities. Fine probes may be made from tungsten wire sharpened with molten sodium nitrate as described by McCrone and Delly (1973) or alternatively some types of dental probes may be used. As the quantity of material available for collection becomes less than a few milligrams special sampling techniques are required. These techniques are not discussed here as they have been described by McCrone and Delly (1973).

2.3   Low-power stereo microscope

In practice the stereo-zoom microscope is simple to use and this is why it finds such wide use in the laboratory. There is no doubt that for manipulative procedures the stereo microscope with its crisp, erect image is an excellent tool. Thus it is an essential item of equipment for the concrete petrographer and will be required if the air content of concrete is to be measured using ASTM C457-90, Standard practice for microscopical determination of air void content and parameters of the air void system in hardened concrete.

The modern, low-power, stereo-zoom microscope employs a 1:6 to 1:8 zoom range dependent on make. With the use of × 10 eyepieces the range of magnification is about × 7 to × 50 but can be extended by accessories. Manufacturers offer a range of eyepieces from × 5 to × 20 but for most purposes the × 10 eyepiece is ideal as the image on a stereo-zoom microscope is clear and crisp up to × 50 magnification. Above × 50 the quality of the image is dependent on the lighting and the type of surface being examined. For many types of concrete samples the image is degraded at × 100. One eyepiece should be fitted with a micrometer scale for length measurement. A detachable stereo-zoom microscope body is essential so that it can be used with a variety of stands as the basic stand commonly supplied cannot accommodate concrete specimens of large size. For this purpose a long swinging-arm stand is required to provide the horizontal and vertical adjustment necessary when examining large core specimens. The stereo-zoom microscope is usually provided with a 60 W halogen lamp illuminator, either mounted on nose rings or on a separate stand and fibre optic ring illuminators are available if a uniform field of light is required. If the eyepiece head is detachable from the body of the stereo-zoom microscope it is possible to mount a camera on a slide track to produce stereo photographic pairs.

There are many models of the stereo-zoom microscope available. When purchasing this type of microscope it is necessary to keep clearly in mind that it is a low-power microscope intended for general viewing and manipulation. Money spent on expensive accessories for a stereo microscope may be better used on higher quality equipment for the petrographic microscope. For the examination of concrete specimens the microscope needs to be of rugged and heavy construction.

2.4   Universal polarising microscope

The polarising microscope is designed to analyse light transmitted through or reflected from substances and this requires strain-free optics aligned with the optic axis of the microscope. Some manufacturers publish short texts on microscopy for the beginner including Grehn (1977), Determann and Lepusch (undated), Patzelt (1986) and Mollring (1976), and there are a number of more detailed texts that are helpful in explaining the practical use of the various components of the microscope. One of the more detailed is Loveland's (1970) book on photomicrography which is still relevant today. The following brief description lists the parts of a typical universal petrographic microscope for transmitted light observations which is suitable for concrete petrography.

1. *Universal stand* (Fig. 2.1).   The stand must be robust enough to provide sufficient stability to support heavy accessories. It will contain graduated focusing controls and provision for the attachment of transmitted and incident lighting systems. These

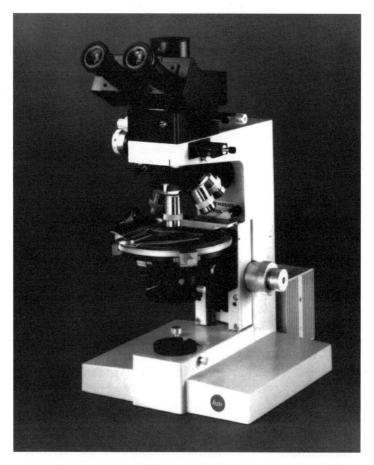

Fig. 2.1 This modern universal type polarising microscope is set up for examination of samples in transmitted light. The black modules which are attached to the white stand by dovetails and clamps may be removed. (Photograph of Ortholux microscope supplied by Wild Leitz).

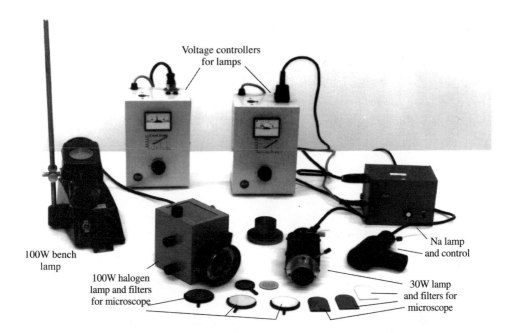

Fig. 2.2 A range of lamps for lighting is illustrated. A light centring disk and neutral density filter are behind the row of lamp filters.

preferably should be attached rather than built-in, to allow flexibility. Dependent on design, a collimating lens, field diaphragm and electrical controls for the lamp may be included in the stand.

2. *Lighting system* (Fig. 2.2).   For transmitted-light work, a 50 to 100 W tungsten halogen lamp, with centring controls and built-in condenser which has an adjustable power supply, is necessary for concrete petrography. Besides the daylight blue, green and diffusion filters usually supplied, a light centring disk and a range of neutral density filters are necessary. A Kodak Wratten 80B filter or its equivalent for colour photography should be purchased as part of the lighting system.

3. *Object stage* (Fig. 2.3).   The petrographic stage is a circular rotating platform, graduated in degrees around its perimeter, which can be clamped and has provision for 45° interval click stops. A detachable ring plate which allows for the fitting of a universal stage is replaced by a solid ring plate in the incident-light mode. There are holes in the top of the stage for stage clips and also threaded holes for attaching mechanical and point-counting stages. An attachable mechanical stage for use with 75 × 25 mm slides can be purchased but is not really necessary for general work. A rack and pinion with yoke is located underneath the stage to which the condenser substage assembly is attached.

4. *Condenser system* (Fig. 2.4).   The condenser system for a petrographic microscope is a strain-free, achromatic two lens Abbé system with a numerical aperture of approximately 0.25. Mounted on the condenser is a swing-in holder into which screws the strain-free top lens. The purpose of the screw-in top lens is to match the numerical aperture of the condenser to that of the higher-power objectives. Condenser top lenses

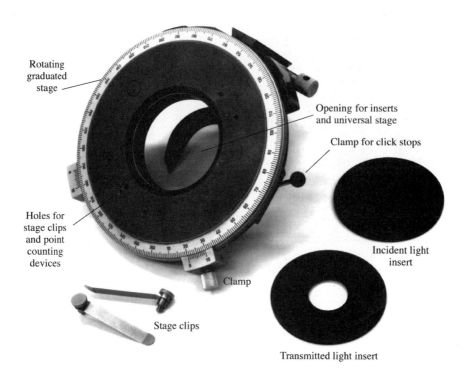

Rotating
graduated
stage

Opening for inserts
and universal stage

Clamp for click stops

Holes for
stage clips
and point
counting
devices

Incident light
insert

Clamp

Stage clips

Transmitted light insert

Fig. 2.3 The object stage with inserts and stage clips.

commonly supplied provide a numerical aperture up to 0.9 for use with dry objectives and 1.3 to 1.4 for use with oil immersion objectives. The alignment of the condenser system must be adjustable. An iris diaphragm for aperture control is mounted between the polariser and the condenser lenses and at the bottom of the main condenser there is a graduated, rotatable polarising plate which is often pivoted so that it can be removed from the light path if required.

Often there is a special filter slot provided just above the polariser for the insertion of a 1/4 wave circular polarising filter. The type of polarising plate used is especially important for concrete petrography and must be of high quality and provide intense light to flood the full aperture of the objective. Nicol prism polarisers and stacked glass plate polarisers designed to protect the optics against heat from strong light sources should be avoided for general work as they limit the numerical aperture. It is necessary for the entire condenser system to be able to rack independently of the stage to allow critical adjustment for Köhler illumination.

5. Objective nose pieces are of two types (Figs 2.5 and 2.1). Rotating nose pieces with individual centring for up to five parfocal objectives are the commonest and most popular. However if a wide range of objectives is in use, some of which may not be parfocal, the nose piece fitted with an objective clutch is preferable. In this case each objective has an individual centring collar. The nose piece includes the tube lens which matches the focal length of the objective to the tube length of the microscope. If

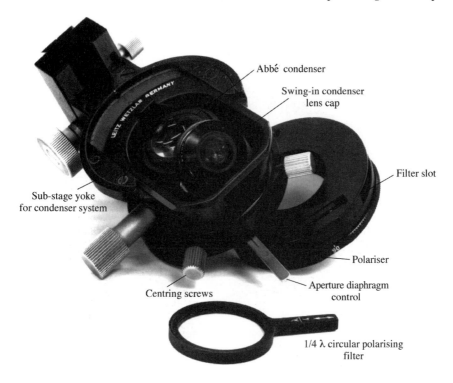

Abbé condenser

Swing-in condenser
lens cap

Filter slot

Sub-stage yoke
for condenser system

Polariser

Aperture diaphragm
control

Centring screws

1/4 λ circular polarising
filter

Fig. 2.4 The condenser system. The polariser is shown swung out from the light path.

objectives with different focal lengths are in use, the appropriate matching tube lens will be required. Extra nose pieces may be purchased with the correct tube lens inserted which avoids having to change the tube lens when objectives of differing focal lengths are used. The tube compensator slot is normally located in the nose piece.

6. A trinocular head (Fig. 2.6) is a common feature of many modern research microscopes and is designed to provide binocular viewing together with provision for the use of special eyepieces, cameras for photography and image analysis, and projection devices. These are all placed in the vertical monocular tube. The inclined binocular tubes are adjustable to match the interpupillary distance of the eyes. When the binocular tubes are adjusted changes in the tube length are automatically corrected so that a camera in the vertical monocular tube remains in focus. On older microscopes the left-hand tube is focusable while the right-hand tube contains specially aligned slots at 45° for the use of an ocular with accurately fixed cross hairs.

In modern microscopes most eyepieces can be focused. To allow the use of the vertical monocular tube the binocular prism can be swung out of the light path but unlike many biological microscopes, the polarising microscope does not use a beam splitter because two special items unique to the polarising microscope are mounted in the body of the trinocular head. These are a focusing and centring Bertrand lens and a polarising plate called the analyser. The Bertrand lens which is used for conoscopic observation of interference figures can be rotated into either the binocular or monocular light paths. The design of the analyser determines the amount of rotation provided.

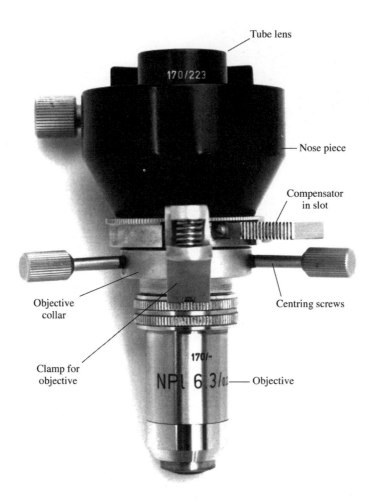

Fig. 2.5 The nose piece with clutch fitting and objective fitted. A compensator is inserted into the slot and the adjusting tools used for centring the objective are shown in place but are usually removed while working. The tube lens marked 170/223 is visible at the top. The more common rotating objective holder is shown in Fig. 2.1.

On many microscopes no rotation of the analyser is available and this limits its use. If the analyser has full rotation through 360° it is mounted on a slide which is not particularly convenient for rapid use (Fig. 2.1).

Some older designs mount the analyser in the tube body so that it pivots easily and quickly into the light path but rotation is restricted to 180° which is sufficient for most purposes (Fig. 2.6). The rotatable analyser is calibrated in degrees and provided with a vernier which reads to 1/10°. Some trinocular heads also contain a tube iris diaphragm for isolating small particles when viewed through the vertical monocular tube when using conoscopic observation.

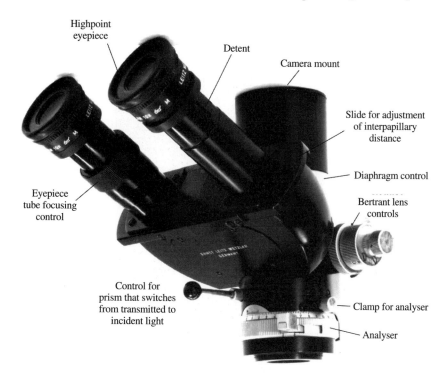

Highpoint
eyepiece

Detent

Camera mount

Slide for adjustment
of interpapillary
distance

Diaphragm control

Eyepiece
tube focusing
control

Bertrant lens
controls

Control for
prism that switches
from transmitted to
incident light

Clamp for analyser

Analyser

Fig. 2.6 The trinocular head with highpoint eyepieces in place. An older style of trinocular head is shown here because the various items of interest can be more easily illustrated than in the modern version shown in Fig. 2.1.

7. The choice of eyepieces is limited to purchasing those matched to the objectives in use because the eyepieces may contain part of the correction designed into the optical system. Since the wearing of spectacles is common, the purchase of some high point eyepieces which allow the user to wear spectacles while operating the microscope, is almost obligatory. These are now available in a wide field option and can be used with or without spectacles by unrolling the soft rubber caps.

On older microscopes one eyepiece of the pair must contain provision for inter-changeable graticules. In modern microscopes both eyepieces are focusable and able to take graticules unless they contain fixed crosshairs. It is convenient if a number of eyepieces for interchangeable graticules are purchased. For precise work an eyepiece containing accurately aligned crosshairs with a registration detent to fit the slots in the ocular tubes is also necessary. A wide range of interchangeable graticules are available. The two that must be purchased are cross lines with a micrometer scale along one axis and another for focusing and delineating photographic fields if this facility is not provided with the camera.

The amount of useful magnification provided by the eyepieces is a subject often discussed in texts where it states that the overall magnification should not exceed × 500–1000 the numerical aperture of the objective . For practical purposes the × 10 or × 12.5 eyepieces give optimum magnification and the use of higher power eyepieces

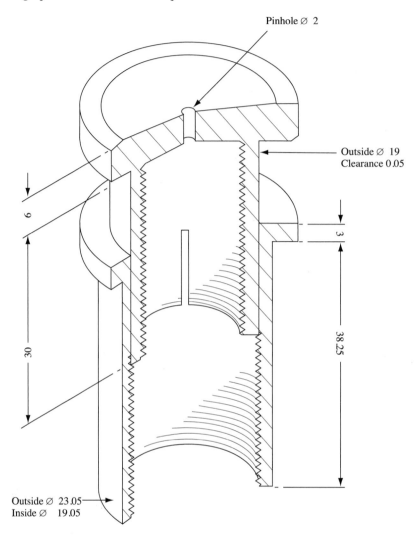

Pinhole ⌀ 2

Outside ⌀ 19
Clearance 0.05

6

30

3

38.25

Outside ⌀ 23.05
Inside ⌀   19.05

Fig. 2.7 The pinhole eyepiece is constructed from black anodised aluminium. The lower portion of the inner barrel contains diametrically opposed slots which enlarge the bottom of this barrel to give a friction fit. The size of the pinhole given is suitable for most conditions but may be varied as required.

is justified only by those with poor eyesight who may need extra magnification to resolve detail. It is better to buy another objective rather than use a higher powered eyepiece. A most useful eyepiece which no longer appears to be available from many manufacturers is the pinhole eyepiece for the observation of interference figures. This eyepiece is not difficult to make and a drawing is illustrated in Fig. 2.7.

8. Since the objective is the heart of the microscope purchasing high-quality objectives is essential. For petrographic work the objectives must be strain free and for photographic work flat-field objectives are desirable. Partially corrected, strain-free flat-field objectives are excellent for general observational work and they are an obvious choice. While such objectives are expensive the money is well spent. However, for the

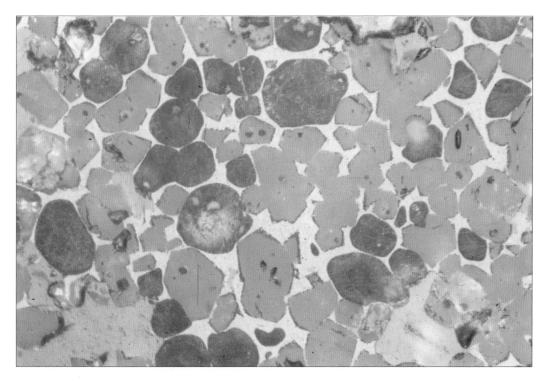

Plate 1 Photomicrograph of unhydrated Portland cement clinker in reflected light after etching with HF vapour. Brown is C_3S, blue is C_2S, the matrix is differentiated into C_3A (light grey) and ferrite (white). Width of field 0.6 mm.

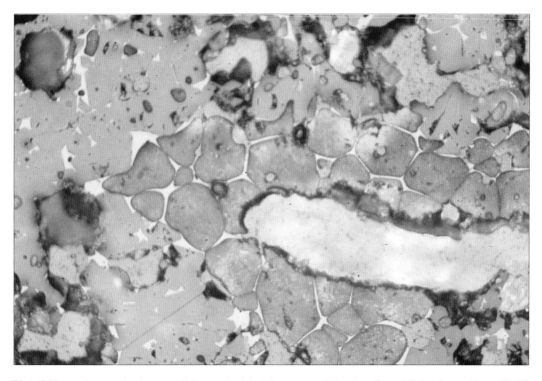

Plate 2 Photomicrograph of nest of C_2S crystals (blue) in cement clinker in reflected light after etching with HF vapour (a resin-filled hole in the centre). Width of field 1mm.

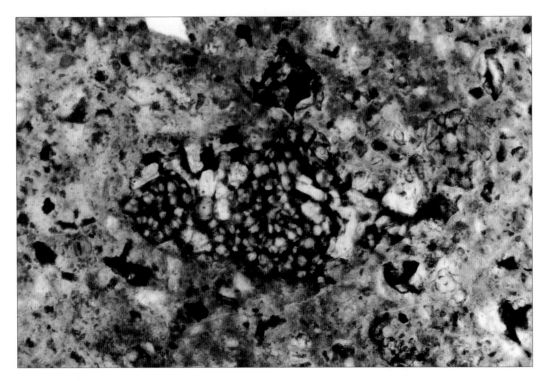

Plate 3 Photomicrograph of cluster of C$_2$S crystals in concrete in thin-section, exhibiting the 'bunch of grapes' appearance. Width of field 0.5 mm. Plane polars. (*Photograph courtesy of Mr Barry J. Hunt, STATS Consultancy.*)

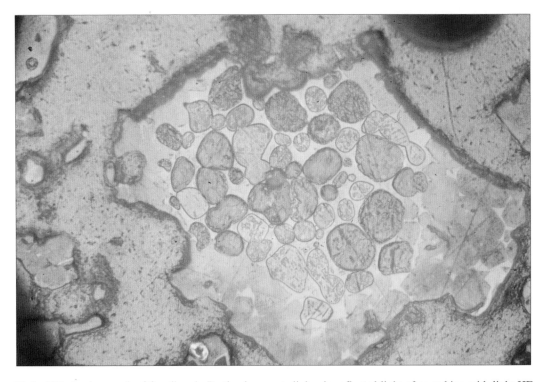

Plate 4 Photomicrograph of free lime in Portland cement clinker in reflected light after etching with light HF vapour. Width of field 0.6 mm.

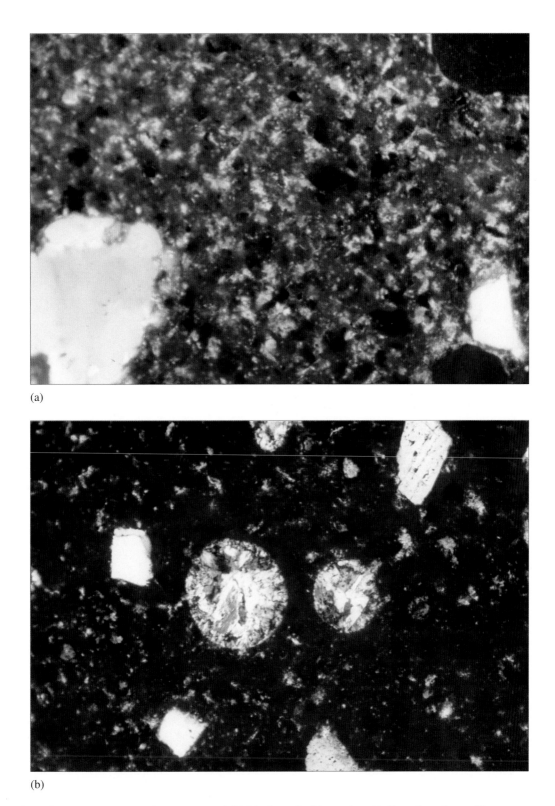

(a)

(b)

Plate 5 Photomicrographs of portlandite as initial hydrates in thin-section: (a) disseminated crystallites, (b) infilled voids. Width of fields 0.4 mm. Crossed polars. (*Photographs courtesy of Mr Mark Knight, Sandberg.*)

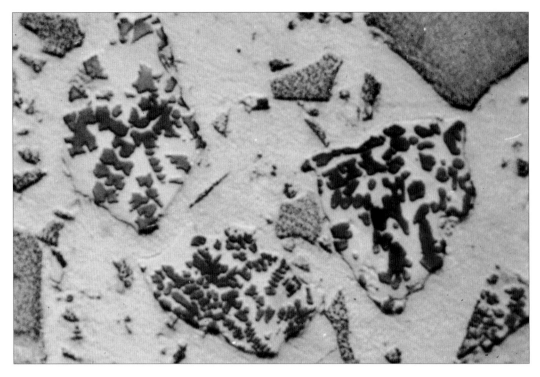

Plate 6 Photomicrograph of unreacted ggbs grains in concrete in reflected light after etching with HF vapour. Width of field 0.3 mm.

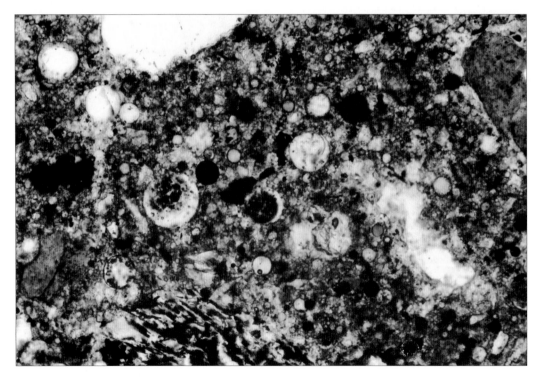

Plate 7 Photomicrograph of pfa cenospheres in concrete in thin-section. Width of field 0.5 mm. Plane polars. (*Photograph courtesy of Mr Barry J. Hunt, STATS Consultancy.*)

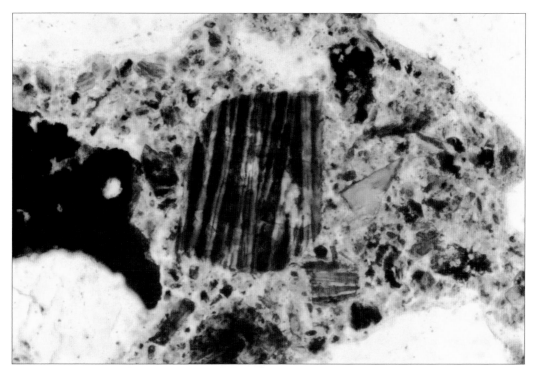

Plate 8 Photomicrograph of pleochroite in high-alumina cement clinker in concrete in thin-section. Width of field 0.2 mm. Plane polars. (*Photograph courtesy of Mr Barry J. Hunt, STATS Consultancy.*)

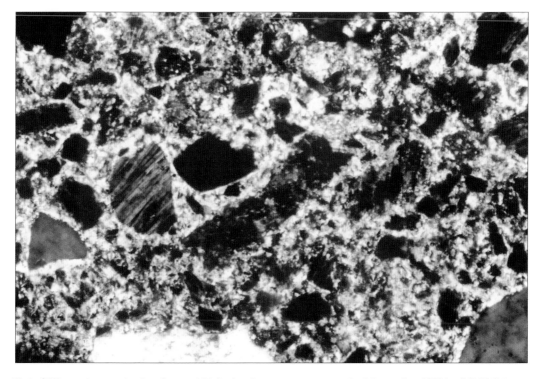

Plate 9 Photomicrograph of carbonated high-alumina cement concrete in thin-section. Width of field 0.4 mm. Crossed polars. (*Photograph courtesy of Mr Barry J. Hunt, STATS Consultancy.*)

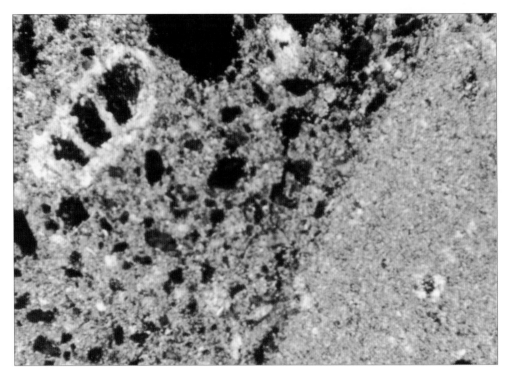

Plate 10 Photomicrograph of Roman lime mortar in thin-section, the matrix comprising finely crystalline calcite. The aggregate particle on the right of the view is chalk Width of field 0.9 mm. Crossed polars.

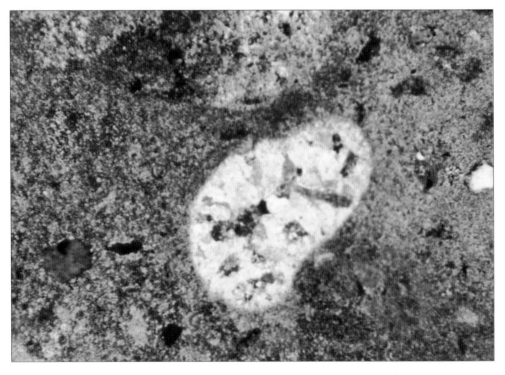

Plate 11 Photomicrograph of Roman lime mortar in thin-section, exhibiting a pocket of coarsely crystallised calcite. Width of field 0.5 mm. Crossed polars.

Plate 12 Photomicrograph of lime-bound mortar in thin-section, containing a relict particle of burnt limestone. Width of field 4 mm. Crossed polars. (*Photograph courtesy of Mr Barry J Hunt, STATS Consultancy.*)

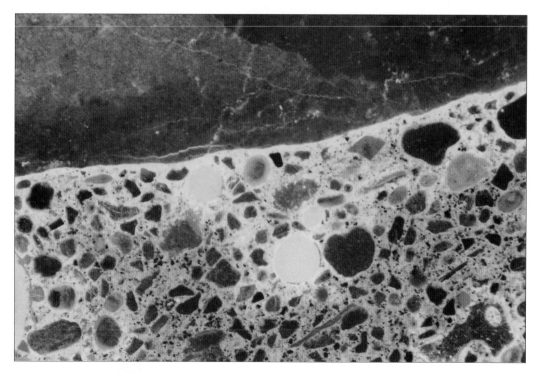

Plate 13 Photomicrograph of internal bleeding in concrete in thin-section, evidenced by voids and porous cement paste beneath an aggregate particle. Fluorescence microscopy. Width of field 4 mm. (*Photograph courtesy of Mr Barry J. Hunt, STATS Consultancy.*)

(a)

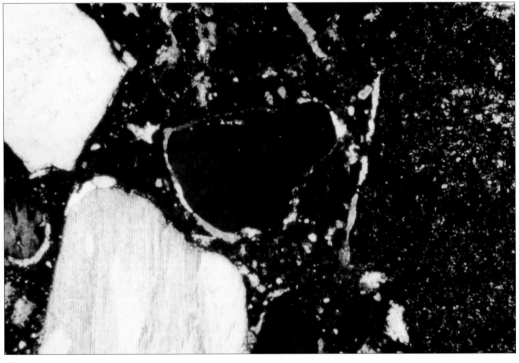

(b)

Plate 14 Ettringite and/or portlandite at the aggregate/cement interface in concrete: (a) SEM micrograph showing ettringite and portlandite in the transition zone, (b) thin-section photomicrograph showing portlandite around an aggregate boundary. Width of fields 0.9 mm. Crossed polars. (*Photograph (a) from Sarkar and Aïtcin (1990) and (b) courtesy of Mr Barry J. Hunt, STATS Consultancy.*)

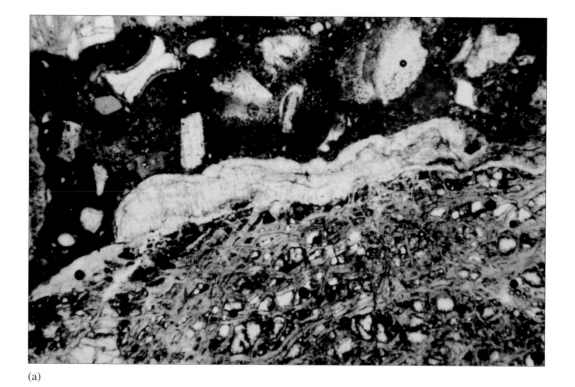

(a)

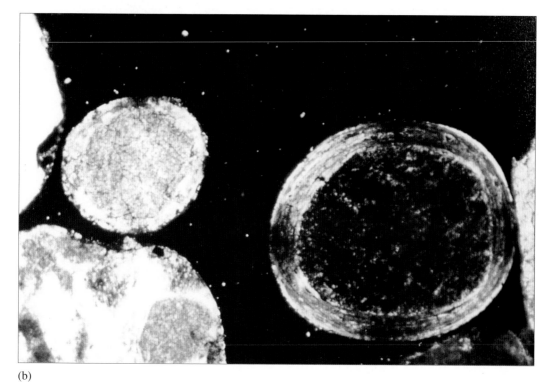

(b)

Plate 15 Photomicrographs of thin-sections showing aggregate particles encrusted with: (a) calcite (encrusting serpentine wadi gravel in concrete), width of field 1.5 mm, plane polars, (b) gypsum (encrusting dune sand grain), width of field 0.5 mm, crossed polars.

Plate 16 Photomicrograph of trachyte aggregate particle in thin-section, showing limonitised micro-brecciation. Width of field 0.5 mm.

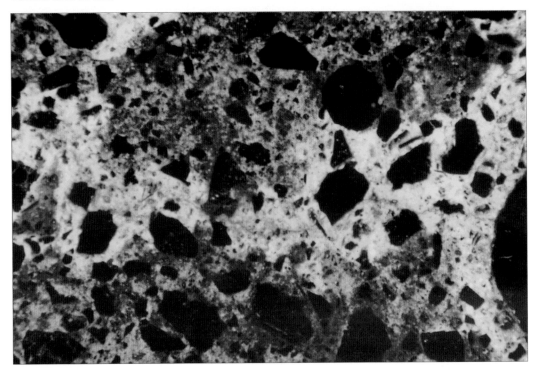

Plate 17 Photomicrograph of bleeding channel (porous zone) in concrete. Fluorescence microscopy. Width of field 0.8 mm.

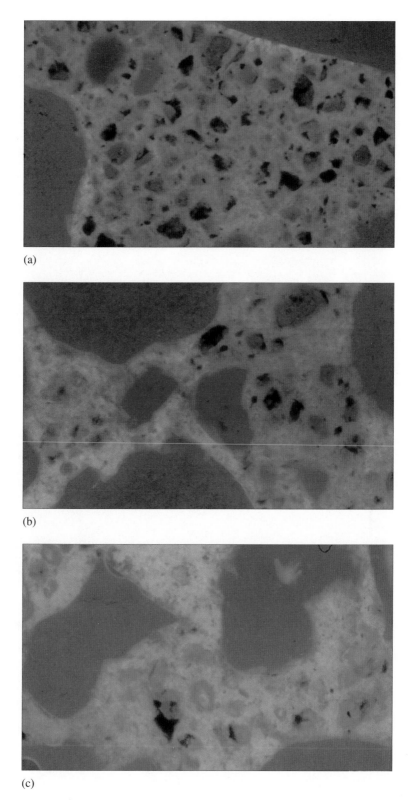

(a)

(b)

(c)

Plate 18 Photomicrographs of fluorescence microscopy in thin-section for concretes of different water/cement ratio: (a) 0.3, (b) 0.5, (c) 0.7. Width of fields 0.3 mm. (*Photographs courtesy of Mr Mark Knight, Sandberg.*)

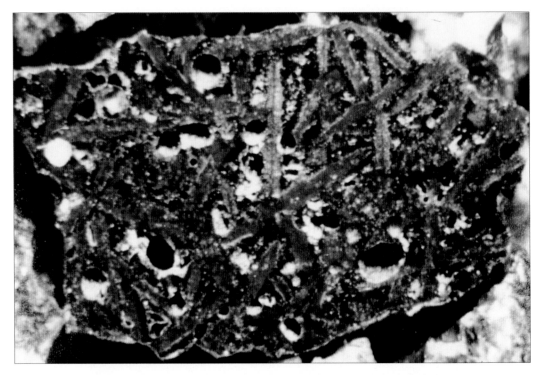

Plate 19 Section through a particle of air-cooled blastfurnace slag, showing large laths of melilite crystals. Width of field 4 mm.

Plate 20 Photomicrograph of layers of hydrate around ggbs kernels in concrete in thin-section. Width of field 0.8 mm.

Fig. 2.8 A lidded box should be used to store objectives to prevent damage and exclude dust.

determination of interference figures a ×50 achromat with a numerical aperture of at least 0.85 is required so this objective should also be purchased.

Magnifications of objectives often follow the geometric progression ×1.25, ×2.5, ×6.3, ×16, ×25, ×40 and the oil immersion objective ×100 although ×10, ×50 and ×60 objectives are still readily available. The most used objectives for concrete petrography are the ×2.5 and ×6.3 (Fig. 2.8). For magnifications greater than ×10, objectives have small working distances and may be provided with sprung noses to prevent damage from accidental contact with the thin-section. Manufacturers now mark their objectives with sufficient data to indicate their correct use. For example, Wild Leitz engrave the following information on the barrel of the ×6.3 objective.

160/–
PL Fluotar 6.3/0.207 P

The first symbol 160/- indicates the objective is designed for an optical path length of 160 mm and does not require the cover glass to be 0.17 mm thick for optimum results. PL Fluotar 6.3/0.207 P indicates the objective is a Leitz PL Fluotar lens, which is their brand name for a corrected flat-field objective, that it has a magnification of ×6.3, and a numerical aperture of 0.207. The symbol P means that it is designed for use with polarised light and is strain free.

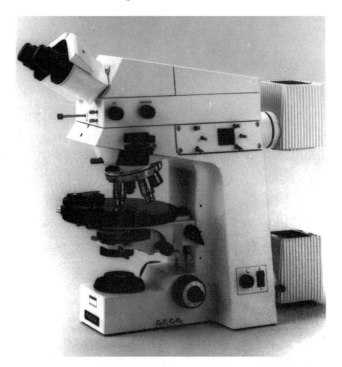

Fig. 2.9 A microscope set up for combined transmitted and incident polarised light. This Zeiss Axioplan is built on a modular principle which allows the use of a wide range of accessories including cameras. The objectives used are designed for infinite path length and separate objectives for transmitted and incident light are not required. (Photograph supplied by Carl Zeiss.)

Another example is the ×100 objective.

160/0.17
PL Fluotar Oel 100/1.3 P

160/0.17 indicates the objective is designed for a 160 mm optical path length and requires to be used with 0.17 mm thick cover glass for optimum performance. The symbols PL Fluotar Oel 100/1.30 P indicates that it is a Leitz PL Fluotar objective for use in oil immersion with a magnification of ×100, has a numerical aperture of 1.30 and has been designed for use with polarised light.

It is to be noted that to obtain the full numerical aperture of this objective, immersion oil must be used between both the top oil lens of the condenser and the section and between the objective and section itself if a condenser with a numerical aperture of greater than 1.0 is used.

The internationally accepted tube length for transmitted light microscopy is now 160 mm. Older Leitz microscopes used a tube length of 170 mm and for incident light microscopy 230 mm. The use of an infinite tube length for incident light is now very common and in some cases this has been extended to transmitted light so that only one set of objectives is required (Fig. 2.9). When buying new objectives for older microscopes care is required to see that objectives with the correct tube length are purchased.

2.4.1   *ADJUSTMENT OF THE POLARISING TRANSMITTED LIGHT MICROSCOPE*

1. Adjust the height of the seat so that the head is comfortable and in a relaxed position when looking down the microscope. The forearms should be able to rest comfortably on the microscope bench and the seat used should be armless and have an adjustable back. To avoid discomfort fully adjustable seating, such as a typist's chair, is essential for extended sessions at the microscope.
2. Adjust the light source to its maximum rated voltage. Remove filters from the substage light path then centre the light source according to manufacturer's instructions and check for evenness of illumination. If incorrect, adjust according to manufacturers instructions. Reinsert filters. These adjustments should be carried out from time to time and always be checked before a photographic session as they are critical for good photomicrography. Some modern microscopes have self-centring lighting systems and may not need adjustment but some checks should always be carried out before commencing photography.
3. Place a thin-section on the stage, open field and aperture diaphragms, and with a low-power objective in place focus the microscope. Set the distance between oculars to your interpupillary distance. Note your interpupillary distance for future use of the microscope.
4. Remove the eyepiece fitted with the micrometer and adjust as follows. Look through the eyepiece against bright daylight and focus the micrometer by adjusting the front lens. Replace eyepiece in right hand ocular tube and using the right eye only critically focus on a recognisable area of the section that contains fine detail. Using the left eye only critically focus the left eyepiece on the same area in the section by adjusting the ocular barrel. The eyepieces are now adjusted to compensate for the individual variation in your eyes. These adjustments are essential to relaxed viewing and should *never* be omitted at the start of a microscope session.
5. Fully close the field diaphragm and adjust height of the condenser system until the edges of the iris are in focus. Centre the iris in the field of view by using the centring screws on the condenser system. Open the field diaphragm until the edges of the iris are just outside the field of view. The lighting is now adjusted to give 'Köhler illumination'.
6. Close the aperture diaphragm. Swing-in the Bertrand lens and focus it on the image of the diaphragm in the back plane of the objective. Adjust aperture diaphragm to leave approximately 80 to 90 per cent of the objective aperture clear. Alternatively, remove an ocular and look directly at the back plane of the objective. This adjustment sets both the numerical aperture of the condenser and objective and the degree of contrast required. Small changes in the setting of the aperture diaphragm cause large changes in the degree of contrast. Beginners at the microscope close down the aperture diaphragm to get strong contrast in the mistaken belief that they are getting a better image. This is incorrect as reducing the numerical aperture to increase contrast degrades the image. The setting of the aperture diaphragm is critical to viewing and should be varied according to the object being viewed. If in doubt use the following maxim. Greater numerical aperture is nearly always superior to greater contrast.
7. Set the lamp to its maximum rated voltage to obtain the correct colour temperature. Adjust this illumination to a comfortable level by placing an appropriate value of neutral density filter into the substage light path rather than reducing the lamp voltage.

This gives a clean white light for optimum viewing. It also gives the correct lighting for colour photography using daylight corrected film provided that a Wratten 80B filter is used in place of the blue filter supplied by the makers.

8. Centre the objective by focusing on a small spot and rotate stage. Centre the objective according to the manufacturer's instructions so that the spot remains stationary while rotating the stage.

9. The adjustments detailed in 5 and 6 need to be carried out whenever an objective is changed and the centring detailed above should be checked from time to time.

10. Place a stage micrometer in the field of view. Take each objective in turn, adjust as above and calibrate the micrometer in the eyepiece against the stage micrometer. Measure both the calibration of the small divisions of the eyepiece micrometer and the diameter of the field of view. Calculate the effective magnification of each objective and assemble a table of effective magnifications, micrometer calibrations and field sizes for the range of objectives in use. This is essential data for quantitative microscopy.

2.4.2   USING THE MICROSCOPE

Before using a new microscope read the manufacturer's instructions on the assembly and operation of the instrument. It is necessary to know how to assemble the microscope and be able to quickly identify how individual items relate to each other. Many microscope accessories are only marked with the manufacturer's name and a code number so that a descriptive list must be assembled for future users of the microscope. This is especially the case for items from older microscopes where the manufacturer may no longer exist and information on code numbers is no longer available.

A good, modern microscope is a precision instrument and will only continue to operate well if it is cared for. Ideal conditions for housing the microscope are a separate room kept at constant temperature as this reduces the pumping of dust into the interior of the optics. Irrespective, a dust cover should always be kept on the microscope when not in use, objective(s) should always be left in place and slot covers closed. Periodically the instrument should be dusted with a soft cloth and exterior surfaces of glass elements cleaned with a soft air brush. If marks or more persistent dirt are present on outer glass surfaces they can be gently cleaned with distilled water and a soft lens tissue. Where marks or dirt on lens surfaces do not respond to this treatment the item should be returned to the manufacturer for repair.

Apart from cleaning the only maintenance required is correct adjustment and an application of a small amount of light polishing wax to the circular stage to ensure that a thin-section slides easily across the surface. Objectives, oculars, condensers, stage mechanisms and other precision adjusted items should be disassembled only by a qualified microscope technician. With time it may be found that movement of the stage or focusing controls becomes uneven or stiff and this is a sign that the microscope requires a routine overhaul. Oil immersion lenses and condenser caps should be carefully wiped clean with a lens tissue *immediately* after use. Periodically they should also be cleaned with a lens tissue moistened with xylene. Once immersion oil has dried on lens surfaces its removal is a job for the specialist.

In the past manufacturers supplied wooden cases (Fig. 2.10) lined with felt which were excellent for the storage of optical items. Today, cases with expanded polystyrene are used which is a retrograde step. As additional items of equipment are purchased it is easy to leave them loose in a drawer without proper storage and this can lead to damage. Optical

Fig. 2.10 Wooden cases to store microscope items appear no longer to be made. This case stores most of the standard items and with the lid removed can be fitted into a drawer.

items should never be stored in rubber or polyurethane foam for any length of time as these materials can decompose and stick to the optics.

The signs of a correctly operated microscope are obvious as the instrument will be clean, well adjusted with adequate storage for ancillary items of equipment. However, all too commonly it will be found that the instrument is used on an open bench in dirty conditions, is out of adjustment and some optical surfaces are damaged. Concrete petrography requires a good quality, well maintained microscope and it is essential that the instrument be treated as a precision item of equipment for this purpose.

2.4.3 LOOKING DOWN THE MICROSCOPE

It is necessary for the beginner at microscopy to see that any defects in vision are corrected before commencing work. This especially applies to eye defects such as astigmatism. When difficulty is encountered in using a microscope it is often because the viewer is straining to look at a plane about 200 mm distant instead of the eyes being relaxed for looking into the distance. Burrells (1961) gives an excellent discussion of the various factors affecting viewing and suggests that the ideal lighting conditions for the field of view should be equivalent to reading a book lit by a 100 W reading lamp. His text is well worth reading as it contains much practical advice on how to use the microscope.

Training the eye at the microscope is a matter of practice. The ability to use continuous small changes of focus to estimate whether relief is positive or negative, operate the analyser,

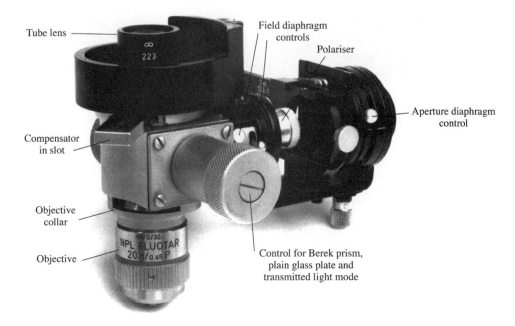

Tube lens

Field diaphragm controls

Polariser

Aperture diaphragm control

Compensator in slot

Objective collar

Objective

Control for Berek prism, plain glass plate and transmitted light mode

Fig. 2.11 The older style of nose piece and condenser system for use with incident light is shown as it illustrates the components better. A modern form of the same module is shown in Fig. 2.12.

rotate the stage and manipulate the section while observing relevant detail is a complex manipulation. With practice the ability can be developed to store details of texture such as colour, shape, relief, twinning, pleochroism, birefringence, anomalous interference colours, zonation and a complex of relationships between components which are recalled for almost instantaneous recognition of minerals and textures when required. It is this ability to identify minerals, rocks and textures quickly which makes optical microscopy a powerful tool in the hands of a skilled petrographer.

2.5   Polarising incident-light microscope

The purchase of a universal microscope stand enables the accessories required for viewing objects by polarised incident-light illumination to be attached. The items required are as follows:

1. *Lighting system.*   A centring and focusing lamp housing, with a 100 W tungsten halogen lamp is required for cement and concrete investigations. As well as the usual daylight filter, a heat filter will be necessary. It is possible to purchase mirror housings which use one lamp that can be switched to either transmitted or incident light or both as desired. Separate lamps, although cumbersome are more versatile.
2. *Vertical illuminator.*   This will include a nose piece with a clutch fitting, diaphragms and collimating lenses (Fig. 2.11). Inside the nose piece is fitted a tube lens, now commonly set for infinite path length, a Berek prism and a plane glass plate. Either the Berek prism or the plane glass plate can be selected to direct the light down the objective. In modern microscopes the Berek prism is often omitted. Since the objective

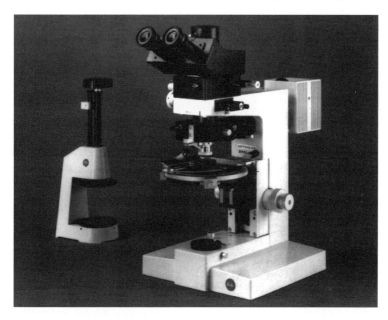

Fig. 2.12 The Leitz Ortholux Pol BK set up for examination in transmitted light employing a clutch fitting for the objective. A mechanical stage is in place and a levelling press is shown to the left. (Photograph supplied by Wild Leitz).

acts as the condenser in this system there is no adjustable condenser system as in transmitted light microscopy. However, an auxiliary adjustable collimator is mounted in front of the field iris diaphragm.

3. *Objectives.* On older microscopes there are separate ranges of objectives for use with polarised incident-light because the tube length for incident-light operation is longer than in transmitted light. In modern microscopes objectives will have an infinite tube length and may also allow their use either in the transmitted or incident-light mode. Objectives intended for metallographic microscopy should be avoided as they will not be strain free. A range of immersion objectives for use in ore and coal microscopy is available but their use is largely restricted to those materials.

4. *Incidental items.* A solid plate to replace the ring plate on the stage and an incident-light stage micrometer are required. The latter is necessary for calibration of objectives. If polished mounts of cement clinkers and other materials are to be examined a levelling press will be necessary (Fig. 2.12).

2.5.1 ADJUSTMENT OF THE POLARISING INCIDENT-LIGHT MICROSCOPE

The adjustment of the microscope in the incident-light mode is similar in many respects to that used in transmitted light. The points of difference are as follows:

1. For adjustment of the light source the lamp housing for the tungsten halogen lamp will require the real and virtual images of the lamp filament to be symmetrically aligned and centred. Follow the manufacturers instructions for this procedure. This may not be required for modern lamp microscopes.

2. There are three modes of viewing possible when the microscope is set up for the incident-light mode. These are with either the Berek prism or the plane glass plate in the light path, or with both removed to allow examination of, for example, polished thin-sections by transmitted light. The Berek prism is better for lower magnifications, gives an image with a greater depth of field and brighter illumination than the plane glass plate. The plane glass plate is better for examining polished surfaces and for the higher magnifications. The Berek prism may not be available on modern microscopes.
3. The field diaphragm has its own focusing condenser and should be adjusted to give a sharp image in the object plane. Centre if adjusting screws are provided. Note that the field and aperture diaphragms are reversed in position in the incident-light mode.
4. The setting of the aperture diaphragm is critical when the Berek prism is used. Very small changes in its setting cause large differences in image quality. The type of specimen being viewed and experience dictates the best setting. When the plane glass plate is used the aperture diaphragm should be wide open unless flare is present. A judicious use of the aperture diaphragm can be beneficial in these cases.

The above adjustments are essentially those required for the examination of polished specimens of cement clinker using a dry objective. There are ranges of immersion objectives for the examination of ores, coals and other materials which use oil and water or glycerine immersion. These objectives are designed to increase the numerical aperture, eliminate unwanted reflections from the sample and allow the determination of bireflectance and absorption. Measurement of these parameters requires special adjustment of the microscope as described by Galopin and Henry (1972) and is outside the scope of this book.

  The operation of the microscope in the incident-light mode is not as simple as in the transmitted-light mode. Equipment tends to be more difficult to adjust, optics do not always operate as well as expected and the final results can be critically dependent on the type of surface being examined. Some objectives will be found to be unsatisfactory for all the purposes claimed and special care should be taken to test them before purchase is made. However, it should be noted that unsatisfactory polishing of mounts is the most common cause of poor results in the incident-light mode and this should be eliminated before condemning an objective. Some good examples of faults in polishing cement clinkers are illustrated by Campbell (1986).

2.6 Analysis of light using the polarising microscope

Before the advent of X-ray diffraction spectrometry, the polarising microscope was one of the principal methods of identifying minerals and giving some indication of their crystal structure. Many of the techniques used for analysing light transmitted through or reflected from crystals now find little use as they have been superseded by other techniques. A knowledge of optical mineralogy is necessary for the use of these methods, which can be found in standard texts such as Hallimond (1953) and Hartshorne and Stuart (1969, 1970). However some of the optical techniques are still used and these will be briefly described.

2.6.1 QUALITATIVE COMPENSATORS

The compensators commonly used for qualitative measurement are as follows. These compensators are all used in the tube slot which is at $45°$ to the polars.

The first-order red compensator or sensitive tint plate, also commonly called the gypsum plate, is an essential item of equipment. It is used for detecting birefringence, determining the directions of fast and slow rays in crystals, and the sign in interference figures up to about second order birefringence. For identification it will be marked with 1λ to indicate that it retards the light by approximately a full wave, which is about 550 nm in practice, and the direction of the slow ray γ will be marked with an arrow. The strong red coloured background produced by this compensator can conceal very weak birefringence and another form of this compensator is available which can be rotated either side of the 45° position to reduce the intensity of the red background and allow better detection of weak birefringence. The form of this plate is known as the 1λ plate in sub-parallel position (Dressler 1980).

The quarter wave compensator retards light by 137 nm giving a grey-coloured background of the first order. It is principally used for determining sign in interference figures especially at higher orders of birefringence, is identified by $1/4\lambda$ and will be marked with the direction of the slow ray γ.

$1/4\gamma$ plate for circular polarisation; this is an arrangement of a $1/4\lambda$ circular polarising plate inserted at 45° above the polariser in the substage condenser combined with the normal quarter wave compensator in the tube slot to produce circular polarisation. With this arrangement and crossed polars, birefringent crystals show the maximum colours for their orientation, irrespective of the rotation of the stage, that is, the birefringence does not wax and wane as the stage is rotated. Its principal use is for strain and fibre studies. To be effective the two $1/4\lambda$ plates must be accurately crossed and have exactly the same retardation.

The quartz wedge is used for measuring the birefringence of crystals and sign in interference figures. Unlike the compensator plates described above it consists of a wedge so that the retardation increases as the wedge thickens and thus it is a more versatile compensator. γ should be marked but it will be found that many old quartz wedges are unmarked so the direction of the slow ray will have to be determined. The problem with the simple quartz wedge is that a portion of the first order is usually missing and the wedge is not uniformly sensitive over its usual range of four to six orders. A number of modifications to move the zero point to the centre of the wedge, the Babinet compensator, or to make the wedge more uniform, have been introduced by Amann, Soleil, Wright & Johanssen. Many of these modifications, especially if a scale is engraved on the wedge, must be used in an ocular (Wright eyepiece) or an objective slot for accurate measurement. If these modified quartz wedges are placed in the tube slot they require the use of a Laspeyres ocular to observe the engraved scale (Hallimond 1953).

2.6.2 QUANTITATIVE COMPENSATORS

A more convenient and accurate method of providing a compensator of variable retardation is the Berek rotary compensator which measures birefringence by rotation of a crystal plate to provide the variable compensation. The amount of rotation is measured on a calibrated drum and the birefringence calculated from calibration charts supplied. The Ehringhaus compensator is similar to the Berek compensator but is modified to give a more uniform scale and is thus more sensitive. A common form of the Berek compensator uses a crystal of magnesium fluoride which has low dispersion and is available to measure up to five orders. Other forms based on calcite will measure up to 10, 20 and 30 orders.

For accurate measurement of weak birefringence or bireflectance the rotary mica or elliptic compensator as proposed by Brace-Kohler is used. This compensator can be purchased to measure in the ranges of up to 1/10, 1/20 or $1/30\lambda$ and is used in the tube slot. To measure retardation quantitatively up to 1λ the de Sénarmont compensator is used. The method of its use for cement clinkers is described by Campbell (1986).

2.6.3 THE WRIGHT EYEPIECE

The Wright eyepiece is a device for accurate quantitative determination of retardation. It contains a rotatable analyser, a slot for compensators together with crosshairs and iris diaphragm for isolating grains. Most of the compensators described above can be used in the eyepiece slot together with devices to determine extinction position precisely which is necessary for accurate measurement of retardation, especially in reflected light. The twinned quartz plates designed by Bravais, Wright and Koenigsburg have generally been superseded by rotary quartz plates for determining extinction position. These are the Soleil's biquartz plate, the rotary sensitive tint plate of Biot Klein and the half shadow plates of Mace de Lépinay and of Nakamura. For most purposes the Nakamura plate is adequate but for retardation of less than $1/20\lambda$ the Koenisburg plate can be used (Hallimond 1953).

Of the above compensators the first-order red plate and the quarter wave plate are intended for general petrographic examinations at the microscope. Where a variable compensator is required the Berek compensator is to be preferred. For the examination of cement grains using Ono's technique the Sénarmont compensator is required to determine the birefringence of alite. The other compensators described above are now little used and restricted to specialist applications.

2.6.4 THE UNIVERSAL AND SPINDLE STAGES

Refractive indices, the optic angle $2V$ and the positions of the optic axes X, Y & Z of a crystal may be determined using these special stages which are attached to the rotating stage of the microscope. The universal stage is particularly applicable to thin-sections but mineral grains may also be examined. It is a complicated device which allows the section to be tilted in three axes and requires skill and experience for its use. This stage is described in most texts on optical mineralogy. An adequate and easy-to-follow description of its use may be found in Wahlstrom (1955). The single axis or spindle stage is a simpler device and still used. It is restricted to grains and cannot be used with thin-sections. A comprehensive description of its use will be found in Bloss (1981). Both these stages are unlikely to be used in concrete petrography. However, if it is necessary accurately to determine the optical properties of a crystal the use of these stages will be required.

2.6.5 FILTERS

The range and types of filters commonly used in microscopy are discussed in some detail by Loveland (1970). Many filters are available as inexpensive gelatin types, of which the Kodak range is an example. Kodak have published full technical data on their range of filters (Kodak 1970). However, dyed gelatin filters are fragile and tend to have a limited life and for general work glass-mounted filters are preferable as they are more permanent. As discussed earlier a blue filter for general viewing at the microscope is essential and this should be a good-quality glass filter. This also applies to the green-sensitive (orthochromatic) filter used for monochrome photography and the diffusion filter.

Neutral density filters are used to reduce the intensity of the light source of the microscope to a comfortable viewing level without materially altering the spectrum in the visible range. As a general rule the light source should always be operated at its maximum recommended voltage to maintain the correct colour temperature and to give an equivalent to white light with the daylight blue filter in place. As neutral density filters are addition it is not necessary to buy a full set. These filters may be designated by optical density instead of optical transmission. Tables to convert from density to transmission are given in the Kodak catalogue (Kodak 1970).

Many of the filters listed in older texts are intended to enhance biological stains and contrast media and will not be discussed. The types of filters of principal interest to the concrete petrographer are the interference, excitation, suppression and colour compensating filters. Interference filters are designed to allow only a narrow band of the light spectrum to pass and are available either as individual band pass filters or as graduated, continuous filters. The size of the band width varies according to the design and requires parallel light for effective operation. Where a well defined band pass is not required dyed gelatin filters or their glass equivalents are adequate. Excitation and suppression filters are discussed later under fluorescence microscopy.

Colour compensating filters are used to make small changes in the colour balance of film during photomicrography. On the microscope it is common to find that the balance of colour film is not acceptable. By some experimentation with colour compensating filters this can usually be corrected. Similarly there is a range of light balance filters for making small changes to colour temperature. If the lamp source is operated correctly these should not be necessary.

For petrographic work filters need to be used with a full knowledge of their purpose. In some cases, as in the accurate determination of refractive indices it is better to use a sodium lamp to give the required spectral lines at 589 nm. In general, apart from neutral density filters, the fewer filters used the better. Where polarised light is used on the microscope it is not permissible to place filters between the polariser and analyser unless they are free of strain and distortion. This restriction does not apply where plain light is being used.

2.7 Fluorescence microscopy

Fluorescence microscopy finds its greatest use in biology (Birk 1984). The technique consists of exciting the specimen to fluoresce by absorption of the shorter wavelengths of light, usually ultraviolet light, and observing this fluorescence under the microscope. If the detail of interest does not fluoresce naturally, which is often the case, it is preferentially stained with a fluorescent dye or stain. Fluorescence microscopy requires a strong light source, an exciting filter inserted before the condenser to ensure only the shorter wavelengths reach the specimen, and a suppression filter placed between the specimen and the oculars to both exclude the shorter wavelengths and isolate the fluorescence for viewing. For observation of fluorescence by reflected light the filters are combined and supplied as dichroic mirrors which must be specifically purchased for the microscope and the wavelengths required.

There are four methods in which observation of fluorescence is applied in concrete petrography. These are to highlight areas of porosity in the fabric, estimate water/cement ratio, delineate crack systems and detect alkali-silica gels. In the first three methods the concrete must be impregnated with a thin epoxy containing the fluorescent dye. For the observation of porosity and estimation of water/cement ratio, thin-sections are required where the concrete has been impregnated with a thin epoxy containing 0.8–1.0 per cent of Hudson Yellow or Fluorescent Brilliant Yellow R. These dyes are listed in the Colour Index under C.I. solvent yellow 43 (Society of Dyers and Colourists 1982). This highlights areas of porosity and other faults which can then be used to evaluate frost and salt resistance of the concrete (Wilk *et al.* 1984, Wilk and Dubrolubov 1984, Dubrolubov and Romer 1985, Jensen *et al.* 1985).

Water/cement ratio is estimated from the intensity of the fluorescence compared with calibration samples (Christensen *et al.* 1984) and a description of the sample preparation together with colour photographs illustrating the appearance of the fluorescence is given by Jensen *et al.* (1985). The excitation filter used is a BG12 which transmits in the blue around 400 nm and a K510 suppression filter which transmits only above 500 nm and isolates the fluorescence. Other fluorescent dyes can be used but these may require different filters and it will be necessary to see that they have reasonable light stability. A 100 W quartz halogen lamp is sufficient to produce the required shorter wavelengths when using these fluorochrome dyes and an ultraviolet lamp is not required. To obtain reliable results from the technique the thin-sections used must be no thicker than 25 μm and be of consistent and reproducible quality. In addition a wide range of calibration samples are required and ideally the calibration samples should bracket the unknown concrete and be reasonably similar in composition. Investigation by image analysis to provide better estimation of fluorescence levels has been carried out by Wirgot and Cauwelaert (1991) and Elsen *et al.* (1995).

The delineation of cracks and detection of alkali-silica gel does not require the use of the microscope and can be easily carried out on cores or large-area ground surfaces. Hobbs (1988) illustrates cracks impregnated with epoxy resin containing a fluorescent dye which have been illuminated by an ultraviolet lamp and then photographed in the dark with a camera using an appropriate barrier filter. In this method protection for the eyes against ultraviolet light may be required.

The detection of alkali-silica gel in concrete differs from the above methods (Natesaiyer and Hover 1988, 1989). It involves washing the concrete surface and then spraying it with a solution containing 10 g of uranyl acetate dissolved in 100 ml of 1.5 per cent acetic acid. After a few minutes to allow adsorption of the uranyl acetate by the alkali-silica gel the surface is rewashed and observed or photographed in ultraviolet light at 240 nm using an appropriate filter. The uranyl acetate which is preferentially adsorbed by the alkali-silica gel is observed as a dull green fluorescence in the dark. St John (1994) investigated the use of uranyl acetate for detecting alkali silica gel and found that Feigl's zinc uranyl acetate solution (Feigl 1958) is more stable and appears to give marginally better results. He also found that if concrete specimens are stored in the laboratory and allowed to dry out thoroughly false negative results for the detection of alkali silica gel can occur. If detailed observation is required ultraviolet sources are available for use with the stereo zoom microscope. Uranyl acetate is toxic and must be handled with care and protection against ultraviolet light is also necessary.

There is no doubt that fluorescence microscopy is useful for estimating the porosity of concrete. Its use for the estimation of water/cement ratio is more controversial and appears to be mainly applicable to water/cement ratios in the range 0.4 to 0.6 (St John 1994). However, many of the other details that the technique highlights are clearly observable in a well made thin-section without the need for fluorescence. The use of a dye in concrete has the disadvantage that it can obscure fine detail of interest in voids and cracks and because of this, dyes should be restricted to estimation of porosity, outlining porous areas and very fine microcracking which may not be otherwise observable. Where facilities for sectioning and microscopy are not available the methods for crack delineation and the detection of alkali-silica gel fulfil a useful purpose. An interesting use of fluorescence in examining ultra-thin-sections has been described by Walker (1979, 1981).

2.8 Fourier transform infrared microscopy (FT-IR)

Whereas fluorescence microscopy uses ultraviolet light to excite fluorescence, at the long wavelength end of the spectrum infrared light can be used to produce spectra from *in situ* areas of samples for analytical purposes. The two techniques are fundamentally different as the microscope in this case is only acting as a method of identifying an area of interest and directing the infrared beam to that area which is then analysed by the attached infrared spectrometer. There is no change in the appearance of the specimen visible through the microscope from the infrared light. Areas of less than 100 μm^2 can be analysed so that FT-IR microscopy is a true micro technique (Reffner *et al.* 1987).

This microscopical technique has become possible because the Fourier transform analysis quickly derives the spectra from all the infrared absorption bands without the need for a lengthy scan of the spectrum by wavelength. The technique can be used in both transmission and reflectance modes. Currently it is mainly used for investigating fibres, coatings and in forensic science. It has possibilities in concrete petrography for examining changes in the surface appearance or properties of concrete especially where these are due to organic materials such as release agents. However, the nature of concrete is such that FT-IR is limited to the reflectance mode. Price and Caveny (1992) have used reflectance FT-IR to study cement crystal phases.

2.9 Phase contrast and related techniques

These illumination systems used for enhancing contrast are mainly applicable to fine particles, colloids and biological materials, and where the specimens are very thin. A good description of the various types of equipment available and their use is given in Loveland (1970). As dark field and interference contrast techniques are rarely applicable to cement and concrete petrography they will not be discussed further. However, interference contrast is used in metallography and some other areas of materials science. The principal application of phase contrast (Fig. 2.13) in concrete technology is in the identification of airborne asbestos fibres associated with the use of asbestos fibre reinforced sheet (AIA 1982).

There are a number of other specialised types of microscopy that are beyond the scope of this book. A general descriptive review has been made by Dziezak (1988) and the subject of chemical microscopy is reviewed in *Analytical Chemistry* periodically. A recent review was made by Cooke (1994).

2.10 Photography in concrete petrography

Cores and other types of concrete samples must be photographed on receipt for record purposes as noted in Chapter 1. The angle of view, type of lighting and background used will depend on the detail to be recorded. In most cases colour photography is to be preferred as it gives a more complete record than black and white. However, it should be remembered that where control of contrast is required black and white film is more flexible than colour film. A standard grey card should be included as part of the subject to eliminate variations

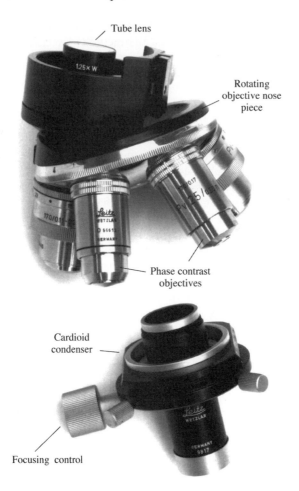

Tube lens

Rotating
objective nose
piece

Phase contrast
objectives

Cardioid
condenser

Focusing control

Fig. 2.13 The rotating nose piece with objectives and cardioid condenser for phase contrast according to Heine. The condenser provides for examination in bright field, dark field and phase contrast to be used. Note that no centring screws are provided for the objectives as this system does not use polarised light.

in batches of colour print paper and the subjectivity of the process operator, and all subjects must also include a scale and sample identification.

Where finer detail of a concrete sample is required this may be recorded with the macro camera. The range of effective magnification using a macro camera to photograph concrete is about ×1 to ×10. Assuming a medium format 120 size film is used and enlarged ×2 an effective magnification of ×2 to ×20 results. This conservative approach gives clear, crisp prints. A good macro camera will consist of a large, rigid, adjustable stand and frame, long adjustable bellows, projection facilities for focusing, a range of close-up lenses (Zeiss Luminar lenses are excellent for this purpose) and a universal back.

It is rarely possible to include a standard grey card in the subject when using the macro camera. However, closely matching colours on the print to the original object is not of such importance when detail is being photographed. Film used should be balanced for tungsten light without the need for filtration and have good reciprocity characteristics for long exposures. Kodak VPL film or other manufacturers' equivalents are suitable. Since scales and identification data cannot be included on the negative unless a databack is available all macro photos must be logged detailing magnification, lighting, subject identification and description of the subject being photographed. It is useful to contact print the negatives both as a record and to allow selection for illustrating reports.

Large-area thin-sections are an unusual photographic subject. Idorn (1965) showed that photographs of complete large-area thin-sections can be made by using them as negatives in a photographic enlarger and enlarging in the range $\times 2$ to $\times 5$. The resulting prints are of course negatives of the thin-section which at $\times 2.5$ show a useful range of detail with microcracks as narrow as 10 μm being visible. These negative prints also assist the plotting of cracks and other detail while working at the microscope. However, the quality of the negative prints are dependent on the thickness of the section. Small variations in thickness across the section cause changes in contrast and density on the print which can be partially corrected by dodging. While negative prints of thin-sections are useful to the petrographer they do contain detail that can mislead. This is because the detail shown is highly dependent on the contrast and opaque aggregates are over emphasised. Attempts to correct these problems by using reversal film to produce a positive leads to considerable loss of detail and information. Similarly the use of crossed polars on the enlarger is not successful. If prints of sections under crossed polars are required it is better to produce these by placing the section between polarising sheets and photographing them on a light box.

With the macro camera good photographs of detailed areas of large-area thin-sections in the range $\times 2$ to $\times 20$ can be made in transmitted light with and without crossed polars. This fills an important gap between studio photography and photomicrography which effectively begins at about $\times 25$ on the microscope. Polarising sheet used for making sunglasses gives adequate results when used with both the enlarger and the macro camera for the procedures described above.

2.10.1 PHOTOMICROGRAPHY

For photomicrography a camera is fitted into the vertical tube of the trinocular head. It is prudent to buy the camera made to fit the brand of microscope being used. In modern microscopy, the 35 mm automatic camera is now standard for routine microscopy. When purchasing the camera it is important to see that any special compensators necessary are fitted at the factory so that the camera can be routinely used on the microscope with crossed polars and compensating plates in the light path.

An alternative system is the photomicroscope. This is a universal type microscope with a built-in camera. Some models have two cameras so that positive and negative film can be used without having to change film cassettes. Provided that the basic microscope is fitted with all the necessary accessories for observations using polarised light the built-in camera makes for ease of use.

While the detailed construction of these 35 mm automatic cameras will vary the following facilities should be available if required. These are, motorised film transport, vibrationless shutter, internal beam splitter to allow continuous integration of light for exposure, whole field and spot light integration, and removable film cassettes to allow the use of alternative films while operating. Control units must contain facilities for setting film speed, reciprocity failure adjustment, shutter controls for exposure, trial, interrupt, bulb and flash synchronisation. End-of-film and over-fast exposure indicators are useful items while a foot control frees both hands for manipulation at the microscope when sequential photography is required.

Automatic cameras are becoming more sophisticated. Spot integrators can be moved to any point in the field and zoom lenses allow the field size to be adjusted between objective magnifications. Identification data are printed on the film and the older type of focusing graticule may be replaced by fibre optic projections. This type of sophistication makes for greater convenience. However, no matter how sophisticated the camera equipment, it is still remarkably easy to take unsatisfactory photographs unless the operator understands the basic principles of photomicrography.

Since the camera will have a fixed focus and will be mounted in the trinocular tube by sliding up to a fixed stop its focus must be carefully checked using a test slide. If it is mounted too far into the tube the problem is easy to solve as the camera can be raised with packing rings to give the correct focus by trial and error. If the camera is mounted above the focal plane it should be returned to the manufacturer for adjustment. It is obvious that if the camera is not accurately in focus when first fitted poor results are built into the system.

Some ancillary items are required for photography. These are, depending on the system used, photographic eyepieces fitted with a graticule with double focusing circle and format outline, a green filter for black-and-white photography, an 80B Wratten filter or its equivalent for daylight colour film and neutral-density filters of which 6.3 and 25 per cent are probably the minimum. Gelatin type colour correction filters can be bought from any supplier of photographic materials and a focusing telescope for determining focus at low magnifications (Hinsch 1979), can be helpful if difficulty in focusing is encountered. The method of taking a photomicrograph should follow a set procedure which should be set down as a laboratory method sheet. Failure to follow this type of procedure is a common cause of unsatisfactory photomicrographs.

2.10.2 RULES FOR PHOTOMICROGRAPHY

1. Load cassettes with film and place a cassette in the camera.
2. Check centring and adjustment of light source. The light must be even and centred.
3. Make sure the microscope is correctly adjusted; see Section 2.4.1. Select the field and see that the area of interest fits inside the format markers in the eyepiece. If a zoom magnification is fitted adjust to give best field.
4. Focus the double circle or other focusing device present in the eyepiece until it is quite sharp. This adjusts the eyepiece to the eye used for focusing. Using this eyepiece alone, rock the fine focus back and forth and estimate the mid-position where the object is in focus. Obviously this movement is large at low magnification and smaller at high magnification. The secret of focusing is to watch the fine detail come in and out of focus in relation to its surroundings. Some operators have problems in focusing at low magnification. Using a focusing telescope to examine the image in the eyepiece can help to overcome this problem (Hinsch 1979).

   At low magnification a considerable depth of field can be focused. The human eye has a remarkable ability to compensate for slightly out-of-focus detail which is not possessed by the camera. Thus a careful adjustment of focus is necessary to get the required detail in focus. At higher magnifications this is not a problem as the depth of field will be smaller. In the ability to select and focus the field, both laterally and vertically, lies much of the secret of good photomicrography. It is much more difficult to photograph at low magnification than is generally realised.

5. Remove the daylight blue filter. No filter is required for film corrected for tungsten light. Use an 80B Wratten type filter or its equivalent for daylight colour film and a green-sensitive orthochromatic filter for black-and-white film. Set the lamp voltage to give the colour temperature of the film being used. Usually 3200° K.

6. Swing out the binocular prism. Using the trial exposure try to set the exposure time between 0.25 and 10 seconds by adjustment of the light source with neutral-density filters. An ideal exposure time is about 0.5 s which avoids any problem with reciprocity failure or blurring due to shutter vibration.

7. Expose film and log necessary data on exposure made *immediately*. Immediate and adequate logging of data of photographs taken avoids many frustrating problems of identification at a later date.

The common problems encountered in taking good photomicrographs of concrete subjects in thin-section are, poorly contrasted, coloured fine detail in the hardened paste with large transitions of contrast between the aggregate and the paste, and flare under crossed polars. Flare is increased by dirty optics. If any of these problems are excessive it is better to look for another field if at all possible. Only experience teaches whether a particular field will make a good subject.

The selection of film brand tends to be a matter of personal taste especially in the case of colour films. Brands tend to have slightly different colour tones and may require the judicious use of colour correction filters. In the case of black-and-white film a DIN 18 (ASA 50) or a film such as Ilford FP4 (ASA 125) or its equivalent is satisfactory for most work. For colour film the most important characteristic required is exposure up to at least 10 seconds without any reciprocity failure and reasonable results up to 30 seconds exposure. For routine microscopy 35 mm film is to be preferred to larger format films for convenience but has the disadvantage that it is not available as negative film corrected for tungsten light. Recent developments in the manufacture of colour film have improved resolution, colour balance and latitude against reciprocity failure and these are giving noticeably improved results in photomicrography.

The processing of colour film for photomicrography presents some problems. In commercial processing laboratories the filtration on colour film processors is normally set for skin tones because photographs of people are one of the commonest subjects. However the colours photographed through the microscope are quite different and require some colour reference for the film processor to match. The answer to this problem is to record the photomicrograph using both positive and negative film. Positive film is processed using the now standard E6 process and requires no colour filtration during processing because it is not printed. Thus it should represent the colours in the object seen by the microscope according to the lighting and colour balance of the film brand used and provides the processor with a reference of how to filter the negative film when printing. It also has the advantage of providing colour slides as a permanent and easily accessible record.

2.10.3 SELECTION OF MICROGRAPHS FOR PUBLICATION

The microtextures of concrete in thin-section are complex and subtle detail visible to the microscopist is degraded or lost when photographed. Colour is generally superior to black-and-white photography and transparencies superior to prints when illustrating detail.

This poses a serious problem when illustrating publications with micrographs as the cost of including coloured illustrations generally rules out their use.

Insufficient information is conveyed by black-and-white illustrations for adequate recognition of textural detail unless the detail portrayed has excellent clarity and contrast. In the literature, black-and-white illustrations of concrete textures are often indecipherable unless areas of interest are clearly marked and all too often the bar scale necessary to indicate magnification is missing. Each micrograph should be carefully chosen to see that when illustrated it does in fact show the details described in the text. This situation will not improve until colour illustration becomes more widespread in publication.

2.11   Quantitative microscopy

Quantitative measurement of dimension and frequency of components in a sample is an essential part of concrete petrography. Bradbury (1991) has discussed the basic measurement techniques for light microscopy. Most microscopes have a fine focus control knob engraved with a scale graduated in microns. This allows the thickness of a section or particle to be estimated provided that it is possible to focus accurately on the top and the bottom of the object. To obtain the correct measurement the distance traversed by the focus must be multiplied by the refractive index of the object. This method works well for particles in liquid immersion but is less applicable to sections where thickness is traditionally and more easily estimated by observing the maximum birefringence of minerals such as quartz.

The essential item required for the measurement of dimension is the eyepiece micrometer. This is a disc of glass engraved with a scale of, for instance, 100 divisions equals 10 mm, preferably including crosshairs for centring, which fits into an eyepiece designed for its use. Using a stage micrometer the eyepiece micrometer is calibrated for each objective. The stage micrometer for transmitted light is engraved on a glass slide and that for incident light is engraved on a polished metal slide. Table 2.1 illustrates typical calibrations.

Table 2.1 Calibration of transmitted light objectives × 10h eyepiece – × 1.25 tube factor.

Objective	Nominal magn.	μm/division	Field size (mm)
× 2.5	× 32	32.5	4.8
× 6.3	× 80	12.5	1.85
× 16	× 200	5.0	0.74
× 25	× 315	3.2	0.48
× 40	× 500	2.0	0.30
× 50	× 620	1.61	0.24
× 100	× 1250	0.80	0.12

× 10h High point eyepiece for use with spectacles

A table such as this needs to be kept prominently displayed close to the microscope. Also included in the table are the diameters of the fields of view which are useful for rapid estimation of the sizes of objects which are in the field of view but too large for the micrometer. The eyepiece micrometer is used for measuring the sizes of crystals and remnant cement grains, crack widths, void diameters and other features giving a level of accuracy adequate for this purpose. If more accurate measurements are required a screw micrometer eyepiece may be used.

2.11.1 MODAL ANALYSIS

To determine the volumes of components in section or numbers of particles in grain mounts requires the use of microscope counting techniques. This is known as modal analysis. The two most commonly used methods for concrete are the Chayes (1956) point-count and the Rosiwal (1898) lineal traverse techniques. ASTM C457 and ASTM C295 briefly describe these methods while ASTM E20 specifies the general technique for measuring particles in the range 0.2 to 75 μm.

Systematic identification and counting of particles, voids or minerals in a finely ground surface or thin-section to estimate volume proportions relies on the assumptions that the surface or section is cut through randomly orientated particles and that the area is large enough to be representative of the material. A further assumption made is that the areas of the components measured per unit of test area also estimate the volume of the components per unit volume. That is, the relationship proposed by Delesse (1848) holds.

The two requirements for point-counting are the ability both to lay down an orthogonal grid on the surface under examination and the identification of the component of the sample that falls under each intersection point of this grid. The following devices are all concerned with generating the orthogonal grid.

Hand-operated point-counters which attach to the microscope stage are readily available but these tend to be limited to working areas of 20 $\times$ 20 mm. That is, they are solely intended for the standard sized, geological thin-section. Similarly, the Swift Automatic point-counter, which electromagnetically operates the specimen stage when any one of thirteen buttons on the tally counting unit is depressed, only has a total movement of 30 $\times$ 30 mm. Dials on the counting unit record the number of times each button is operated. The Swift Automatic point-counter can be adjusted to traverse in intervals of 50, 100, 200 and 500 μm in the horizontal direction. It is manually reset orthogonally at the end of each traverse with click stops that are any multiple of 200 μm. In this method the point-counting stage is moved incrementally along a lineal traverse and the component observed under the crosshairs in the ocular identified and tallied at each increment. A series of such lineal traverses can be measured in this way so that the points examined comprise a series of regular grid intersections.

There is no doubt that instruments such as the Swift Automatic point-counter are to be preferred over hand-operated stage counters for the quantitative estimation of components in rock thin-sections. Another method is to use a counting grid inserted in the microscope eyepiece. The component at each intersection is counted and then a mechanical stage is used to move a set distance to a new field until the specimen surface has been traversed. This method as applied to cement clinkers is discussed by Hoffmänner (1973).

However concrete, because of its inherent variability, and use of aggregate which can range from dust which is less than 100 μm in size to aggregates whose maximum size commonly is in the range of 20 to 40 mm requires specimens with large surfaces to be truly representative. ASTM C457 (1990) specifies a counting table, which is available commercially, suitable for point-counting or making lineal traverses of large, ground surfaces of concrete. Using a stereo-zoom microscope in the range of $\times$60 to $\times$80, the volume of air voids and hardened cement paste can be determined. The problem with this method is that at this magnification the stereo-zoom microscope does not image the ground surface of the concrete with sufficient clarity to be able reliably to distinguish remnant cement grains from small grains of aggregate in the cement paste. This is not a problem with air voids which are usually clearly distinguishable and may require as much as $\times$120 to be able to count the smallest bubbles. In this case accuracy of identification is a function of the sharpness of the void margins.

2.11.2 ALTERNATIVE METHODS OF COUNTING

It is relatively easy to design a counting stage to examine a ground surface. This is much more complicated for large-area, thin-sections where they are examined by transmitted light, as large lateral movements are required while maintaining the correct relationship of the section to the condenser and objective of the microscope. Some automatic stages are available as components of image analysis systems but if these are not available it is necessary to use improvised methods. One method is being successfully used to count 1750 points over an area of 120 × 95 mm in approximately 30 minutes.

A tally unit such as that of the Swift Automatic counter is used for recording the results. A large, orthogonal grid is drawn on a piece of cartridge paper with Indian ink using the finest drawing pen possible. This grid is then photographically reduced using Kodalith film to give the required intercepts. Intercepts of 1.25 and 2.5 mm are the most convenient but they do not have to be exactly these dimensions provided they are constant across the negative. Once the negatives are processed there is no limit to the number of grids that can be produced by contact printing from the original film. The most convenient size of finished Kodalith negative is 125 × 100 mm but larger sizes can be produced with a process camera if required. The 125 × 100 mm size with a 2.5 mm grid contains 1750 points and for a 1.25 mm grid gives 3000 points. If a 5 mm grid is required the 2.5 mm grid is used omitting alternate intersections during counting.

To use one of these grids the appropriate sized Kodalith is taped, emulsion side down, on top of the cover glass of the thin-section (Fig. 2.14). Using a magnification in the range × 30 to × 70 the section is slid with the fingers from intersection to intersection quickly traversing the section backwards and forwards until the grid has been counted. For counting a very large section the grid is retaped over adjacent areas of the section. The size of the thin-section is now limited only by the ability of the hand to span its width for manipulation. Some blurring of the section and the grid occurs but this is not sufficiently serious to interfere with the counting.

Modal analysis, which forms part of the general subject known as stereology, has long been accepted as a standard method in petrography and materials science. It is a complex subject and for further information DeHoff and Rhines (1968) should be consulted. However, for the petrographer who may not count a great number of samples during a year, the methods given above should suffice. The question of how many samples are required and how many points to count depends on the adequacy of sampling and the accuracy required. A working rule for concrete is that it is best to count over as many samples and the largest surface area possible because of the inherent variability of concrete.

Several sources of error are inherent in these point-counting procedures of modal analysis in addition to the size of the surfaces sampled. The Delesse relationship assumes that the ratio of the area occupied by a single component in a randomly cut surface to the total measurement area is a consistent estimate of the volume percentage of the component in the whole sample. Since an analysis made on a randomly cut surface is essentially the analysis of an area, the relation holds, but the size of the volume concerned will be dependent on the homogeneity of the material. Preferential orientation, segregation and banding of components, grain size variation and other forms of heterogeneity will introduce errors to

Fig. 2.14 This large-area thin-section of portions of DSP mortar bars is surrounded by aggregate used as filling and has a Kodalith grid taped across its surface for point counting. The 25 μm thick section measures about 150 × 100 mm and is mounted on 4 mm thick float glass.

the estimates of volume proportions if the size and shape of the sample volume is not considered carefully. A further discussion of modal analysis is given in Section 4.8.

Observer bias can be introduced when identifying minerals in thin-section during point-counting. When coloured and transparent minerals or components occur adjacent to each other in the section the coloured materials tend to be overestimated and the transparent ones underestimated as illustrated diagrammatically in Fig. 2.15

The precision of point-counting methods depends on the number of points counted, the grain size or sizes of the component particles in addition to size of the area traversed which must be representative of the volume to be sampled. Van der Plas and Tobi (1965) devised a chart, Fig. 2.16, for estimating the error to be expected in the volume percentage of a particular component given the total number of points counted. As an example of the use of this chart suppose that some 2500 (n) points had been tallied and that the component of interest represented half the points. That is 50 per cent (P) of the volume. From Fig. 2.16 for 95 per cent probability the relative error (2σ expressed as a fraction of P) shown by the dashed lines is 4 per cent. The solid lines give the error at 95 per cent probability (2σ) as 50 ± 2 per cent. The Delesse relationship applies only if the step size of the measuring grid is larger than the individual particle sizes of the components being measured.

When carrying out measurements on concrete it is not possible to comply with the above requirement because of the large size of the coarse aggregate in relation to the smaller components being counted. The counting errors involved are usually not serious. Tables

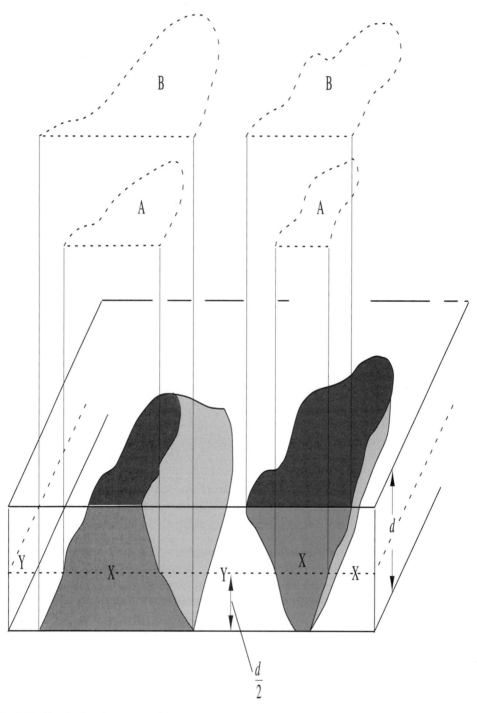

Fig. 2.15 The dark coloured particles in the section thickness d should be estimated as covering areas A whereas an observer will see them as covering areas B.

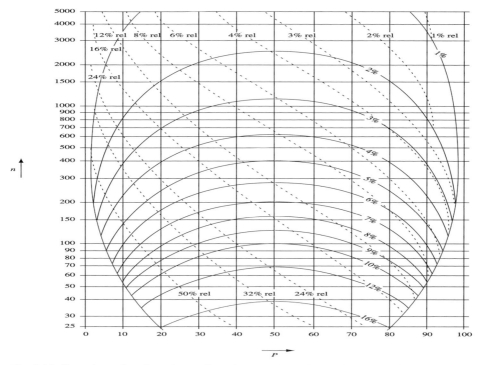

Fig. 2.16 Chart for estimation of error in point counting. The total number of points n counted on the specimen are shown on the y axis and the percentage P of a component on the x axis. At the intersection point of n and P the relative error is read from the dotted lines and standard deviation at 95 per cent probability from the solid curved lines.

1 to 4 in ASTM C457-90 specify the number of counts and areas to be traversed when determining the air void content of concrete and this specification may also be used for counting other components in concretes.

A comparative technique for quantifying the microscopical evidence of ASR and other durability aspects had been suggested by Sims *et al.* (1992).

2.11.3 IMAGE ANALYSIS

Mounting a television camera on the microscope and processing the image from the camera with a computer is called image analysis. Where large numbers of samples have to be measured the image analyser is highly successful provided that sufficient image contrast is present to allow unequivocal identification. For biological specimens this contrast is often provided by staining or other contrast enhancement techniques.

In the case of concrete there are many problems in providing the necessary contrast. There appears to be no image analysis system that gives entirely satisfactory results on thin-sections of concrete. In the case of polished sections of cement clinker some of these problems are now being overcome and the image analysis is being used as a semi-routine technique (Marten *et al.* 1994). Some aspects of ground surfaces of concrete such as air voids or cracks, are imaged more successfully as contrast enhancement is easier. Chatterji and Gudmundsson (1977), Roberts and Scali (1984), Laurencot *et al.* (1992), Buckingham and Spaw (1988) and Cahill *et al.* (1994) have proposed methods where the air voids in a ground surface of concrete are filled with contrast material and other methods such as filling the voids with fluorescent paint are now available commercially. A method for determining paste volumes where the contrast of the paste is increased by acid etching has also been proposed by Gudmundsson and Chatterji (1979) and Efes (1988) and Shi (1988) have investigated the use of image analysis for determination of composition. Petrov and Schlegel (1994) have investigated the automatic image analysis of autoclaved aerated concrete and Lange *et al.* (1994) the characterisation of the pore structure of cement-based materials.

Attempts have been made to estimate the fluorescence of impregnated thin-sections for the determination of water/cement ratio using image analysis. This has been attempted by Wirgot and Van Cauwelaert (1991). Following up this work Elsen *et al.* (1995) used automated image analysis to determine the water/cement ratio of hardened cement paste and concrete and concluded that more research is needed to assess the reproducibility of the method and the variability between laboratories.

Computer programs are becoming increasingly sophisticated in manipulating data derived from images (Russ 1990) and can now be used on a personal computer with sufficient capacity. Also there has been a big improvement in the video camera devices available which can be fitted to a microscope. The more expensive of these cameras is capable of producing images which approach the resolution of photography and there is no doubt that within the foreseeable future these systems may become viable alternatives to the traditional film camera. However, the basic problem of providing sufficient detail and contrast between the components to allow reliable counting remains irrespective of the system being used. A short course on image analysis in mineralogy has been presented by Petruk (1989).

A method is described by Kaneda *et al.* (1995) for obtaining three-dimensional images by using a colour video camera mounted on a light microscope which electronically merges a series of images from different depths of focus.

2.12   Examination of particles

Where it is necessary to examine large particles such as aggregates used in concrete, the experienced petrographer may carry out a visual examination. From observation of grain size, mineral species present, colour, texture and specific gravity it is possible to identify many common rocks. If a more definitive identification of the rock and mineral types is required the aggregates may be crushed to approximately 100 μm and examined with transmitted light under the microscope in an immersion medium. The refractive index oil chosen should, if possible, not differ too markedly from the refractive indices of the particles under examination, to prevent excessive relief. Once again the experienced petrographer is often able quickly to identify the common rock types using this technique and by point-counting give a quantitative estimate of the various rock or mineral components present, (Fleischer *et al.* 1984).

Where the identification of particles is in doubt they may be subjected to X-ray powder diffraction analysis which is usually sufficient to identify most crystalline minerals. It is the facility of X-ray diffraction analysis that has made so many of the old microscopic techniques superfluous. Infrared spectroscopy can also be used to identify minerals present by grinding the particles and mounting in Nujol between KBr discs. This has the advantage that only a few mg of sample is required.

It is possible to identify most minerals by determining their refractive indices in immersion liquids and determinative tables based on refractive indices and other optical data have been published by Fleischer *et al.* (1984). A range of suitable immersion liquids has been described by Larsen and Berman (1934) and closely calibrated ranges of liquids are available commercially. To determine its refractive indices a crystal is rolled to give the correct orientation for measurement and the immersion liquid replaced until a close match between the refractive index of oil and the crystal is obtained. This technique, which is so often described in standard texts and made to sound easy, is in fact quite difficult and requires experience before it can be used with ease. If full optical data are required the spindle stage techniques as described by Saylor (1966) and Bloss (1981) may be used.

When photographing particles the rules given in Section 2.10.2 are generally applicable. However, for purposes of effective illustration it may be necessary to resort to some of the special procedures given by McCrone (1983).

2.12.1 DISPERSION STAINING

This is a method for identifying fine particles and has special application to dusts, asbestos, industrial materials and cases where it is necessary to identify a very small amount of a particular mineral or material, even as little as one particle, among a large number of other particles. The methods, determinative tables and graphs are given by McCrone and Delly (1973). A special objective with a central stop and annulus is required which may be purchased from the McCrone Institute.

The technique consists of placing the fine particles in immersion liquids of high dispersion which are commercially available from Cargill (R.P. Cargille Laboratories Inc., New Jersey, USA), and observing the particles under the microscope using the special objective. Dependent on the differences between the refractive indices of the immersion liquid and the particles, characteristic colour fringes appear around the margins of the particles which can be used for identification. If one particle with a different refractive index is present it will have different colour fringes and will stand out in the field of view. In this respect dispersion staining is a powerful technique and should always be kept in mind by the petrographer especially when dealing with fine, heterogeneous materials.

2.13 Examination of thin-sections

The standard method for observing rock texture is the examination of thin-sections. Traditionally these are cut to a thickness of approximately 30 μm and all optical data for rocks in thin-sections are based on this thickness. In addition the mounting media used in making the section usually have a refractive index close to 1.54.

For concrete a thickness of approximately 25 μm often better reveals texture and for cement clinker a thickness of less than 20 μm is necessary to prevent crystal overlap obscuring texture (Campbell 1986). Provided that the area of the section does not materially exceed 25 $\times$ 25 mm the top surface may be finished to produce a polished thin-section which can be examined in both transmitted and reflected light. This technique is important for the examination of cement clinkers. For general petrographic purposes it allows the section to be not only used on the microscope but also for examination by the scanning electron microscope. Numerous descriptions of thin-sectioning techniques have been made. For the concrete petrographer the descriptions given in Allman and Lawrence (1972) and Campbell (1986) for preparing the traditional sized sections are recommended.

Where concrete is to be thin-sectioned larger sections are required because of the variability of the material. Sections as large as 150×100 mm to represent the full cross-section of a core may be prepared although many concrete petrographers work with 75×75 mm sections and obtain acceptable results. Methods for producing these large-area thin-sections by grinding have been described by St John and Abbott (1983) and using lapping techniques by Wilson (1978). The mounting media used to prepare these sections is invariably an epoxy resin, many of which have refractive indices of 1.7 to 1.8, which has to be taken into account when viewing minerals.

The manipulation of the traditional sized, thin-section presents few problems. It is not necessary to use a mechanical stage and the experienced petrographer is able to position the section on the circular stage with his fingers, even at high magnifications. Large-area thin-sections are more difficult to manipulate. Firstly, they may be mounted on glass plates as thick as 4 mm and in these cases the use of a 4 mm focal length top lens in the condenser is required. Secondly, when examining the edges of the specimen it is necessary to use a stage clip to anchor the section. It is still possible to slide the section under a stage clip for positioning. Thirdly, very large sections collide with the microscope stand so that only part of the stage rotation can be used. However, manipulation is basically similar to the smaller sections and with practice few difficulties are encountered. A stage for use with large-area thin-sections has been designed by Hjorth (1965).

There are problems of locating points of interest in these large sections. It is necessary to either use orthogonal coordinates or mark points of interest on negative prints of the sections. Because of the large amount of detail present in a large-area thin-section of deteriorated concrete it can be difficult to obtain an overview of all the data. Plotting cracks, carbonation, areas of alkali silica gel and other data on a negative print helps to clarify relationships and assists diagnosis.

2.13.1 STAINING MINERALS FOR IDENTIFICATION

It is possible to treat thin-sections during their preparation with a variety of solutions that distinguish between types of feldspars, carbonates and some sulfates as well as other minerals. This has application to aggregates where there may be an alkali carbonate problem or unsoundness due to the presence of sulfate minerals. The procedures are described by Allman and Lawrence (1972) and a general review of methods has been given by Reid (1969).

2.14 Spectroscopic methods of identification

X-ray diffraction spectrometry (XRD) is one of the few methods available for the direct identification of crystalline phases in substances and is widely used in the material sciences (Klug and Alexander 1974). It is to be noted that phases must be crystalline as a unique diffraction pattern is not obtained from amorphous substances. The usual method of analysis is to grind a sample, approximately 0.5 g is required, to less than 50 μm in particle size. The sample is then mounted and inserted into a diffractometer where the X-ray diffraction data is collected electronically and stored in a computer or recorded on a strip chart. Identification of the phases present is made by comparison with a library of known patterns.

In the case of cement clinkers and anhydrous cement powders the principal mineral phases can usually be readily identified from the diffraction trace. Quantitative estimates of the cement phases are possible by using selective extraction of the silicates by salicylic acid in methanol (SAM) and extraction of the alumina bearing phases by potassium hydroxide and sugar solutions (KOSH) (Gutteridge 1979, Takashima 1958, 1972).

Identification of the cement phases present in concrete and mortars is more complicated because the calcium silicate hydrates are poorly crystalline and give weak diffraction peaks which tend to be swamped by the diffraction peaks from calcium hydroxide. In addition peaks from minerals present in the aggregates may also interfere.

The practical limits for detecting phases depend on the crystallinity and ordering of the phase. For quartz as little as 0.1 per cent may be detected but for many cement hydrates as much as 5 per cent may be required for detection. As a rough guide, if less than 2 per cent of a phase is present X-ray diffraction analysis may have difficulty in detecting the phase. In general, results obtained from X-ray diffraction techniques are at the best only semi-quantitative unless careful standardisation procedures are used.

If only small amounts of sample are available the diffractometer can still be used. Very thin layers of powder can be placed on double-sided adhesive tape or special backing plates. If only a few mg of material are available a photographic technique such as the Gandalfi camera may be used. In this way identification of a few grains of material, perhaps selected from a larger sample, may be achieved.

The value of X-ray diffraction methods for research studies of cement hydration have been extended by the use of synchrotron energy dispersive X-ray diffraction techniques (Barnes *et al.* 1992). This powerful technique enables real-time investigations of early hydration reactions to be undertaken such as the synthesis of ettringite (Muhamad *et al.* 1993) and the rapid conversion of calcium aluminate cement hydrates (Rahid *et al.* 1992, 1994).

Infrared spectroscopy may also be used to identify minerals (Farmer 1974), but finds its greatest use in the identification of organic materials including admixtures in concrete. In the case of minerals the spectra are usually only fingerprints which may not always be unique. Information on hydroxyl groups and water molecules can be derived from the spectra so that it is possible to obtain information on both hydrates and glassy materials. In this respect infrared spectroscopy remains a useful technique for some aspects of cement and concrete investigations and should not be neglected as a technique.

An infrared spectrum is the measure of the vibrations of chemical bonds in a material which can be diagnostic of functional groups especially in the case of organic molecules. While it may not always be possible to identify an organic material the infrared spectrum will usually provide information on the presence of functional groups allowing the nature of the material to be derived.

The older spectrometers mechanically scan the infrared spectrum using a grating which is slow and cumbersome. Modern instruments now use Fourier transform methods (FT-IR) to analyse the infrared spectrum which is very rapid. When this method is combined with computer databases built into the instrument, consisting of a library of spectra, searching for and identifying a spectrum is considerably improved. The older method of manually searching very large libraries of spectra such as that produced by Sadtler Research Laboratories is being superseded.

Powder samples are prepared by set procedures of grinding and encapsulating in potassium bromide discs as specified by the instrument manufacturer. Liquid samples are mounted as a film between potassium bromide or sodium chloride plates. Only a few mg of sample are required so that infrared analysis can be used as a microanalytical technique.

There is a considerable range of spectrometric techniques that are used for analysis which are outside the scope of this book. Sibilia (1987) discusses most of the techniques required for materials characterisation if further information is required.

2.15   Scanning electron microscopy

The scanning electron microscope (SEM) is now used extensively in materials science and has many applications in cement and concrete petrography. For a general text reference should be made to Hayat (1989). It has two modes of operation that are of importance. The first of these is the ability to produce images of surfaces in the range $\times 50$ to $\times 50000$ with sufficient depth of field to give a three-dimensional effect. The second is to use the electron beam's production of X-rays which allow analysis of volumes as small as 1 μm in diameter (see Section 2.15.1). However, the electron beam only penetrates a few microns below the surface so that analyses are essentially restricted to surface material and accuracy is vitally dependent on the quality of the polished surface. It is to be noted that this is not an analysis of mineral phases as in X-ray diffraction but the analysis of elements.

Many modern scanning microscopes can take samples as large as $150 \times 100 \times 30$ mm. When large samples are examined they may require degassing overnight to obtain the required vacuum in the SEM and there will be some restriction on the degree of sample rotation. Samples can be in any form provided that they can withstand the high vacuum required in the specimen chamber for the operation of the microscope. Electrically non-conducting materials such as concrete must be given a thin conductive coating to prevent charge build-up on the surface of the sample. This coating procedure will normally be carried out by the microscope operator.

If reliable analytical results from X-ray analysis of cementitious materials are to be attained a flat, finely polished surface without any smearing is necessary. Powders may be encapsulated in mounting resin and must be polished using non-aqueous lubricants. Provided the surface being analysed is well polished the modern SEM can analyse for many elements down to about 0.1 per cent from a volume of 1 to 2 μm in diameter and with suitable standardisation the results are reasonably quantitative. However, there is a cut-off point for analysing for the lighter elements at sodium, unless the SEM is equipped with a light element detector which then allows for analysis down to carbon. There are other problems with the X-ray analysis of the lighter elements which restricts both detection limits and accuracy so analytical results for these lighter elements should be treated with caution.

Other X-ray analytical modes of operation are possible. Areas may be mapped in terms of elemental composition and line traverses can be made to plot changes in element concentration. These techniques are useful for delineating reaction zones and areas where the concentration of an element has changed from its normal distribution in the material. It is also possible to image a surface using back-scattered electrons which broadly reflects the atomic number. The higher the atomic number the brighter the area in the image.

Where the SEM is used to examine the polished surface of conventional light microscope thin-sections it becomes a powerful technique as it is possible to analyse areas of interest previously defined optically, for elemental composition. In the scanning image mode which gives such clear pictures of fracture surfaces the relationship to microscopy is tenuous as the techniques are imaging different attributes of the sample. There is a danger with SEM imaging of fracture surfaces of producing dramatic pictures which can be difficult to relate to the broader features of a sample as seen by thin-section microscopy and other techniques.

2.15.1   ELECTRON PROBE MICROANALYSIS

The electron micro-probe (EMP) is similar in many respects to the SEM and in modern instruments these differences are diminishing. However, the electron beam is larger, about 1 μm in diameter as opposed to 100 nm on the typical SEM and higher beam currents are used. In addition, the X-ray detection system uses wavelength dispersive detector systems as opposed to the energy dispersive detectors normally used with the SEM. Because the electron beam is larger and often operated at higher voltages the penetration of X-rays below the surface of the sample can be as deep as 15 μm. When all these factors apply the accuracy of analysis tends to be better on the EMP and detection of the lighter elements is superior to the SEM.

For EMP work the sample surfaces to be analysed must be polished and coated for satisfactory results to be obtained. Areas of interest must be pre-marked as the optical microscopes fitted to many instruments for viewing the sample surface have poor resolution and contrast so that it may be difficult to locate areas of interest. As in the SEM mapping elements by area or line traverse are standard techniques of analysis. Because the electron beam tends to be operated at higher energies in the EMP, non-conducting elements can rapidly heat if the beam is focused in one spot for any length of time. This can lead to movement and volatilisation of elements with low melting points and affects the accuracy of analyses unless care is exercised.

The modern EMP is becoming more versatile and sophisticated. It can take large samples so that sections of concrete cores can be examined, may have up to six high-speed channels for analysis and also be equipped with image scanning facilities. It still remains the premier method for *in situ* microanalysis in geology and materials science and thus is applicable to concrete petrography.

2.15.2   SPECIALISED SCANNING ELECTRON MICROSCOPES

A range of specialised SEMs are now available which do not require the specimen to be coated and can perform examinations in the wet state, Danilatos (1991). However, because of their very nature there are some limitations on the resolution and analytical capability of these instruments. To negate the effects of charging in low vacuum mode (>270 Pa) the sample chamber contains air and this causes beam skirting. To overcome this problem, EDAX modifications to the machine's X-ray detector are required. In respect of hydrating cement, work has been carried out by Jennings and Sujata (1992) and Caveny (1992). The principles of variable pressure SEMs have recently been reviewed by Mathieu (1996).

2.16   Thermal analysis

This technique consists of programmed heating of a sample in a special furnace where the atmosphere can be controlled if required and then measuring energy or weight changes which take place. The method where energy changes are measured is known as differential thermal analysis (DTA) and that where weight changes are measured as thermogravimetric analysis (TGA). Separate instruments are normally used for the two techniques although with improvements in instrumentation combined models are now available. In both techniques as little as a few mg of sample can be analysed. Commonly, samples are prepared in powder form but liquids and some types of solids can also be analysed. The techniques spans the temperature range of $-195\,°C$ to $1600\,°C$ but for the type of materials likely to be analysed in cement and concrete technology the range ambient to 1150°C is applicable.

Thermogravimetric analysis is a quantitative method of analysis while differential thermal analysis is usually qualitative. If quantitative differential thermal analysis is required another related technique known as differential scanning calorimetry (DSC) is used. Currently commercial differential scanning calorimetry instruments are limited to a maximum temperature of approximately 700°C. However, there are variants of differential scanning calorimetry which can be made semi-quantitative if required which will operate to higher temperatures.

Thermal analysis techniques have been widely used in the clay, ceramic and materials sciences. Its principal use in cement chemistry is for measuring dehydration reactions, detecting phases such as gypsum and organic materials. Bhatty (1991) has reviewed the application of thermal analysis to cement-admixture systems. Reviews of thermal analysis applied to cement and concrete have been made by Ramachandran (1969), Barta (1972) and Ben-Dor (1983). DTA finds special use in detecting the conversion of the hexagonal aluminate hydrate to the low-strength cubic phase in concrete containing high-alumina cement (Midgley and Midgley 1975) and TGA is one of the few methods for estimation of calcium oxide in cements. The difference between the amount of calcium hydroxide found by TGA and the free lime determined by chemical analysis is equivalent to calcium oxide.

Differential scanning calorimetry is now the preferred thermoanalytical method because of its high sensitivity especially for the detection of organic materials in cements (Portilla 1989). It can be used for detection of the small changes that cause pack and false sets in cements and easily detects the first stages of hydration, ettringite, gypsum and syngenite. However, like all highly sensitive thermoanalytical systems that use low mass sample holders it does not satisfactorily discriminate between gypsum and hemihydrate. The older type of massive sample holder with thermocouples that protrude into the sample are required for this purpose. Smillie *et al.* (1993) have applied differential thermal analysis to the detection of syngenite.

2.17   Chemical analysis of cement

There are a number of different aims in analysing a sample. In many of the methods given above the aim of the analysis is to identify an unknown component. This may be carried out directly by diffraction X-ray, microscopy, thermal analysis and other techniques depending on the nature of the sample or indirectly by determining the elemental composition. Another aim is to determine precisely the composition of a sample whose identity is already known, for the purposes of controlling quality and properties. As there is usually plenty of sample available the methods used are largely dictated by the degree of accuracy required. Recommended methods for the chemical analysis of cementitious materials will be found in most national standards. Wet chemical methods are in many cases being superseded by X-ray fluorescence analysis which is capable of producing results of a similar reliability (Norrish and Chappell 1977).

2.17.1   *CHEMICAL ANALYSIS OF BLENDED CEMENTS*

The analysis of blended cements presents analytical problems many of which still have no satisfactory solution. Materials such as limestone, granulated blastfurnace slag, fly ash, silica fume and a range of natural pozzolana are added to cement either by intergrinding or addition at the concrete mixer. The use of these materials is becoming common. If the specifier wishes to determine that the claimed blending material has been used at the specified addition this can be reliably determined only if samples of the cement and the blending material are also supplied to the analyst. Where the analyst is supplied a blended cement without any data on the materials which it contains the probability of obtaining a reliable analysis is not good. Identification of the blended material presents few problems. Either examination of residues after acid extraction or examination of the cement by optical microscopy or SEM will usually identify the blended material. A general discussion of methods has been given by Papadoulos and Suprenant (1988).

The methods required to determine the quantity of blended material will vary according to the material added. In the case of Portland/limestone cement it may be possible to estimate the limestone content by using a combination of TGA and XRD methods to determine calcium carbonate and then checking these results by a lithium gluconate extraction of the cement fraction. Where granulated blastfurnace slag is used a selective dissolution technique has been proposed by Demoulian *et al.* (1980) and further investigated by Luke and Glasser (1987). Investigation by Goguel (1995) found that Demoulian's technique was not applicable to all types of granulated slag and that selectivity ratios have to be increased by the use of large cations in the leachant. Similar techniques may possibly be applied to Class F fly ash but for the Class C high lime fly ashes their appears to be no reliable analytical technique for their determination in blended cements without the original material being available.

The greater portion of silica fume and natural pozzolana are not soluble in either hydrochloric or nitric acid and a reasonable estimate may be made on the basis of the amount of residue left after acid extraction of the blended cement. Where the use of blended cements is specified it is prudent for the specifier to obtain representative samples of the cement and blending material if check analyses of the blended cement are to be carried out. Otherwise reasonably accurate estimation of the amount of blending material present may not be possible.

2.17.2   CHEMICAL ANALYSIS OF HARDENED CEMENT PASTE IN CONCRETE

The identification and estimation of the materials present in hardened concrete is not a simple matter and gives quantities which may not be the same as those originally used in the concrete mix. Even concrete mixed under close control is a somewhat variable material, and this variability may be further increased by incomplete compaction and areas of local inhomogeneity due to segregation and settlement in the plastic state. The setting of concrete is a series of chemical reactions which can continue far into the life of the structure. Components, such as chemical and mineral admixtures also take part in these reactions and in the process may be partially or completely consumed as they are incorporated into the hardened matrix of the concrete. Concrete also reacts with its environment. Moisture and alkalis can move both into and out of the concrete while carbon dioxide from the atmosphere and ions such as chloride and sulfate can penetrate the outer layers. Thus it may be impossible to determine all of the components used in the original mix.

Methods for the analysis of concrete were detailed and discussed by Figg and Bowden (1971) and their publication remains a useful reference. A Working Party of the Concrete Society and the Society of the Chemical Industry (Concrete Society 1989) updated and discussed methods of analysis but some of the useful pictorial information given in Figg and Bowden has not been included. These two publications are good sources of information for the reader who does not wish to consult the specialised literature. BS 1881: Part 124 (1988) also details some methods for the analysis of hardened concrete.

The component most likely to require determination is the cement content as this information is vital where there is a dispute over the quality of the concrete. The classical method involves acid dissolution of the concrete to determine the calcium and silicon oxides in the hardened paste and then uses these results to determine the cement content from an assumed composition of the cement. Acid dissolution also attacks aggregate, the amount of attack varying considerably dependent on aggregate type. The problem is that it may be difficult to estimate satisfactorily the unwanted contribution made by the aggregate to the analytical results and large errors can occur. Determination of hardened paste volumes by microscopic point-counting in thin-section avoids the problem of contamination from aggregate and allows estimation of the cement content provided the water/cement ratio can be determined. It is claimed by French (1991) that microscopical point-counting methods are quicker and more accurate than chemical analysis in the determination of the cement content of concrete. Errors in the chemical determination of cement content have been estimated at the 95 per cent confidence level as ± 25 kg/m^3 for sampling, ± 30 kg/m^3 in analysis and 40 kg/m^3 for sampling and analysis combined (Concrete Society 1989). Confidence levels for microscopic point-counting do not appear to be available.

Where high-alumina cement has been used in the concrete the cement content is usually estimated from the alumina content which is prone to error. Chinchon *et al.* (1994) have proposed a method based on the X-ray fluorescence analysis of the iron content of the concrete corrected for the iron in the aggregates which they claim reduces the errors.

The older method for the determination of water/cement ratio, BS 1881: Part 124 (1988), is a complex method and does not always give satisfactory results. Determination of the porosity of the hardened paste-based fluorescence techniques as described in Section 2.7 is used in some European countries to determine water/cement ratios. Another method of relating porosity to water/cement ratio as measured by porosimetric data has been proposed by Tenoutasse and Moulin (1990).

A method for determining the origin of the cement used in a hardened concrete, providing limestone aggregates are not present, has been described by Goguel and St John (1993). Once the identity of the works which manufactured the cement is known it is then possible to use historical chemical analyses to estimate the composition of the cement used. Because the composition of cement from a works does not vary greatly the original alkali content of the concrete and other compositional details can be calculated if the cement content is estimated as detailed above. Where construction records are no longer available this is the only method of estimating the alkali content in the concrete as originally mixed and is an effective method of resolving disputes on this issue.

The determination of the alkali content in a concrete that has been subject to weathering is possible but the results may not relate to the original alkali content as mixed. Not only can acid dissolution of concrete extract alkalis from some aggregates but the alkalis themselves are mobile and may be deeply leached from the concrete by weathering and in some cases exchange into the aggregates. These are competing effects which can be partially resolved by dissolution of the hardened paste with alkaline EDTA instead of acid to determine the alkalis, and also by determining the silica and alumina contents to interpret the results (St John and Goguel 1992).

Chemical admixtures, such as plasticisers, super plasticisers, air entrainers, retarders, accelerators and integral waterproofers may be identified in concrete (Concrete Society 1989), provided that it is not more than a few days old. In older concretes this is not always possible because chemical admixtures based on organic materials may be broken down by the alkaline pore solution (Milestone 1984). Investigation of some of the published methods (Rixon 1978) indicates that they do not appear to work for older concretes (St John 1992).

Identification of mineral additions in old concretes is usually possible by dissolution of the hardened cement paste and examination of the residues. Sufficient of the mineral addition has to remain in the concrete to allow identification of the residue by microscopic or spectroscopic analysis. Accurate estimation of the quantity of the mineral addition used is possible only for very young concretes and even then may be difficult. In older concretes a considerable portion of the mineral addition is no longer present because it will have reacted to form hydrates of a similar composition to those formed from the cement and as a result estimates will be unreliable. French (1991) has proposed an electron microscope method for the determination of slag and fly ash in concrete.

When concrete is attacked by sulfate or chlorides the amount of attack can be estimated by analysis of the concrete. The methods commonly used are detailed in BS 1881 Pt 124: (1988). It is necessary to have some idea of the sulfate content of the cement so that this can be subtracted from the results to give the amount of external sulfate that has penetrated the concrete. Where estimation of ettringite is required an XRD method has been reported by Ludwig and Rudiger (1993). It should be noted that ettringite is only one of the sulfate compounds formed in sulfate attack on concrete.

Total soluble chloride can be determined by acid extraction of a powdered concrete sample. Work by Dhir *et al.* (1990) shows that to extract chlorides fully the sample should be boiled in the acid for a longer period than specified in BS 1881. On chemical grounds extraction would also be assisted by using acid concentrations of 1M or less and for the lower chloride levels colorimetric determination is superior to titration. Dhir *et al* (1990) also concluded that the method used to determine water-soluble chlorides is unimportant provided that the extraction time exceeds 24 hours. A rapid method for measuring acid-soluble chloride in powdered concrete samples has been proposed by Weyers *et al.* (1993).

2.18 The petrographer's bookshelf

Although some of the selected texts are now difficult to obtain they are invaluable as guides for concrete petrography. A broader listing has been given by Delly (1986) for a basic microscopy library.

Adams, A.E., MacKenzie, W.S. and Guilford, C. 1991: *Atlas of sedimentary rocks under the microscope.* UK, Longman Scientific & Technical.

American Geological Institute. 1962: *A dictionary of geological terms.* Prepared under the direction of the American Geological Institute. New York, Dolphin Books, Doubleday.

Bloss, F.D. 1966: *An introduction to the methods of optical crystallography.* New York, Holt, Reinhart & Winston.

Brindley, G.W. and Brown, G. (eds) 1980: *Crystal structure of clay minerals and their X-ray identification.* London, Mineralogical Society Monograph No. 5

Campbell, D.H. 1986: *Microscopical examination and interpretation of Portland cement and clinker.* Skokie, Illinois, Construction Technology Laboratories, a Division of the Portland Cement Association.

CRC Press. *CRC handbook of chemistry and physics.* Florida, CRC Press.

Dana, J.D. *The system of mineralogy.* 7th edn Palache, C., Berman, H., Frondel, C. 1944 *Elements, sulfides, sulfosalts, oxides, I.* 1951 *Halides, nitrates, borates, carbonates, sulfates, phosphates, arsenates, tungstates, molybdates, etc, II.* 1962 Frondel, C. *Silica minerals, III,* New York, Wiley.

Hartley, W.G. 1962: *Hartley's Microscopy.* England, Senecio Publishing.

Heller, L. and Taylor, H.F.W. 1956: *Crystallographic data for the calcium silicates.* London, DSIR Building Research Station, HMSO.

Hewlett, P. (ed.) 1997: *Lea's Chemistry of Cement and Concrete* 4th edn, London, Edward Arnold.

Insley, H. and Fréchette, van D. 1955: *Microscopy of ceramics and cements including glasses, slags and foundry sands.* New York, Academic Press.

JPCDS, International Centre for Diffraction Data. Published annually. *Powder diffraction files.* USA.

Kerr, P.F. 1977: *Optical mineralogy.* New York, McGraw-Hill Books.

Lea, F.M. 1970: *The Chemistry of Cement and Concrete* 3rd edn, London, Edward Arnold, London.

MacKenzie, W.S., Donaldson, C.H. and Guilford, C. 1991: *Atlas of igneous rocks and their textures.* UK, Longman Scientific & Technical.

Mackenzie, W.S. and Guilford, C. 1980: *Atlas of rock forming minerals in thin-section.* London, Longman Group.

Merck and Co. *The Merck index of chemicals and drugs.* Rahway, New Jersey, Merck.

Rigby, G.R. 1948: *The Thin-Section Mineralogy of Ceramic Materials.* British Refractories Research Association.

Taylor, H.F.W. 1990: *The chemistry of cements.* London, Academic Press.

Tröger, W.E. 1971: *Optische bestimmung der gesteinsbilderenden Minerale. Pt I Identification tables.* Newly rev. edn by Bambauer, H.F., Taborsky, F. and Trochim, H.D. Stuttgart, E. Schweizerbart'sche Verlagsbuchhandlung.

Verein Deutscher Zementwerke E.V. (ed.) 1965: *Microscopy of cement clinkers. Picture atlas.* Düsseldorf, Beton-Verlag GmbH.

Wahlstrom, E.E. 1979: *Optical crystallography.* New York, Wiley, 5th edn.

Winchell, A.N. and Winchell, H. 1952: *Elements of Optical Mineralogy. Part II. Description of minerals.* New York, Wiley.

Winchell, A.N. 1943: *The optical properties of organic substances.* Madison, The University of Wisconsin Press.

Winchell, A.N. and Winchell, H. 1966: *The microscopical character of artificial inorganic substances and optical properties of artificial minerals.* New York, Academic Press.

3
Sampling and Specimen Preparation

3.1 Sampling problems

It is necessary to collect samples for petrographic examination which will be representative of the material under investigation and also provide enough material for all the laboratory tests required. In the case of particulate materials sampling does not constitute a problem as there are methods based on statistical principles to ensure that the samples are representative. Since particulates are loose materials, normally there are few difficulties in obtaining the necessary size of sample for this purpose.

Since concrete is a mixture of materials many of the same principles can be applied providing the constituents to be measured are reasonably distributed in the concrete. However, there may be engineering or economic constraints to obtaining the necessary number of samples to be truly representative. In addition, petrographic examination often involves not only determination of well distributed constituents such as volumes of hardened paste, aggregate and air content but also other aspects of the concrete such as highly localised deterioration. While strategies can be devised to cope with sampling a segregated aggregate stock pile the variability common in concrete and the localisation of deterioration in structures can present considerable problems if representative samples are required.

In order to deal satisfactorily with these problems the petrographer must first give careful consideration to the kinds of information and the level of detail required of a petrographic examination before designing an appropriate sampling programme. The information required is of three types. The first of these is concerned with the characterisation of a concrete in the whole structure or part of it. This will entail identifying and measuring constituents such as the hardened cement paste, air content, aggregate types and other materials or properties that are distributed through the mass of the concrete. In these circumstances a random or systematic programme of sampling will need to be designed in order to obtain results representative of the whole concrete. However, petrographic examination is usually restricted to concrete that is located away from areas of high variability, which in practice means that measurements are restricted to the internal portions of cores which are considered to represent the bulk of the concrete and the variable areas such as those present near the surface are often ignored. The precision of the result obtained will depend on the density of sampling, and both experience and judgement will be required to decide on the level of precision necessary for their correct interpretation.

The second type of information required is concerned with the variability associated with the natural or induced zonation of the concrete. These zones can be broad and diffuse, for example the variation of water/cement ratio between the bottom and top of lifts, or relatively narrow such as the zonation close to external surfaces. In most cases zones can be considered as discrete separate areas to be randomly sampled but there is a danger with this type of sampling that it may miss critical areas of inhomogeneity. In practice, there is usually little attempt made to randomly sample these areas and instead samples are located on the basis of detailed inspection and appraisal of the concrete and a reliance on the experience and skill of the inspector.

The third type of information involves identifying the nature and cause of specific faults and types of deterioration. Once again this can range from a broad distribution through the structure to a highly localised area which needs to be sampled. Invariably the selection of sample sites is based on judgement derived from inspection of the structure and on experience. Random or systematic sampling would be counter-productive when investigating the causes of faults and deterioration of concrete. It cannot be over emphasised that effective sampling in such cases requires considerable skill and judgement based on a thorough evaluation of the structure backed up by practical experience. Ineffective sampling is a common cause for unsatisfactory petrographic examination.

One of the most common reasons for carrying out a petrographic examination of concrete is that faults and deterioration are causing the concrete to perform below expectation. The identification of the nature and causes of such faults or deterioration requires sampling to be directly related to the fault(s). However, in order to relate the fault or deterioration to the whole structure it is also necessary to characterise the quality of the concrete in the structure in order to determine whether it was mixed and cured according to specification for example. Ideally this will require an additional programme of random sampling but because of the costs involved and practical difficulties the petrographer is usually restricted to samples from the faulty concrete and is then forced to estimate the quality of the concrete from these samples. This is unsatisfactory and the limitations should be clearly recognised.

In cases of dispute it is essential that the structure is sampled both on random and selective bases. Where non-random samples are examined results of measurement of constituents used to characterise the concrete should be carefully recorded as being derived from samples that may not be truly representative of the bulk of the concrete. The sampling strategies necessary to provide these various kinds of information are summarised in Table 3.1.

Table 3.1 A sampling strategy for the petrographic investigation of concrete structures.

Information requirement	Dimensions of area/volume to be investigated (m^2/m^3)	Typical example	Sampling scheme
1. Characterisation	1–50	Structural component	Systematic/random
2. Variability			
Large scale	1–2	Aggregate segregation	Systematic selection
Small scale	0.2–0.01	Surface zone alternation	Selective
3. Defects	0.5–0.01	Surface spalling	Selective

Notes
1.   The number of core samples required depends on the size of the area/volume to be sampled and the level of precision required (see also Section 3.5).

2. and 3. Although cores from selected areas are used to provide information directly related to the problem being investigated, additional cores will be necessary in order to characterise the bulk concrete.

3.2   Inspection of structures

Petrologists and engineers will appreciate the importance of visiting and carefully examining any site which is the subject of an investigation. If samples are to be taken for petrographic examination it is important for the petrographer to be involved in the sample selection. Where precast units are to be investigated it is often very helpful to observe the manufacturing process and to evaluate the raw materials used.

As already indicated the level of detail and the time spent on a particular site inspection will be dictated by economic considerations and the size and complexity of the structure. Problems with concrete structures may range from defects in curb-stones and roof tiles to deterioration of major viaducts or dams. Nevertheless, the importance of carrying out an initial investigation on site cannot be over emphasised and whenever possible, details of the materials used and the methods and date of construction should also be obtained. Study of similar structures or other structures in the local area constructed of similar materials can also be helpful in providing 'case-study' evidence, particularly if these other structures vary in age from the one under investigation. Account must also be taken of climatic and other external environmental factors at the location, since factors such as freeze-thaw conditions for example, may be of considerable importance when assessing the causes of deterioration.

Since cracking is a very common feature on concrete structures, the nature and extent of any cracking is best observed when surfaces are drying from recent rain since the differential drying picks out and highlights fine cracks or other defects. It is necessary to record the weather conditions at time of inspection as it may have an influence on the observations made.

An assessment may also need to be made of the particular environmental conditions to which each part of the structure has been exposed. In particular the wetting and drying frequency and temperature variation that an element is subjected to should be recorded because these factors influence various mechanisms of deterioration common in concretes. Where a large structure is to be examined it may be appropriate to produce a plan or series plans indicating 'climate exposure severity' to overlay the engineer's plans of the structure. If more exact information is not available, it may be useful to use an arbitrary scale on a scale of say 1 to 10 to indicate the severity of environmental effects.

Other extraneous factors such as de-icing salts or water run off may also vary for different parts of a structure and can be recorded in a similar manner.

A careful and detailed record of all observations should be made as the inspection proceeds. Drawings can be marked, coloured or shaded to indicate the local severity of each feature. Features which commonly need recording include cracking which can vary widely in nature and style depending on the causative mechanism, surface pitting and spalling, surface staining, differential movements or displacements, variation in algal or vegetative surface growths, surface voids, honeycombing, bleed marks, constructional and lift joints and exudations or efflorescence.

Certain parts of a structure may be considered to merit a more detailed inspection than others prior to sampling. It is at this stage that recourse may be made to detailed mapping and measurement of cracks; photographic recording and surveying techniques might be used to check any apparent misalignment.

Reference should be made to the appropriate guides and codes of practice if further details concerning the inspection of particular types of concrete structure are required.

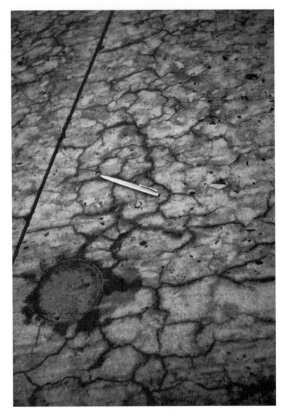

Fig. 3.1 Map cracking of concrete road slabs due to alkali-aggregate reaction, South Africa. Note a concrete core sample has been taken for laboratory study and the hole made good.

The completed site survey with its detailed observations should allow a preliminary assessment of the structure to be made and will have identified areas requiring more detailed examination and testing.

This further work may include on-site non-destructive testing or longer-term monitoring of parts of the structure. In all but minor causes of deterioration, it is probable that some laboratory work will be required, and consequently samples will be required.

3.2.1   THE SURFACE EXPRESSION OF CONCRETE DETERIORATION

The commonest defects observed on concrete surfaces are cracking, weathering and discoloration. Cracking can range from hairline superficial cracks or crack networks to large open cracks several millimetres wide which exhibit differential movements (Fig. 3.1).

The reasons why cracks develop vary widely, but because they are the surface expression, and perhaps the only indication of some form of distress in the structure they have become the subject of numerous articles, for example Fookes (1982), reviews in textbooks and special publications such as the Concrete Society Technical Report 22 (1982). The petrographer will be required to examine both macroscopic and microscopic cracks and crack patterns during the course of the investigations. Some of these cracks exhibit characteristic features which relate to the original cause of cracking, though their interpretation should be tempered with caution. Section 5.8 of Chapter 5 deals specifically with various forms of cracking in concrete, their appearance and causes, while the relationship between concrete deterioration and crack development is covered in Chapter 6.

Cracking which develops sub-parallel to the concrete surface may progress to an extent where spalling occurs. Other features of deterioration are the development of exudations or efflorescence. By far the most common is calcium carbonate derived from the gradual leaching of calcium hydroxide from the cement paste, its subsequent evaporation at a surface and the carbonation of the hydroxide solution by the air. Exudations can be used to provide evidence of the nature of a deterioration mechanism since they may contain salts produced within the concrete or as in the case of deterioration due to alkali-silica reactions, gels produced by the reaction.

The porosity and permeability of a concrete can also provide information about the quality of the original concrete or the subsequent leaching of the material. The calcium hydroxide formed by the hydration of cement rapidly carbonates on exposure to the air. Carbonation of the surface of a concrete is inevitable, and the carbonation will progress inwards with time. This inward progression is rarely more than a few millimetres in a good quality concrete but its rate of progression will be greater if the concrete is porous or the surface is cracked.

Variation in algal and vegetative growth on old concrete surfaces can often provide indirect evidence of the alkalinity of the surface layers and the moisture availability over localised areas.

3.3   Specimen selection and investigative objectives

Planning of the site investigation, selection of sample sites and petrographic examination should be aimed to meet the objectives of the investigation in the most efficient and economic way. This plan should be agreed between the client, petrographer and other relevant technical experts before the work is started. The petrographer must be in a position to evaluate the possible nature of the suspected problem and to assess the levels of precision necessary which will in turn dictate the types and number of samples required. In this respect a good knowledge of the costs involved in the various stages of sampling, preparation and examination are also required so that a balance between cost and reliability of the proposed investigation can be made to the client.

The larger and more complex investigations are usually broken into a series of stages with each succeeding stage being modified by the information obtained by the previous one. As indicated in Section 3.1 some parts of an investigation may require different levels of precision from others. Careful planning is required to see that unnecessary sampling and examination is avoided because of the costly nature of petrographic examination.

3.4   Particulate materials

Particulate building materials can vary in size from boulders to sub-micron sized powders so that a range of sampling techniques needs to be used. Common particulates which may be examined by the petrographer are aggregates, cement, cement replacements such as blastfurnace slag, pozzolana and limestone and other additions such as pigments.

Sampling of particulate materials is based on the concept of random sampling to ensure the variability of the particulate material is represented in the sample taken. True random sampling is amenable to mathematical analysis as discussed by Harris and Sym (1990) who demonstrate how the reliability of the data improves with increasing sample size. The maximum size of the particles present will be the controlling factor in the determination of the size of the representative bulk sample. Harris and Sym (1990) provide tables and nomograms relating individual particle size, sample size and sampling error. However, segregation within a stockpile or variation in a conveyor feed must also be taken into account when sampling. Specific sampling procedures given in National Standards have been devised to minimise the effects of such variations and to ensure that representative samples are obtained.

Theoretical reviews concerned with the size of random particulate samples and the probabilities associated with selection of a particular constituent from them normally deal with ideal single-sized spherical particles which are all of the same density. In practical sampling procedures allowance must be made for deviation from such ideal material. If there is a wide distribution of particle sizes for example, it may be necessary to devise a separate sampling procedure for each size fraction.

British Standards Institution recommendations for minimum size samples of aggregates and fillers is given in Table 3.2 which are similar to the recommendations given in ASTM D75 (1992), Table 1.

Table 3.2 Minimum masses of bulk samples for normal density aggregates according to BS 812: Part 102, Clause 5.

Nominal particle diameter range (mm)	Minimum mass of bulk sample (kg)
28 and larger	50
5 to 28	25
5 and smaller	10

In the case of cements, BS EN196-7 (1987) recommends 7 kg and ASTM C183 (1995) 5 kg as a minimum representative bulk sample from a defined batch. Taken together these recommendations suggest that for the purposes of sampling building materials the minimum size bulk sample of powders (i.e. particulate materials with maximum particle diameters less than 100 μm) should be 5 kg. Specific sampling procedures are described in detail in National Standards and a good review of sampling procedures for aggregates is given by Smith and Collis (1993).

In most cases bulk representative samples are supplied to the petrographer who should ensure that unambiguous identification is marked on the sample bags and in addition loose identification tags are present inside the bags. On receipt at the laboratory each sample should be logged and given a laboratory reference number.

The next step involves reduction of the bulk sample to sizes suitable for laboratory testing and petrographic examination. In the case of powders this reduction may involve the separation of a few mg of material from as much as a 5 kg sample. The most reliable method of carrying out this type of sample reduction is to use a sample splitter, such as a riffle box or preferably a spinning riffler. Allen (1990) reviewed the various methods of sample reduction for powders and concluded that the cone and quartering method is subject to considerable sampling error.

To accommodate factors such as particle shape, composite particles and particle size distribution a comprehensive sampling method has been developed by Gy (1982) which is widely used in the mining industry for estimating the minimum mass of sample required. Gy's methods are considered in detail by Pitard (1989) and reviewed in Harris and Sym (1990).

In BS 812: Part 104 (1989) an application of Gy's formula is applied to estimating approximate sample size for petrographic analysis. This simplified version of the formula assumes likely values for the numerical factors dealing with shape, size distribution, composite grains (liberation factor) and density which are used in the original formula. Also the approximate proportion of the constituent of interest must be known or inferred from previous results to use the formula. The minimum test portion M (kg) is related to the estimated proportion of a particular constituent p (%) whose nominal maximum diameter is d (mm) by the following formula:

$$M = 0.0002(100 - p)d^3/p$$

Alternatively, for general petrographic analysis smaller samples (approximately half those calculated from the above formula) may be used following values given in Table 3.3 after Harris and Sym (1990) who assume the material to be narrowly graded and calculate the results to have a precision of ± 10 per cent relative.

Table 3.3 Minimum mass of test proportion required where constituents of interest may be present as a small percentage of the sample Harris and Sym (1990).

Nominal maximum particle size (mm)	Minimum mass of test portion (kg)
40	50
20	6
10	1
<5	0.1

The size of the final subsample used is determined by the analytical technique to be used. In the case of chemical analyses 0.5–1 g may be required while for powder diffraction X-ray analysis only 50–100 mg is necessary. However, where constituents are to be identified by microscopic point-counting the size of the subsample is controlled by the smallest concentration among the constituents present that is of interest. For small percentages of constituents this necessitates examination of a large number of particles. The 95 per cent probability limit for the detection of a constituent is given in Table 3.4. Thus it is necessary to examine at least 600 particles to insure with 95 per cent probability that a particular constituent present as 0.5 per cent of the total sample will be identified.

In order to quantify the proportions of the different constituent particles present in a sample they may be spread out as a layer for examination or if small enough cemented into a tablet with resin (Section 3.4.1) and prepared as a thin-section suitable for the petrographic microscope. The volume proportions of the individual constituents identified may be determined using an appropriate modal analysis technique. The precision of the estimates of the proportions of the constituents present in a sample will depend on the number of particles examined, the accuracy of their identification and whether the sample

Table 3.4 Number of particles to be identified to ensure detection of a particular constituent with a 95 per cent probability.

Percentage of constituent in the sample	Number of particles to be examined to ensure detection
3	100
1.5	200
0.7	400
0.5	600
0.2	1500
0.1	2000

is truly representative of the bulk material. A discussion of modal analysis methods and their precision is given in Section 2.11.1.

In some circumstances, for example, estimating the proportions of different rock types in an aggregate stockpile and where visual identification is possible, the sample can be sorted on site. However, an appropriate method of obtaining the representative bulk sample must be used, an accurate record of the sampling and sorting procedures supported by photographs must be kept and representative sub-samples of the different constituents returned to the laboratory to confirm field identifications by petrographic or other means. Although such on-site preselection of materials may be an attractive and possibly the only option available in remote or difficult sites care and skill is required in assessing the precision of the results obtained.

3.4.1   PREPARATION OF PARTICULATE SAMPLES FOR EXAMINATION

Where the identification and estimation of constituents in particulate materials can be determined by external appearances of the particles, examination of the loose material may be carried out using a low-power stereo microscope. The only specimen preparation required in these cases is to ensure that a sufficient number of particles are examined as discussed above and in Section 2.11.1. To assist identification selected particle grains may be crushed and placed in immersion oil mounts for examination by the petrographic microscope. These are routine petrographic procedures and for aggregates the methods required are discussed in ASTM C295 (1990), BS 812: Part 104 (1989) and Dolar-Mantuani (1983), Klein and Hurlburt (1993) and Larsen and Berman (1934).

When the material is in the form of a powder most standards are restricted to detailing methods of measuring fineness and surface area, for example, as specified in ASTM C184 (1985), ASTM C204 (1996), ASTM C786 (1994) and BS EN 196-6 (1992). General methods for the microscopic determination of particle sizes in fine powders is given in ASTM E20 (1994).

The preferred method for examining sands and powders is to cast them into blocks of resin which are thin-sectioned for microscopic examination by transmitted light. If the powder is too fine for examination then X-ray fluorescence analysis of polished mounts and thin-sections may be required. Campbell (1986) describes the use of polished thin-sections for the examination of cement and clinker by incident, transmitted light and scanning electron microscopy.

If the sample is to be polished, as opposed to fine grinding, the diameter is effectively limited to 30 mm due to the considerable problems involved in satisfactorily polishing larger samples. Reusable polyethylene moulds such as those used for casting metallurgical specimens are suitable. If these are not available polyethylene beakers may be substituted. For the larger specimens, where finishing will be restricted to fine grinding on a metal lap, reusable polyethylene moulds are unlikely to be available and moulds can be constructed from cardboard, sheet metal or any other suitable disposable material. It is necessary to coat these moulds liberally with the release agent recommended for the resin used so that they do not stick to the cast resin.

The way in which the particulate sub-sample is embedded in the resin is dictated by the amount of material available, whether it is porous and whether it contains fine material that can separate out. Firstly, all materials must be thoroughly dried, generally at 105°C, except hydrated cementitious materials which should be restricted to 60°C. Relatively coarse powders, sands and aggregates may be mixed with polyester or epoxy resin attempting to include as much material as possible so that the maximum number of particles will be intersected by the plane of grinding and polishing. The resin used should then be mixed with its hardener according to the manufacturer's instructions, and poured into the mould leaving at least 5 mm free space for frothing. Note that if large amounts of resin are used the specimen may have to be cooled during initial hardening to prevent overheating and flash hardening from the heat of polymerisation. The mould containing the mixture is placed in a vacuum desiccator and a vacuum of at least 10 mm Hg is applied. The surface of the resin should be allowed to froth and degas for a number of minutes and the vacuum released to force the resin into the pores of the particles. This procedure is repeated until the air has been removed from the sample.

The mould is removed from the vacuum desiccator and placed in a warm place and the resin allowed to harden. In the case of polyester resin it can, if required, be cured in about 30 minutes by placing the sample in an oven at 60°C. Epoxy resins should be left overnight to harden in a warm room and then fully cured by heating to 60°C for 1–2 hours. This additional curing is important to harden the resin so that it takes a good polish. If vacuum moulding has been performed correctly no air bubbles will be visible in the cast and the more accessible pores in the particles will be filled. In general, polyester resin is more suitable for the smaller specimens as it is easier to handle and impregnates better because of its lower viscosity. Epoxy resin should always be used for larger specimens as it is stronger and harder.

Where porous particles are being cast the above procedure is modified. The dry particles are packed into the mould which is then transferred to the vacuum desiccator. A vacuum is applied and the sample left to evacuate for a few minutes. In this case the desiccator is fitted with a device to allow the liquid resin to be run into the mould under vacuum so that the resin is not displacing air and is better able to fill the voids. After a few minutes the vacuum is released and the procedure described above of applying and releasing the vacuum is followed until the resin is degassed.

If very fine material which can separate out is present in the sample or the amount of material is limited a modification of one of the above procedures is used. In this case only a thin layer of resin and sample is placed on the bottom of the mould. The mould is taken from the vacuum desiccator and allowed to stand until the resin is just starting to gel. Further resin is then added over the top of the gelled layer with which it will not intermix and the sample again degassed as described above. This gives a specimen with a thin layer of embedded sample backed by clear resin.

Both Araldite and polyester resins can be coloured by dyes which will delineate cracks and pores in the particles. A range of dyes are available from resin manufacturers but some of these may interfere with staining techniques or be difficult to detect. Yellow fluorescent dyes are gaining increasing popularity for general petrographic work because of their high visibility and difference in colour from colours produced by the majority of staining techniques used for petrographic identification of minerals. These dyes also have the advantage that they can be made to fluoresce as described in Section 2.7 and used for the determination of water/cement ratio and porosity in concrete. All dyes used in thin-sections should be used with discretion as they tend to obscure fine textural detail. Further details concerning fluorescent and other dyes together with the staining techniques used in the petrographic thin-sections are given in Section 3.9.1.

3.5   Sampling concrete

3.5.1   BULK SAMPLING

General descriptions of the inspection, investigation and sampling of concrete structures are given by Allen (1993), Baker (1992), ACI Committee 364 (1993), ACI Committee 201 (1992), British Cement Association (1992) and ASTM C 823 and the use of cores to assess the strength in concrete structures is specified in BS 6089 and the Concrete Society (1987).

One difficulty in determining the type of sampling required for petrographic examination of concretes from structures is the wide range of conditions that are encountered. As there are only a few standards which are solely concerned with the petrographic examination of concrete an attempt will be made to discuss the various aspects of sampling so that the reader will be aware of the implications of any decisions made.

The preferred method of sampling concrete for petrographic examination is to drill a number of cores from the structure. Descriptions of coring the techniques for coring structures can be found in Andersen and Petersen (1961) and US Department of Interior (1981).

Obviously the diameter of the core is vitally important in relation to the size of the constituents present in the concrete which in practice means the core diameter should be not less than three times the maximum size of aggregate. In the majority of structural concretes the maximum size of aggregate does not exceed 37.5 mm and 100 mm diameter cores are adequate. In mass concretes larger aggregates are used and 150 mm or larger diameter cores may be necessary. However, scattered through the literature are many statements which claim that 75 mm and even 50 mm or smaller diameter cores may be used. The justifications for these statements are that potential damage to critical reinforcement may prevent the drilling of larger diameter cores and the high costs involved.

It is also argued that since deterioration is often confined to the hardened cement paste it is not necessary to use larger diameter cores. This does not take account of the fact that, as pointed out in ASTM C 856 (1995) for deteriorated concretes, core recovery is much poorer for smaller diameter cores. This suggests that in the smaller diameter cores recovered there is a greater possibility of damage occurring from the coring process. There are strong arguments for the use of larger diameter cores and as a general rule cores of less than 100 mm in diameter should not be used for petrographic examination of structural concretes. If circumstances force the petrographic examination of smaller diameter cores the report should include the reasons for the use of the smaller cores and their likely impact on the results obtained.

The length of the concrete cores will be controlled by the type of examination and tests required. The outer zone of a reinforced concrete, the 'covercrete', is usually less than 75 mm in depth. Even where severe AAR has occurred the altered outer zone rarely exceeds 100 mm in depth beyond which any effects of reaction become more uniform. Thus below a depth of 100 mm the core usually represents the bulk concrete. For petrographic purposes this suggests that the outer 100 mm must be examined plus 100 to 150 mm of further depth to characterise the bulk concrete. Core depths of 250 mm are usually adequate which accords with ASTM C 856 (1995). For general sampling of concrete ASTM C 823 (1995) recommends lengths between 150 and 600 mm dependent on the type of structural element being sampled. Where physical tests will be required sufficient length of core should be taken to allow for both petrographic examination and the test specimens. Preferably separate cores should be taken for the petrographic examination and for the physical tests.

In summary, adequate petrographic examination requires that deteriorated structural concrete should be sampled by drilling 100 mm or greater diameter cores to a depth of not less than 250 mm. Where these sampling conditions cannot be met the petrographic report must take account of the effect of the sampling conditions on the results of the examination.

As noted in Section 3.1 the number and positioning of samples will depend on the type of test or examination required. Where concrete is to be tested for a physical property or attribute, representative sampling as specified in ASTM C 823 (1995) or provisions of an appropriate National Standard should be used. This type of sampling assumes that the concrete or identifiable portions of the concrete are of a similar condition and that a probability sampling plan can be formulated. Guidance for the formulation of probability sampling is given in ASTM E 105 (1958). For compliance with construction specifications ASTM C 823 (1995) specifies that not less than five samples shall be taken from each category of concrete for each test procedure stipulated. It will be apparent that if representative sampling is carried out, the coring, testing and examination of concrete from a large structure is a costly exercise.

If the assumption of the concrete being in a 'similar condition' cannot be made, the number and positioning of samples for petrographic examination becomes a matter of judgement based on skill and experience. Where deterioration is being investigated samples must be critically sited, based on careful observation of cracking, exfoliation, efflorescence, exudation, staining and other changes visible on the concrete surfaces. ASTM C 856 (1995) specifies that for petrographic examination 'the minimum size sample should amount to at least one core for each mixture or condition or category of concrete . . .'.

In practice this translates into the following type of scenario. As an example, when investigating suspected alkali-aggregate reaction it is usual to define three sample areas consisting of most severely cracked, averagely cracked and uncracked concrete from which one core will be extracted. The exact siting of each core in the three designated areas is solely up to the judgement of the engineer or petrographer. If this judgement is faulty the results of the petrographic examination will not be reliable and this is the reason why the petrographer should be present when sample sites are selected. It is not unusual to find when investigating alkali-aggregate reaction that the severity of external cracking does not correlate well with the severity of the reaction internally. Thus it is not surprising that complaints are voiced about unreliable diagnosis of alkali-aggregate reaction. There are also other contributory reasons which are discussed in Section 6.7.10.

It should be apparent that taking one core sample from each designated condition of concrete is not a satisfactory method of sampling. The possibility of error is large and the use of one core to characterise unaffected concrete is not acceptable. A more desirable sampling regime would be to drill at least two critically sited cores from each designated area of cracking and the number of cores drilled from the unaffected concrete to be controlled by the attributes of the concrete to be measured as designated by ASTM C 823 (1995) or an appropriate National Standard.

Under some circumstances, fragments such as spalled or broken pieces of parts of the structure may be the only samples available. It is possible to provide limited information from broken samples such as aggregate identification but samples of this type must be avoided because mechanical breakage may mask or complicate *in situ* deterioration processes. In addition samples of this type usually represent only external portions of the concrete and if they have lain around the structure will have been subject to considerable weathering. Wherever possible the petrographer should reject spalled or broken samples.

Inspecting engineers need to accept that the reliability of results from petrographic examination are subject to the same sampling methodology as applies for instance to the measurement of compressive strength. It is regrettable that in too many cases the petrographer is expected to produce meaningful results based on examination of a limited number of undersized samples without any consultation over the selection of the sample sites.

3.5.2   SAMPLING METHODS

Water-cooled diamond-bit coring is the recommended method of sampling concrete for petrographic examination. Where smaller diameter cores are drilled, that is 75 mm or lesser diameter, the core bits used should be designated by internal diameter as the loss of core diameter on a nominal 50 mm diameter coring bit is considerable. A recent development in coring, which is suitable for unreinforced concrete makes use of a 5–6 atmosphere/pressure of air flush to cool the bit with a water mist spray which keeps the swarf as a paste and prevents clogging and glazing of the cutting edge.

Concrete cores can also be dry cut using either diamond or tungsten carbide bits. Dry cut diamond bits as large as 150 mm are available but drilling depths are limited to about 350 mm. The diameter of tungsten carbide bits is restricted to about 50 mm and depths to about 100 mm. These coring bits are useful when no water is available or leaching of core surfaces by cooling water is unacceptable such as in the determination of chloride levels. However, they all suffer from the disadvantage that there is no cooling and considerable heat can be generated during drilling.

A coring bit larger than 75 mm in diameter cannot be used in a hand-held drill. The amount of torque necessary to drive the 75 mm and larger coring bits requires a motor fixed in a frame that is rigidly clamped to the structure. This ensures a clean core with minimum damage due to chattering of the bit and is another good reason for using larger diameter cores for petrographic examination.

For concrete of restricted dimension, such as pavements and slabs or precast products, wet diamond sawing of samples is preferable to coring as it is less likely to cause damage. The use of dry silicon carbide saw blades is not recommended except for subsampling. There are cases where wet cut core or prism samples need to be analysed for alkalis by obtaining a dry portion from the centre of the sample where the concrete will not have been leached. This subsampling can be carried out by using a dry silicon carbide blade mounted on a masonry saw.

Where powder samples are required for analysis of chlorides dry drilling is essential. It is necessary to use a tungsten carbide tipped masonry drill to collect dry powder from successive depths as the drill penetrates the concrete. However, since there is little control whether the drill encounters large aggregate particles, a large number of samples from different locations are required. The technique is described by Roberts (1986) and may also be used for the determination of other elements such as alkalis where leaching is suspected.

When examining concrete where the presence of a damaged, possibly friable, or heavily cracked outer surface is important to the investigation, the surface itself must be examined using a portable field microscope, sampled and photographed *in situ*, unless it can be sawn from the structure without damage. However, cross-sections through the damaged surface can be cut from normal cores provided the surface is first protected and supported by coating with a low viscosity epoxy resin which may then be reinforced using fibreglass matting and further coats of resin. Once the resin has cured the concrete can be cored through the protective layers in the normal way. Specialist concrete repair companies have the necessary expertise for applying epoxy coatings to difficult areas such as inverts.

In the special case of taking core samples from *in situ* concrete pipes such as sewers, coring done from the outside should be stopped 3–5 mm short of the inner surface. An appropriate handle should then be epoxy cemented to the core sample which should then be broken free from the pipe using a chisel or by drilling successive holes round the cut using a hand-held drill with a small diameter tungsten carbide bit.

Once the core is broken out of the structure it should be labelled and its orientation marked. After inspection it should be wrapped immediately in several layers of plastic film, sealed into a polythene bag to preserve its 'on site' moisture condition and the details of the core recorded. A suitable record sheet is illustrated in Fig. 3.2 which should accompany the core to the receiving laboratory.

3.5.3   SUBSAMPLING

The coring and cutting processes open up large surface areas of fresh concrete which, if they remained wrapped for any length of time may show an increase in reaction products and allow local leaching of the concrete. Thus the cores should be transported to the receiving laboratory which should then unwrap, inspect, photograph and 'log in' the cores without delay. It is important that the petrographer is present when the cores are unwrapped so that an initial petrographic assessment can be made and the appropriate positions of planes for cutting can be marked onto the sample prior to the preparation of the ground surfaces and thin-sections.

The initial petrographic examination should be made both visually and by using a low power stereo microscope, preferably with a 'zoom' magnification facility, and all features of particular interest photographed and recorded.

Features such as exudations and hygroscopic patches on surfaces may be indicative of alkali-aggregate reaction. They are most clearly apparent as a core dries out from its initial moist condition and may be sampled at this time. Void or crack infillings and other surface deposits of interest are also best sampled during this examination. Particular attention should be paid both to the external cut surface and the fracture surface where the core has been broken out from the drill hole. These surfaces may provide useful information relating to aggregate-cement bonding, microcracking and to other detailed textural features, so a comparison of cut and fractured surfaces should not be overlooked during the examination. Once the initial inspection and any wet cutting is complete the cores for petrographic analysis should be dried at 40–60 °C. This process may require several days for larger samples.

CONCRETE CORE RECORD

Client	Site	Job No.	Site Ref.
Date	Type of rig	Operator	Core No.

Location of core in structure (including grid refs)	PHOTOGRAPH (TICK BOX) Core ☐　Hole ☐　Sample Area ☐

DRILL ORIENTATION (TICK BOX) Vertical up　☐ Vertical down　☐ Horizontal　☐ Inclined　☐　　Angle ± α + α – α	METHOD OF FLUSH (TICK BOX) Water flush　☐ Air flush　☐ Dry　☐ Antifreeze　☐	DIA. OF DRILL BIT IN MM

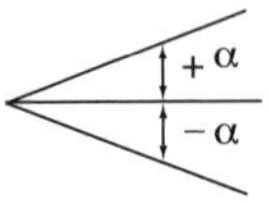

DEPTH OF CARBONATION IN MM	NUMBER OF PIECES	REINFORCEMENT
Max　　　Min		Diameters in mm
	ASSEMBLED LENGTH	Distance from surface in mm
	Max.	Condition
	Min.	Vertical or horizontal

Comments on CORE SAMPLE (condition of concrete, condition of steel, cracks, delamination of core etc.)

Comments on CORE HOLE (cracks, voids, poor compaction, reinforcement, loss of water during coring etc.)

Sketch of core hole LOCATION on the structuire (if necessary)

Notes:
1. Mark core with site reference number and core number
2. All core locations to be photographed or sketched
3. All cores to be labelled as below

End　　　　　Length　　　　　Broken
UP (↑)　　OUT [◄———]　　[1│2│3]

4. If outside of core is destroyed by coring,
   mark core with estimate of depth of new surface e.g.

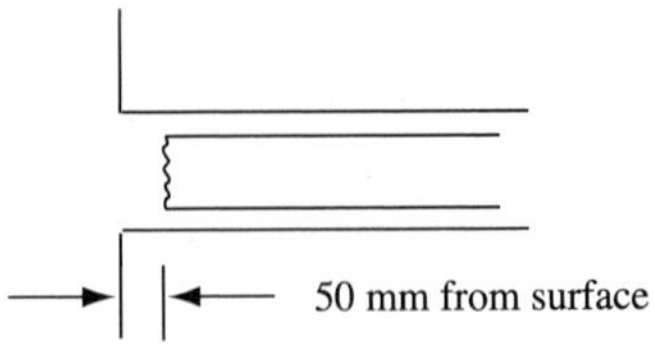

———► │◄— 50 mm from surface

Fig. 3.2 A typical core log record sheet for use on site.

The subsampling for petrographic examination of a drilled core or saw cut sample consists of the production of a plane surface and thin plates for the preparation of thin-sections. The location and size of these surfaces as subsamples are no less important than the original bulk sampling of the structure. The size of sample plane required for the determination of air content in hardened concrete is specified in ASTM C 457 (1990) and reproduced in Table 3.5. This table indicates that as the aggregate size increases and the paste volume decreases the effect is compensated by the requirement for an increased area of examination. It is to be noted that this table is based on the concrete being reasonably homogeneous and well compacted. It recommends that if the concrete is heterogeneous in distribution of aggregate or large air voids the measurements should be made on a proportionately larger area of ground surface. It also specifies that the distance between section planes shall exceed the largest diameter of aggregate present to ensure that each plane intersects new material.

Table 3.5 Minimum area of finished surface for microscopical determination of air content as specified by ASTM C 457.

Nominal maximum size of aggregate in the concrete (mm)	Minimum area of surface for measurement (cm^2)
150	1613
75	419
37.5	155
25	77
19	71
12.5	65
9.5	58
4.8	45

These recommendations for minimum surface area are often used as a guide for petrographic examination in general so it is important that their basis is understood. In a typical concrete containing an air content of 5 per cent, assuming a uniform distribution of air voids of 50 μm in diameter, a 150 cm^2 sample plane cuts about 300,000 voids. This sample population is large enough to provide a realistic estimate of the percentage of air voids in the structure providing the concrete is reasonably homogeneous. Obviously the whole population is not counted but providing the statistical requirements of modal analysis are met reliable estimates of the air content present in the sample can be made. In this respect ASTM C 457 (1990) includes an excellent discussion of the types of errors involved in the measurements.

If the same sample plane is now used to examine the coarse aggregate, which is assumed to comprise say 50 per cent of the concrete volume, to be spherical, and to average 12.5 mm in diameter, the 150 cm^2 sample plane will cut a population of less than 50 aggregate particles which is clearly inadequate as a representative sample of the coarse aggregate in the structure. Even for 5 mm aggregates the population sampled is only 300. While the argument used is simplistic in some of its assumptions it clearly indicates a fundamental problem in the petrographic examination of concrete where examination is typically carried out on constituents ranging in size from 37.5 mm to 5 μm. These constituents may be aggregate, a crack system or any other quantifiable feature of the concrete. As the size of the constituent increases, the extent of the sample surfaces required for adequate characterisation increases dramatically.

As can be seen from the above discussion the sizes of the sample surfaces recommended by ASTM C 457 (1990) are only applicable to those cases where the constituent being measured is less than about 1 mm. Where the constituents to be quantified have average sizes much greater than 1 mm the area of surface required for examination needs to be increased significantly if a precise, quantitative and representative modal analysis is to be attempted. In many cases the requirement for a representative modal analysis implies that as much of the sample as is practical must be prepared as finely ground surfaced or thin-sections and attention must be given to the precision of the measurements made. As an example, in a typical sampling of a concrete structure affected by alkali aggregate reaction three 100 mm diameter by 300 mm length cores are drilled and three 150 × 100 mm thin-sections are prepared from plates cut along the diametral core axes for petrographic examination. This subsampling provides an adequate area for detecting the presence of reaction associated with the fine aggregate, but it is not satisfactory if reaction is restricted to coarse aggregate particles. While the petrographic examination may detect the presence of alkali-aggregate reaction other observations on the reaction will be limited and may be biased. As a minimum, where the concrete needs to be fully characterised three extra sections should be cut so that a complete diametral axial plane of each core is sampled.

Three approaches to providing sufficient sample area are commonly used in petrographic laboratories which are largely dictated by the facilities available (see Sections 3.6 and 3.7). The more satisfactory method is to prepare large-area thin-sections with edge dimensions of 100 mm or larger. This allows full and uninterrupted study of any variations or segregation from the surface to the interior along the diametral axis of the core to at least a depth of 150 mm.

Alternatively, several thin-sections of up to 75 × 50 mm in size are prepared but this may be unsatisfactory for a full petrographic study of normal concrete because of edge losses due to cutting and insufficient area being available for modal analysis. Where the objective of the petrographic study is more limited or specialised, for example when only aggregate identification is required or steel corrosion products adjacent to reinforcement are the subject of the investigation this type of thin-section may be appropriate. The use of the 25 × 25 mm sized thin-section traditionally used by geologists is not acceptable for the general petrographic examination of concrete. However, there are specialised situations where polished thin-sections are required for examination by transmitted light supplemented by SEM study where this size of thin-section is applicable.

The third approach, favoured by some laboratories in North America, is to prepare finely ground surfaces with edges of 100 mm or greater for examination by incident light as described in ASTM C 856 (1995). As much information as possible is obtained by examination using a low-power stereo microscope. Areas of interest are located and the matching opposing surface is then used to prepare thin-sections and specimens for optical, chemical, X-ray diffraction and other examinations. No indication is given about the dimensions of the thin-sections to be used but it is known that in practice these were usually 25 × 25 mm in size. This approach has the advantage that the ground surfaces are relatively easy and cheap to produce and the costs of thin-sectioning are restricted. The problem is that only a limited amount of information can be gained from examination of ground surfaces and the use of small sections cut from the matching surface presents many problems of interpretation.

There is general agreement that where possible the orientation of the sample plane in the core should be controlled by the zonation present in the concrete. The outer zones of concrete are variable because of inhomogeneities arising during placement and subsequent weathering and thus the main axis of sectioning must always include the outside surface. The other common inhomogeneity in concrete is due to plastic settlement causing voids beneath aggregates which are best revealed by sections cut in a vertical plane. Where the core is drilled vertically the orientation of the diametral axis is not important, unless dictated by a surface feature, as all diametral planes will be vertical. For horizontal cores whose orientation should always be marked, the vertical diametral plane should be chosen unless surface features, such as the orientation of a crack, dictate otherwise.

All edges, especially the external surface, must be fully retained on the thin-section so that surface features such as carbonation and any gradation from the surface are available for study. With modern methods of thin-sectioning given in Section 3.7 this is readily achievable.

The last type of surface that should always be examined is a fresh fracture surface prepared by breaking open a portion of the core. When the examination of the thin-sections and other petrographic studies have been carried out and diagnosis completed these should always be confirmed by examination of fresh fracture surfaces. Such examinations may provide valuable information, for example crack or void infillings are readily observed on a new freshly exposed surface, but with time they dry out and become almost invisible.

There are a wide range of products such as pipe, sheet, tiles and cement renderings as well as laboratory specimens which may require sampling for petrographic examination. Some of the dimensions of these products may only be a few millimetres. However, the same sampling principles discussed above are applicable although they will be modified by the nature of the product. Cross-sections of thin sheet products can be cut and glued together in a stack to produce an acceptable area for sectioning. Similarly 25 mm mortar bars can be cross-sectioned so that up to 24 cross-sections are included in one 150 × 100 mm thin-section for examination. Surface alteration is often important in thin-walled products so that as much of these surfaces should be included in the sample as possible.

In other situations, for example in forensic studies, the petrographer may only have fragmentary samples to work with. In such circumstances the combination of ground surface and thin-section examination will provide the maximum amount of information which may be augmented by other techniques such as those referred to in Sections 2.14 to 2.17.

3.6 Preparation of finely ground and polished surfaces

There is a distinction between finely ground and polished surfaces which is important. Finely ground surfaces are produced when the smallest grinding medium used is limited to about 5 μm whereas polished surfaces require that the final grinding media is submicron in size. It is relatively easy to produce finely ground surfaces of up to 100 × 100 mm in size but difficult and extremely time consuming to polish areas greater than 30 mm in diameter.

When finely ground and polished surfaces are compared under the petrographic microscope at × 200 magnifications the differences in the roughness of the surfaces observed are very large. The finely ground surface is dark and rough due to light scattering with blurring, smearing and poor differentiation of constituents. On the other hand a well-polished surface is bright and crisp, almost like a mirror, with sharp edges and good differentiation between components. This indicates that observation of a finely ground surface is best restricted to low-power stereo microscopy using magnifications not exceeding × 100. Where surfaces are required for examination by incident light using higher magnifications, such as for etched cement clinker or SEM and X-ray fluorescence analysis of samples, a high-quality polished surface is essential.

Finely ground surfaces are specified for the determination of air content in hardened concrete according to ASTM C 457 (1990). It should be noted that the level of detail on ground surfaces that is observed is critically dependent on the quality of the optical system of the microscope. Ground surfaces are also useful for determining broad textural detail of aggregates, gross inhomogeneity and crack systems and for staining alkali-silica gel. In some respects it is easier to observe these broad details on a ground surface than in thin-section and as such they can be an aid to the petrographic examination of thin-sections. Where facilities for preparing the necessary sized thin-sections are not available the petrographic examination of ground surfaces is preferable to using inadequately sized thin-sections. The number and location of section planes has already been discussed.

3.6.1   *LABORATORY METHODS OF CUTTING CONCRETE*

The petrographic sectioning of concrete requires the cutting of large slabs and surfaces from samples. Concrete can only be cut satisfactorily with diamond impregnated saws. Silicon carbide abrasive blades produce poor quality, uneven cuts and tend to damage surfaces. Their use is restricted to the dry removal of portions of sample which will be later crushed and ground for chemical analysis.

There is a wide range of diamond cutting and grinding equipment available, its design depending on the particular application. A review of diamond cutting equipment and a discussion of its application is given by Allman and Lawrence (1972) and more recent information may be obtained from the technical literature available from diamond tool suppliers and firms specialising in petrographic and metallographic supplies.

The technicalities of cutting samples will not be discussed here as this is the province of the skilled technician. However, there is one aspect that the petrographer should be aware of that is peculiar to concrete. Water is always used as the cutting fluid for diamond sawing unless the blade dips into a bath of cutting fluid. In this configuration the water will either contain a rust inhibitor or an oil/water emulsion may be used. It should be noted that some oil/water based cutting fluids are demulsified by concrete swarf. It is also possible to fill the bath with a straight cutting oil cut back with a kerosene. This is preferable for the larger saws where the rate of cutting is slow and prevents leaching from the sample. This cutting fluid may be flammable and carcinogenic, particularly in the form of vapour or aerosol. Therefore it is essential that appropriate fire precautions and extraction systems are installed which comply with national safety regulations. With dense concretes the extent of water leaching rarely extends beyond 1 mm but in more porous concretes this may become a problem especially for chemical analyses.

It is the petrographer's responsibility to discuss with the technician how the sample is to be handled. Where the concrete is particularly friable, highly porous or broken up it will require consolidation by gluing, impregnation or casting with an epoxy resin and there should be agreement at what stage and how this is to be carried out. Similarly, it is necessary for the petrographer to indicate how the core should be cut especially where a range of tests are to be carried out. Where it is desired to slice a core along the complete length of its diametral axis this may be restricted by the length of cut available from the fixed blade saw available. Where cores exceed this length the petrographer will be required to indicate

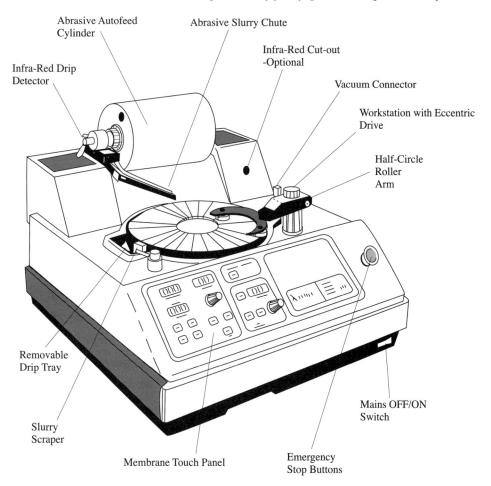

Abrasive Autofeed Cylinder

Abrasive Slurry Chute

Infra-Red Cut-out -Optional

Infra-Red Drip Detector

Vacuum Connector

Workstation with Eccentric Drive

Half-Circle Roller Arm

Removable Drip Tray

Mains OFF/ON Switch

Slurry Scraper

Membrane Touch Panel

Emergency Stop Buttons

Fig. 3.3 A diagram showing the main features of a modern lapping machine (courtesy of Logitech Ltd).

where transverse cuts are to be made and the extent to which continuity of cutting planes to be maintained. In practice, the slabs of concrete cut for later thin-sectioning should not be less than 10 mm and preferably be about 15 mm thick to ensure sufficient mechanical strength to allow subsequent handling and processing.

Provided that the petrographer clearly indicates the requirements, there are few problems for the skilled technician in a well-equipped preparation laboratory in cutting and preparing a wide range of concrete samples.

3.6.2 *PREPARATION OF GROUND SURFACES*

The cut concrete slice which is to be ground must be thoroughly dried in a thermostated drying cabinet at a temperature that does not exceed 60 °C. The minimum drying time is overnight and for larger specimens longer times should be allowed.

There are two methods of grinding surfaces depending on the end use of the specimen. In the first method the sample is ground in retaining rings by successively decreasing sizes

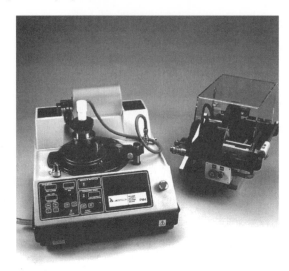

Fig. 3.4 Modern large-format thin-sectioning equipment: cutting and grinding machine (right) mechanical lap (left) (courtesy of Logitech Ltd).

of loose abrasive on a rotating, cast-iron lap until the desired quality of finish is achieved. The only consolidation of the sample carried out is surface impregnation with wax to maintain the sharpness of the margins of air voids. This method is described in ASTM C 457 (1990).

Where the sample is to be subsequently polished, thin-sectioned or required for delineation of crack systems and porosity, impregnation must be carried out with a low viscosity epoxy resin before grinding. Dye can be incorporated in the epoxy resin which makes original cracks, voids and porosity clearly visible and differentiates them from any damage which may occur during subsequent preparation. The vacuum impregnation procedures required are described in Section 3.4.1.

A variety of grinding and polishing machines are manufactured. Most rely on the basic principle of a rotating lap but there is also a range of equipment which uses a diamond cup wheel to traverse the specimen. The main features of a mechanical lapping machine are illustrated in Fig. 3.3 and modern large-format thin-sectioning equipment in Figs. 3.4 and 3.5. The details of grinding and lapping methods have been well described in the literature of Allman and Lawrence (1972) and Hutchinson (1974). For cementitious materials all lapping and grinding must be carried out with non-aqueous fluids. As it is preferable that the procedures be carried out by experienced technicians the detailed methods will not be discussed.

Whatever grinding method is used it is essential that no surface relief is induced due to differences in hardness between aggregate particles and the cement paste. The flatness of the surface can be checked by viewing it at a low angle of incidence in a strong light or under a low power stereo microscope. The clarity of the observations will depend on the quality of the finished surface and the optical system of the microscope used to examine it. It will be necessary for the petrographer to indicate to the technician the quality of the surface finish required.

Fig. 3.5 A large-format jig for sections up to 150 × 100 mm for use with a mechanical lap (courtesy of Logitech Ltd).

3.6.3 *PREPARATION OF POLISHED SURFACES*

Polished surfaces are required where the surface will be examined with a high-power microscope or by microanalytical techniques using energy or wave dispersive X-ray methods. Specimens may consist of polished mounts or polished thin-sections. In a polished mount the sample is embedded in a cylindrical block of resin of which one face is polished. A polished thin-section is merely a thin-section of which the top face has been polished and the mounting of the protective cover glass omitted. When examined by transmitted light a temporary cover glass is fitted with an immersion oil which is later removed with a solvent before coating for SEM examination.

The area of surface which can be successfully and economically polished is limited to a maximum diameter of about 30 mm. It is possible to polish larger area thin-sections (personal communication from Kelvin Pettifer, Building Research Establishment, UK, retired) using a Logitech type machine but the procedure is difficult and may take days. It can be justified only where detailed *studies* of concrete *are* required using analysis by transmitted light to obtain textural relationships backed up by electron microanalysis. As the surface area to be polished increases the time required for polishing increases exponentially. It becomes increasingly difficult to maintain a flat, scratch-free surface without relief or smearing occurring. In the polishing of architectural stone the final polish achieved depends on smearing of minerals induced by localised heat during polishing. Large areas can be polished but this type of polished surface is useless for petrographic purposes, and this distinction needs to be clearly understood.

The various methods for polishing cement clinkers, concrete and building materials have been described by Ahmed (1991) and Walker (1979, 1981) describes methods for the production of polished ultra-thin-sections. Polishing is usually not a difficult technique providing that the operator has the necessary skill and experience. However, concrete is one of the more difficult of materials because of its heterogeneous nature and the range of hardness of the components. One problem during sample preparation is the introduction of cracks into the hardened cement paste. A method for overcoming this problem has been proposed by Bager and Sellevold (1979) and Hornain *et al.* (1995). It is strongly recommended that polishing be carried out by a skilled and experienced technician. This type of skill is required because the quality of the polished surface is important for both reflected light and SEM studies.

In electron probe microanalysis the basis of the method relies on the comparison between X-ray intensity obtained from the selected area of the specimen surface with the intensities obtained from a standard. To minimise sources of error it is important both specimen and standard are polished to an equally high quality of finish.

3.7   Preparation of petrographic thin-sections

The preparation of thin-sections of rock for study with the petrological microscope has been an important part of geological laboratory work for more than a century, Reed and Mergner (1953). Essentially a thin-section is produced by cutting about a 5 mm slice from a rock with a diamond saw and after suitable impregnation to stabilise the slice one face is ground flat and smooth. This face is then glued to a slide and the rock slice is then reduced in thickness until it is approximately 30 μm thick. The upper surface is finished by fine lapping before gluing a 0.17 mm thick cover glass over the top surface for protection. In special cases the upper surface is polished and the cover glass omitted (Moreland 1968).

It is possible to produce a rock thin-section entirely by manual methods. However, since the end of the Second World War most of the manual procedures used in preparation have been supplanted by mechanical methods. This has been possible because of advances in the technology of adhesives and precision cutting and grinding equipment. The problem with hand finishing a thin-section is that it is extremely difficult to retain the original edges of the rock slice in the section. Traditionally the ragged thinned edges produced by the hand finishing are cut away to produce a clean edge. Also it is difficult to produce a section by hand much greater in size than 75 × 50 mm. In practice the 'standard' thin-section measures about 25 × 25 mm. This loss of edges and limitation on size has generally been acceptable to the petrographer when examining rocks but is completely unacceptable for concrete samples.

Most manuals on petrology and optical mineralogy have included some type of description of thin-section preparation and there are numerous other descriptions in the literature. Two references that include good descriptions of the techniques as well as a range of ancillary techniques are Allman and Lawrence (1972) and Hutchinson (1974). Campbell (1986) has described techniques for sectioning cement clinkers and Wilson (1973) for producing sections on semi-automatic lapping equipment.

Concrete and mortars present special problems in their preparation as petrological thin-sections. Their mineral components differ in their relative hardness, some are water soluble or may be damaged by water and the material itself may be fragile due to pre-existing cracks, voids and porous areas. These problems may be overcome by the use of non-aqueous grinding fluids during preparation and the use of epoxy resin adhesives which effectively stabilise the fabric of the concrete. However, the examination of a thin-section of concrete requires much larger areas than those used for rocks and retention of the original section edges is essential for adequate examination. These requirements place the preparation of concrete thin-sections in a different category to that normally used in geology.

The first semi-routine preparation of large area thin-sections of concrete was described by Poulsen (1958) and Andersen and Petersen (1961). By a combination of hand and mechanical techniques they prepared a wide range of large-area thin-sections for the investigation of the alkali-aggregate reaction in Denmark. Jones *et al.* (1966) used a modified surface grinder with a vertical spindle and diamond cup wheel to produce large-area thin-sections of concrete on a semi-automatic basis. Further development of this technique was made by St John and Abbott (1983) to enable it to be used to prepare 25 μm thick sections of concrete up to 150 × 125 mm in size on a routine basis. Concurrently with these developments Logitech Ltd mechanised the traditional lapping techniques for the preparation of thin-sections Wilson (1973). Initially the Logitech machines were intended for multiple preparation of routine rock thin-sections but with further development (Wilson and Milburn 1978) sections of up to 100 × 75 mm may be prepared with full edge retention so the technique is now applicable to concrete. Recently this size has been extended to 150 × 100 mm (Fig. 3.5).

Lapping sections with alumina grinding powder as opposed to grinding with fine diamonds embedded in a bronze cup wheel produces thin-sections with differing characteristics. The diamonds embedded in the cup wheels are usually about 60 μm (260–300 mesh) in size and while this produces a reasonably fine surface finish it is not as fine as that produced by lapping with 5 μm alumina. While the resin used to mount the section and its cover glass effectively hides the rougher finish it does place a limitation on the use of high magnifications for examination. On the other hand, lapping frosts the surface of the mounting resin which can obscure detail in cracks and voids. This effect is absent where a diamond cup wheel is used and fine detail is more easily observed in cracks and voids. Unless the diamond cup wheel is in good condition some crystal shattering can occur especially where the section thickness is reduced to 25 μm. Lapping can leave fine alumina embedded in the sample unless especial care is taken to clean the section surfaces before mounting.

The two grinding techniques can be combined. The thin-section is first cut on a surface grinder and then the upper surface lapped on a Logitech machine to produce a surface which is ground finely enough to be used for examination by SEM. Using a demountable cover glass the section is examined by the petrological microscope and then after removal of the cover glass and application of a suitable conductive coating detailed examination under the SEM is possible.

It cannot be over emphasised that concrete thin-sections must be produced by precision mechanical methods if adequate examination is to achieved. It is the authors' experience that hand-prepared smaller-sized thin-sections of concrete make it almost impossible to carry out adequate examination of material deterioration in structures. The preparation of thin-sections is a task for skilled technicians using precision equipment. For this reason detailed descriptions of the sectioning techniques have been omitted. The extra cost in having large-area thin-sections cut by specialists is more than justified by the increased reliability of results obtained.

However, the concrete petrographer should be familiar with the preparation techniques used as these affect the petrographic examination, for instance, the effective use of epoxy resin is dependent on drying the concrete of free moisture. Temperatures greater than 60°C must not be exceeded as dehydration of ettringite and microcracking due to temperature gradients can occur. Thus drying of a 10–15 mm thick slice of concrete will require at least sixteen hours and if lower temperatures are used even longer times will be required.

The use of a dye in the impregnating resin, as described in Section 3.9.1, needs to be specified. The routine use of a dye in the impregnating resin to outline cracks and pores is controversial as it can obscure detail and ideally should not be required if sectioning is carried out correctly. On the other hand the yellow fluorescent dyes are effective in highlighting areas of porosity and determining water/cement ratio (Christensen *et al.* 1984 and Jensen *et al.* 1985). One compromise is to cut additional smaller sections, 75 × 50 mm in size, to include the fluorescent dye so that the large section is retained for general examination.

The epoxy resins used are critical to the sectioning process. Impregnating, casting and adhesive types of resins are required each with specific properties. The problem in specifying these resins is that formulations are usually confidential to the manufacturer and code designations change with time. The casting resin is the easiest to specify and should be readily available from all manufacturers as it is one of the basic resins used for the manufacture of a range of products. It is an unmodified liquid resin derived from epichlorhydrin and bisphenol A with an epoxy gram equivalent of 180–200 and a viscosity of 9000–13000 mPa s. When used with an appropriate amine type hardener and cured briefly at 60°C after hardening at room temperature, a hard cast results that is excellent for embedding slices of concrete in preparation for machining.

The adhesive resin used should also be an unmodified epoxy, should not contain solvents and may contain a wetting agent. It needs to be less viscous than the casting resin and its setting shrinkage should not be excessive. Laminating epoxies used for making fibreglass products may be suitable. The choice of impregnating resins presents many problems not the least being the fact that concrete is a difficult material to impregnate unless the viscosity of the resin is close to that of water. The depth of impregnation of the fabric of even highly porous concretes is limited using epoxy intended for crack injection. When concretes impregnated with these resins are examined in thin-section they often appear as crumpled, heavily shrunk coatings inside pores and microcracks and it is questionable whether more than superficial impregnation of the fabric penetrates beyond a depth of 0.5–1 mm. Because impregnation of concrete is difficult, only highly porous and friable specimens should be impregnated before casting and this should be regarded as a consolidation process rather than impregnation. In practice, impregnation to consolidate the fabric of the layer of concrete in the thin-section occurs when the ground slice is glued to the glass mounting plate. While this impregnation is superficial this does not matter as the final section is only 25–30 μm thick.

The types of epoxy resin discussed above have refractive indices around 1.58 which are higher than the 1.54 of the traditional mountants for rock thin-sections. Special epoxy resins with a refractive index of 1.54 are available but these are not necessarily suitable for large-area thin-sections because of high setting shrinkage. Other adhesives, such as UV curing cyanoacrylate types are also used but their applicability to large area thin-sectioning needs to be demonstrated. Whatever resins are used the requirements for thin-sectioning are stringent. The resins must be able to be used at temperatures up to 60°C, not have excessive setting shrinkage, be hard and have good adhesive properties and also be resistant to the oils and solvents used during sectioning. Experience to date indicates that unmodified epoxy resins are still generally superior for this purpose.

Where sections exceed 75 × 50 mm in size thicker glass mounting plates must be used. Dependent on size, 3 or 4 mm float glass is used and one side only is ground or lapped to provide a flat, parallel surface for mounting the concrete slice. If both sides of the glass mounting plate are ground the glass plate loses most of its strength and breaks easily. The use of the thicker glass mounting plates will require the use of a longer focus condenser cap on the substage assembly of the petrographic microscope.

The choice of large-area thin-sectioning technique may also be dictated by the nature of the investigation. While the production of sections on surface grinders with diamond cup wheels has some advantages it is not applicable to the examination of steel reinforcement embedded in concrete. If the steel/paste interface and associated alteration products are to be retained for examination the section must be prepared by a lapping technique. This is because the type of diamond and bond used in the cup wheels will not grind steel without smearing and overheating occurring. A method for overcoming this problem has been proposed by Garrett and Beaman, (1985).

3.8   Specimen preparation for special purposes

Concrete and its related materials are complex assemblages of mineral materials ranging from mm to submicron in size. Many of the minerals are themselves chemically complex, present as solid solution or amorphous materials which may not be resolved under the microscope. To supplement optical examination the petrographer may find it necessary to use other techniques such as chemical analysis, powder X-ray diffraction, infrared spectroscopy, thermal analysis and electronmicroscopy (see Chapter 2).

3.8.1   THE POWDERED SPECIMEN

If chemical, X-ray, infrared or thermal analysis is required, specimens will need to be in the form of a powder. The particle size of the powder is not very critical for thermal analysis but for the other analytical techniques the sample may require grinding very finely.

The method for the preparation of test specimens for the analysis of concrete is given in BS 1881:Pt 124 (1988) and the modes of operation of crushing and grinding equipment are detailed in Allman and Lawrence (1972) and manufacturers' literature.

The heterogeneity of concrete and aggregates requires that several kilograms be crushed in order to obtain a representative subsample from one sample site of 100 to 200 g passing a 150 μm sieve. For example BS 1881:Pt 124 (1988) requires the minimum linear dimension of the sample to be at least five times that of the nominal maximum dimension of the aggregate particles and also to be not less than 1 kg for the analysis of concrete. It also points out that a major source of error in the analysis of hardened concrete is inadequate sample preparation.

The first stage in crushing a large concrete sample, such as a core, involves breaking it down into 20 to 30 mm sized fragments suitable for feeding to the crusher. Initial breaking is performed using either a mechanical knife edge, hydraulic splitter or hammer and bolster. In the further reduction of these concrete fragments to a powder passing a 150 μm sieve it is necessary to limit contamination, restrict loss of material, principally as dust, from the sample and carry out the reduction as quickly as possible to limit exposure to atmospheric carbon dioxide.

Reciprocating jaw crushers are commonly used for crushing materials to pass a 5 mm sieve as they can handle large quantities of material rapidly. Iron is the most likely contaminant from a jaw crusher but this should be minimal if the jaws are made from specialised hard steel alloys. Cross contamination from previous samples is a much more serious problem which is minimised by thorough wire brushing and cleaning with compressed air between samples.

If carried out carefully jaw crushing produces material that can be taken directly to the hand pestle and mortar, and for small amounts of material of less than about 1 g hand grinding is the simplest and most effective method. However, there is a range of mechanical equipment that can be used to avoid hand grinding to reduce the minus 5 mm sample to a coarse powder. The commonest of these are various types of disc mills. BS 1881:Pt 124 (1988) specifies that before this is done the sample be split down to 500 to 1000 g using a riffle box and that as the material is reduced in size further sample splitting be carried out for material passing the 2.36 mm and 600 μm sieves. The final sample required is 100 to 200 g of material passing a 150 μm sieve which is sufficient for the determination of cement, aggregate and original water contents. Due consideration must be given to the possibility of contamination of the sample by the materials of the mill, disc mills in particular can contaminate samples significantly if the sample material is hard and grinding protracted.

Where a powder finer than 150 μm is required a small subsample, often only a few grams, is ground in a specialised laboratory mill. There is a wide range of this equipment available with the type and method of grinding being dictated by the analytical technique to be used. There are no general references available but several manufacturers of milling equipment produce informative catalogues. In brief, X-ray diffraction analysis requires <45 μm, X-ray fluorescence <100 μm preferably 45 μm, infrared analysis 2 μm sized particles for analysis. The details and associated problems of the preparation of samples for X-ray diffraction analysis is discussed by Klug and Alexander (1974), X-ray fluorescence analysis by Norrish and Hutton (1969), infrared analysis by Farmer (1974) and solution chemical analysis of minerals by Sulcek and Povondra (1989).

3.8.2   SMALL SELECTED SAMPLES

During the inspection of structures, or the examination of cores and other laboratory samples, samples of efflorescence, exudations and crack infillings are collected for later detailed study. These samples are often only a few mg in size and should first be examined without pre-treatment. The initial steps are examination using a stereo binocular microscope followed by examination of part of the sample in immersion oil by the petrographic microscope. These initial steps will often provide valuable information about the number and type of constituents present and will indicate whether further detailed analysis is required.

Electron microscopy and electron probe microanalytical investigations can be carried out on the original unmodified material but where reliable chemical data is required the sample must be prepared as a polished mount. As the samples are usually less than 10 g they are best crushed using a small percussion mortar and then ground to a fine powder by hand in a small agate pestle and mortar. The particle sizes required for X-ray, infrared and chemical analysis have already been discussed in Section 3.8.1.

It is not possible to completely separate the fine aggregate particles from the hardened cement paste in mortars and concretes but the proportion of hardened paste can be increased by selective sampling. After breaking the sample with a hydraulic splitter fragments rich in hardened paste are selected and further processing is restricted to a percussion mortar. By a process of selection and sieving the hardened paste can be enriched by about 100 per cent with minimal contamination from the crushing apparatus or degradation of phases due to heating. This is the preferred method for minor and trace analysis and has been used successfully for this purpose by Goguel and St John (1993). Selective sampling may also be used for the detection of organic admixtures by Fourier transform infrared spectroscopy in contrast with the laborious chemical extraction of large powdered samples used by Hime (1974), Rixom (1978) and Concrete Society (1989).

3.9 Preparation involving dyes, stains and etches

Techniques using chemical solutions or vapours to selectively stain minerals as an aid to their identification have been used for many years. Geologists have made use of such techniques to identify a variety of rock- forming minerals and clays (Allman and Lawrence 1972, Reid 1969) while cement chemists have used etching techniques to identify phases present in polished sections of clinkers and slags (Campbell 1986). More recently, dyes have been incorporated into concrete to enable small cracks, voids and porous areas of the section to show up clearly.

The basic mechanism for both stains and etches is the same in that a chemical reagent reacts selectively with specific minerals. In the case of etches the surface layers of the minerals are removed in solution and in the case of stains a reaction product remains on the surface and is either coloured or a colour can be developed by further treatment. The complexities of the individual recipes arise from the need to select appropriate reactant concentration, reaction time and temperature conditions so that the desired effect is produced. Some skill and experience is usually necessary so that the mineral is neither under nor over etched or stained.

3.9.1 DYES FOR RESINS

There are two types of colorants suitable for incorporation into the epoxy resins which are used for specimen impregnation. The first type consists of dry pigments in a form which may be mixed directly with the resin prior to impregnation but unless the mixing is thorough a patchy uneven colouring is produced and the coloured pigment particles may not fully permeate the finer pores in the cement paste. A better approach is to use pre-mixed colouring pastes marketed by major producers of epoxy resin which dissolve in the resin when heated to 100°C. These pastes are available in red, green and blue and provided they are thoroughly intermixed with the resin at 100°C produce an even coloration.

The second type consists of dyes that are soluble in the resin. These are superior to pigments and impart a uniform colour to the resin. The colorants of greatest value to concrete petrographers are dyes both for highlighting cracks and voids in white light and for their fluorescent properties in ultraviolet light which can be used to detect porosity (see Section 2.7). These dyes need to be reasonably stable to resist fading. A yellow dye known as Albisol Brilliant Yellow R has been found particularly suitable and is manufactured under a number of names that will be found listed in the Society of Dyers and Colourists (Annual).

Table 3.6 An outline of some useful staining techniques for concrete petrography.

Specimen	Method outline	Observation	Reference
Sulfates Polished and ground surfaces. Thin-sections. Powders.	2 min. immersion in 2:1 mixture of $BaCl_2$: $KMnO_4$ 6% solution, wash first with water then saturated oxalic acid.	Ettringite, gypsum, anhydrite, plaster stain pink to purple.	Poole and Thomas (1975).
Alkali-silica gel Broken and ground surfaces thin-sections.	15 min. immersion in 10% uranyl acetate 1·5% acetic acid solution, washed in water.	UV light at 240 nm wavelength gives green fluorescence.	Natesaiyer and Hover (1988, 1989).
Broken and ground surfaces. Thin-sections.	72 hour absorption in 4M cuprammonium sulfate, washed in water.	Gel exudations in voids and cracks stain blue.	Poole, McLachlan and Ellis (1988).
Carbonates Broken and ground surfaces.	Etch for 10 secs. in 15% HCl. 30 sec.immersion in 1:1g Alizarin Red S + 0.9g Potassium	Calcite stains pink. Ferroan calcite stains purple-blue.	Dickson (1966).
Thin-sections. Powders.	Ferricyanide in 100 ml 1·5% HCl. 10 sec. immersion in 0.2g Alizarin Red S in 100 ml 1·5% HCl, washed in water.	Ferroan dolomite stains turquoise. Dolomite is unaffected.	

3.9.2   STAINING PROCEDURES

Aggregates in concrete are usually easily identified without recourse to staining. Some carbonate aggregates do present problems because distinctions between limestones, dolomites and dolomitic limestones are not always clear, particularly if the rock is fine grained. Identification of carbonates is important when investigating the alkali carbonate reaction (see Section 6.7.12). Allman and Lawrence (1972) detail several methods for separating these rock types using selective stains. One of the best of these techniques published by Dickson (1966) allows differentiation of all the common carbonate rocks and if used carefully will pick out zoning and overgrowths on individual particles. Sulfate minerals are also sometimes difficult to identify when they are present as small particles mixed with a fine aggregate and also as fine-grained sulfate phases such as ettringite in cement paste. An effective stain which may be adapted for use on ground or polished surfaces, or on thin-sections is described by Poole and Thomas (1975) and has been used to confirm the presence of sulfate bearing phases in studies of sulfate attack on concrete (Harrison 1992).

The development of alkali-silica gel in concrete is discussed in Section 6.7. This gel is quite difficult to identify unequivocally on broken and sawn concrete surfaces but readily absorbs the uranyl ion which fluoresces strongly in ultraviolet light (see Section 2.7). The method described by Natesaiyer and Hover (1988, 1989) is useful as a method for detecting the presence of alkali-silica gel when in the field. A summary outline of these methods is given in Table 3.6, but the original references should be consulted before a method is used for the first time.

3.9.3   ETCHING PROCEDURES FOR CEMENT CLINKERS

The identification of the mineral phases in cement clinkers using reflected light (incident light) microscopy requires the use of selective etching techniques and will also reveal structural features such as twinning and zoning (see Table 4.2). A comprehensive review of these techniques is given by Campbell (1986), together with colour photomicrographs illustrating the results obtained. Many of the techniques require very brief exposure of the specimen to the etching solution or vapour. The common method used is to invert the section and dip the polished face into a thin layer of the etchant contained in a shallow petri dish. In the case of HF vapour the hydrofluoric acid, often diluted to slow the etching down, is placed in a platinum crucible and the inverted polished face is held above it. The etched surface is washed with alcohol to stop the reaction and the surface quickly dried in a current of warm air. Over etching is a common cause of unsatisfactory results but lowering the temperature of etching has been found to give better results for some of the etches (Fundal 1980). A clean, high-quality polished specimen surface is essential if unambiguous results are to be obtained. Some of the methods used are outlined in Tables 3.7 and 4.2, but for full details reference should be made to the original published method before a particular procedure is attempted for the first time.

Table 3.7 Mineral identification: stains and etches for cement minerals and slags.

Specimen type	Method outline	Observation	Comment
1. Aluminates			
Polished surface.	5% KOH in ethyl alcohol, immerse for 20 secs., wash with 1:1 ethyl alcohol, water, then isopropyl alcohol buff for 1 sec. with isopropyl alcohol.	C_3A turns blue.	Campbell 1986 (variants using NaOH)
Polished surface.	10% boiling NaOH 20 sec. immersion, wash with alcohol.	CA turns blue or brown.	HAC.
Polished surface.	1% borax solution, etch for 30 secs., wash with alcohol.	$C_{12}A_7$ turns grey.	HAC.
2. Silicates			
Polished surface.	0.2g salicylic acid + 25 ml isopropyl alcohol + 25 ml water, 20–30 sec. immersion, alcohol spray wash.	C_3S and C_2S turn blue-green.	C_3S stains faster than C_2S.
Polished surface.	0.5g salicylic acid + 50 ml methyl alcohol, 45 sec. etch.	C_3S and C_2S etch, C_2S shows lamellar structure.	Methyl alcohol more reactive than isopropyl.
Polished surface.	1.5 ml HNO_3 + 100 ml isopropyl alcohol, 6–10 sec. etch.	C_3S blue to green, C_2S brown to blue.	'Nital' etch.
Polished surface.	KOH etch for 20 sec. followed by 6 sec. Nital etch.	C_3A turns light brown.	
Polished surface.	1g Ammonium Nitrate + 20 ml water + 20 ml ethyl alcohol + 10 ml acetone + 150 ml isopropyl alcohol, 25–30 sec. etch.	C_3S first, then other silicates brown → blue → green → yellow green.	Colour depends on length of etch.
Polished surface.	Hydrofluoric acid vapour at 20°C for 5–10 secs.	C_3S brown, C_2S blue.	Hazardous substance.
Polished surface.	Water pH 6.8–7 on a saturated microlap cloth, 3 secs. isopropyl alcohol wash, warm air dry.	MgO, hard high physical relief. CaO, dark → green/blue, C_3A, dark blue. C_3S, brown. C_2S, lamellar structure.	
3. Hydrates			
Polished surface.	Naphthol green B dissolved in water/alcohol mixture, etch time up to 12 hours.	$Ca(OH)_2$ turns green.	Lea (1970) p. 238.
Thin-sections.	$BaCl_2$ $KMnO_4$ 2:1 mixture as 6% solution in water, 2 min. immersion, wash with water and then saturated oxalic acid solution.	Ettringite, gypsum anhydrite, plaster stain pink to purple.	Poole and Thomas (1975)
Thin-sections, ground surfaces and powders.			
Ground surfaces and powders.	4M cuprammonium sulphate solution, 72 hour immersion at 25°C, water wash.	Alkali silica gels turn pale blue.	
4. Slags			
Polished surface.	10% $MgSO_4$ solution, etch for 60 secs. at 50°C.	C_2S shows striation.	Unsoundness BS 1047.

4

Composition of Concrete

4.1 Scope

There is frequently the need to establish the identity of the mix constituents within hardened concrete and sometimes also the mix proportions. Petrographical examination is a direct and effective method of achieving these objectives. Typical causes of the need for such analyses include disputes during construction, materials assessments during the life of a structure for condition or for predicting performance in respect of a change of use and matching constituents for repair concretes. In one region of the UK, petrographical examination has been adopted as the method for classifying concrete materials during routine building surveys (Stimson 1997).

Cement types, aggregates and mineral additions (including pigments) are usually identifiable using a range of petrographical techniques. However, the many admixtures, which are increasingly used in small dosages to modify the properties of concrete, are more difficult to characterise by either petrographical or chemical techniques. The principal mix parameters may each be readily evaluated by petrographical methods, especially water/cement ratio, aggregate/cement ratio, coarse/fine aggregate ratio and void content (including entrapped and entrained air voids), providing adequately large and representative samples are available.

4.2 Cement types

4.2.1 ANHYDROUS PORTLAND CEMENT PHASES AND CLINKER

The sintered product of burning limestone and an aluminosilicate rock (clay or shale) at temperatures of up to 1500°C is known as Portland cement 'clinker'. The grey powder generally called 'cement' is produced by inter-grinding the clinker and ~ 5 per cent gypsum to act as an early set retarder. Lea (1970) gives a detailed account of the history of the development of Portland cement.

The properties of Portland cement can be modified by altering its chemical composition by varying the raw feed materials and mix proportions, also the rate of cooling of the clinker, in order to produce different clinker mineralogies and hence different Portland cement types (Table 4.1). Fineness (or specific surface) can also be varied, with effects on the rates of hardening and strength gain, and there has been a tendency over time for cements to be more finely ground. Generally, pre-1950s Portland cements will be found to be more coarsely ground than their modern equivalents.

Table 4.1 Principal Portland cement types (as specified in draft European and equivalent British Standards).

Cement type		prENV 197-1	BS	Clinker %	Others %	Secondary constituents
Portland cement	I	I	BS12 BS4027[1]	95–100	—	—
Portland slag cement	II	II/A-S II/B-S	BS146	80–94 65–79	6–20 21–35	Blastfurnace slag
Portland silica fume cement		II/A-D	—	90–94	6–10	Silica fume (microsilica)
Portland pozzolana cement		II/A-P II/B-P	—	80–94 65–79	6–20 21–35	Natural pozzolana
		II/A-Q II/B-Q		80–94 65–79	6–20 21–35	Industrial pozzolana
Portland fly ash cement		II/A-V II/B-V	BS6588	80–94 65–79	6–20 21–35	Siliceous fly ash (e.g., UK pfa)
		II/A-W II/B-W	—	80–94 65–79	6–20 21–35	Calcareous (high lime) fly ash
Portland burnt shale cement		II/A-T II/B-T	—	80–94 65–79	6–20 21–35	Burnt shale
Portland limestone cement		II/A-L II/B-L	BS7583 —	80–94 65–79	6–20 21–35	Limestone
Portland composite cement		II/A-M II/B-M	—	80–94 65–79	6–20 21–35	Composite of two or more of the above
Blastfurnace cement	III	III/A III/B III/C	BS146 BS4246[2]	35–64 20–34 5–19	36–65 66–80 81–95	Blastfurnace slag
Pozzolanaaaic cement	IV	IV/A IV/B	— BS6610	65–89 45–64	11–35 36–55	Pozzolanaaaa or fly ash
Composite cement	V	V/A V/B	—	40–64 20–39	36–60 61–80	Composite of two or more of the above

Notes
[1] Sulphate-resisting Portland cement will eventually be covered by a separate part of ENV 197.
[2] BS 4246 covers a cement with a blastfurnace slag content of 50–85 per cent.

Nearly all hydrated cement paste within concrete contains residual kernels of unhydrated cement or clinker, although the proportion and grain size vary greatly according to a number of factors, and these pieces of unhydrated clinker can be used to characterise the cement type by mineralogical analysis.

The microscopical study of Portland cement clinker to determine mineralogy commenced in the last years of the nineteenth century (Le Chatelier 1905). Törnebohm (1897) gave the names alite, belite, celite and felite to four distinctive crystalline components which he and Le Chatelier had observed within Portland cement clinker, plus an isotropic residue phase. Insley (1936, 1940) later demonstrated that Törnebohm's 'alite' was tricalcium silicate (C_3S), that 'belite' and 'felite' were two different habits of usually beta-dicalcium silicate (βC_2S), that 'celite' was a calcium aluminoferrite phase and that the isotropic residue contains calcium aluminates (mainly C_3A) and glass. The calcium aluminoferrite phase was first thought to be C_4AF (brownmillerite), but is now known to be a solid solution series ranging from C_2F to just beyond C_6A_2F (Midgley 1964), although according to Lea (1970) the median value is close to C4AF, which is the compound assumed by Bogue (1955) for his normative calculations from chemical analyses of cement.

The identification of clinker mineralogy has been described in detail elsewhere (Insley and Fréchette 1955, Midgley 1964, Gille *et al.* 1965, Lea 1970, Bye 1983, Campbell 1986). In general, petrographical examination of Portland cement clinker reveals a texture of relatively large crystals of alite (C_3S) and belite (C_2S) set in a matrix mainly comprising C_3A and calcium aluminoferrite. The principal characteristics of the Portland cement minerals are detailed in the Glossary.

Briefly, the texture of a cement clinker is dominated by pseudo-hexagonal crystals of C_3S, varying from 25 μm to 65 μm in size, together with less frequent rounded crystals of C_2S usually with polysynthetic twinning, varying from 10 μm to 40 μm in size, set in a poorly differentiated matrix. The crystals of C_3S are usually well formed and dispersed, but C_2S can vary from well scattered crystals to clusters or nests and is nearly always present as small relict inclusions in C_3S crystals.

Differentiation of the aluminate and ferrite phases in the matrix is primarily a function of the rate at which the clinker has been cooled below about 1250 °C, which is approximately the liquidus temperature. In modern cement manufacture clinker is cooled moderately fast, so that even with adequate etching, details of the matrix may not be well differentiated. It is only when the clinker has been more slowly cooled in the manufacture of older cements that the aluminate and ferrite phases stand out clearly.

Microscopical examination of polished and etched surfaces in reflected light is the most useful method, but thin-section examination can also be helpful (Campbell 1986). The techniques are equally applicable to portions of unground clinker, samples of cement or pieces of concrete matrix containing residual kernels of clinker. Table 4.2 summarises the appearance of the main cement minerals in reflected light using various etchants. Fundal (1980) has demonstrated that clinker microstructures can also be used to assess clinker properties and quality.

Calcium silicates (C_3S, C_2S)

Accounting for nearly 80 per cent of a typical ordinary Portland cement clinker, C_3S and C_2S form crystals which appear as agglomerated phenocrysts, with the other cement constituents forming the fine-grained interstitial component. The dominant C_3S (alite) usually forms large euhedral pseudo-hexagonal crystals, which are colourless in thin-section but appear brown in reflected light after HF etching (Plate 1). Inclusions of C_2S, lime (CaO) or periclase (MgO) are not uncommon within C_3S crystals; also alite is typically impure, with small proportions of magnesium and aluminum substituting for silica.

The C_2S (belite and felite, but now usually just referred to as belite), by contrast, usually forms anhedral or sub-hedral crystals and in some cases these appear well rounded in cross-section. It is believed that C_2S in Portland cement clinker is usually the βC_2S polymorph, although Insley (1936) demonstrated at least three different types of βC_2S according to crystal twinning styles. French (1991a) has referred to four types of belite, according to the number of sets of inversion lamellae and the presence of inclusions. C_2S is typically slightly coloured in thin-section (yellow, brown or green) and generally appears blue in reflected light after HF etching (Plates 1, 2).

Table 4.2 Appearance of clinker phases in reflected light using various etchants.

Clinker phase	Etchant or stain and etchant	Immersion/ exposure time (s)	Appearance in reflected light
Free lime, CaO	Distilled water	3–5	Etches rapidly
Tricalcium silicate C_3S (Alite)	10% KOH		Does not etch
	HF vapour	2–5	Buff brown
	HNO_3 in alcohol[1]	2–15	Greyish brown
	NH_4NO_3/Salicylic acid[2]	25–30/30	Yellow-green[3]
Dicalcium silicate C_2S (Belite)	10% KOH		Does not etch
	HF vapour	2–5	Blue
	HNO_3 in alcohol[1]	2–15	Blue
	NH_4NO_3/Salicylic acid[1]	25–30/30	Brown (when C_3S is yellow-green)
Tricalcium aluminate C_3A	Water	3–5	Etches
	10% KOH	10–20	Bluish grey
	HF vapour	2–5	Light grey
	HNO_3 in alcohol[1]	2–15	Light grey
Ferrite C_4AF	10% KOH	10–20	White
	HF vapour	2–5	White
	HNO_3 in alcohol[1]	2–15	White

Notes

1 Gille *et al.* (1965). Nitric acid in alcohol, using 1:100 or 1:1000 solutions, prepared at least a fortnight prior to use.
2 Ahmed (1994). Stain using 1 g NH_4NO_3, 150 ml isopropyl alcohol, 20 ml water, 20 ml ethyl alcohol and 20 ml acetone; then etch using 0.2 g salicylic acid, 25 ml isopropyl alcohol, 25 ml water.
3 Depending on time, the colour ranges from light brown to brown to purplish brown to blue to blue-green to yellow-green.

   In slowly cooled clinker, the C_2S can occur also as very small rounded particles dispersed in the interstitial component and/or as 'rims' of small particles around the edges of the large C_3S crystals. Clusters of rounded C_2S crystals have sometimes been termed 'bunch of grapes' or 'basket of eggs' texture because of their appearance in thin-section (Plate 3) and, when large, this texture can indicate an unacceptable degree of inhomogeneity in the kiln raw feed (Fundal 1980, Bye 1983).

Calcium aluminates and calcium aluminoferrites (C_3A, C_4AF)

Forming in the order of 10 per cent of ordinary Portland cement, C_3A typically forms one component of the interstitial material between the C_3S and C_2S crystals. This 'dark' interstitial material has been sub-divided into three types: small rectangular outlined crystals believed to be C_3A, long prismatic crystals (said by Insley and Fréchette (1955) to be 'the more usual form') thought to be a C_3A polymorph modified by alkali in solid solution, and an amorphous (or glass) phase. The precise nature of the amorphous interstitial material appears uncertain, but it appears to be logical and agreed that the amount of glass increases for rapidly cooled clinker.

In thin-section the C_3A crystals are colourless to brown and frequently isotropic. In reflected light after HF etching the C_3A material is distinguished by being light grey in appearance and clearly interstitial in occurrence.

The other element of the interstitial component (the 'light' interstitial material) is the calcium aluminoferrite (or simply 'ferrite') phase: the old term 'celite' has fallen out of use. The ferrite material is distinctive in both thin-section and reflected light. Ferrite has a much higher refractive index than the other clinker constituents (1.9–2.0 compared with about 1.7 for C_3S, C_2S and C_3A), is highly birefringent and is pleochroic, typically ranging from pale amber to dark reddish brown (French (1991a) reports a colour range from dark green to almost opaque). Apart from being truly interstitial, thus taking the form imposed upon it by the neighbouring constituents, ferrite can occur as small prismatic or dendritic crystals or as fibrous material. In reflected light, the ferrite phase is unaffected by most etching agents and appears nearly white and highly reflectant.

Lime and periclase (CaO, MgO)

Significant amounts of 'free' lime (CaO) may occur in Portland cement clinker if the raw feed mixture was overloaded with limestone or if that mixture was incompletely burned in the kiln but a normal range would be from 0.5 to 1.5 per cent. Excessive proportions of CaO lead cements to be 'unsound' and standard physical tests are routinely carried out on cement to avoid this problem (BSI 4550 1978), but Vivian (1987) has reported delayed expansion in concrete caused by 'hard burnt' uncombined free lime which was not detected by routine testing.

When present in clinker, CaO is almost always found as distinctive spheroidal particles and never exhibits its cubic crystal faces (Plate 4); CaO might also occur as small inclusions within C_3S or C_3A crystals. CaO grains are often very small but occasionally form rounded crystals, or clusters of small crystals, which might be as large as the C_3S crystals. In reflected light, CaO particles are distinguished by their rapid etching with water (whereas etching to distinguish other clinker constituents typically involves KOH or HF reagents) and their spherical shape.

Periclase (MgO) can form in clinkers in which the magnesium content is higher than that which can be taken into solid solution (about 1.5 per cent according to Bye 1983) and if the clinker is not cooled too rapidly. Usually such discrete MgO averages only around 1 per cent of clinker, but occasionally might amount to 5 per cent or more. Although MgO appears to have some value as a flux in cement making, excessive proportions of free MgO can sometimes lead to long-term unsoundness caused by expansion resulting from very slow hydration (Lea 1970): this led the Americans to introduce an autoclave test for assessing cement soundness (ASTM C151 1993).

MgO forms in association with the other interstitial components (e.g. C_3A and ferrite), as small angular crystals or clusters of crystals. In reflected light, these crystals are readily identifiable as highly reflecting grains on a polished but unetched surface, when they are emphasised by the relief caused by their relative hardness and by distinctive dark borders (Fig. 4.1).

Other phases and gypsum

The other significant but minor constituents of Portland cement clinker are the alkali sulfates and double sulfates, which are the source of most of the alkalis associated with Portland cement. Excess sodium and potassium in the clinker mix appear to combine with

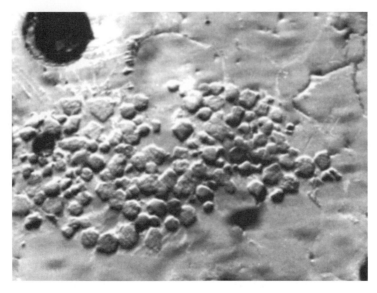

Fig. 4.1 Photomicrograph of periclase in Portland cement clinker in reflected light, unetched (from Fundal 1980). Width of field *c*. 150 μm.

calcium silicates and calcium aluminates in a complex manner (Lea 1970). According to Bye (1983), water etching of clinker specimens removes the alkali sulfates and creates pores, suggesting that these alkali sulfates might be observed in clinker as later-stage pore infillings.

Gypsum (or occasionally anhydrite) is not a component of the clinker, but is instead added to the clinker at the grinding stage and is thus an essential but discrete constituent of the cement power produced. If milling temperatures are too high, the gypsum ($CaSO_4.2H_2O$) can become dehydrated to form the 'hemi-hydrate' ($CaSO_4.nH_2O$), giving rise to problems of 'false setting' (i.e. apparent setting of cement which can be removed by remixing) caused by rapid rehydration on exposure to moisture (cf. setting of 'Plaster of Paris' which is hemi-hydrate, $CaSO_4.0.5H_2O$). The gypsum reacts more or less immediately with C_3A during Portland cement hydration (Lea 1970) and is rarely observed as a residual unreacted cement constituent within hardened concrete or mortar.

4.2.2   IDENTIFICATION OF CEMENT TYPE IN CONCRETE

As the type of cement (Table 4.1) is often specified for concrete, disputes can arise concerning whether or not the correct cement was used. More importantly, the type of cement might have an important influence over the durability of a concrete being assessed for condition or for future use. This distinction has most often concerned the difference between ordinary and sulfate-resisting Portland cements, but increasingly questions arise concerning blended cements (see Section 4.2.4) or the presence of separately batched mineral additions (such as ggbs or pfa, see Section 4.6). These cement types or blends of materials can be identified from their mineralogical characteristics.

General principle of microscopical methods

Hardened concrete contains residual unhydrated particles of cement (or clinker), although the quantity and the particle size both vary widely, depending upon composition, curing and exposure. Concretes in which mineral additions (e.g., ggbs, pfa, natural pozzolanas) have been deliberately used will always contain, often substantial, proportions of unreacted mineral addition material. In most cases, microscopical examination of these various residual particles enables the cements and additions to be identified from their mineralogy and statistical analysis offers prospects for quantifying blend proportions. A discussion of the commonly employed method used for distinguishing between OPC and SRPC provides an effective introduction to the approach required for identifying any cement and/or mineral addition content of hardened concrete.

When ground Portland cement is examined in transmitted light the majority of the particles appear monomineralic. The most obvious particles are crystals of C_3S which are often broken and rarely have any adherent matrix. Discrete crystals of C_2S are much less common and are more likely to have some matrix material attached to them. The largely monomineralic character of most of the cement particles is not surprising, considering that 80–90 per cent of the particles are less than 45 μm in size in a typical modern cement. As discussed further in Chapter 5, the residues on a 90 μm sieve vary from as little as 2 per cent for modern cements to as much as 10 per cent for older cements. These oversize particles, which range from 90 μm to 300 μm or greater in size, consist of a matrix in which are commonly embedded C_2S and less commonly C_3S crystals. In hardened cement paste these remnant grains are surprisingly resistant to hydration and their examination in polished section can be used to differentiate some cement types.

Microscopical procedure

A method for distinguishing OPC from SRPC is given in BS 1881 (1988) and a good account of the procedure is given by Grove (1968). In considering the reflected light techniques, it is important to realise that different effects are produced by different etchants and, commonly, these are selected to highlight a constituent of particular interest (Table 4.2). In Section 4.2.1, for example, reference has been made to the appearance of various clinker phases when viewed after HF vapour etching, which is frequently used today as a general etchant; it was also explained that CaO was best detected by water etching and that MgO was most obvious on the polished surface prior to etching. The technique described by Grove (1968) employs a 100 g/l aqueous solution of KOH as the etching agent, which highlights the aluminate phase and is thus particularly applicable to the differentiation of OPC and SRPC.

Sulfate attack on concrete is most usually associated with the expansive formation of ettringite within hardened concrete as the result of reaction between externally derived sulfates in solution and the calcium aluminate and calcium monosulfoaluminate hydrate phases in the hydrated cement paste (see Chapter 6). Sulphate-resisting Portland cement (SRPC) is based on the principle of reducing the C_3A content of the interstitial clinker phase in manufacture by correspondingly increasing the content of ferrite (notionally C_4AF), although it remains uncertain why the calcium aluminoferrite is not also susceptible to sulfate attack (Lea 1970). SRPC is thus defined by limiting the permitted content of C_3A: to 3.5 per cent in BS 4027 (1996) and to 5 per cent in ASTM C150 (1994). As research suggested that the effectiveness of low-C_3A cements was reduced at higher C_4AF contents, ASTM C150 also places a 20 per cent upper limit on the total content of C_3A and C_4AF

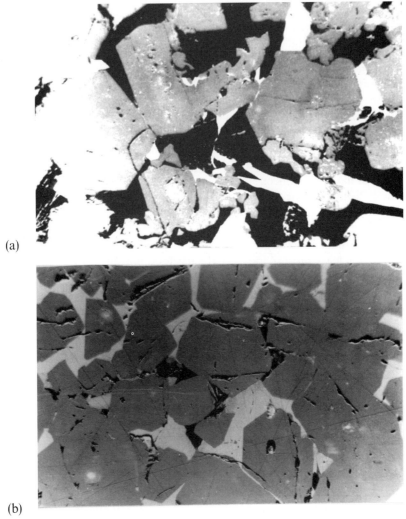

(a)

(b)

Fig. 4.2 Photomicrographs: comparison of OPC (a) and SRPC (b) clinkers in reflected light after etching with 10 per cent KOH. Width of fields *c*. 250 *μm*. (Photographs courtesy of Mr Martin Grove (as for Grove 1968)).

together. A recent review of sulfate attack and the effectiveness of SRPC has been carried out by Lawrence (1990).

In distinguishing SRPC from OPC, therefore, the ratio of C_3A to C_4AF (ferrite) is thus critical. In Grove's method, a highly polished surface of the concrete matrix (see Chapter 3) is treated with the KOH solution for 10 or 20 seconds at room temperature, then washed in ethanol. It is important to avoid over-etching the specimen. When the residual unhydrated cement particles in the etched surface are examined under a reflected light microscope, the C_3A exhibits a distinctive blue or blue-grey colour, whilst the ferrite is unetched and appears bright and highly reflectant; the other clinker phases are dull by comparison.

In most cases the difference between representative clinker grains of OPC or SRPC are visually apparent (Fig. 4.2), with the volume ratio of ferrite to C_3A rarely exceeding 2:1 for OPC, but being more than 5:1 for SRPC. In fact the C_3A can be difficult to find in many SRPCs, such is the dominance of ferrite. Grove (1968) states it is usually necessary 'to search' for C_3A and French (1991) agrees that ferrite 'makes a very large part' of the matrix in SRPC. Therefore, although the relative proportions of C_3A and ferrite could be quantified by point-counting in Grove's method (Weigand 1994), in practice this is rarely necessary for the experienced cement microscopist.

Interpretation of the findings and some difficulties

In many cases the KOH etching method is a reliable and comparatively straightforward means of confirming whether the cement used in a concrete was OPC or SRPC. The optional method using HF vapour is usually also effective, except that the visual contrast between C_3A (light grey) and ferrite (white) in the interstitial component is rather less obvious; however the HF method also provides information about the other clinker minerals, notably C_3S and C_2S.

Optical studies of Portland cement clinker prior to grinding illustrate its often variable texture and composition, and analysis of these textural variations can assist cement technologists in perfecting manufacturing techniques (Fundal 1980). It therefore follows that, after grinding, the individual cement grains will be variable in composition and only the exceptionally large grains are likely to approach being representative of the original overall clinker. When examining specimens of hardened concrete matrix, the cement type determination has to be based upon whatever residual unhydrated particles of cement remain and the limitation on representativeness of individual grains must be taken into account. At least twenty such residual particles should be examined and the mineral composition of each assessed.

The authors also recommend that the approximate size of each particle should be recorded and the compositional findings classified according to size band (<20 μm, 20–40 μm and >40 μm bands have been found satisfactory). Then, in making final judgement in apparently marginal cases, reliance weighting should be placed upon the findings for the coarser particles (>40 μm) and, if necessary, further analyses should be carried out, perhaps using additional specimens of the concrete matrix in question. One form of practical reporting of results is shown, with an example, in Table 4.3.

Mixtures of Portland cement types in concrete are uncommon and are very difficult for the microscopist to detect with certainty. Such mixtures may result from contamination, for example within cement silos, or from inadvertent use of different types of cement (or different sources of the same type) at the concrete-making stage. However, clinkers of either the same or different cement type from different sources are sometimes deliberately blended. One UK manufacturer has employed blends of OPC and SRPC in order to create a 'reduced alkali' cement (the alkali content of SRPC is typically lower than that of the related OPC), for possible use with aggregates considered to be potentially alkali-reactive (see Chapter 6). Such mixtures of cement types can sometimes be identified and possibly even quantified by statistical treatment of the microscopical findings, although a comparatively large number of residual unhydrated particles would need to be found in the concrete matrix and examined.

The microscopical method obviously depends upon the hardened concrete matrix containing a sufficient number of residual unhydrated cement particles of microscopically resolvable size and this is not always the case. In some modern very finely milled cements (e.g., rapid-hardening Portland cement, RHPC), such unhydrated particles might sometimes be hard to find.

Table 4.3 Determination of Portland cement type by reflected light microscopy – suggested form and a typical example (sulfate-resisting Portland cement in a mortar with building lime).

Sample details and methods of treatment

Laboratory ref:	11525	Site ref:	50
Location:		Block A, South elevation	
Type of material:		Brickwork jointing mortar	
Preparation & procedure:		Polished specimen etched for 3 seconds in HF vapour	

Test results – apparent cement type

Unhydrated grain size:	$>40\ \mu m$	40–$20\ \mu m$	$<20\ \mu m$
Number of grains:			
OPC appearance	–	2	3
SRPC appearance	3	12	10
HAC appearance	–	–	–
WPC appearance	–	–	–
Total:	3	14	13
Percentage of grains:			
OPC appearance	–	14	23
SRPC appearance	100	86	77
HAC appearance	–	–	–
WPC appearance	–	–	–
Total:	100	100	100

Test results – other observations

Mineral additions:	Building lime
Other constituents or features:	Yellow pigment

Conclusions

Apparent cement type:	Sulfate-resisting Portland cement
Other comments:	Mortar matrix was soft

French (1991a) has claimed that the content of residual unhydrated cement in concrete is related to the water/cement ratio and the temperature during curing. In concretes made at 20 °C, for example, he found that the cement powder was 'virtually completely hydrated' for water/cement ratios of 0.6 or more. Similarly, for concrete cured at between 40 °C and 50 °C, French found that hydration was 'generally complete', although others have reported a 'limited' presence of residual clinker for concretes cured up to ~80 °C (Patel *et al.* 1995). In the authors' experience, for the bulk of concretes made at temperatures below 20 °C and/or at water/cement ratios of less then 0.6, there is usually a sufficiency of unhydrated cement particles in the matrix for the cement type to be identified microscopically.

Although OPC and SRPC are by far the most commonly used cements, there are other specialised varieties of Portland cement, blended cements (see Section 4.2.4) and non-Portland cements such as aluminous cements. A good summary of the range of modern cement types is given by Neville and Brooks (1987). The possibility of these other cement types, or the possible presence of separately batched mineral additions, must not be forgotten by the microscopist when examining concrete.

Microscopy is particularly useful, for example, for confirming the presence of white Portland cement, which is frequently employed for concrete for architectural usage, sometimes with various pigments which mask the white cement coloration. White Portland cement is made by using limestone and aluminosilicate raw materials, such as bauxite, chalk and china clay, which are low in the iron which usually gives OPC its distinctive grey colour. The resultant white clinker has less than 1 per cent of ferrite and a substantially higher C_2S content than OPC, whilst the aluminate content is comparable with OPC or even a little higher than average (Bye 1983). Thus the near absence of ferrite and the high content of C_2S makes the cement relatively easy to identify microscopically. White Portland cement has a lower alkali content and typically also lower strength properties than conventional OPC.

Other methods

Unhydrated cement particles are visible in thin-sections of concrete, especially if the sections are ground thinner than usual for petrographical purposes (say to 10 or 15 μm, rather than the standard 25 μm for concrete and related materials), but it is extremely difficult to distinguish between types on a mineralogical basis, mainly because the grain size of the critical interstitial components are close to the practical resolution limits of the optical system. In most Portland cement concretes, it is the distinctive clusters of round C_2S (belite) crystals set in the clinker matrix (Plate 3) which are most readily noticed as residual cement material in thin-section (Parker and Hirst 1935).

Extreme cases might be discernible in thin-section when present as comparatively large residual grains. Examples would include sulfate-resisting Portland cement which is notably ferrite-rich, or white Portland cement which is equally notably ferrite-poor and also high in C_2S, or a non-Portland type of cement (such as HAC, see Section 4.2.4). The presence of mineral additions, especially ggbs or pfa, either as a blended cement or separately added, is usually detectable in thin-section, but silica fume might be difficult to resolve unless present in agglomerations (see Section 4.6.3).

It is possible to judge the likelihood of rapid-hardening Portland cement (RHPC) being present in a concrete from thin-sections, by studying the maximum size of the unhydrated cement particles. The bulk of remnant cement grains in RHPC will be found to be less then 20 μm in size and difficult to find in older concretes.

A potentially effective alternative to the range of optical techniques for identifying cement type in concrete is microanalysis of the residual cement particles using an electron probe microanalysis in connection with scanning electron microscopy (SEM). French (1991a) has described three approaches to the determination of 'apparent' cement type by electron microprobe on polished sections of concrete:

1. identification of the interstitial phases in residual cement particles (this is comparable with the optical methods);
2. overall chemical analysis of the larger residual cement particles; and
3. chemical analysis of the hydrated paste.

In the first approach, analysis of a 'spot' 5 μm in diameter enables the proportions of C_3A and C_4AF to be determined using the Bogue (1955) equations, but the analyses can be unreliable owing to the small size of the interstitial areas and consequent problems of interference from surrounding phases. Clearly the overall microanalysis of residual cement particles, the second approach, ought to be effective at characterising the cement, providing, as with the optical methods, that a sufficiently large number of particles is analysed. The method also offers a prospect of identifying different sources of the same type of cement, as each cement source might be expected to be chemically distinctive. The third approach, area microanalysis of the hydrated cement paste, is complicated by the additional presence of aggregate dust. French (1991a) states that the electron probe method yields one of three findings: high C_3A implying OPC, low C_3A implying SRPC, or ambiguous apparently intermediate C_3A content which a statistical analysis of the data might help to clarify.

Hammersley (1980) found few other techniques to be helpful in identifying cement type. SEM, X-ray diffraction, infrared absorption spectroscopy, differential thermal analysis and chemical analysis were each found to be useful in studying hydration or in distinguishing between Portland and non-Portland types of cement in concrete, but were of limited application in identifying the type of Portland cement. One of these, overall chemical analysis of concrete, is frequently used in commercial practice to suggest the probable cement type present, but, where the aspect of cement type is important, confirmatory optical or electron microprobe identification is indispensable.

Where the ranges of possible cement and aggregate sources are limited and known, it can be possible to 'fingerprint' the cement present by selective chemical analysis of the concrete material. Goguel and St John (1993a, b), for example, devised a simple procedure for New Zealand concretes whereby analysis for Ca, Sr and Mn enabled the cement source to be identified.

4.2.3   HYDRATED CEMENT PHASES

The four principal cement constituents (C_3S, C_2S, C_3A & C_4AF, see Section 4.2.1) are each hydraulic and thus react with water on mixing to form a range of hydrated phases. Detailed accounts of the complex chemistry and mineralogy of cement hydration are given elsewhere (Copeland and Kantro 1964, Lea 1970, Czernin 1980, Bye 1983, Taylor 1990). The initial post-hardening result of hydrating Portland cement at normal temperatures is a groundmass of poorly crystalline and microporous hydrated compounds with scattered inclusions of crystallised Portlandite (calcium hydroxide), other crystalline phases and residual kernels of unhydrated cement. Carbonation and general recrystallisation of this hardened 'cement paste' takes place gradually with time (see Chapter 5).

C-S-H and the microstructure of cement paste

As the calcium silicates (C_3S and C_2S) dominate the composition of Portland cement, it follows that the constitution of hydrated Portland cement paste is largely dictated by the reactions between these calcium silicates and water (idealised chemical equations are given in Table 4.4). Tricalcium silicate (C_3S) reacts with water and hydrated calcium silicate is quickly formed, together with a supersaturated solution of calcium hydroxide from which crystals of calcium hydroxide are subsequently precipitated, finally leaving a 'pore solution' which is saturated in respect of calcium hydroxide. Dicalcium silicate (βC_2S) exhibits a broadly similar hydration reaction, albeit much less rapidly and producing two-thirds less Portlandite.

The hydrated calcium silicate, which is principally responsible for the binding properties of Portland cement, appears to exhibit a highly variable chemical composition (lime:silica ratios typically range from 1 to 2) and is practically amorphous. Two types of hydrated cement paste (or 'gel' as it is often called) have generally been recognised, C-S-H (I) and

Table 4.4 Hydration reactions for Portland cement mineral phases (after Czernin 1980).

Tricalcium silicate:

$$2(3CaO.SiO_2) + 6H_2O \rightarrow 3CaO.2SiO_2.3H_2O + 3Ca(OH)_2$$

Dicalcium silicate:

$$2(2CaO.SiO_2) + 4H_2O \rightarrow 3CaO.2SiO_2.3H_2O + Ca(OH)_2$$

Tricalcium aluminate and Gypsum:

$$3CaO.Al_2O_3 + 3(CaSO_4.2H_2O) + 26H_2O \rightarrow 3CaO.Al_2O_3.3CaSO_4.32H_2O$$
$$\text{ettringite}$$

then

$$2(3CaO.Al_2O_3) + 3CaO.Al_2O_3.3CaSO_4.32H_2O + 4H_2O \rightarrow 3(3CaO.Al_2O_3.CaSO_4.12H_2O)$$
$$\text{monosulfate}$$

then

$$3CaO.Al_2O_3 + Ca(OH)_2 + 12H_2O \rightarrow 4CaO.Al_2O_3.13H_2O$$

Tetracalcium aluminoferrite:

$$4CaO.Al_2O_3.Fe_2O_3 + 4Ca(OH)_2 + 22H_2O \rightarrow 4CaO.Al_2O_3.13H_2O + 4CaO.Fe_2O_3.13H_2O$$

C-S-H (II), and both are broadly analogous in structure to the naturally occurring mineral tobermonite. Indeed, some cement mineralogists have referred to the hydrated calcium silicate phase of cement paste as 'tobermonite gel' (Brunauer *et al.* 1958). More recently, Diamond (1976) has reported four types of C-S-H by SEM study.

Hydration of the aluminate and ferrite phases (C_3A and C_4AF) also involves the alkali sulfate and gypsum components of Portland cement (in the absence of gypsum, the extremely rapid hydration of C_3A would lead to 'flash setting'). The precise series of reactions which occur are seemingly dependent upon a number of factors, including the relative proportions present of C_3A, sulfates and calcium hydroxide, but generalised chemical equations are included in Table 4.4.

Initially, C_3A and calcium sulfate react rapidly to form ettringite, hydrated calcium sulfoaluminate. The ettringite envelopes the unhydrated C_3A and effectively prevents the cement from setting too quickly. With time the ettringite is replaced by calcium monosulfoaluminate ('monosulfate'). As reaction occurs between residual C_3A and the initially formed ettringite coating layer, once the primary sulfates have been consumed, normal setting of the cement proceeds. If there is residual C_3A present, in the presence of calcium hydroxide later stage hydrated calcium aluminate can also be formed, usually of the form C_4AH_{13}, but the 'hydrogarnet' (C_3AH_6) might be formed under some circumstances. The much slower hydration of C_4AF is roughly analogous to that of C_3A after the consumption of the sulfates (Table 4.4), leading to the compound C_4FH_{13}.

The considerable intermixing of the various hydrated phases in the main period of hydration results in an extremely complicated mixture of phases. Its microstructure is difficult to resolve even using electron microscopy. Under the SEM, the C-S-H phases are often said to appear as filaments or tubular structures, although Jennings and Pratt (1980) have suggested that this appearance results from the rolling up of thin amorphous sheets

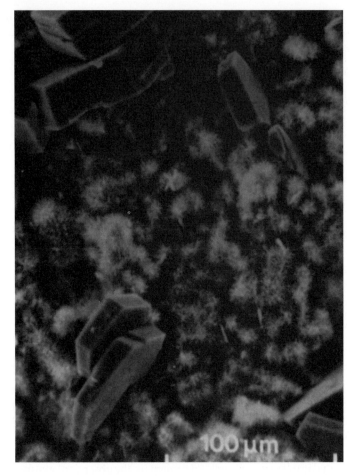

(a)

Fig. 4.3 SEM micrographs of hydrated Portland cement paste. (a) 3-day-old paste containing well-defined portlandite crystals. (Photograph from Bye 1983.)

caused by the drying involved in specimen preparation for electron microscopy. C-S-H also occurs pseudomorphing the original cement grains. The Portlandite and hydrated calcium aluminate form hexagonal platelets, whilst crystalline ettringite and monosulfate form distinctive clusters of acicular crystals. Some typical SEM views of hydrated cement paste are shown in Fig. 4.3. Diamond (1976) has estimated a typical composition of almost fully hydrated paste to be (see also Chapter 5):

C-S-H	70%
Portlandite	20%
ettringite/monosulfate	7%
minor phases	3%

Hydrated cement paste is porous, exhibiting a specific surface area in the region of $200 m^2/g$, implying pores of $<10nm$ diameter which are not easily observed by SEM (Bye 1983). Powers and Brownyard (1948) recognised two types of cement paste porosity: gel pores contained within an area of CSH particles and relatively coarser capillary pores representing space originally occupied by air or by mix water not used in hydration. The latter are

(b)

Fig. 4.3 SEM micrographs of hydrated Portland cement paste. (b) fracture through dense 'ground mass' in 8-month-old paste. (Photograph from Bye 1983.)

clearly related to the cement and water contents of the concrete mix and are used as the basis of several methods for estimating the original water/cement ratio of concrete (see Section 4.4).

The hydrated cement paste in samples of concrete is difficult to study by any optical system. As most of the cement 'gel' is virtually amorphous, it appears featureless and isotropic in thin-section, although the crystallinity, grain size and distribution of Portlandite is usually readily apparent under the polarising microscope (see later in this section). The composition of cement paste can be qualitatively determined by X-ray diffraction (Fig. 4.4), providing interference from aggregate constituents can be avoided. Over an extended period of time, cement paste gradually alters and crystallises and carbonation occurs as a result of reaction with the atmosphere (see Chapter 5).

Degree of hydration

The extent to which the cement in concrete becomes hydrated principally depends upon the water/cement ratio and the degree of curing. Degree of hydration is important as

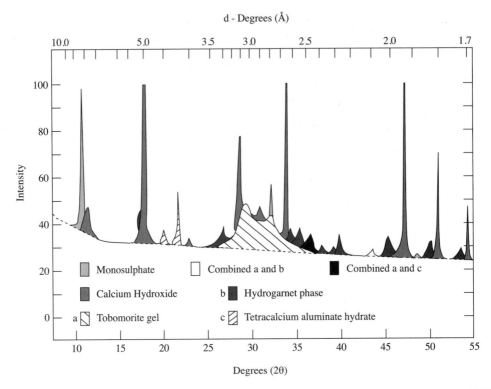

Fig. 4.4 Schematic X-ray diffractogram of Portland cement paste (from Copeland and Kantro 1964).

concrete strength is influenced not only by the content of cement in the mix but also by the degree of hydration of that cement. A very low water/cement ratio is always desirable providing the concrete can be compacted, but can also lead to under-hydration of the cement in some circumstances and consequently weak concrete. Conversely, an excessively high water/cement ratio can lead to almost complete hydration, but cause a large capillary pore volume in the cement paste (even 'bleeding') and thus impair both strength and impermeability of the concrete.

Thus, for any combination of constituents, required mix properties and conditions, there is a critical degree of hydration associated with the maximum attainable strength and minimised permeability. Of course, except for concretes kept in completely dry conditions, the degree of hydration is liable to increase with time after the initial period of hardening and strength gain, as residual cement particles very gradually become hydrated.

Determination of the degree of hydration might therefore sometimes assist in the assessment of concrete strength and permeability. Such estimates are feasible using chemical analysis, whereby the determined cement and CO_2 contents are compared with the content of combined water, but the method requires generalised assumptions and is complicated by the presence of aggregates. As the anhydrous cement compounds are crystalline, X-ray diffraction analysis is potentially effective, but interference from the poorly crystalline hydrates and aggregate materials make reliable interpretation difficult.

Microscopical examination in reflected light of polished specimens of concrete matrix (see Section 4.2.2) offers a direct method of determining degree of hydration. In concretes of unknown mix design, the volumetric ratio between the contents of hydrated and residual unhydrated materials in the cement paste is clearly related to the degree of hydration, although many factors affect the volume change accompanying hydration, so that such an approach can only be qualitative.

Ravenscroft (1982) has described a microscopical quality control method for determining the degree of hydration for concretes of known original mix proportions. In this method the volume proportions of residual cement particles and fine aggregate in portions of the hardened cement matrix are compared with those of the original mix proportions. The volumetric determinations by point-counting under the reflected-light microscope are converted to gravimetric results by assigning appropriate specific gravity values to the cement and aggregate constituents.

In this way, the Degree of Hydration (*Dh*) is determined as follows:

$$Dh = \frac{(Cm/Sm - Cs/Ss) \times 100}{Cm/Sm}$$

where Cm/Sm = cement to sand ratio by mass in original concrete mix, and Cs/Ss = unhydrated cement to sand ratio by mass in the sample. An example of this determinative method is shown in Table 4.5.

Portlandite ($Ca(OH)_2$)

It was explained earlier that Portlandite ($Ca(OH)_2$) is formed during the hydration of the calcium silicates in Portland cement and it is thus characteristic of uncarbonated cement paste. Typically, $Ca(OH)_2$ forms as hexagonal crystals with a perfect basal cleavage. In hardened cement paste it appears as anhedral or euhedral crystallites, plates or as short prisms, up to about 100 μm in maximum crystallite length when precipitated during the main hydration phase. The crystallites are distinctive in thin-section under the polarising microscope, being colourless and highly birefringent (see Glossary).

Table 4.5 Determination of the degree of hydration (a hypothetical example).

The formula proposed by Ravenscroft (1982) is

$$Dh = \frac{(Cm/Sm - Cs/Ss) \times 100}{Cm/Sm}$$

where Dh = degree of hydration
  Cm/Sm = cement/sand ratio in mix
  Cs/Ss = unhydrated cement/sand ratio in sample

If a concrete had mix proportions of 1:2:4

the $Cm/Sm = 1/2 = 0.5$

If point-counting of the mortar component had given

unhydrated cement 105 points
sand grains 895 points

the $Cs/Ss = 105/895 = 0.12$

thus, the degree of hydration (*Dh*) may be calculated from

$$Dh = \frac{(0.5 - 0.12) \times 100}{0.5} = 76\%$$

Christensen *et al.* (1979) reported that, using optical microscopy, Portlandite appears to occur in hardened concrete in three forms (Plate 5):

1. small crystals uniformly distributed in the cement paste (Plate 5(a))
2. coarser crystals at aggregate particle boundaries and/or at the edge of voids (Plate 5(b))
3. recrystallised relatively very coarse crystals in cracks and cavities.

Also, Berger and McGregor (1973), using electron microscopy, suggested that sub-microscopic ($\sim$0.1 μm) globular particles scattered within the CSH were probably amorphous $Ca(OH)_2$. In comparing different techniques for determining the content of $Ca(OH)_2$ in set Portland cement, Midgley (1979) found that X-ray diffraction, which only detects crystalline material, underestimated the amount present. For example, in one sample containing $\sim$16 per cent $Ca(OH)_2$ determined using differential thermal analysis, X-ray diffraction indicated only $\sim$13 per cent and Midgley attributed this to the presence of $\sim$3 per cent amorphous (i.e., non-crystalline) Portlandite in the sample.

Referring to the conspicuous Portlandite which occurs at the interface between aggregate particles and the cement paste, French (1991a) noted that the quantity of such material appears to be dependent upon the aggregate type. He found that, for a given water/cement ratio, relatively thick layers of Portlandite develop on siliceous particles, such as chert or quartz, but that the layers on carbonate rock surfaces were thin and impersistent. Also, the amount of this boundary Portlandite evidently increased with aggregate surface area, being apparently more abundant in concretes made with finer sand.

The observed distribution and crystallite size of the Portlandite in the cement paste can sometimes be used as indicators of the likely properties of the concrete mix or as evidence of the conditions to which the concrete has possibly been exposed, although any such interpretations should be treated with circumspection and must be corroborated by other factors. For example, Portlandite crystals formed in concretes made with high water/cement ratios tend to be well-defined, relatively coarse and predominantly situated along aggregate boundaries and void edges; whilst those formed at lower water/cement ratios are often irregular in shape, smaller and distributed uniformly through the paste (French 1991a). According to Jepsen and Christensen (1989), the 'unusually large' $Ca(OH)_2$ crystals which form when 'the concrete is too wet' (i.e., high water/cement ratio) can be indicative of bleeding.

Portlandite size and form can also indicate frost or freeze-thaw damage to concrete. The solubility of Portlandite decreases with rising temperature (Taylor 1964) and the corollary is that the solubility is increased at lower temperatures. For example, Bassett (1934) demonstrated that the solubility was 1.3g/l as CaO at 0°C, dropping to 1.2 g/l at 18°C or 0.5 g/l at 100°C (see also Chapter 5). According to Idorn (1969), microcrystalline $Ca(OH)_2$ can thus redissolve and be reprecipitated during freezing and thawing cycles, allowing large crystalline accumulations of Portlandite to form (Fig. 4.5). The occurrence of such accumulations, plus evidence of chemically unaltered cement paste (Idorn cites the presence of unhydrated C_2S) 'is a good indication' that freeze-thaw action might have caused damage.

Again, however, caution must be exercised in making such interpretations and any apparent freeze-thaw action indicated by Portlandite appearance should be verified by the presence of other characteristic features, such as macro- and micro-cracking patterns. French (1991a) has reported that very coarse Portlandite crystals occur in factory-made concrete subjected to steam curing, when they might be intergrown with ettringite (see next section). Patel *et al.* (1995) have also shown that relatively coarse Portlandite can be characteristic of concretes cured at elevated temperatures.

Fig. 4.5 Photomicrograph of accumulation of portlandite in concrete affected by frost. The field of view is about 0.4 mm across. (From Idorn 1967 and 1969.)

Calcium hydroxide can be dissolved by percolating water, often being redeposited in open spaces elsewhere within the concrete. Thus, often comparatively very coarse secondary development of Portlandite within voids and open cracks within hardened concrete (i.e., type 3 in Christensen *et al.* (1984)) are fairly reliable indicators of such leaching by the migration of water through the concrete. Commonly such secondary Portlandite appears either as a lining to the void or crack (Fig. 4.6), or as a more or less complete infilling by coarse platy crystals. Such Portlandite leaching products often occur in close association with secondary ettringite deposits. Portlandite is also subject to carbonation, but this is dealt with in detail in Chapter 5.

Ettringite and some other complex phases

As already explained, ettringite ($3CaO.Al_2O_3.3CaSO_4.26$–$32H_2O$) is the initial cement hydration product of C_3A and water in the presence of calcium sulfate. When the hydration of C_3A continues after the supply of sulfate ions diminishes, then the first-formed ettringite decomposes and is progressively replaced by the 'monosulfate' form ($3CaO.Al_2O_3.$-$CaSO_4.12H_2O$). These are often referred to as the 'high' (ettringite) and 'low' (monosulfate) forms of calcium sulfoaluminate hydrate.

However, these hydration products are generally submicroscopic and it is usually secondary ettringite which is observed under the optical microscope during the examination

Fig. 4.6 Photomicrograph of secondary portlandite deposits lining a void. Width of field *c*. 1.5 mm.

of concrete. An exception is the more coarsely crystalline ettringite which can develop in concretes cured at very high temperatures (French 1991a). The secondary ettringite results either from one type of sulfate attack (see Chapter 6) or, perhaps more commonly, from leaching through the percolation of water.

In the late nineteenth century, the needle-like crystals found to be associated with the type of deterioration now recognised as sulfate attack were termed 'cement bacillus' and Candlot (1890) identified these 'bacilli' as ettringite. Acicular crystals of ettringite are rather common in concrete, often lining or filling voids or cracks (Fig. 4.7), and it is important to realise that these have most usually resulted from the redistribution of sulfates within the concrete by migrating water and thus might indicate a degree of leaching: expansive sulfate attack is comparatively rare and, if suspected, should be corroborated by other observations.

Detailed descriptions of the complex compositions, structures and properties of ettringite and monosulfate are given by Turriziani (1964) and Lea (1970). Typically ettringite forms long, thin, colourless 'needles', of low to medium birefringence, commonly occurring in radiating clusters or sometimes as crustiform linings. The optical properties are detailed in the Glossary and summarised in Table 4.6. Ettringite is relatively easy to detect by differential thermal analysis (Midgley and Rosaman 1960), when it is characterised by a large endotherm at low temperature (110–150 °C), or by X-ray diffraction (Midgley 1960, Crammond 1985a).

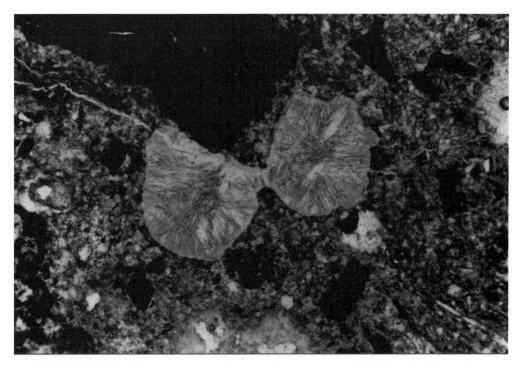

Fig. 4.7 Photomicrograph of ettringite filling voids in concrete in thin-section. Width of field *c.* 0.9 mm. Plane polars. (Photograph courtesy of Mr Barry J. Hunt, STATS Consultancy.)

Table 4.6 Some secondary deposits in concrete (adapted from ASTM C856 1995; see glossary for full details).

Compound/ mineral	Chemical formula	Frequency of occurrence	Notes
Calcium carbonate/ calcite	$CaCO_3$	Very common	Fine-grained, white or grey masses or coatings in the paste, in voids or cracks, or on exposed surfaces
Calcium carbonate/ aragonite	$CaCO_3$	Rare	Minute white prisms or needles in voids or cracks
Calcium carbonate/ vaterite	$CaCO_3$	Common	Spherulitic encrustations on laboratory specimens and also identified by XRD
Calcium sulfoaluminate hydrate/ ettringite	$3CaO.Al_2O_3.3CaSO_4.32H_2O$	Very common	Fine, white fibres or needles or spherulitic growths in paste, voids or cracks
Calcium sulfoaluminate hydrate/ monosulfate	$3CaO.Al_2O_3.CaSO_4.12H_2O$	Very rare	Minute, white to colourless plates in voids or cracks
Calcium aluminate hydrate	$4CaO.Al_2O_3.13H_2O$	Very rare	Mica-like, colourless twinned plates in voids

Table 4.6 (*continued*).

Compound/ mineral	Chemical formula	Frequency of occurrence	Notes
Hydrous sodium carbonate/ thermonatrite	$Na_2O.CO_2.H_2O$	Rare	Minute inclusions in alkali-silica gel
Hydrated aluminium sulfate/ paraluminite	$2Al_2O_3.SO_3.15H_2O$	Very rare	In cavities in extremely altered concrete
Calcium sulfate dihydrate/ gypsum	$CaSO_4.2H_2O$	Unusual	White to colourless crystals in paste, in voids or along aggregate surfaces in concrete affected by sulfate or seawater attack
Calcium hydroxide/ Portlandite	$Ca(OH)_2$	Ubiquitous	White to colourless plates in paste, voids and cracks
Magnesium hydroxide brucite	$Mg(OH)_2$	Unusual	White to yellow, fine-grained encrustations and fillings in concrete attacked by seawater
Hydrous silica/ opaline silica	$SiO_2.nH_2O$	Unusual	White to colourless, fine-grained amorphous; resulting from intense leaching or carbonation of paste
Alkali-silica gel	$Na_2O.K_2O.CaO.SiO_2.nH_2O$	Quite common	White, yellowish or colourless; viscous, fluid, waxy, rubbery or hard; in voids or cracks, within aggregates or as exudations
Hydrated iron oxides limonite	$Fe_2O_3.nH_2O$	Common	Opaque or nearly so, brown stain in cracks, in paste around some iron-bearing aggregate grains (e.g., pyrite) or on surfaces
Calcium carbo-sulfo-silicate hydrate/ thaumasite	$CaSiO_3.CaCO_3.CaSO_4.15H_2O$	Quite common	Needles of similar general appearance to ettringite, with which it can occur; forms in concrete exposed to damp cold conditions
Syngenite	$(K_2Ca(SO_4)_2).H_2O$	Very rare	Fibrous material found in cavities and zones peripheral to slate aggregate particles
Hydrotalcite	$Mg_6Al_2(OH)_{16}CO_3(H_2O)_4$	Very rare	Foliated platy to fibrous masses

In recent years, it has been discovered that early prolonged high temperature (say >65–$75°C$) curing of precast concrete can interfere with the normal hydration reactions of C_3A and sulfates, initially suppressing the formation of calcium sulfoaluminate hydrate,

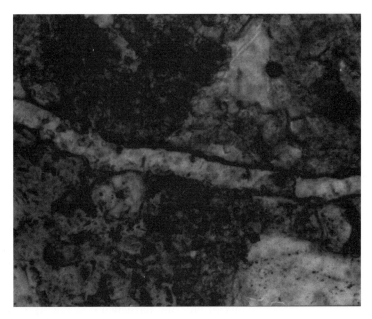

Fig. 4.8 Photomicrograph of ettringite formed in a fissure at a cement/aggregate boundary in delayed ettringite formation. The concrete was estimated to have experienced an early peak temperature in excess of 85°C. Width of field *c.* 0.4 mm. (Photograph courtesy of Dr D. W. Hobbs, British Cement Association.)

so that later expansion and damage can be caused by 'delayed ettringite formation' (DEF) when the concrete is subsequently exposed to moist conditions or wetting (Heinz and Ludwig 1987, Lawrence *et al.* 1990). Heinz and Ludwig (1987) reported that the expansive secondary ettringite formation occurred mainly at the interfacial zone between cement paste and aggregate particles (Fig. 4.8).

Lawrence *et al.* (1990) gave some preliminary recommendations for minimising the risk of delayed ettringite formation, variously controlling the premature application of high temperature (i.e., not before 3 or 4 h after casting), the rates of both raising and later reducing the curing temperatures (i.e., not greater than 20°C/h), and the maximum level of curing temperature (i.e., not greater than 60 or 70°C). Cement chemistry also appears to be a critical factor (Lawrence 1995).

Although the occurrence of secondary ettringite observable to optical microscopy, particularly when infilling peripheral cracks around aggregates, is often a feature of concrete damaged by delayed ettringite formation, caution should be exercised in attributing damage to this cause on the sole basis of ettringite presence, unless substantiating evidence is available including information relating to the conditions of manufacture. Deng and Tang (1994) suggested that only ettringite which forms in certain ways within the cement paste can cause expansion, whilst that precipitating within existing voids or cracks exerts little or no expansive stress. Similarly, Poole *et al.* (1996) concluded from their experiments that the growth of ettringite within the cement paste was the principal cause of expansion in cases of DEF and that ettringite rims around aggregate particles were not always formed.

In 1966, Erlin and Stark reported on concrete deterioration caused by the secondary formation of the complex compound, thaumasite ($CaSiO_3.CaCO_3.CaSO_4.15H_2O$). A useful review of the thaumasite literature is given by Van Aardt and Visser (1975), who demonstrate that thaumasite forms preferentially at lower temperatures (i.e., 5°C rather than 25°C).

According to Crammond (1985b), thaumasite forms in conditions that are very damp as well as cold, and also when there are abundant proportions of sulfate and carbonate ions available. Typically thaumasite and ettringite form together and both can contribute to expansion and degradation of the matrix (see Chapter 6). Berra and Baronio (1987) described a case in which portions of a concrete tunnel lining had been 'transformed into an incoherent whitish mass' consisting chiefly of thaumasite resulting from attack by water containing both sulfates and aggressive CO_2. Crammond and Halliwell (1995) have described cases of the thaumasite form of sulfate attack in which the required carbonate ions were derived from limestone and dolomite dust within the concrete.

Thaumasite is superficially rather similar in appearance to ettringite under the microscope, and indeed often occurs as intimate mixtures with ettringite, but thaumasite has strong birefringence in contrast to the weak birefringence of ettringite (Varma and Bensted 1973). Although the X-ray diffraction patterns for thaumasite and ettringite are quite similar, Crammond (1985a) has demonstrated a quantitative procedure for analysing mixtures of thaumasite, ettringite and gypsum in concretes and mortars. French (1991a) has suggested that there might be solid solutions between thaumasite and ettringite.

Chlorides added to a concrete mix or entering hardened concrete as a contamination can combine with C_2A or C_4AH_{13} to form complex calcium chloroaluminate hydrates (e.g., $3CaO.Al_2O_3.CaCl_2.10H_2O$ – Friedel's salt). According to Figg (1983), virtually all of the calcium chloride ($CaCl_2$) added to concrete as an accelerator usually becomes combined into this low-chloride chloroaluminate, although the high-chloride form has been reported from the USSR when exceptionally large amounts of $CaCl_2$ were used for placing concrete at sub-zero temperatures. Damage to concrete arising from the formation of calcium chloroaluminate hydrates appears to be uncommon.

Similar chloroaluminate compounds can form when sodium chloride is present, either in the mix water (e.g., sea water or saline mix water) or as an externally derived contaminant (e.g., sea water, de-icing chemicals, sabkha ground conditions in some arid desert regions). Free chloride in concrete is a major cause of reinforcement corrosion (see Chapter 6), so that the amount of calcium aluminate available to react with chloride salts is an important consideration when setting tolerance limits for chloride content in reinforced concrete. For example, BS 8110 (1985) sets a limit of 0.4 per cent chloride ion for OPC concrete with a relatively high C_3A content, but a limit of only 0.2 per cent chloride ion for SRPC concrete with a restricted C_3A content. In reinforced concretes exposed to conditions of both sulfate and chloride, therefore, it may sometimes be deemed counterproductive to specify a type of Portland cement to resist the sulfates which might be less able to inhibit the effect of the chlorides on the embedded steel.

The calcium chloroaluminates also form needle-like crystals and under the microscope are difficult to distinguish from ettringite. However, the X-ray diffraction pattern permits distinction (Crammond 1985a) and, of course, electron probe microanalysis can be carried out under the SEM (Fig. 4.9). Reaction between sulfates or chlorides and the aluminoferrite compounds of cement can also give rise respectively to various sulfoferrite and chloroferrite analogues (Lea 1970).

4.2.4   BLENDED AND SPECIAL CEMENTS

In the UK and North America it is increasingly common for the cement in concrete mixes

Fig. 4.9 SEM micrograph of hexagonal plates of calcium chloroaluminate hydrate (from French (1991a).

to be partially replaced by mineral additions of cement fineness (or finer in the cases of microsilica and metakaolin) (Malhotra 1987). These materials can modify the properties of both fresh and hardened concrete. The uses and potential benefits of using either ground granulated blastfurnace slag (ggbs) or pulverised-fuel ash (pfa), which are the predominant materials used, are well documented and one useful brief resumé is given by Dewar and Anderson (1992).

Amongst other advantages, subject to good concrete-making practice, either ggbs or pfa might be expected to reduce costs (as a cement replacement), to reduce the heat of hydration which is especially important in large concrete pours and usually to improve resistance to a range of durability threats, such as sulfate attack or alkali-aggregate reactivity (see Chapter 6).

On the mainland of Europe, in contrast with the UK and North America, it has been traditional practice for such mineral additions to be blended with the Portland cement component during the cement manufacturing process (Corish 1989). The use of such blended cements obviously simplifies the site batching of concrete, but permits less flexibility of mix design. Apart from rare cases of inadequate concrete mixing leading to a markedly uneven distribution when the addition has been added at the mixer, it is usually not possible for petrographical examination to differentiate between concretes made using separate addition and cement components and those made using factory blended cements.

Ggbs and Portland blastfurnace cements

The nature of ggbs is described in more detail in Section 4.6.2. A by-product of the iron-making industry, granulated or pelletised blastfurnace slag is produced by the rapid

Fig. 4.10 Photomicrograph of unreacted ggbs grains in concrete in thin-section visible as greyish particles often with dark edges. Width of field *c.* 0.4 mm. Plane polars. (Photograph courtesy of Mr Barry J. Hunt, STATS Consultancy.)

quenching of molten blastfurnace slag as it passes through water sprays, followed by either water granulation or pelletisation on a spinning drum (Hooton 1987). The granulated material produced is largely glassy, typically at least 95 per cent being glass with a chemical composition close to that of the bulk composition: around 40 per cent lime, 30 per cent silica, 15 per cent alumina and variable proportions of magnesia and other constituents (Lee 1974). The minority crystalline components are dominated by melilite or merwinite (see Section 4.6.2) with some oldhamite and residual native iron (Scott *et al.* 1986). Ground pelletised slag is essentially similar but tends to contain more vesicles (gas bubbles).

No hydration products are formed when ggbs is exposed alone to water, but the 'latent hydraulicity' of ggbs is activated by the calcium hydroxide and alkalis liberated during the hydration of Portland cement (Lea 1970, Hooton 1987). It seems that the hydration products of a binder containing both Portland cement and ggbs are broadly similar to those of Portland cement alone, except that the content of Portlandite ($Ca(OH)_2$) might be reduced at higher ggbs levels because of the reduced lime/silica ratio. Small and infrequent proportions of hydrated gehlenite, C_2ASH_8 (gehlenite is an end member of the melilite solid solution series) and hydrogarnets, $C_4(A, F)H_{13}$, have also been reported. However, it is the usually considerable quantity of residual unreacted ggbs in hardened concrete that will be apparent to the petrographer and which enables the presence of ggbs to be identified.

Fresh concrete made using Portland blastfurnace cement has a fairly distinctive dark greenish appearance but with time, following oxidation, the colour becomes similar to that of normal concrete. Unreacted remnants of ggbs appear in the cement paste matrix as angular, even shard-like, particles of silicate glass. They are clearly visible either in thin-section under the petrological microscope (Fig. 4.10) or in highly polished surfaces,

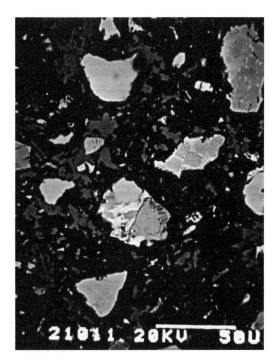

Fig. 4.11 SEM micrograph of unreacted ggbs grains in concrete visible as grey/white particles in the hardened matrix.

etched with HF vapour, under the reflected-light microscope (Plate 6). It is possible that hydraulic reactions affect only the smaller, sub-microscopic ggbs particles as the coarser pieces observable under the microscope typically exhibit sharply defined boundaries and yield no apparent evidence of any interfacial reaction, even after many years. French (1991b) has suggested that, because ggbs particles as small as 1 μm can be seen by SEM to be completely unreacted (Fig. 4.11), ggbs particles might be divided into those which react entirely to produce hydrates and those which are essentially inert.

Whilst identifying the presence of significant ggbs in a hardened concrete presents few difficulties, quantifying the relative proportions of Portland cement and ggbs is much more problematic. The chemical procedure given in BS 1881 (1988) relies upon the determination of sulfide and is thus inappropriate if any other of the concrete constituents contain sulfide. It is also ineffective if all or part of the sulfide has been oxidised to form sulfate. It is possible microscopically to point-count the residual unreacted ggbs and Portland cement clinker grains within the cement paste matrix, but the result needs to assume a similar cement:ggbs ratio in the hydrated groundmass, which need not be the case. If French (1991b) is correct, the relationship between the hydrated ggbs and the unreacted ggbs could be highly variable.

French (1991b) has advocated a method using electron probe microanalysis, whereby the cement:ggbs mixture in the hydrated paste is estimated from chemical composition data separately determined by spot analyses of unreacted ggbs, residual cement clinker and the hydrated groundmass. However, even this could be complicated by his postulated division between inert and reactive types of ggbs particle. It is possible that a combination of the microscopical point-counting and microanalytical methods could enable a reasonably dependable estimate of cement and ggbs mix proportions to be determined.

Pfa, fly ash and Portland pozzolanic cements

Pulverised bituminous coal (pulverised-fuel) is burnt at some electricity power stations, producing 'pulverised-fuel ash' (pfa) as a by-product, about 20–25 per cent of which becomes fused together like clinker and is known as 'furnace bottom ash', whilst the remaining 75–80 per cent is a fine pfa dust or 'fly ash' which may be collected from the combustion gases (Owens 1980 Part 2, Berry and Malhotra 1987).

The chemical composition of pfa or fly ash is dependent upon the rank of coal burnt and also the nature of the clay impurities in the coal, but all are dominated by silica (45–50 per cent) and alumina (25–30 per cent). Two main types are recognised (ASTM C618 (1996): 'Class C' types are produced from sub-bituminous coal (or even lignite) and are relatively high in calcium content (> 10 per cent), whereas 'Class F' types are produced from bituminous coals and are lower in calcium content (< 10 per cent). UK power stations produce only 'Class F' pfa.

Mineralogically, most pfas are dominated (45–70 per cent) by aluminosilicate glass (Hubbard *et al.* 1985), principally in the form of solid spheroidal particles less than 20 μm in diameter (RILEM 1988) and occasionally forming hollow 'cenospheres' (Raask 1968). Hubbard *et al.* (1985) found the crystalline component of UK pfas variably to consist of mullite ($Al_6Si_2O_{13}$), haematite, quartz and magnetite, with about 5–10 per cent unburnt coal. The product from any one power station was found to be rather consistent and the variations between stations was largely correlated to the clay impurities in the coal used, aluminosilicate glass being generated by the fusion of illite, and mullite being formed by the recrystallisation of kaolinite. McCarthy *et al.* (1984) reported a different and wider variety of crystalline phases for 'Class C' pfas from Western USA, perhaps accounting for 'Class C' materials being cementitious as well as pozzolanic, whereas 'Class F' ashes are only normally pozzolanic (RILEM 1988).

The presence of pfa is readily identified in thin-sections of concrete under the petrological microscope because of the distinctive spheroidal or ellipsoidal glassy particles or even relatively coarser hollow 'cenospheres' (Plate 7). In their studies of pfa concrete structures, Thomas and Matthews (1991) routinely used SEM of broken surfaces to confirm the presence of pfa, as betrayed by either unreacted spherical particles or relicts of mullite crystals (Fig. 4.12). However, like ggbs, the quantitative determination of pfa in concrete is complicated by the small particle size, which ranges down to sub-microscopic, and the uncertain extent to which some proportion of the pfa in hardened concrete has possibly been consumed in the formation of hydrates.

The aluminosilicate glass phase in pfa is considered to be genuinely pozzolanic and may react in the presence of water with calcium hydroxide liberated by the normal hydration of Portland cement to form stable hydrated cementitious compounds (Eglinton 1987). Such pozzolanicity is likely to vary with both chemical composition and particle size, but the extent to which pozzolanically derived hydrates exist in concretes made using pfa remains uncertain. Abundant residual and apparently completely unreacted pfa particles are a characteristic feature of such concretes. Semi-quantitative techniques, including the point-counting of residual pfa and cement clinker particles in optical microscopy or the visual assessment of those residual particles in SEM, enable the original cement and pfa proportions to be estimated.

Chemical techniques are fraught with difficulty, especially in the absence of reliable reference constituent samples, although some success has been claimed for complicated procedures using differential solubility rates (Gomà 1989). As it seems likely that the

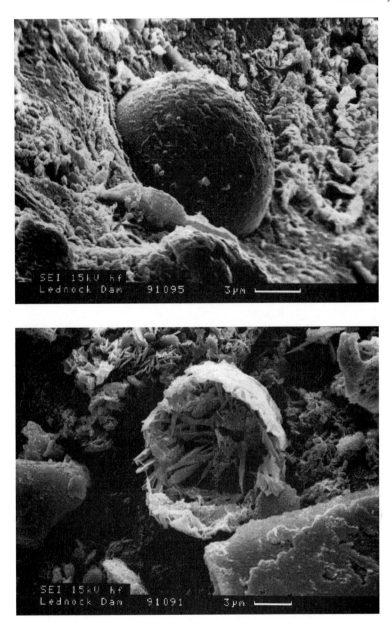

Fig. 4.12 SEM micrographs of pfa in concrete: (a) unreacted cenosphere; (b) relicts of mullite crystals present as needles in the corroded sphere (from Thomas and Matthews 1991).

hydrates produced by any pozzolanic activity of pfa will be different in composition and character from those generated by Portland cement hydration, it is thought possible that a microanalytical technique using an electron probe could possibly be developed.

Pfa is the main pozzolana currently used in the UK, either as a 15–35 per cent Portland cement replacement in 'Portland pulverised-fuel ash cement', BS 6588 (1985) or as a 35–40 per cent replacement in 'Pozzolanaic cement with pfa as pozzolana', BS 6110 (1985).

However, on the mainland of Europe and elsewhere, a range of other natural and artificial pozzolanas are available and sometimes used in concrete (Table 4.7). The natural pozzolanas are mainly volcanic ash or tuff deposits, including for example the German ground 'trass' materials, the 'Santorin Earth' from Greece and of course the Roman source of zeolitic tuff from Pozzuoli near Mount Vesuvius in Italy which gave pozzolana its name. The diatomaceous earths are the other main group of natural pozzolanas. Apart from pfa, the artificial pozzolanas mainly include burnt clay or shale, but other materials have been used, for example 'rice husk ash'. Further information on pozzolanas may be obtained from Lea (1970), Mehta (1987) and RILEM (1988).

Table 4.7 Some natural and artificial pozzolanas used in concrete.

Type of source	Type of pozzolana	Pozzolanic classification after RILEM (1988)
Natural	Volcanic tuff (e.g., trass)	—
	Volcanic ash/earth	—
	Diatomaceous earth/diatomite (e.g., Moler in Denmark)	—
	Bauxite	—
Artificial – waste/by-product	Fly-ash/pulverised-fuel ash (pfa) – high-calcium[1]	Cementitious & pozzolanic (II)
	Fly-ash/pulverised-fuel ash (pfa) – low-calcium[1]	Normal pozzolanic (IV)
	Ground granulated blastfurnace slag (ggbs)	Cementitious (I)
	Condensed silica fume (csf)/Microsilica	Highly pozzolanic (III)
Artificial – synthetic	Burnt clays and shales	—
	Burnt moler/diatomite	—
	Rice husk ash – controlled incineration	Highly pozzolanic (III)
	Metakaolin	Highly pozzolanic (III)[2]

Notes
1   High-calcium > 10 per cent CaO, Low-calcium < 10 per cent CaO, as defined in RILEM (1988).
2   Not included in RILEM (1988), but here allocated to Class III.

High-alumina cement

High-alumina cement (HAC), or calcium aluminate cement, differs from Portland cement in being manufactured by the fusion of limestone and bauxite, instead of limestone and clay or shale (Robson 1962, 1964). As a consequence, HAC clinker consists chiefly of calcium aluminates with much smaller amounts of ferrite and silicate phases. Patented in 1908, production of HAC commenced in France in 1913 and in Britain in 1925. In addition to its rapid strength gain properties, Bied (1926) discovered the superior sulfate-resistance properties of HAC.

   Reference to the phase equilibrium diagram for the $CaO-Al_2O_3-SiO_2$ system shows that HAC falls almost entirely within the stability field of monocalcium aluminate (CA), which is found to be the main phase present and the principal cementitious material (Table 4.8).

Other aluminate compounds include $C_{12}A_7$, CA_2 and C_3A (rare), whilst larnite (βC_2S) and/or gehlenite (C_2AS) may also be present depending upon silica content. Calcium aluminoferrites form with compositions in the range C_4AF–C_6AF_2, whilst most of the ferrous iron appears to be contained within wüstite-(FeO) bearing glass.

Table 4.8 Anhydrous mineralogy of high-alumina cement.

Phase	Cement 1[1]	Cement 2[1]	Cement 3[2]
		% by mass	
CA	60	60–65	58
C_2AS	15–20	2	—
C_2S	10–15	10	—
Pleochroite	—	2	20
Iron-bearing glass	10	5	—
Wüstite (FeO)	—	—	12
C_6AF_2	—	—	10
Other iron compounds	Very low	High	—

Notes
1   Robson (1964).
2   Midgley (1967).

One characteristic constituent of HAC, which is fibrous and pleochroic and variously known as 'pleochroite' or 'Q', has proved difficult to identify chemically (Sourie and Glasser 1991). However Kapralic and Hanic (1980) have established the general formula for 'Q' as $Ca_{20}Al_{32-2x}Mg_xSi_xO_{68}$ (where x can vary from 2.5 to 3.5), which is believed to be isostructural with 'pleochroite' wherein FeO substitutes for MgO and Fe_2O_3 substitutes for some Al_2O_3.

HAC clinker is generally dominated by short prismatic crystals of CA, mixed with ferrite, other calcium aluminates and sometimes gehlenite. The crystal sizes are very dependent upon cooling rate (Fig. 4.13). The more slowly cooled (and hence coarser textured) HAC clinker can be studied using optical microscopy, when the cement type is easily distinguishable using either etched polished specimens or thin-sections (Plate 8, Fig. 4.14, Plate 9). Pleochroite distinctively occurs as long lath-like crystals with a striking pleochroism (Plate 8).

Residual pieces of such clinker are nearly always observable in the matrix of HAC concrete, especially at the recommended low water/cement ratios typically less than half of the cement hydrates. Figure 4.14 illustrates the typical appearance of HAC concrete matrix in thin-section. Chemical tests have also been described for the rapid differentiation of HAC from Portland cement in concretes (Roberts and Jaffrey 1974).

The hydration and ageing of HAC is summarised by Midgley and Midgley (1975) and Sims (1977) (Fig. 4.15). The dominant CA reacts rapidly with water to produce the components which give HAC concrete its high early strength. Pleochroite and βC_2S react more slowly but appear to contribute to strength gain in due course. The iron compounds and any C_2AS are either inert or only slowly reacting. The initial hydration products at normal temperatures are principally hexagonal CAH_{10}, with some C_2AH_8 depending upon cement chemical composition and the temperature, and alumina gel. It appears that mineralogical changes might occur at or near concrete surfaces exposed to the atmosphere,

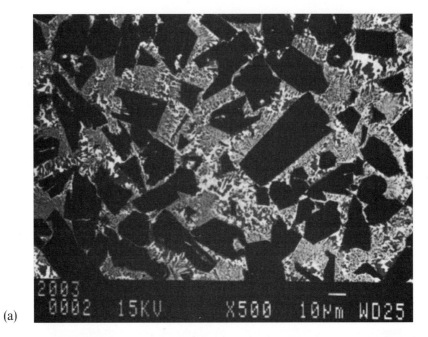

(a)

(b)

Fig. 4.13 SEM micrographs of a high-alumina cement clinker sample (a) fast cooling ($\sim 300\,°C/min$): crystals of CA in a dendritic matrix including C_2AS, (b) slow cooling ($\sim 5\,°C/min$): coarser dark CA crystals, plus grey C_2AS and white ferrite (from Sourie and Glasser 1991).

such that CAH_{10} dehydrates to CAH_5 (Midgley 1967) or even perhaps decomposes to alumina gel and calcite (Robson 1964).

These early-formed hexagonal calcium aluminate hydrates are metastable and spontaneously invert (or 'convert') with time to the cubic hydrogarnet (C_3AH_6) and monoclinic

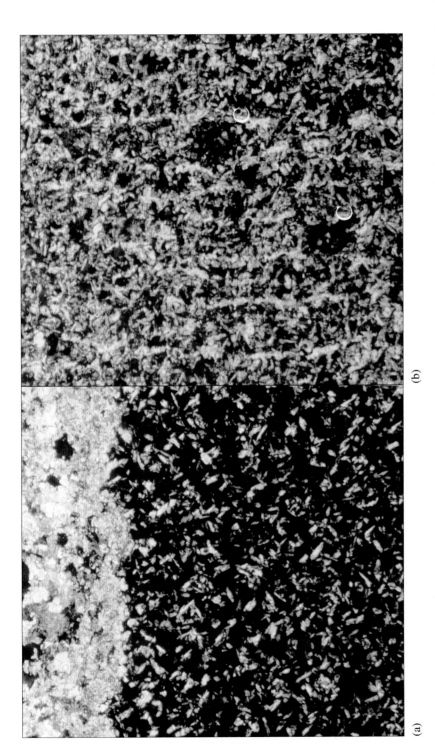

(a)

(b)

Plate 21 The appearance of α-dicalcium silicate hydrate in geothermal cement grout. Test grout, water: powder = 0.62, containing 3 per cent bentonite, exposed for three months to geothermal fluids at 160°C. Width of fields 1.1 mm. (a) The small white laths of α-dicalcium silicate hydrate can be clearly seen in the area below the zone of golden yellow carbonation at the top of the picture. The area photographed is completely leached of calcium hydroxide so that the laths are clearly visible against the dark background. Crossed polars. (b) An area from the centre of the same specimen. Here the α-dicalcium silicate hydrate crystals form a background in which are embedded aligned calcium hydroxide and the remnants of a few grains of cement (C) and bentonite relics which have managed to survive the extreme conditions. Circular polarisation.

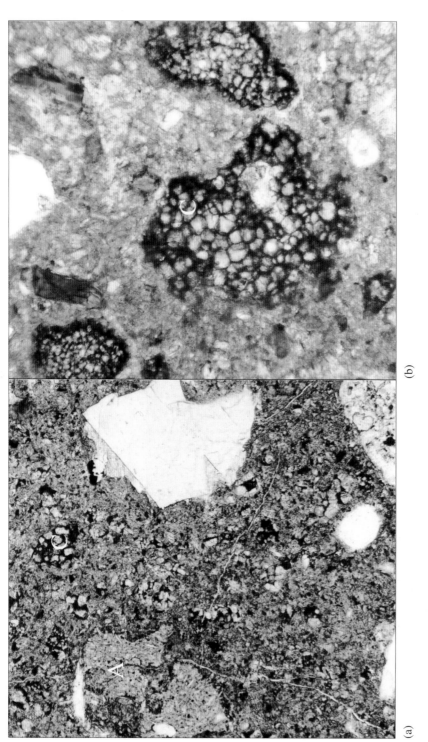

(a)

(b)

Plate 22 Examples of remnant grains of cement from two bridges built in the 1920s. Width of fields 0.9 mm. (a) The remnant cement grains (C) in this bridge concrete consist of clusters of belite together with some alite set in a dark interstitial matrix. Hydration rims are minimal. The microcracks are due to the reaction of the andesite aggregate (A) with cement alkalis. (b) The concrete from this hydro structure contains large remnant grains (C) consisting of nests of belite, some of which have dark hydration rims. Fragments of clinker as large as this are not common in modern concretes.

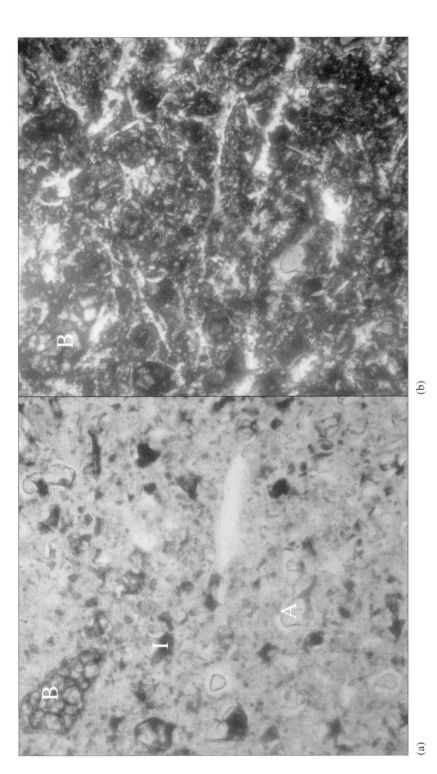

(a)

(b)

Plate 23 Partially hydrated grains of cement. A high cement content of 740 kg/m³ combined with a water/cement ratio of 0.36 has resulted in self-desiccation and insufficient space for hydrate formation in spite of moist curing this test concrete at 38°C for 7 years. This has further resulted in numerous partially hydrated grains of cement still being present. Width of fields 0.44 mm. (a) The fragments of alite (Al) are surrounded by clear hydration rims while the fragments of dark interstitial phases (I) are irregular in shape and have no visible hydration rims. The belite cluster (B) at upper left is a remnant cement grain. A sliver of chert is visible at middle right of photo. (b) Under crossed polars the hardened cement matrix is filled with calcium hydroxide which shows as irregular white areas. The yellowish birefringence of the belite (B) in the remnant cement grain is clearly visible and the biregringent texture of the sliver of chert is just discernible.

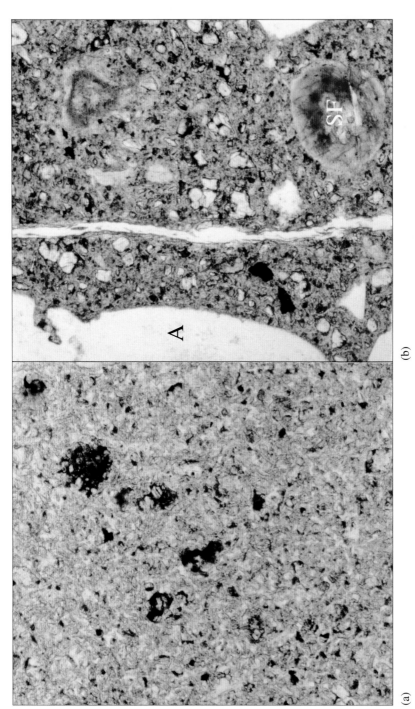

(a)

(b)

Plate 24 A comparison between the extremes of hydration. Width of fields 0.9 mm. (a) This geothermal cement grout has been hydrated at 160°C for 3 months. The result is near complete hydration with a few remnants of dark interstitial matrix of the cement and possibly some bentonite relics set in a background of α-dicalcium silicate hydrate. Same area as shown in Plate 21(b). (b) This super-high-strength DSP mortar sample contains sufficient unhydrated cement to act as a fine filler. Grains as small as 5-10 µm are visible showing hydration ceased early due to internal desiccation. The two roughly circular areas in the texture are balls of undispersed silica fume (SF). Quartz aggregate is also indicated (A). DSP sample mixed at water/cement = 0.23 with 20 silica fume and quartz sand as aggregate. Moist cured for 28 days and then exposed to wetting and drying for 3 months.

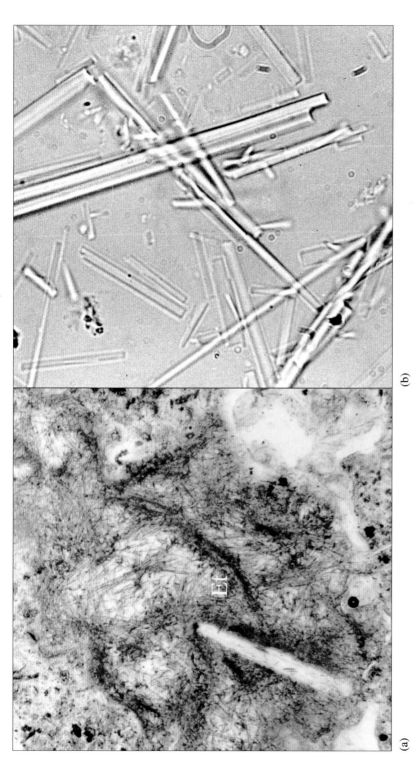

(a)

(b)

Plate 25 Ettringite. (a) The low refractive index plus low birefringence under crossed polars, identify the fine needles which fill the pore in this old concrete as ettringite (Et). There is no evidence of sulfate attack or alkali-aggregate reaction but some indication that moisture was moving slowly through the concrete in this seventy-year-old hydro structure. Width of field 0.75 mm. (b) At high magnification, ettringite needles are resolved as rods with a thickness of 1-3 μm. This sample was taken from a massive efflorescent deposit of ettringite which formed over a ten-year period in a submerged hydro dam gallery. The mechanism for the unusual formation of ettringite is discussed by Kennerley (1965). Width of field 0.05 mm.

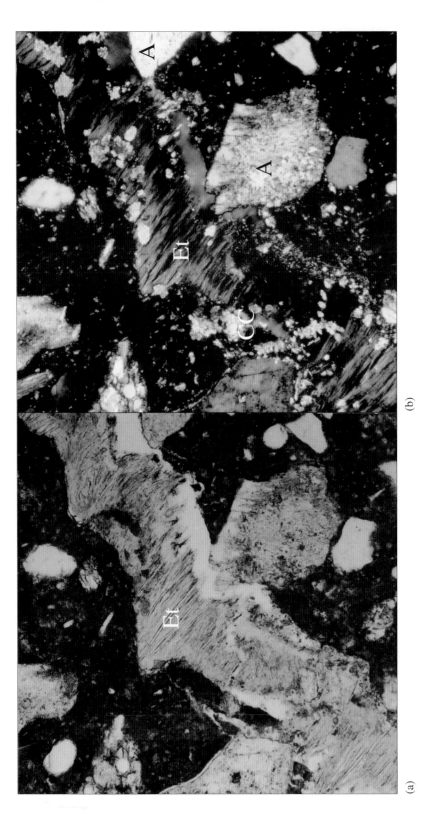

(a)

(b)

Plate 26 Massive ettringite in the cracks of an eighty-year-old concrete from Lady Evelyn Lake Dam in Canada. Width of fields 0.9 mm. (a) The ettringite (Et) deposited in the crack is massive and appears as curtain-like material. The concrete is undergoing an advanced alkali-aggregate reaction and the combination of long severe exposure and advanced reaction appears to have promoted this rather unusual formation. (b) Under crossed polars the typical low birefringence of the ettringite (Et) is clearly apparent together with unidentified dark material. There are also some isolated clusters of carbonation (CC) present in the texture. Aggregates labelled A.

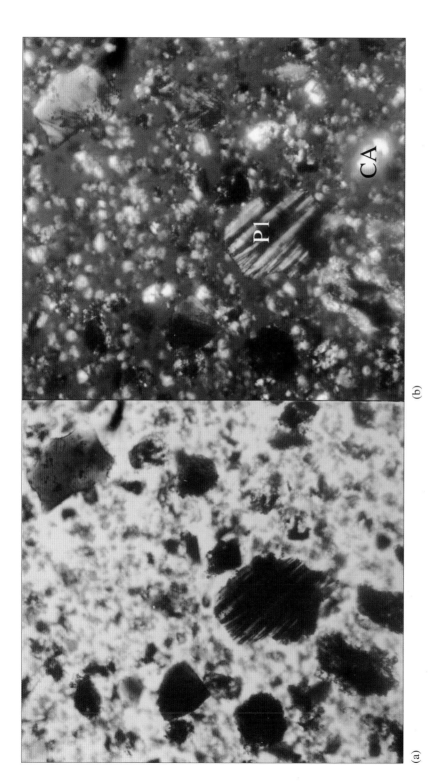

(a)

(b)

Plate 27 Anhydrous high-alumina cement mounted in oil with a refractive index of 1.54. Width of fields 0.45 mm. (a) The large particle in the centre has the fibrous character and pleochroism typical of pleochroite ($6CaO.Al_2O_3.4MgO.SiO_2$). It is not possible to identify most of the other particles which are glass materials varying in colour from opaque to brown. (b) Under crossed polars the fibrous character of the pleochroite (Pl) stands out. The white birefringent particles are calcium aluminate (CA). Some of the brown-coloured birefringent particles may be a ferrite phase.

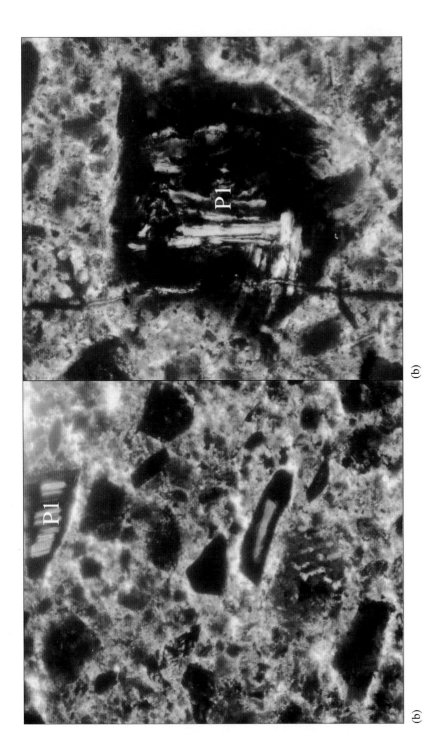

(b)

Plate 28 The texture of hydrothermally cured high-alumina cement. Width of fields 0.9 mm. (a) After 7 days at 150°C in this geothermal grout (water/cement ratio = 0.4) only the most resistant particles remain unhydrated. At the top, the fibrous particle of pleochroite (P1) with its distinctive lilac-coloured pleochroism stands out. Other large particles are opaque or have well developed hydration rims. (b) The large particle of pleochroite (P1) has fibres lying at right-angles to each other and shows the range of pleochroism that varies from white to lilac in colour. Most of the remaining particles visible are either isotropic or opaque.

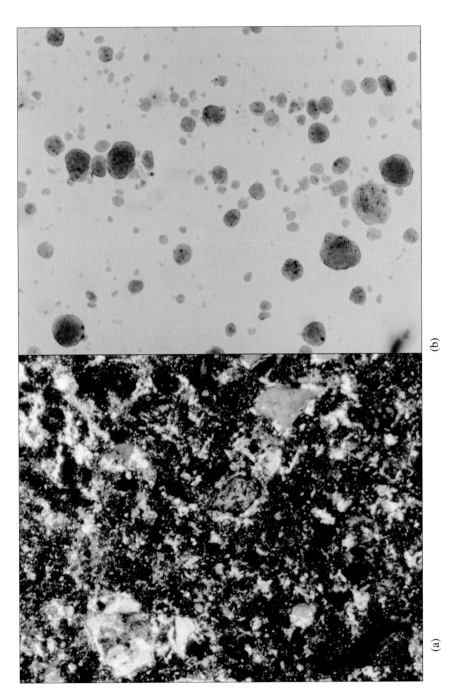

(a)

(b)

Plate 29 (a) The texture of Portland limestone cement . Width of field 1 mm. The limestone interground with the Portland cement used in this test sample (317 kg/m³ of Portland/10 per cent limestone cement, water/cement ratio = 0.6) appears as small birefringent specks of light which on careful examination often have a rhombic shape. The texture of the limestone filler is similar to post-sampling carbonation and care will be required in sample preparation to prevent carbonation which can be confused with the limestone. (b) The appearance of silica fume. Width of field 1.2 mm. The actual particles are below the resolving power of the microscope but they aggregate into lumps which are visible at this magnification.

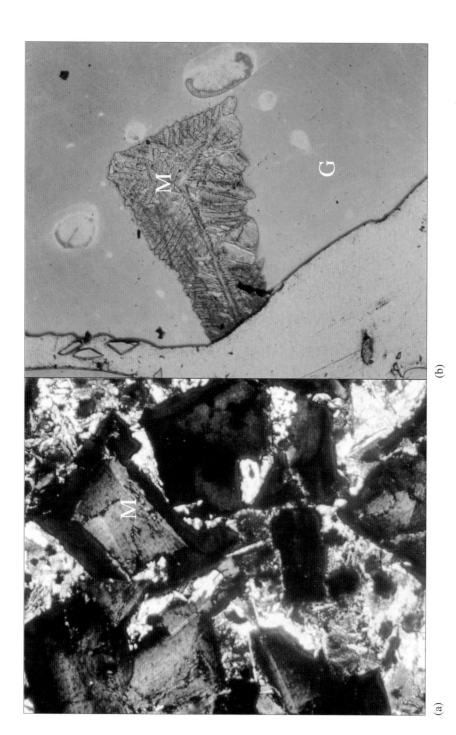

(a)

(b)

Plate 30 The appearance of melilite and granulated blastfurnace slag. Width of fields 0.8 mm. (a) Melilite crystals (M) are very distinctive in appearance. They have rectangular sections, zonation and a distinctive anomalous blue interference colour. The melilite is set in a matrix of crystals of dicalcium silicate, pyroxene and other minerals in this fragment of crystalline slag. (b) When slag is granulated, i.e., rapidly cooling from the molten state, it forms a glass in which are interspersed incipient crystals of melilite and other material. In this polished section, which has been etched with HF vapour, an incipient crystallisation of melilite (M) is embedded in the glass (G). The reason for the different etching colours of the glass is not known.

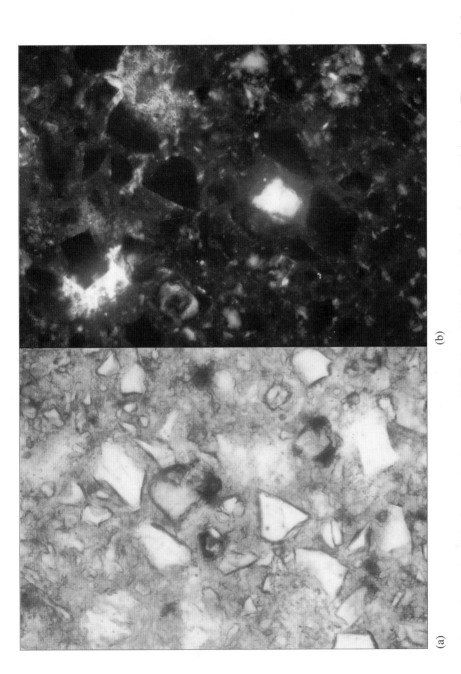

(a)

(b)

Plate 31 The appearance of ground, granulated blastfurnace slag in concrete. Width of fields 0.25 mm. (a) The colourless angular fragments of the slag are clearly visible set in the brown matrix of the hardened cement paste. Diffuse greenish patches of calcium sulfide interacting with the ferrite phase are scattered through the texture which imparts a bluish tinge to fresh surfaces containing slag. The concrete in the test sample used contained 300 kg of slag, 100 kg OPC and was mixed at a water/ cement ratio of 0.51. The concrete was moist cured for 28 days. (b) Under crossed polars the slag stands out as dark fragments because its glassy character makes it isotropic.

(a)

(b)

Plate 32 The appearance of well formed calcium hydroxide crystals in two sixty-year-old concretes. Width of fields 0.16 mm. (a) The fissure present at the margin of the argillite aggregate (A) at the left is due to plastic settlement of the paste (P). It is filled with well crystallised calcium hydroxide. More commonly, fissures like this are only partly filled because they do not remain saturated. The interconnected pore (V) contains a tablet of calcium hydroxide and the paste at its margin is also filled with calcium hydroxide. Crossed polars. (b) Between crossed polars the large pore contains well formed laths of calcium hydroxide set in a fibrous mass of ettringite (Et). This pore will have been filled with fluid to allow crystals of this size to form. The range of interference colours of the calcium hydroxide are well illustrated.

(a)

(b)

Plate 33 The effect of water/cement ratio on calcium hydroxide. Width of fields 0.8 mm. (a) Under crossed polars, this test concrete, (water/cement ratio = 0.4), contains well dispersed calcium hydroxide throughout the paste texture which gives the appearance of only limited crystallisation at aggregate (A) margins. Remnant and partially hydrated cement grains are noticeable when the polars are uncrossed. (b) In contrast another test concrete, (water/cement ratio = 0.8), contains larger aggregations of calcium hydroxide and crystallisation at aggregate margins is now obvious. There are far fewer remnant and partially hydrated grains in this sample compared to the one in Plate 33(a).

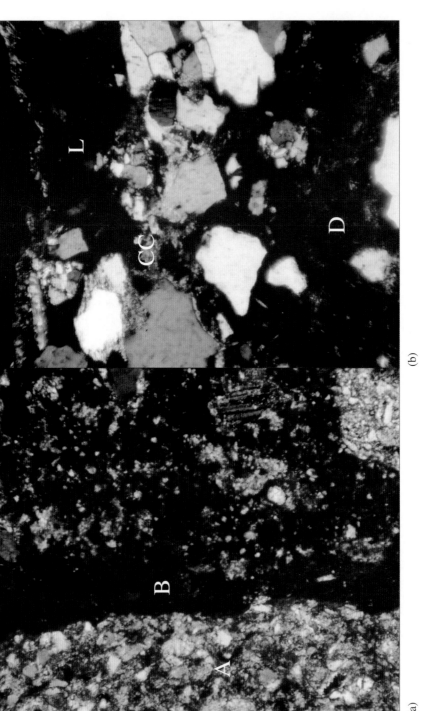

(a)

(b)

Plate 34 The absorption of calcium hydroxide into aggregate margins and the leaching of calcium hydroxide from crack margins. (a) Under crossed polars the dark band (B) of paste along the margin of the greywacke aggregate (A) in a thirty-five-year-old bridge is clearly deficient in calcium hydroxide. In cases such as this it is not clear whether the calcium hydroxide has been absorbed into the aggregate or the paste has been desiccated by the aggregate absorption. The lack of partially hydrated cement grains in the band suggests the former. Width of field 1.5 mm. (b) Under crossed polars the area (L) along the margins of this wide crack has been completely leached of calcium hydroxide and has undergone mild corrosion. There is a diffuse band of calcium carbonate (CC) backed by another area where calcium hydroxide is deficient (D). At the bottom of the picture normal paste texture is just visible. This is the effect of soft water flowing for sixty years through a large crack in an irrigation dam. Width of field 0.9 mm.

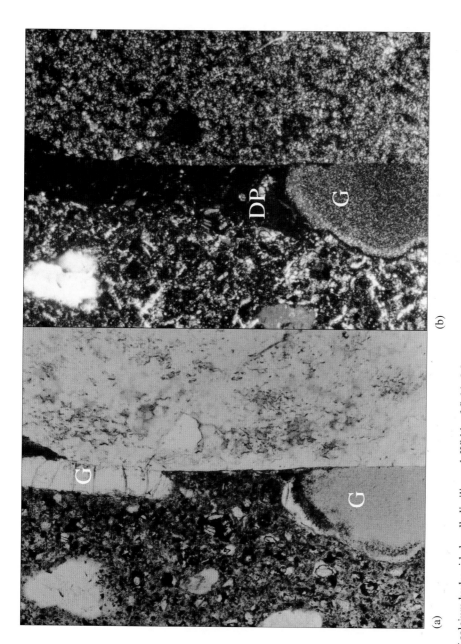

(a)

(b)

Plate 35 Absorption of calcium hydroxide by alkali-silica gel. Width of fields 0.8 mm. (a) The chert aggregate at the right has two patches of gel (G) along its margins. The lower patch is light brown with a darker margin while the upper is clear in colour with transverse cracks. The cement paste is full of remnant and partially hydrated cement grains with the brown interstitial phase being prominent. There are two grains of fine aggregate embedded in the paste. (b) Under crossed polars the texture of the chert aggregate shows clearly. The lower patch of gel (G) is crystallised while the upper patch has disappeared because it is a true gel and hence isotropic. Along parts of the aggregate margin the cement paste is dark (DP) due to gel having soaked into the paste and absorbed the calcium hydroxide. In the remaining paste a network of calcium hydroxide is clearly visible.

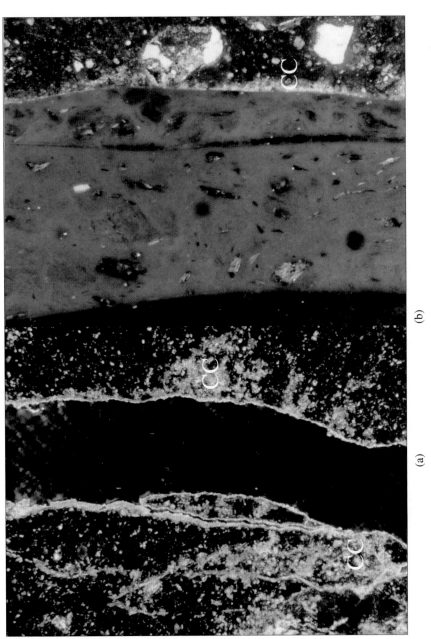

(a)

(b)

Plate 36 Calcium carbonate along a crack margin and at the outer surface of a very young concrete. Width of fields 0.8 mm. (a) Under crossed polars this 250 µm wide crack located 50 mm deep in this fifty-year-old concrete illustrates the typical appearance of the initial stages of carbonation (CC) which has not proceeded. The crack margins can be regarded as surfaces available to the atmosphere but protected from the weather. The depth and intensity of carbonation in such cases is largely dependent on the permeability of the concrete. (b) Under crossed polars an early stage of carbonation (CC) is visible. In this case the young test specimen had to be briefly dried to allow an epoxy coating to be applied. It can be clearly seen that the epoxy has been applied in two layers and contains a filler. Adhesion between the epoxy and concrete is excellent which is typical for these types of resin systems.

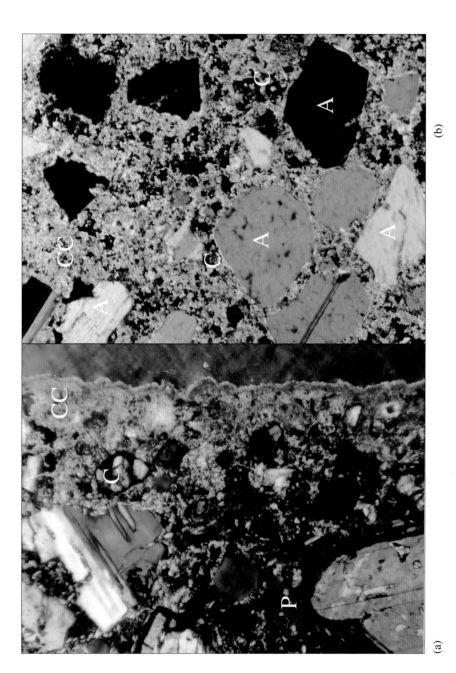

(a)

(b)

Plate 37 Typical carbonation at the surface of concrete and an example of intense carbonation deep inside a concrete. Width of fields 0.8 mm. (a) Under crossed polars. Classical carbonation of a ten-year-old structural concrete. The surface is weathered and carbonation (CC) forms a band that varies irregularly between 250 and 500 µm in width grading into the sound paste (P). The surface has dried out so that the rate of carbonation has exceeded the rate of hydration resulting in the presence of prominent partially hydrated cement grains (C). (b) 50 mm deep inside this sixty-year-old concrete the paste is intensely carbonated (CC). The calcium carbonate crystals appear discrete and the texture has a fretted appearance due to remnant grains (C). It is probable that most of the paste texture has been destroyed by intense carbonation which is due to the high porosity of this concrete. Aggregates labelled A.

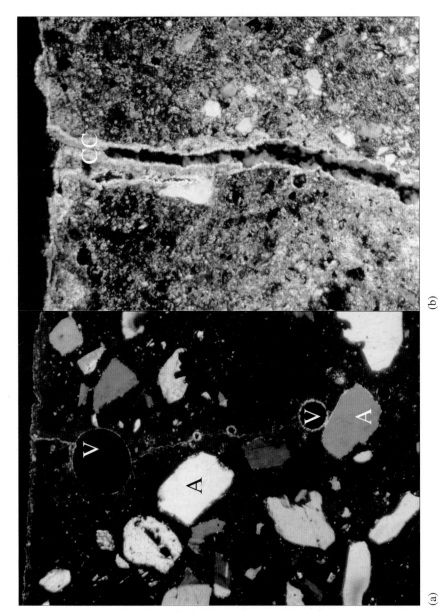

(a)

(b)

Plate 38 Carbonation associated with surface cracking. Width of fields 0.8 mm. (a) Under the crossed polars the surface of this three-year-old bridge concrete is only lightly weathered and carbonated although some of this is hidden as the section is a little thick. The drying shrinkage crack, about 2 mm deep and <10 μm wide, intersects some pores (V), and is filled with carbonation for its entire length. Aggregates labelled A. (b) Under crossed polars this tension crack, which is about 30 μm wide is plugged with carbonation (CC). The surface is slightly weathered and the depth of carbonation indicates that locally the thirty-year-old bridge concrete is of only moderate quality. This is a very typical surface texture.

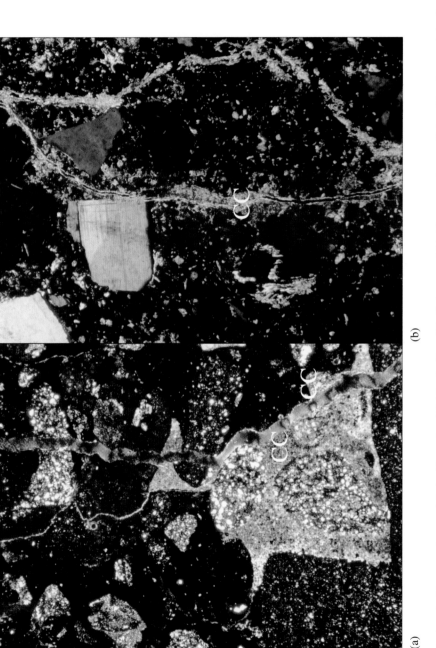

(a)

(b)

Plate 39 Carbonated paste deep inside concrete associated with cracks. (a) At a depth of 50 mm in this fifty-year-old bridge concrete, a 250 μm wide crack is not carbonated along its margins but intersects areas of paste that are clearly carbonated (CC). This indicates that the large crack is a later occurrence but other evidence strongly suggests it is not an artefact. The reasons for the uncarbonated margins are not easy to explain but may be related to environmental conditions. This complex behaviour of carbonation is not uncommon. Width of field 2.3 mm. (b) At a depth of 30 mm in this forty-year-old bridge concrete, a splintered tension crack is almost wholly filled with carbonation (CC). Although it does not exceed 25 μm in width for its whole length it is carbonated to a depth of nearly 40 mm. Areas of paste adjacent to the crack are also spotted with carbonation. This is in contrast to the adjacent photomicrograph where carbonation is largely confined to discrete areas of paste along the margins of the crack. Crossed polars. Width of field 0.9 mm.

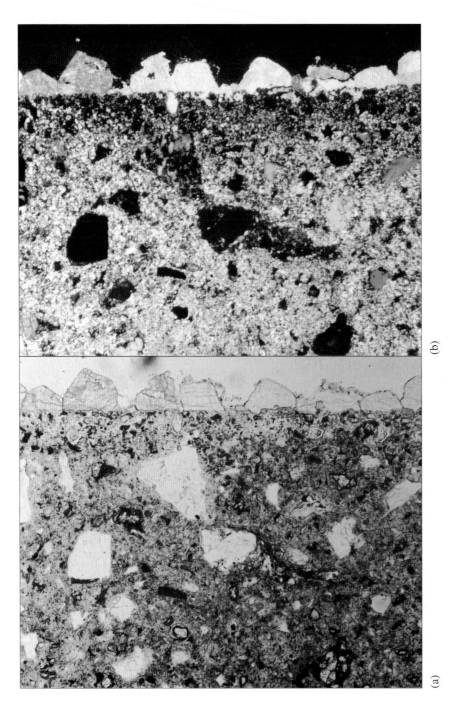

(a)

(b)

Plate 40 The appearance of well crystallised calcium carbonate present as efflorescence on the surface of concrete. Width of fields 0.8 mm. (a) A mixture of irregularly shaped and well formed rhombs of calcium carbonate is present on the surface of this test specimen (317kg/m^3 cement, water/cement ratio = 0.6, cured 28 days in fog room). The crystals stand out because of their relief and the distinctive pressure twinning which is commonly present in thin-section. (b) Under crossed polars the range of interference colours of the calcium carbonate crystals shows clearly. The outer layer of the concrete is well carbonated with some signs of leaching having occurred which is the probable reason for the presence of the efflorescence.

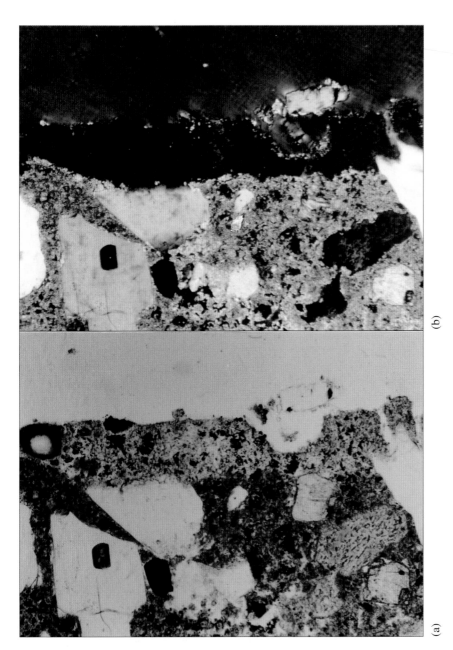

(a)

(b)

Plate 41 The effect of weathering on the surface of a fifty-year-old bridge concrete. Width of fields 0.12 mm. (a) Along the outer surface there is a band of lighter coloured paste which contains aggregates and diffuse, dark remnants of the cement paste. The outer surface is often ragged but it is not very apparent in this photomicrograph. This light band of material largely consists of gelatinous silica which is the remnant of corrosive attack on the cement paste. (b) The band of gelatinous silica is now clearly apparent because it is isotropic under crossed polars. The remainder of the cement paste is shown to be heavily carbonated. This type of texture is typical of attack by mildly acidic atmospheric conditions which in this case have taken place over fifty years. Circular polarisation.

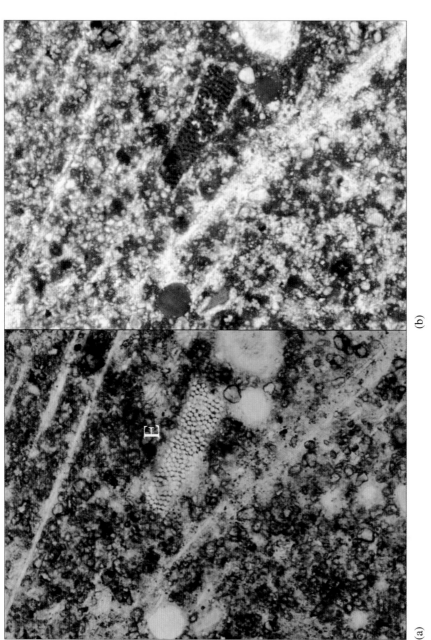

(a)

(b)

Plate 42 The fibre glass/paste interface. Width of fields 0.8 mm. (a) A cross-section of a strand of an E glass fibre lies in the middle of the photomicrograph. Other individual fibres lie approximately in the plane of the picture of this cement-lime test prism. Numerous relics of partially hydrated cement grains are clearly visible. (b) The cross-section of the glass fibre is surrounded by calcium hydroxide which has only been able to penetrate between the strands in isolated areas. Other individual fibres lying in the plane of the section are liberally coated with calcium hydroxide. Relics of both the cement and lime used are visible. Circular polarisation.

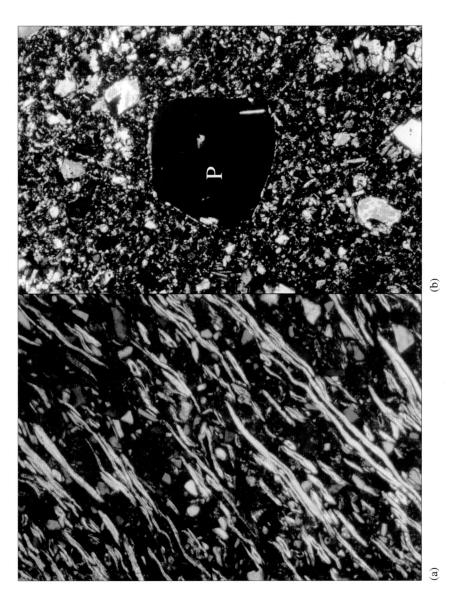

(a)   (b)

Plate 43 Paste interfaces with kraft pulp fibre and a low porosity aggregate. Width of fields 0.8 mm. (a) The kraft pulp fibres have been aligned at 45 degrees to show maximum birefringence and the section cut in a plane to intersect the maximum number of fibres. Autoclaving at 180°C has reacted most of the calcium hydroxide with the silica flour of which the coarser relics are still visible. There appears to be no visible calcium hydroxide at the fibre interfaces. Polarised light. (b) Around this pyroxene aggregate (P) is a semi-continuous layer of calcium hydroxide which is typical of low porosity aggregates in mortars of high cement content. The pyroxene aggregate has been positioned at extinction to highlight the calcium hydroxide at its interface. All the particles in this sample are derived from a crushed pyroxene olivine basalt. Polarised light.

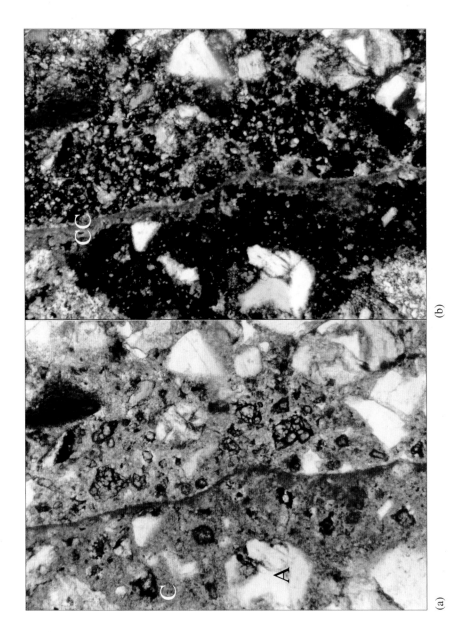

(a)

(b)

Plate 44 The interface between layers of shotcrete. Width of fields 0.8 mm (a) The interface between the two layers of shotcrete is delineated by a diffuse dark band. In the area shown contact is excellent but voids and fissures can be present at these interfaces. Remnant cement grains are dotted through the paste (C) among the greywacke aggregate (A). (b) The dark band is now seen to be the lightly carbonated surface (CC) of the previous layer of shotcrete. Note the difference in the amount of calcium hydroxide present between layer 1 and layer 2. This is probably a segregation effect that occurs as the layers are built up during spraying. Crossed polars.

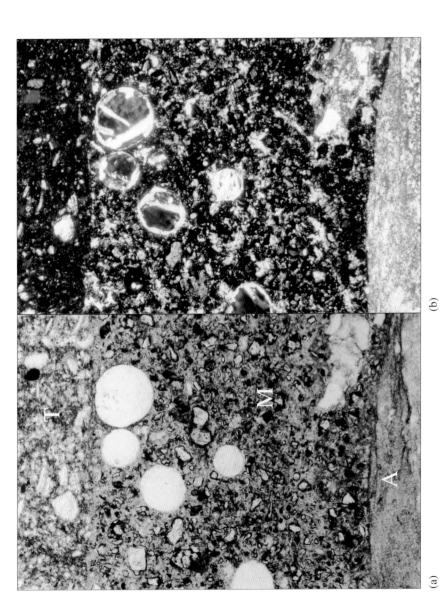

(a)

(b)

Plate 45 An interface between a ceramic tile and bedding mortar. Width of fields 0.8 mm. (a) The area of the rich bedding mortar (M) between the bottom surface of the tile (T) and a sand particle (A) contains some entrapped air and numerous partially hydrated and remnant cement grains. The paste is in close contact with the aggregate and the tile but in many other areas the interface contained voids and had parted from the tile. (b) At the tile interface a thin line of calcium hydroxide is just visible and to a lesser extent at the aggregate margin. The entrapped air voids contain laths of calcium hydroxide. It is possible the styrene-butadiene latex used in the mortar has had some effect on the way the calcium hydroxide has crystallised at the interface. Crossed polars.

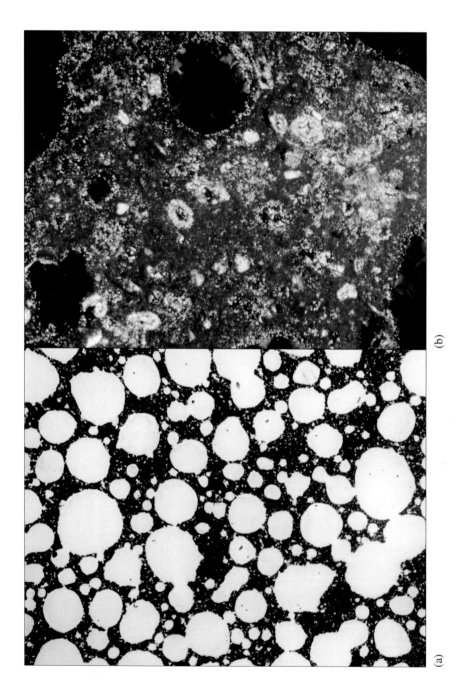

(a)

(b)

Plate 46 The texture of an aerated concrete. (a) The texture of an aerated brick. The hardened paste has carbonated due to many years lime silica brick (possibly Siporex) has probably been produced by the use of of storage in the laboratory. Relics of the silica flour and lime used are still visible. aluminium powder. The pore structure is predominantly in the 1.5-2 mm range but Width of field 0.9 mm. pores as small as 0.25 mm are present. Width of field 16 mm. (b) Detail of the lime

(a)

(b)

Plate 47 The texture of no-fines concrete. (a) In no-fines concrete large irregular voids are formed in the cement paste (C) which in this laboratory sample has cracked due to drying and/or carbonation shrinkage. The rounded quartz aggregate used (A) is clearly visible. Width of field 16 mm. (b) Detail of A. The cement paste has carbonated and cracked due to long storage in the laboratory. Shrinkage has also caused cracking at the interface of the aggregate. Cement clinker remnants are clearly present in the hardened paste. Width of field 0.9 mm.

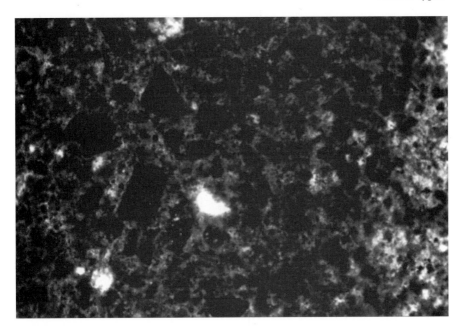

Fig. 4.14 Photomicrograph of high-alumina cement concrete matrix in thin-section. The dark crystals are mainly CA. Width of field *c.* 0.3 mm. Crossed polars. (Photograph courtesy of Mr Mark Knight, Sandberg.)

gibbsite (AH_3). When HAC is hydrated at an elevated temperature (40°C), Scrivener and Taylor (1990) found that CAH_{10} and an 'inner product' were still the initially formed phases, which within a matter of hours converted into C_2AH_8 and AH_3, then into C_3AH_6 and more AH_3. The 'conversion' reaction may be idealised as:

$$3CAH_{10} \rightarrow C_3AH_6 + 2AH_3 + 18H$$

or

$$3C_2AH_8 \rightarrow 2C_3AH_6 + AH_3 + 12H$$

The factors controlling the rate of conversion and the magnitude of its effect on the properties of HAC concrete are most importantly temperature and water/cement ratio: other factors include humidity, possibly stress and notably alkalis possibly deriving from adjacent Portland cement products, some aggregates and even mould release agents. These controlling factors actually influence the nature of the mineralogical phase change, including for example the higher mineral density of C_3AH_6 as compared with that of CAH_{10}, the increase in crystallite size and the complete change in habit on conversion which effects crystal packing (Sims 1977).

The progress of the conversion reaction is usually assessed by measuring the 'degree of conversion', whereby the relative proportions of CAH_{10} and C_3AH_6 (or AH_3) are determined by mineralogical analysis, most commonly using differential thermal analysis, or DTA (Chemical Society 1975) (Fig. 4.16). However it should be appreciated that this concept of 'degree of conversion' is an over-simplification and that the use of DTA for this purpose is subject to considerable imprecision.

Generally, the conversion of HAC in concrete leads to a progressive loss of strength, affecting structural integrity, and a simultaneous increase of porosity. The relationship

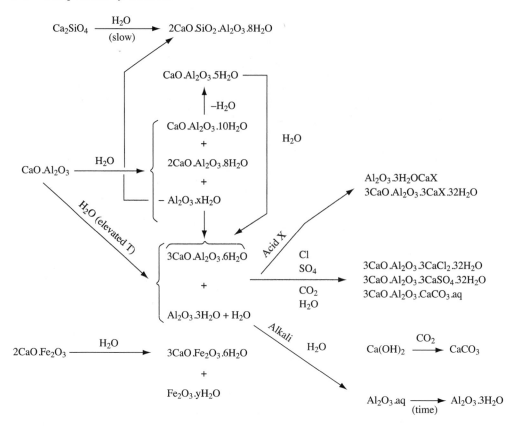

Fig. 4.15 Hydration and ageing reactions of high-alumina cement (from Midgley and Midgley 1975).

between conversion and strength loss is complex and practically unpredictable; in any case, in practice strength loss frequently continues even after the attainment of full conversion, especially when the concrete is not constantly dry. Although some structural collapses occurred in the UK (Building Research Establishment 1973, Bates 1964, Neville 1974) and more recently in Spain, assuming no building faults irrespective of concrete type, a majority of structures containing HAC concrete survive the detrimental effects of conversion alone. The structural use of HAC concrete ceased in the UK in the mid-1970s.

Converted and hence more permeable HAC concretes become vulnerable to further decay of the concrete itself and reduces protection for any embedded steel. Further deterioration of converted HAC may variously result from carbonation, alkaline hydrolysis, sulfate attack, chloride attack, or even combinations of these agencies (Building Research Establishment 1981, Collins and Gutt 1988). HAC conversion is discussed further in Chapter 5.

The petrographical examination of any HAC concrete must therefore address the conversion state of the cement matrix, as well as identifying the presence of HAC, and the extent to which any secondary chemical actions might have further impaired the strength and integrity of the material. In particular, now that most HAC concrete in existing structures in the UK can be assumed to be highly converted, concern has been expressed over the possibly detrimental effects of carbonation of the HAC concrete cover to the depth

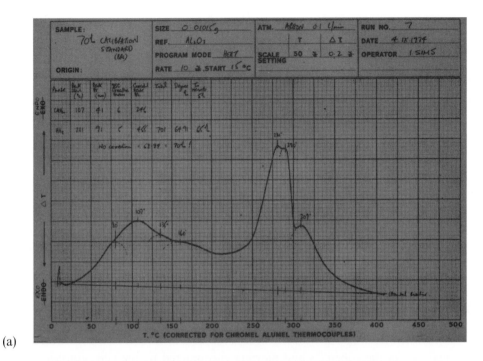

(a)

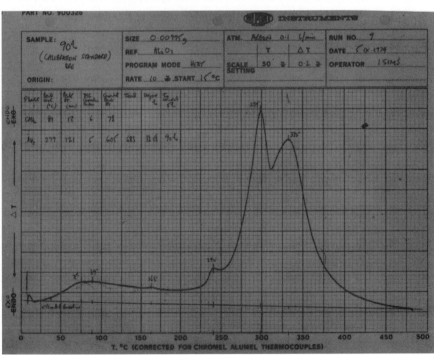

(b)

Fig. 4.16 Example differential thermal analysis (dta) charts for estimating the degree of conversion in high-alumina cement concrete: (a) 65/70 per cent converted; (b) 90 per cent converted.

of steel reinforcement or prestressing wires (Crammond and Currie 1993, Currie and Crammond 1994). It is feared that such steel is at greater risk of corrosion. According to Building Research Establishment Digest 392 (1994), 'petrography (optical microscopy) is the only cost-effective way to establish the depth of carbonation' of HAC concrete.

Carbonation of HAC is mineralogically quite different from that of Portland cement (see Chapter 5). The mechanism may be envisaged as C_3AH_6 reacting with CO_2 from the atmosphere to form an intermixture of calcium carbonate (mainly calcite) and aluminium hydroxide (mainly gibbsite). It seems that damp concrete is particularly likely to undergo this carbonation reaction, so that particularly samples from continuously or intermittently damp locations should be examined. Wholly or partially carbonated HAC is easily identified in thin-section (Plate 9).

A particularly damaging form of HAC carbonation is associated with a process known as 'alkaline hydrolysis', whereby alkaline solutions ingress HAC concrete as a result of water leakage via adjacent Portland cement concrete materials (e.g., concretes or screeds overlying HAC concrete beams). The alkaline solutions dissolve and decompose the calcium aluminate hydrates, to form alkali aluminate and calcium hydroxide, which can then be carbonated by atmospheric CO_2 to form calcite and gibbsite, so regenerating the alkali hydroxide to react again. A case study, concerning alkaline hydrolysis, possibly combined with sulfate attack, affecting HAC beams in a school in north-west England built in the 1960s, has been described by Dunster and Weir (1996).

Supersulfated and other special cements

A great majority of the concretes and mortars encountered by the petrographer will be found to be made using one of the variants of Portland cement and most commonly OPC. It is becoming quite common for ggbs, pfa or another pozzolanic material to be present as a binding material in combination with Portland cement and this was discussed in an earlier section. HAC is occasionally encountered and is quite distinctive as described in the previous section. Comparatively rarely, concretes or mortars will be found to contain an atypical binder, sometimes based upon or containing Portland cements, and sometimes being quite different. Whilst the petrographer obviously must be aware of such special or proprietary cements, apparently unusual binders in concrete are, at present, always more likely to be a modified or altered form of Portland cement than a completely different type of material.

'Supersulfated cement' is the name given to a form of cement based upon ggbs (80–85 per cent) activated by calcium sulfate (i.e., 10–15 per cent anhydrite or burnt gypsum) and a small proportion of Portland cement (5 per cent). The cement develops only a low heat of hydration and is particularly resistant to sulfates and other aggressive chemicals. It has been used to only a limited extent in the UK, but has been utilised more widely elsewhere, notably in Belgium. Apart from the high content of residual ggbs (see earlier sections and Section 4.6), the hydration products are said to be mainly ettringite and a tobermorite-like phase (Lea 1970).

A list of some special cements and their principal characteristics is given in Table 4.9. Of these, the petrographer is liable to encounter masonry cement, particularly when examining mortar samples. Masonry cements based on a mixture of building lime and OPC are discussed in Section 4.2.5, but other types of 'masonry cement' are encountered which comprise OPC with a finely ground inert material and an air-entraining agent. In the UK, the inert filler is usually ground limestone, but at least one variety is made using a finely ground quartz filler.

Table 4.9 Some special non-Portland types of cement and their principal characteristics.

Cement type	Manufacture/composition	Advantageous properties
High-alumina cement (HAC)	Fusing limestone and bauxite/mainly monocalcium aluminate (CA)	Rapid strength development, good refractory and sulfate resistance
Slag cement	Ground granulated blastfurnace slag activated by hydrated lime	Chemical resistance
Supersulfated cement	80–85% ground granulated blastfurnace slag activated by 10–15% inter-ground anhydrite (calcium sulfate) and with a small proportion of Portland cement or lime	Chemical resistance
BRECEM	Blend of high-alumina cement with ground granulated blastfurnace slag	Benefits of HAC without the problems of conversion
Masonry cement	Finely inter-ground Portland cement with limestone, or another inert filler, sometimes plus a plasticising and/or waterproofing admixture	Intended for use in mortars for masonry jointing, producing a more plastic mortar than that using ordinary Portland cement
Magnesium oxychloride or Sorel cement	Magnesite (magnesium carbonate) is calcined, ground and mixed with a magnesium chloride solution/brucite (magnesium hydroxide) bound by magnesium oxychloride	Produces a hard strong material, used for flooring when mixed with inert filler and a pigment

Experimental programmes have indicated promising results for blends of HAC with ggbs and other materials, but it is not believed that such cements have yet been widely used in practice. The addition of ggbs to HAC has been reported to inhibit conversion by the formation of C_2ASH_8 (strätlingite) instead of C_3AH_6 as the main product of hydration (Majumdar *et al.* 1990, Richardson and Groves 1990, Majumdar and Singh 1992). Other additions might also be effective in preventing or inhibiting HAC conversion, including microsilica and metakaolin (Majumdar and Singh 1992) or even powdered limestone or quartz (Piasta *et al.* 1989).

4.2.5 BUILDING LIME AND CEMENT/LIME MIXTURES

Prior to the invention of Portland cement, most mortar and concrete materials were made using lime as the binder. The use of lime has been traced to the early Egyptian and Greek civilizations, and the Romans became technically sophisticated in the manufacture of lime concretes (Granger 1962, Davey 1965, Malinowski 1982). Lime mortars are occasionally still used today, especially with stone masonry. More frequently lime is used in tripartite mixtures with sand and Portland cement, particularly for brickwork jointing mortar, and the petrographer must be aware of this possibility, especially when examining samples of mortar.

Lime is manufactured by the burning of limestone, when, at a temperature of around 900°C, the calcium carbonate decomposes, so that CO_2 is driven off leaving only calcium oxide (known as quicklime). This process is known as 'calcination'. For building purposes, the lime is then hydrated by slaking to produce calcium hydroxide (known as slaked lime). Hardening and strength development is achieved by carbonation to re-form calcium carbonate when slaked lime is exposed to the atmosphere:

$$CaCO3 \rightarrow CaO + CO_2 \tag{1}$$
Calcite   Lime

$$CaO + H_2O \rightarrow Ca(OH)_2 \tag{2}$$
Lime   Calcium hydroxide

$$Ca(OH)_2 + CO_2 \rightarrow CaCO_3 + H_2O \tag{3}$$
Calcium hydroxide   Calcite

The burning of limestones containing a proportion of clay leads to the production of 'hydraulic' limes, which contain calcium silicates, calcium aluminates and other phases in addition to calcium oxide. Depending on the proportion of clay in the original limestone, these hydraulic limes range in silica and alumina content from as low as 1–2 per cent ('fat' limes) up to 50 per cent ('eminently hydraulic' limes), with intermediate materials being variously termed 'lean' or 'semi-hydraulic' limes. The fat limes harden only by the carbonation process shown in the above equations, whereas the hydraulic and semi-hydraulic limes harden wholly or additionally by the hydration of the silicate and aluminate compounds.

Building lime

Building limes are classified according to their composition and whether or not they exhibit any hydraulic properties (i.e., an ability to set under water).

BS 890 (1972) gives four types of hydrated lime (either as fine dry powder or as lime putty):

1. hydrated high-calcium lime (white lime)
2. hydrated high-calcium by-product lime
3. hydrated semi-hydraulic lime (grey lime)
4. hydrated magnesian lime.

BS 890 excludes the class of 'eminently hydraulic' limes, which are stated to be 'not widely available in the UK'. The by-product lime (2) is derived from the wet-process manufacture of acetylene from calcium carbide, and, in BS 890, the only specified difference from white lime (1) is an additional limitation on soluble salts.

In the UK, 'white limes' (or fat limes) are manufactured from Chalk, Jurassic oolitic limestone and from some of the purer Carboniferous limestones. Semi- (or moderately-) hydraulic limes are made by burning limestones which contain a proportion of clay or shale, such as the greystone from the Lower Cretaceous of the UK Thames Basin. According to Roberts (1956), after burning an argillaceous limestone at 1000°C, the products include βC_2S, C_2AS, C_4AF plus other silicates and aluminates. The hydraulic properties are mainly attributed to the hydration of βC_2S. White lime can be converted into hydraulic lime by the addition of pozzolanic material at the mortar or concrete mixing stage. The Romans made particular use of this technique, employing various volcanic ash materials or even ground brick (Davey 1965).

The eminently hydraulic limes can contain up to 50 per cent silica and alumina (Lea 1970) and have been made in the UK from burning Blue Lias, Chalk Marl and the shaley limestones from the Carboniferous in Scotland (the aptly termed 'Cementitious Group'). There is an uncertain boundary between these eminently hydraulic limes and the so-called 'natural cements', made historically by calcining natural mixtures of calcareous and argillaceous material; for example the 'Roman cement' made from septarian nodules occurring in the London Clay.

The hydrated magnesian limes are broadly similar to white lime, except that the carbonate rock parent contained the mineral dolomite ($CaMg(CO_3)_2$) in analogous addition to calcite ($CaCO_3$). An amount of magnesium oxide (MgO or periclase) is formed on calcination, subsequently slaked to magnesium hydroxide ($Mg(OH)_2$ or brucite) and finally carbonated to $MgCO_3$ or magnesite.

In contrast to the wealth of literature on the texture and composition of hydrated Portland cement, there has been little study in recent times of the nature of set lime binders. The cementing mechanism has been studied by Moorehead and Morand (1975) and Moorehead (1986). The rate and extent of hardening is governed by the diffusion of CO_2 from the atmosphere to the hydrated lime reaction site, where carbonation occurs by a solution mechanism, leading to a 35 per cent increase in mass and a 12 per cent increase in solids volume.

Significant expansion does not occur on carbonation. However, as the internal pores accommodate the increase in volume and the whole composite becomes denser and less permeable. The heat generated by the carbonation reaction, 74 kJ/mol according to Moorehead (1986), is sufficient to evaporate the water liberated by the reaction (see equation 3 above) and sometimes also the residual capillary water, so that in some circumstances carbonation may be terminated prematurely.

Ancient lime mortars and concretes are quite distinctive in thin-section under the petrological microscope, with the matrix being predominantly finely crystalline calcite (Plate 10), together with residual particles of partially burnt limestone, charcoal contamination from the original lime kiln and patches of more coarsely crystallised calcite (Plate 11) deriving from the slow carbonation of pockets of trapped slaked lime (Sims 1975). Malinowski and Garfinkel (1991) have likened the textural appearance of such material to the natural rock travertine.

Old mortars and concretes originally made using an hydraulic type of lime might initially appear similar to those containing non-hydraulic lime, because of the ultimate carbonation of calcium silicate hydrates, but relict particles of unhydrated dicalcium silicate and other potentially hydraulic constituents are liable to be observable. Sims (1975) used X-ray diffraction to search for evidence of the presence of calcium silicates or the formation of calcium silicate hydrate phases in some Roman concretes made using ground brick, but in those particular samples the brick did not appear to have taken part in a pozzolanic reaction with the lime.

Moorehead (1986) has studied the microstructures of laboratory-prepared lime-cemented materials using electron microscopy. In the mixture prior to carbonation, the slaked lime binder was visible as distinctive platy crystals of calcium hyroxide (Fig. 4.17a). After carbonation, cryptocrystalline and amorphous calcite were found to have formed and to have infilled the larger voids and spaces between aggregate particles (Fig. 4.17b). The calcite grain size was generally less than 1 μm, with a significant proportion less than

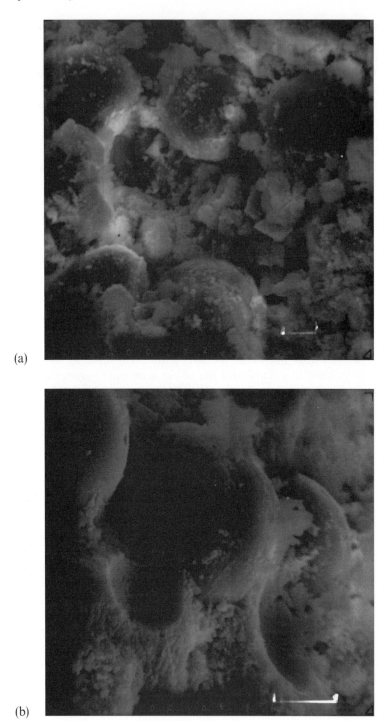

(a)

(b)

Fig. 4.17 SEM micrographs of building lime (a) before and (b) after carbonation (from Moorhead 1986). The white bars are equal to 10 μm.

0.1 μm and sometimes with more than 50 per cent of the calcite being amorphous or nearly so in respect of X-ray diffraction detection. Moorehead also found up to 8 per cent residual uncarbonated calcium hydroxide.

It is expected that, with time, such a texture will recrystallise to form the type of microcrystalline groundmass apparent in ancient lime mortar and any residual calcium hydroxide is likely to become carbonated. However, such a maturation time will vary very widely, depending upon the compactness of the lime mortar material itself and the environmental conditions to which it is exposed. The authors have encountered old lime mortars which exhibit a fully carbonated hard outer crust and a softer interior of incompletely carbonated material. Building lime is discussed further in Chapter 7.

Cement/lime mixtures

Portland cement mortars for masonry work are today generally preferred over lime mortars for their much more rapid strength gain characteristics. However, masonry mortars must not be over-strong and a reduction in Portland cement content to control the ultimate strength typically leads to the production of 'harsh' and unworkable mixes. Combined lime-cement mortars overcome this problem and are widely used, often by adding Portland cement to a pre-mixture of sand and lime. It is common for mortars to be gauged such that the lime and Portland cement components are present in equal volumes: traditional mix proportions of 1:1:6 by volume are most frequently encountered.

In the chemical analysis of hardened mortars, the presence of building lime will be suggested by the significant excess of calcium over that required for combination with soluble silica in calculating the Portland cement content. However, such calcium could also derive from calcium carbonate in the aggregate (i.e., particles of limestone, chalk or shell fragments) and therefore only microscopy can unequivocally identify the presence or absence of a substantial content of building lime. Once microscopical examination has established the nature of the non-Portland cement source or sources of calcium, the chemical analysis can be used to estimate the original mix proportions.

Identifying the presence of building lime in a cement/lime/sand mortar by microscopy is not always straightforward, especially if it is likely that the hydrated Portland cement has been subjected to carbonation. The absence of calcareous aggregate material in a sample found to contain significantly excess calcium by chemical analysis is clearly indicative of the presence of building lime. Pockets of trapped calcium hydroxide, whether or not subsequently carbonated, can also indicate the presence of building lime, providing it can be shown that the 'pocket' is not a void secondarily infilled with Portlandite as a result of leaching.

The most positive evidence of building lime is the identification of relict burnt limestone particles (Plate 12). Spheroidal pfa particles might be observed sporadically in the matrix, but these could be derived either from building lime or from the Portland cement (modern kilns for both lime and Portland cement manufacture frequently use pfa and traces of residual fuel occur in the products).

4.3 Aggregate types and characteristics

4.3.1 *PETROGRAPHIC IDENTITY OF AGGREGATE*

Aggregate particles typically comprise about three-quarters of the volume of a concrete and their petrological type, maximum size, size grading, surface texture and shape have influences on both the engineering and architectural properties of the concrete. The petrographic identity of the aggregate (i.e., the rock or mineral type or types of which it is composed) is an important consideration in establishing the composition of any concrete. In many cases the petrographic identity of the aggregate will indicate provenance, at least to within particular regions and in rare cases it is possible to pinpoint specific quarry or pit sources.

However, the petrographer should concentrate aggregate description on those features which have a potential bearing upon concrete properties; for instance, a petrogenetic analysis of a granite is not relevant to concrete performance, whereas assessing the extent of secondary alteration of the granite could provide valuable information pertinent to concrete durability.

Various undesirable or 'potentially deleterious' constituents might also be identified, including *inter alia* clay lumps, discrete (or 'free') mica, gypsum, pyrite and alkali-reactive materials. The presence of such constituents in the concrete will frequently need to be quantified and any evidence of detrimental activity recorded (see Chapters 5, 6). Further information on undesirable aggregate constituents may be found in Sims and Brown (1997) and a list of some of the more common ones is given in Table 4.10.

Table 4.10 Some potentially deleterious constituents found in aggregates. Compiled from information in Sims and Brown (1997).

Potentially deleterious constituent	Possible adverse effect in concrete[1]				
	i	ii	iii	iv	v
Clay coatings on aggregate particles	–	✓	–	–	–
Clay lumps and altered rock particles	–	–	✓	✓	✓✓
Absorptive and microporous particles	–	–	✓	✓	✓✓
Coal and lightweight particles	–	–	–	–	✓✓
Weak or soft particles and coatings	–	✓✓	✓	✓	✓✓
Organic matter	✓✓	–	✓	–	–
Mica	–	–	✓✓	–	✓
Chlorides[2]	✓	–	–	✓	–
Sulfates	✓	–	–	✓✓	✓
Pyrite (iron disulfide)	–	–	–	✓✓	✓✓
Soluble lead, zinc or cadmium	✓✓	–	–	–	–
Alkali-reactive constituents	–	✓	–	✓✓	–
Releasable alkalis	–	–	–	✓✓	–

1   i   chemical interference with the setting of cement
   ii   physical prevention of good bond between the aggregate and the cement paste
   iii modification of the properties of the fresh concrete to the detriment of the durability and strength of the hardened material
   iv interaction between the cement paste and the aggregate which continues after hardening, sometimes causing expansion and cracking of the concrete
   v   weakness and poor durability of the aggregate particles themselves.
2   The main problem with chlorides in concrete is associated with the corrosion of embedded steel.
✓✓ = main effect   ✓ = addition effect

The aggregates can be studied on broken concrete surfaces or in the sides of drilled cores. However, the aggregates are most easily identified and described in finely ground large-area surfaces, using unaided visual and low-power microscopical procedures. Fine-grained coarse aggregates and the fine aggregates are most easily examined in thin-section under a petrological microscope.

Typical aggregate combinations

A concrete mix design usually aims to achieve a reasonably continuous aggregate particle size distribution (or grading), so that inter-particulate space, to be occupied by cement paste, is minimised after compaction. This is typically achieved by blending a coarse aggregate (say >5mm particle size) with a sand or fine aggregate (say <5mm particle size). Occasionally, a natural sand and gravel material will exhibit a suitably continuous grading and may be used as a single 'all-in' aggregate. It is more common for the sand and gravel components of a natural material to be processed separately and then recombined in a controlled way to achieve a continuous grading. Crushed rock coarse aggregates are frequently combined either with natural sand fine aggregates or with a controlled blend of the crushed rock fines produced as a by-product (i.e., 'crusher-run fines') with a natural sand.

It is therefore necessary to recognise that the aggregates within a hardened concrete sample might represent a combination of materials from various different sources. It is usually convenient to describe both the coarse and fine fractions, with 5mm being a common threshold particle size between the two, unless another demarcation is made apparent by a clear compositional change.

Crushed rock coarse aggregates

The presence of crushed rock coarse aggregates in concrete is usually indicated by a relatively angular particle shape (see Section 4.3.3) and an essentially single-component (i.e., 'monomictic') composition. The rock type is usually straightforward to classify within broad generic groupings, such as those listed in BS 812: Part 102 (1989) (Table 4.11) or defined in ASTM C294 (1991).

French (1991a) has highlighted the classification difficulties which can arise because the rocks seen as aggregate particles within concrete are separated from their field association and thus have to be identified from their mineralogies and textures alone. This maximises the possibilities of different interpretations by different petrographers and generally leads to a lack of confidence in petrography by engineers who do not always appreciate the imprecise and widely overlapping nature of petrological nomenclature.

There would be much to be gained from a universally standardised and unambiguous petrological classification scheme for engineering purposes, but this is easier stated than achieved: this problem is discussed in Sims and Brown (1997) and more fully in Chapter 6 of Smith and Collis (1993). French (1991a) advocates 'full descriptions of the rocks ..., illustrated with photomicrographs, rather than the simple use of "standard" terminology'. However, for purely compositional analyses, this might lead to unnecessarily detailed work and clearly the descriptive detail justified must be tailored to the objectives of the petrographical examination (see also Chapters 5 and 6).

Even a crushed rock aggregate will not necessarily exhibit an entirely uniform petrographic composition. A crushed sedimentary rock, for example, might comprise a variety of types reflecting the bedded sequence within the source quarry. Crushed limestones, for example, might exhibit a range of carbonate rock types, including dolomitic varieties and silicified material, and sporadic chert particles are very common. Crushed greywacke might comprise a complete range from grain-dominant to matrix-dominant, virtually mudstone, types.

Table 4.11 Names and simple definitions for rock types commonly used for aggregates in the UK (after BS 812: Part 102: 1989).

Petrological term	Description
Andesite	a fine grained, usually volcanic, variety of diorite
Arkose	a type of sandstone or gritstone containing over 25 per cent feldspar
Basalt	a fine grained basic rock, similar in composition to gabbro, usually volcanic
Breccia	rock consisting of angular, unworn rock fragments, bonded by natural cement
Chalk	a very fine grained Cretaceous limestone, usually white
Chert	cryptocrystalline (resolved only with a high-power microscope) silica
Conglomerate	rock consisting of rounded pebbles bonded by natural cement
Diorite	an intermediate plutonic rock, consisting mainly of plagioclase, with hornblende, augite or biotite
Dolerite	a basic rock, with grain size intermediate between that of gabbro and basalt
Dolomite	a rock or mineral composed of calcium magnesium carbonate
Flint	cryptocrystalline silica originating as nodules or layers in chalk (see Chert)
Gabbro	a coarse grained, basic, plutonic rock, consisting essentially of calcic plagioclase and pyroxene, sometimes with olivine
Gneiss	a banded rock, produced by intense metamorphic conditions
Granite	an acidic, plutonic rock, consisting essentially of alkali feldspars and quartz
Granulite	a metamorphic rock with granular texture and no preferred orientation of the minerals
Greywacke	an impure type of sandstone or gritstone, composed of poorly sorted fragments of quartz, other minerals and rock; the coarser grains are usually strongly cemented in a fine matrix
Gritstone	a sandstone, with coarse and usually angular grains
Hornfels	a thermally metamorphosed rock containing substantial amounts of rock-forming silicate minerals
Limestone	a sedimentary rock, consisting predominantly of calcium carbonate
Marble	a metamorphosed limestone
Microgranite	an acidic rock with grain size intermediate between that of granite and rhyolite
Quartzite	a metamorphic rock or sedimentary rock, composed almost entirely of quartz grains
Rhyolite	a fine grained or glassy acidic rock, usually volcanic
Sandstone	a sedimentary rock, composed of sand grains naturally cemented together
Schist	a metamorphic rock in which the minerals are arranged in nearly parallel bands or layers. Platy or elongate minerals such as mica or hornblende cause fissility in the rock which distinguishes it from a gneiss
Slate	a rock derived from argillaceous sediments or volcanic ash by metamorphism, characterised by cleavage planes independent of the original stratification
Syenite	an intermediate plutonic rock, consisting mainly of alkali feldspar with plagioclase, hornblende, biotite, or augite
Trachyte	a fine grained, usually volcanic, variety of syenite
Tuff	consolidated volcanic ash

Igneous rock sources can also give rise to mixtures, for example as the result of vein swarms, or small dykes within the main rock body, or, in the case of volcanic rocks, variations within or between successive lava flows or tuff deposits. All types of rock might exhibit additional variations as the result of patchy or zonal alteration and weathering. Variants considered potentially to be of engineering significance should be separately identified and semi-quantified.

The petrographer must be aware of the heterogeneity likely to arise from these variations in aggregate composition and, whenever practicable, should ensure that adequately large sample areas are examined for the results to be sensibly representative.

Natural gravel coarse aggregates

Natural gravel coarse aggregate in concrete is usually indicated by rounded or sub-rounded particle shape (see Section 4.3.3) and a mixed or multi-component (i.e., 'polymictic') composition. Oversized gravel particles will be crushed, giving rise to the added presence of angular particles and also particles which are partly well-rounded and partly angular. Determination of petrographic composition should follow the same principles as those for crushed rock aggregate, except that a variety of completely different rock types might be encountered and each should be separately identified and semi-quantified.

Marine gravel coarse aggregate can usually be distinguished from an equivalent terrestrial material, by the presence in the former of recent calcareous shell debris, although in coarse aggregate this might form only a small proportion and then must be distinguished from fossil shell debris which is sometimes present in terrestrial sand and gravel deposits.

The compositional presentation of the aggregate observations will depend upon the rock constituents identified and their relative proportions. For example, a gravel aggregate which is found to comprise 80–90 per cent flint and 10–20 per cent of a varied assortment of other materials, might reasonably be labelled a 'flint gravel'. Another gravel aggregate which is found to comprise 20 per cent greywacke, 20 per cent chert, 20 per cent limestone, 20 per cent quartzite and 20 per cent varied others, might better be described as a mixed or 'polymictic' gravel mainly containing greywacke, chert, limestone and quartzite.

Visual estimation of relative proportions is usually adequate for concrete compositional analysis purposes, when the percentage estimate charts devised for sediments are usually helpful (Terry and Chilingar 1955) (Fig. 4.18). Quantitative determinations by point-counting are arduous and time consuming and will not always be necessary; certainly they will not be justified unless the concrete sample area is large enough to be adequately representative.

Crushed rock and natural sand fine aggregates

The identification and description principles for coarse aggregates largely apply similarly to fine aggregates, except that examination is usually carried out under a microscope and monomineralic grains will often be present in addition to rock particles. Particle shape and uniformity of composition are again the main features which distinguish crushed rock and natural sand materials, although blends can be difficult to differentiate if the natural sand particles are not well rounded. Natural sand particles can sometimes exhibit secondary surface growths, or superficial erosion or weathering zones, when microscopically examined in thin-section.

Mixed or polymictic sand compositions are readily quantified by point-counting in thin-section under the microscope. Particle size differentiation of various constituents is quite common with natural sands and any quantified analysis must take this into account. For example, the flint-bearing natural sands of south-eastern England typically contain their flint within the coarser particle size range of 1mm to 5mm, whereas the <1mm fractions are dominated by quartz. Marine sands will usually be distinguished from terrestrial sands by the presence in the former of recent shell debris, sometimes in considerable quantity.

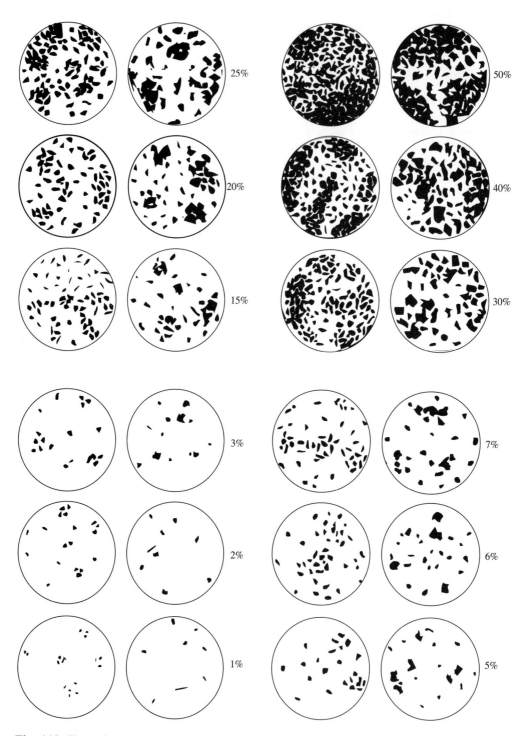

Fig. 4.18 Charts for estimating percentage composition of rocks and sediments (from Terry and Chilingar 1955).

4.3.2   PARTICLE SIZE AND AGGREGATE SIZE GRADING

The overall particle size distribution (or 'grading') of aggregate exerts an important influence over the properties of concrete, particularly by controlling with other factors the mix workability and the feasibility of achieving a satisfactorily high density by normal compaction methods. The aggregate maximum particle size and overall gradings are therefore necessary features to be established as part of any petrographical assessment of concrete composition.

In practice, modern design mix techniques and the availability of a range of admixtures for modifying concrete properties can enable successful concrete to be made using virtually any grading, providing that grading remains reasonably consistent during a period of

Table 4.12a Coarse aggregate grading limits specified by BS and ASTM standards (adapted from BS 882 (1992) and ASTM C33 (1993)).

BS 882: 1992				Type				
	Graded aggregates (mm)			Single-sized aggregates (mm)				
	40–5	20–5	14–5	40	20	14	10	5
Sieve (mm)				% by mass passing				
50.0	100	—	—	100	—	—	—	—
37.5	90–100	100	—	85–100	100	—	—	—
20.0	35–70	90–100	100	0.25	85–100	100	—	—
14.0	25–55	40–80	90–100	—	0–70	85–100	100	—
10.0	10–40	30–60	50–85	0–5	0–25	0–50	85–100	100
5.0	0–5	0–10	0–10	—	0–5	0–10	0–25	45–100
2.36	—	—	—	—	—	—	0–5	0–30

ASTM C33: 1993						Size (mm)							
	90–37.5	63–37.5	50–25	50–4.75	37.5–19	37.5–4.75	25–9.5	25–12.5	25–4.75	19–9.5	19–4.75	12.5–4.75	9.5–2.36
Sieve (mm)						% by mass passing							
100	100	—	—	—	—	—	—	—	—	—	—	—	—
90	90–100	—	—	—	—	—	—	—	—	—	—	—	—
75	—	100	—	—	—	—	—	—	—	—	—	—	—
63	25–60	90–100	100	100	—	—	—	—	—	—	—	—	—
50	—	35–70	90–100	95–100	100	100	—	—	—	—	—	—	—
37.5	0–15	0–15	35–70	—	90–100	95–100	100	100	100	—	—	—	—
25	—	—	0–15	35–70	20–55	—	90–100	90–100	95–100	100	100	—	—
19	0–5	0–5	—	—	0–15	35–70	20–55	40–85	—	90–100	90–100	100	—
12.5	—	—	0–5	10–30	—	—	0–10	10–40	25–60	20–55	—	90–100	100
9.5	—	—	—	—	0–5	10–30	0–5	0–15	—	0–15	20–55	40–70	85–100
4.75	—	—	—	0–5	—	0–5	—	0–5	0–10	0–5	0–10	0–15	10–30
2.36	—	—	—	—	—	—	—	—	0–5	—	0–5	0–5	0–10
1.18	—	—	—	—	—	—	—	—	—	—	—	—	0–5

supply and use. The petrographical analysis of a population of samples from the same or supposedly similar mixes must therefore address the aspect of aggregate grading consistency.

Aggregate particle size and grading are traditionally characterised and quantified by sieve analysis (BSI 812: Part 103 1985a), wherein the proportions by mass passing through each of a series of sieve sizes are computed. The set of sieve sizes chosen for this purpose varies between different methods (those specified in British and American Standards are shown in Table 4.12a), although graphically presented results ought to be similar. The nominal particle sizes used to describe standardised aggregate gradings similarly vary (Table 4.12b).

Table 4.12b Fine aggregate grading limits specified by BS and ASTM standards (adapted from BS 882 (1992) and ASTM C33 (1993)).

| Type Sieve size | BS 882 | | | | ASTM C33 |
| | Overall limits | Coarse (C) | Medium (M) | Fine (F) | General[1] |
		% by mass passing a sieve			
10 mm	100	100	100	100	—
9.5	—	—	—	—	100
5	89–100	89–100	89–100	89–100	—
4.75	—	—	—	—	95–100
2.36	60–100	60–100	65–100	80–100	80–100
1.18	30–100	30–90	45–100	70–100	50–85
600 μm	15–100	15–54	25–80	55–100	25–60
300	5–70	5–40	5–48	5–70	10–30
150	0–15/20[2]	0–15/20[2]	0–15/20[2]	0–15/20[2]	2–10

Notes
1   Additional criteria are given in ASTM, depending on the concrete mix and to ensure evenness of grading.
2   The upper limit is increased from 15 to 20 per cent for crushed rock fines, except for heavy-duty floors.

Sieve analysis results are influenced by a number of factors (Smith and Collis 1993), including *inter alia* sieve aperture shape, aggregate particle shape (see Section 4.3.3), whether sieving is carried out dry or wet and the energy and duration of the sieve shaking. Broadly speaking, however, sieved size fractions are determined by the largest dimension of a cuboidal particle or the larger width dimension of an elongated particle. It is therefore these dimensions which must be measured and recorded when evaluating aggregate grading in hardened concrete.

The nominal size and grading of aggregates in concrete can be determined by petrography, preferably from sawn and finely-ground surfaces or using large-area thin-sections (St John 1990). It is usually more helpful to attribute the maximum particle size observed to the nearest standard size (i.e., measured 22mm might be better reported as nominal 20mm), except that anomalously large particle (e.g., 'plums') should be additionally recorded.

It is important to recognise that random cross-sections through the centres of aggregate particles almost always exaggerate the dimensions (see Fig. 4.19): only those rare cross-sections exactly perpendicular to the dimension concerned will provide an accurate measurement, all otherwise oblique sections are likely to suggest a larger than actual dimension. Conversely, sections which transect only the tip of a particle might *under*state the size of that particle. Thus, in determining aggregate particle sizes, the petrographer

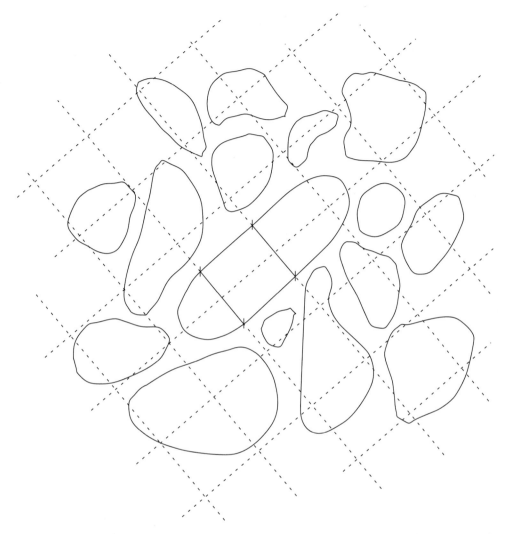

Fig. 4.19 Sketch illustrating that particle dimensions are usually misrepresented by cross-sections. A variety of particle shapes are overlaid by a grid of traverse lines. Only the solid intercepts in the two-dimensional sketch would accurately reflect the particle size as measured by sieve analysis, whilst the dashed intercepts would all either under-estimate or over-estimate particle size. Even the solid intercepts in the sketch might not be accurate in three dimensions.

must make intuitive allowance for these 'apparent' particle dimensions in all sectional specimen surfaces.

Grading analyses can be quantified from either sawn, ground or large-area thin-section specimens, using linear-traverse or particle counting techniques; this is exceptionally arduous, in view of the large number of measurements needed to satisfy the demands for adequate precision, and rarely justifiable for normal concrete petrography. However, French (1993) has described a procedure based upon the geometrical relationship between the 'grading curve' of chord intercepts along lines of traverse and the sieved grading curve for an idealised aggregate of spherical particles.

In this chord method, lines of traverse are described across a large sample area, spaced at 10 mm intervals and totalling ~4 m of traverse. By examining these lines of traverse under a microscope, the total length of traverse, the number of aggregate particles encountered and the individual chord intercept lengths are recorded. The chord lengths are corrected to compensate for their typical underestimation of particle diameter (with spherical particles the correction factor is suggested by French to be ~1.5). A chord correction factor can be estimated for a particular concrete by comparing the average chord length against the average particle size, calculated from the total length of traverse and the total number of particles encountered, assuming an overall aggregate volume for the concrete (say 70 per cent). Corrected chord lengths are then allocated into convenient and appropriate size fractions (e.g., 20–14 mm, 14–10 mm and 10–5 mm) and a grading curve can be obtained from the relative proportions of these size fractions.

The method requires the painstaking examination of large areas of ground concrete surface and is potentially subject to errors arising from its assumptions: the aggregate particles will not be spherical and in many cases will be strongly asymmetrical, the distribution of aggregate particles may be uneven, the particle densities may vary (affecting the relationship between the volumetric data and the originally gravimetric grading) and the total aggregate volume may be unknown and/or atypical. Although the procedure may have merit in critical investigations, once its accuracy and precision have been demonstrated, the authors do not consider that quantified aggregate gradings should be undertaken as part of routine concrete petrography.

However, with experience, petrographers will be able subjectively to assess the evenness of the overall aggregate grading (see Fig. 4.20). In this way, it should be possible to identify any particle size excesses or deficiencies in the overall grading which might have occurred either as an unintended result of blending aggregates with slightly incompatible gradings, or as a deliberately designed effect 'gap-graded' concrete. Often it will be appropriate to describe the maximum particle size and grading separately for the coarse and fine aggregates, but it will not usually be possible for the petrographer to differentiate an overall coarse aggregate grading which was originally compiled using 'single-sizes' processed from the same aggregate source.

The grading of the fine aggregate (i.e., particles mainly passing a 5 mm sieve, or 'sand') is particularly critical, as this can have significantly influenced the strength and durability of the hardened concrete. The current edition of BS 882 (1992) stipulates a fine aggregate grading system which comprises a wide overall grading envelope, plus three overlapping additional sets of grading limits designated C, M or F (i.e., coarse, medium or fine) (see Table 4.12b).

Although it is less arduous to quantify the grading of a fine aggregate using a counting technique in thin-section which is standard practice for sedimentary petrographers in analysing the particle sorting of natural sandstones, it is nevertheless rarely considered essential in concrete examination. However, the nearest appropriate BS 882 (or other as appropriate) sand grading should be determinable, when it should be appreciated that many aggregates will comply with two of the three overlapping limit sets.

In UK concretes made before 1983, the fine aggregate grading should be assessed with reference to the now superseded four-zone system first introduced in the 1954 edition of BS 882. This system, which was based on a survey of available UK sand gradings and was not intended to imply any relationship with quality or performance, described four zones from the coarsest (Zone 1) to the finest (Zone 4) in which the proportional ranges of material passing 600 μm did not overlap (Table 4.13). This offers a prospect to the petrographer of

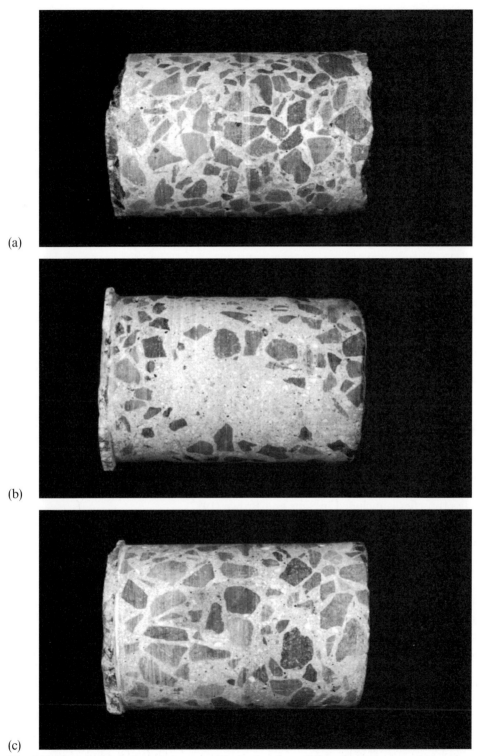

Fig. 4.20 Photographs of concrete cores, illustrating three different appearances: (a) evenly graded and distributed coarse aggregate; (b) unevenly distributed coarse aggregate; (c) apparently gap-graded coarse aggregate (although this concrete has the appearance of being gap-graded, the coarser sand sizes are similar in colour to the cement matrix and thus not easily discernible).

identifying the sand grading zone with some reliability in critical cases, by preferential particle counting in thin-section of the 600 μm–300 μm size fraction.

Table 4.13 Superseded British Standard fine aggregate grading limits – note the non-overlapping 600 μm limits (after BS 882: Part 2: 1973).

Type	Zone 1	Zone 2	Zone 3	Zone 4
Sieve Size		% by mass passing[1]		
10.0 mm	**100**	**100**	**100**	**100**
5.0	**90–100**	**90–100**	**90–100**	95–100
2.36	**60–95**	75–100	85–100	95–100
1.18	**30–70**	55–90	75–100	90–100
600 μm	**15–34**	**35–59**	60–79	**80–100**
300	**5–20**	8–30	12–40	15–**50**
150	0–10[2]	**0–10**[2]	**0–10**[2]	**0–15**[2]

Notes
[1] A cumulative total tolerance of 5 per cent is permitted on the limits not given in bold type.
[2] For crushed stone (rock) sands, the permissible limit is increased to 20 per cent.

The quantities in aggregates of very fine and ultra-fine material (i.e., those passing the 75 μm or 63 μm sieves according to different standards authorities) are particularly critical to the properties of concrete, but these constituents are usually difficult for the petrographer to assess using only optical microscopy. However, the actual particulate size of this material, especially whether it is mainly silt-sized (i.e., $<63/75$ μm >2 μm) or mainly clay-sized (<2 μm), and the mineralogical composition, especially whether or not clay minerals are present, are important considerations and these aspects can sometimes be established by optical petrography. Clay fines will sometimes occur coagulated as coarser clay aggregations or 'clay-lumps'. A more detailed discussion of the factors influencing the effect of fines on concrete properties is included in Sims and Brown (1997).

Scanning electron microscopy is increasingly available to practising concrete petrographers and can be considered, as an extension to the optical examination, to study the content, characteristics and distribution of these very fine and ultra-fine aggregate constituents.

As an aid to the petrographer, destructive chemical techniques can sometimes be employed to determine aggregate grading, using representative but disposable samples. Figg (1989) describes a 'mechanical' technique whereby hardened concrete is first carefully disaggregated by physical means, then the residual cementitious materials are removed mainly using acid treatment and finally the resultant loose aggregate is subjected to a conventional sieve analysis. This procedure is potentially quite accurate but somewhat operator-dependent in respect of reliability and of limited usefulness when the aggregate itself includes or comprises acid-soluble components.

In time it might be possible for the size and shape characteristics of aggregates in concrete to be rapidly determined using image-analysis techniques, but at present these methods remain at the research stage.

4.3.3 PARTICLE SHAPE

Aggregate particle shapes and their variations can influence the concrete mix characteristics and hence the properties of the resultant hardened concrete. Particle shape also affects the achievable degree of cement/aggregate bonding and component interlock within concrete, which are each of critical importance to the mechanical properties. A careful description of aggregate particle shape should therefore be included in any petrographical determination of concrete composition.

The shapes of aggregate particles are typically described using the two parameters 'sphericity' and 'roundness', long established for use in characterising the grains in clastic sedimentary rocks (Fig. 4.21). A simplified version of such a scheme was included in BS 812 (1975), except that the low sphericity particles were differentiated into 'flaky' (i.e., thickness small relative to other dimensions) and 'elongated' (i.e., length large relative to other dimensions). BS 812: Part 105 (1989) includes methods for quantifying the contents of flaky and elongated particles as flakiness and elongation indices.

Generally speaking, it is considered that, in terms of shape, 'cuboidal' aggregate particles characterised by high sphericity and low roundness are to be preferred, with particles which are both flaky and elongated being the least desirable. Although this is a fair overview, many aggregates which are far from ideal in terms of particle shape are used in concrete without significant impairment to its properties, especially if the shape characteristics are consistent and compensatory measures are taken in the mix design.

The aggregate particle shapes can be described, using the above-mentioned morphological terms, as part of a petrographical examination of hardened concrete. The sphericity aspects are liable to relate to rock type: for instance, closely cleaved or bedded rocks tend to yield flaky and sometimes also elongated particles. Roundness, by contrast, is more usually related to the type of source: for example, crushed rocks tend to form angular particles, whereas natural gravels are more likely to contain variably eroded and thus rounded particles.

Mixtures of aggregate particles are quite common, as shapes may be modified by processing, such as crushed over-sized gravel particles, which can be in part rounded and in part angular. Particle shapes in some crushed aggregates are also dependent to an extent upon the crusher type: for example, jaw and cone crushers can exacerbate a tendency of some rocks to form flaky particles. When a degree of quantification of aggregate particle shape in concrete is required, a visual estimate rather than a time-consuming particle count is usually adequate.

When required, schemes exist for a more critical consideration of particle shape parameters which are most likely to have an effect on the aggregate performance. Lees (1964), for example, devised a system for categorising aggregate particles in terms of axial ratios obtained from the longest, intermediate and shortest orthogonal dimensional axes. This gives a plot of flatness and elongation ratios divided into the four fields, cuboidal (equidimensional), elongated, flaky and flaky-elongated.

Alternatively, French (1991a) proposed the measurement of 'aspect ratios' (i.e., longer dimension of the particle divided by the shortest), which range from unity for highly spherical particles to about 1.7 for regular cubic grains. He considered that when all of the particles have aspect ratios of 3 or less, the shape factor has little influence on the quality of the concrete.

(a)

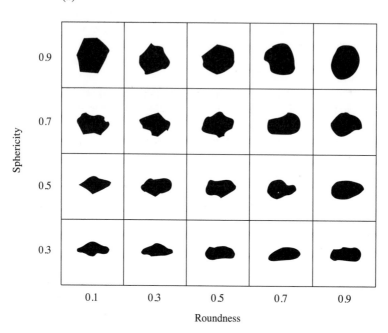

Sphericity = nominal diameter / maximum intercept
Roundness = average radius of corners and edges /
radius of maximum inscribed circle

(b)

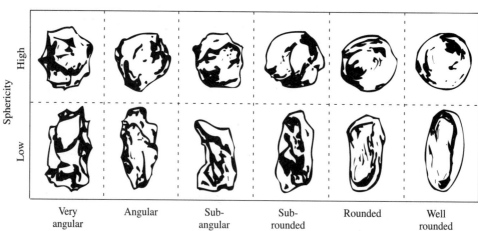

Fig. 4.21 Classification of aggregate particle shape (from Sims and Brown 1997).

4.3.4 OTHER PARTICLE CHARACTERISTICS

When examining samples of hardened concrete, the petrographic composition and the aggregate particle sizes and shapes are readily described. Other factors relating to the particle surfaces or their internal micro-structures are less straightforward, but can often have a significant effect on concrete properties.

The interface between aggregate particle surfaces and cement paste is one of the most important parameters in determining the mechanical performance of concrete. Although aggregate-paste interfacial zones are often studied in some detail during failure investigations (see Chapter 6), their potential importance is frequently overlooked during routine assessments of composition and quality.

It is therefore recommended that the aggregate/paste interface should be regarded and described as if it were an additional *component* of a concrete. According to Sarkar and Aïtcin (1990), theoretical models for the elastic behaviour and mechanical properties of concrete prove inadequate when they fail to address the aggregate:cement interface.

Physical strength and integrity of the bond achieved at the aggregate/paste interface variously depends upon the character of the aggregate particle surface and the nature of the immediately adjacent hydrated cement. The micro-topography of aggregate particle surfaces should be examined for 'texture' and for coatings or encrustations. The immediately adjacent cement paste zone should be examined for entrapped voids, microporosity and the concentrated development of mineral phases, such as Portlandite or ettringite.

Many attempts have been made over the years to find an objective method for measuring the roughness of aggregate particle surfaces, most recently using Fourier and fractal techniques, but the simple descriptive schemes, such as that included in BS 812 (1989), remain the most widely used. However, it is difficult to apply these textural concepts to aggregates in hardened concrete, particularly when viewed in section, when only the general roughness or otherwise of the aggregate:cement interface will be readily apparent.

It is reasonable to assume, within limits and with other factors held constant, that relatively rougher or more rugose particle surfaces will facilitate stronger aggregate:paste bonds, because of improved physical keying and enlarged contact areas. However, other factors are not held constant in practice and extremes of surface texture are likely to be disadvantageous: for example, unduly rough aggregates will adversely influence workability and might also consume quantities of cement paste without any commensurate binding benefit.

Dolar-Mantuani (1983) considered that there is a complex interrelationship between the main textural features of aggregate surfaces and the quality of the bond to cement paste. In particular, she stressed the importance of surface absorption, in comparison with surface roughness, with the strength of bond being potentially increased by the penetration of cement slurry into permeable aggregate surfaces.

One of the fundamental activities in placing *in situ* concrete, or in moulding precast concrete products, is some form of compactive effort, such as internal or external vibration, whereby the constituents are consolidated and entrapped air is reduced to a minimum. This compactive activity tends to promote segregation and bleeding. Excessive compaction can lead to impaired concrete quality, although the inherent susceptibility of concretes to segregation and bleeding also depends upon a variety of mix factors.

The most familiar evidence of bleeding is the formation of a layer of water on the upper surface of a freshly emplaced concrete and, within normal limits, this is a feature which helps to prevent premature surface drying and consequential cracking. However, localised internal bleeding also occurs throughout the body of a concrete, typically manifesting itself by the concentration of water at the aggregate/paste interfaces.

This can lead to the formation of unusually porous hardened paste at the transition zone, because of the locally high water/cement ratio. Otherwise, if water has completely segregated to form either a pocket or a continuous film at the aggregate particle surface, voids or spaces can be created if the water remains trapped during the plastic phase and only evaporates from the concrete after hardening.

Such evidence of internal bleeding is often apparent at the aggregate/paste interface on the underside of an aggregate particle (see Plate 13). Entrapped air voids might also be found in such locations, which tend to occur most readily with elongated aggregate particles that have become orientated parallel to the concrete upper surface.

The film of fluid which can be formed at the aggregate/paste interface as the result of internal bleeding will naturally have the chemical composition which is characteristic of the pore fluid in freshly placed concrete, being rich in calcium hydroxide and sulfates (Lea 1970). It is therefore quite common for layers of Portlandite and/or ettringite to be deposited at the aggregate/paste interface as the fluid evaporates (Plate 14). Sarker and Aïtcin (1990) describe the well-crystallised nature of these developments. French (1991a) suggests that these layers of Portlandite particularly form at the surfaces of siliceous aggregate particles and the amount of Portlandite precipitated usually increases as the particle size of the aggregate decreases.

The efficacy of the bond between aggregates and hardened cement paste can obviously be influenced by any material which interposes between these two components, including, for example, the layers of water and crystalline deposits discussed above. Dust coatings of aggregate particles are liable severely to reduce the aggregate:paste bond, although these might be difficult to detect during the examination of concrete using purely optical techniques.

Thicker encrustations on the surfaces of aggregate particles will be more visible to the petrographer, but need not be detrimental. One of the most common encrustations will be calcite, which will sometimes have been formed *in situ* within sand and gravel deposits as the result of precipitation from percolating groundwaters, or might be the residual natural cement from an eroded conglomeritic sedimentary rock. Such partial calcitic encrustations are quite common with the wadi gravels widely used for aggregate in the Arabian Gulf (see Plate 15a) and these have not usually been considered to be detrimental, unless contaminated by potentially deleterious salts.

In some parts of the Arabian Gulf region, prospective aggregates can be seriously contaminated by potentially detrimental salts (Fookes and Collis 1975). Although it is feasible for water soluble salts to be removed by washing, a greater long-term threat to concrete durability is posed by gypsum, which is only slowly soluble in water and might well survive aggregate processing either as discrete particles or as surface coatings (see Plate 15b). Crammond (1984, 1990) has demonstrated that expansion approaching 1 per cent over about 5 years can be caused in mortar-bar test at room temperature by as little as 5 per cent crystalline gypsum in an aggregate.

Some sand and gravel aggregates can be contaminated with clay, such that particles are coated with thin layers or even films of clay. This will inhibit the formation of an effective bond and might even cause disruption of an initial bond as the result of moisture activating the clay material.

Although the factors affecting the aggregate/paste interface are usually the more critical, the internal characteristics of aggregate particles are occasionally important, especially in the case of high and very high strength concretes. There are obviously certain constituents which are themselves usually regarded as undesirable at best and deleterious at worst. These are fully described by Sims and Brown (1997) and summarised here in Table 4.10. Otherwise this discussion refers to aggregate particles which are compositionally acceptable, but which might exhibit questionable microstructural features.

It is usually a reasonable assumption that rock particles that survive the rigours of aggregate processing will possess adequate strength to perform their role as a constituent of concrete. However, this might not apply to higher strength concretes (Sarker and Aïtcin 1990), or to concretes which are subjected in service to exceptional shear stresses, when otherwise stable aggregate particles might prove to have potentially vulnerable flaws. Crushed rock aggregate particles, for example, can exhibit potentially vulnerable micro-fractured surface zones.

Sometimes rocks will have been fragmented or become micro-brecciated as part of their geological history or as a result of blasting in the quarry. Aggregates produced from such materials might comprise particles which contain potentially numerous weak points. The trachyte shown in Plate 16, for example, was found to exhibit a limonitised natural micro-brecciation and, whilst the rock performed admirably in a range of aggregate strength tests prior to its use, it failed seriously in service and in subsequent investigation using the magnesium sulfate soundness test method (ASTM C88 1990, BSI 812: Part 121 1989).

Rocks can also exhibit natural fabrics or textures which might adversely influence their performance as aggregates. Metamorphic rocks, for instance, frequently display a schistose or gneissose fabric, often causing aggregate particles to be strongly anisotropic in their mechanical behaviour. Because such rock fabrics tend also to create flaky and/or elongated aggregate particle shapes, which can then adopt a preferred orientation, effectively 'planes' of potential weakness can be created within a concrete containing these types of aggregates.

The integrity of rock aggregate particles might also be impaired where crystalline textures have been 'loosened' by weathering or by stress relaxation following quarrying. Certain types of rocks, particularly those comprising crystalline calcite, such as some marbles and limestones, are occasionally also liable to thermal hysteresis, whereby thermal cycling can lead to irreversible expansions and deformations of aggregates particles within hardened concrete (Senior and Franklin 1987).

4.4   Water/cement ratio

4.4.1   DEFINITIONS AND RELATIONSHIP TO CONCRETE PROPERTIES

The relationship between cement and water is a matter of crucial importance to concrete, because it is the hydration of cement and its consequent hardening into a strong composite with aggregates which affords the material most of its advantageous properties. Hydration of Portland cement is, in detail, a complicated chemical process and the resultant microstructures, particularly the pore structures, are related to the strengths of hardened cement paste and concrete.

Aggregate/paste bond is probably the primary limiting factor regarding the mechanical behaviour of most concretes. Yet, even if the aggregate is disregarded, the strength typically achieved by hardened cement paste is up to 1000 times lower than the 'theoretical strength' which may be calculated, according to Neville and Brooks (1987). This difference is accounted for by Griffiths' theory of fracture mechanics, whereby minute cracks or other flaws in brittle materials act as stress-raisers by concentrating stress at their tips. There

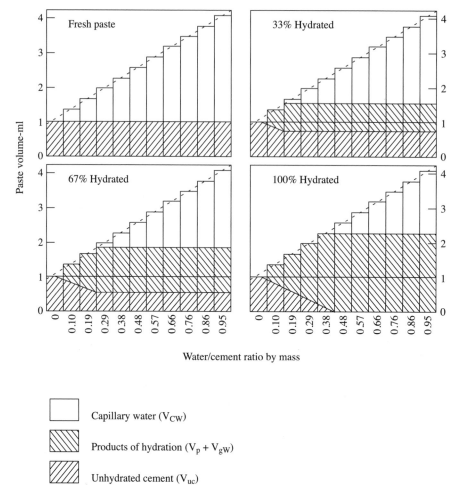

Fig. 4.22 Composition of cement paste at various stages of hydration. The percentage indicated applies only to pastes with enough water-filled space to accommodate the products of hydration at the degree of hydration indicated (from Powers 1949a).

are various micro-defects and discontinuities within the complicated structure of cement paste, but around half of the volume consists of micro-pores and these are the principal 'flaws' which critically control the strength potential.

The content of pores in the cement paste is determined by the quantity of water used in relation to the amount of cement, traditionally expressed as the 'water/cement ratio'. Thus, Neville and Brooks (1987), discussing the factors influencing concrete strength, stated: 'The most important *practical* factor is the water/cement ratio, but the *underlying* parameter is the number and size of pores in the hardened cement paste.'

The hardened cement paste comprises the solid hydrates plus the adsorbed water held in very tiny 'gel pores' around 2 nm in diameter. Because this hydrated cement occupies a smaller volume than that of the original dry cement and water, there are residual spaces or 'capillary pores' much larger than gel pores, at around 1 μm or 1000 nm in diameter. If the cement was fully hydrated and there was no excess of water over that required for cement hydration, these capillary voids would amount to about 18 per cent of the original volume.

However, in practice there is always an excess of water over that required for cement hydration, essentially acting as a lubricant to enable concrete to be mixed and placed. As the amount of excess water increases, so the volume of capillary voids, initially filled with water, increases. Powers (1949a) developed the diagram shown in Fig. 4.22 to illustrate the relationships between the water/cement ratio, the degree of hydration and the relative proportions of unhydrated cement, hydrated cement and water-filled capillary voids.

It is apparent from the above discussion that the water/cement ratio is an important criterion in the composition of concrete, controlling the microporosity of the cement paste and hence having a critical influence on concrete strength properties. Furthermore, because durability is generally related to the water permeability of concrete, which is in turn largely dependent upon cement paste porosity, water/cement ratio also has an important bearing upon durability.

4.4.2   INDICATORS OF WATER/CEMENT RATIO

Variations in the relative proportions of cement and water in the mix have significant influences on the physical and mineralogical characteristics of the resultant concrete. It therefore follows that study of the relevant features of hardened concrete can provide an insight into the original water/cement ratio of the mix. Although procedures have been devised for using these factors to quantify the original water/cement ratio for a given concrete sample, in truth the result can only ever be *indicative*, because the determinable parameters which are influenced by water/cement ratio are also affected by various other factors in complex and practically unpredictable ways.

In practice, accurate estimation of the water/cement ratio in a concrete is rarely necessary. Often the petrographer will only be required to establish whether or not a particular concrete had a water/cement ratio which significantly exceeded a given value. Also, many structural concrete mixes will have been made using water/cement ratios within a narrow range and the imprecision of the determinative methods would not allow these mixes to be differentiated. As cement paste volume increases with water/cement ratio, for a given cement content, it is often possible to assess the likely level of water/cement ratio from the paste volume (excluding entrapped and entrained air) and this will frequently suffice (Fig. 4.23).

Water-voids and bleeding

During and after the placement of concrete, there is a tendency in the plastic phase for the constituents to segregate, with coarse aggregate particles settling and excess water and entrapped air rising towards the upper surface. This effect is minimised in good quality cohesive mixes, but in poor concrete leads to materials segregation. The segregation of water from the other constituents within concrete is known as bleeding and, within limits, is a typical feature of all normal concrete mixes. Excessive bleeding, however, can detrimentally affect both the strength and durability of concrete and has a variety of causes.

The likelihood of excessive bleeding increases with water content, so that the occurrence of excessive bleeding is one sign of a possibly high water/cement ratio. Bleed water either reaches the surface of the concrete, where it evaporates, or becomes trapped within the concrete interior at the time of setting. It is common for such entrapped bleed water to accumulate beneath aggregate particles (Ash 1972), leading either to water voids or to local

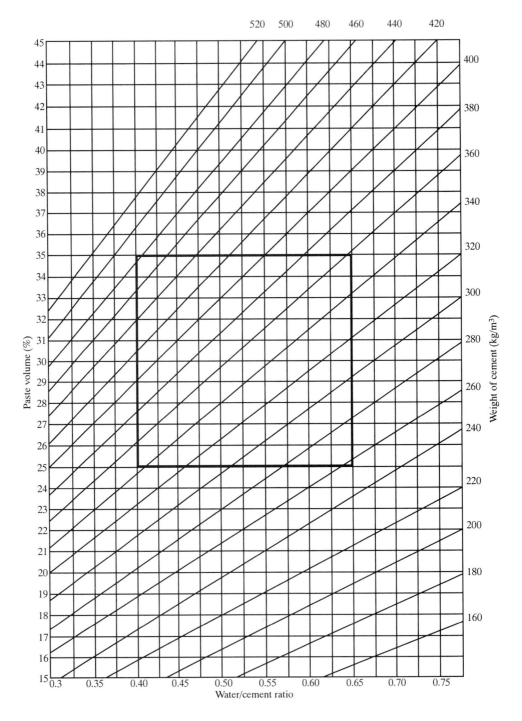

Fig. 4.23 Relationship between water/cement ratio, hardened cement paste volume and cement content, assuming the cement specific gravity to be 3.12. The values in the large square represent a typical range of concretes.

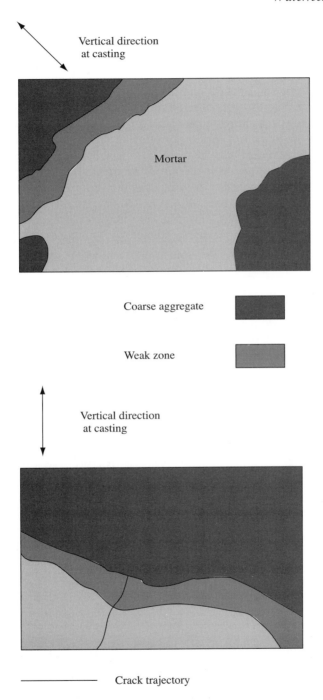

Fig. 4.24 Sketches to illustrate water voids and porous paste caused by bleeding beneath aggregate particles (from Ash 1972).

zones of conspicuously porous cement paste (Fig. 4.24). Bleed water may also occupy cavities and voids left by incomplete compaction, leading to smooth internal coatings of secondary deposits within the void which help to distinguish water voids from air voids. In some cases, bleed water can migrate along particular channels within the concrete, which later manifest themselves as conspicuously porous zones or tortuous channels within the cement paste (Plate 17).

The presence of water voids beneath aggregate particles in hardened concrete is readily observable during the macroscopical examination of the sides of drilled cores. Even zones of conspicuously porous cement paste beneath aggregate particles are frequently betrayed by relatively greater erosion during drilling, producing slight grooves under the aggregates in the core side surfaces. Otherwise the more porous zones beneath aggregates or along bleed channels can be detected with the aid of differential absorption techniques. The simplest method is to wet the drilled or sawn concrete surface and to make observations during drying: the more porous zones are the last to dry. Ash (1972) described a method using red ink, but today impregnation with a low-viscosity resin containing a fluorescent dye is probably the most efficient technique, whereby porous zones absorb more resin and thus fluoresce more in ultraviolet light.

Capillary porosity

It was explained in Section 4.4.1 that hydrated cement paste inevitably contains capillary pores around 1 μm in size and that the total volume of capillary pores increases with increasing water. The correlation shown in Table 4.14 indicates that the relationship between the capillary porosity of cement paste and the water/cement ratio is particularly pronounced at the critical lower end of the range, the capillary porosity trebling between water/cement ratios of 0.40 and 0.55 (Christensen *et al.* 1979).

Table 4.14 Relationship between the capillary porosity of hardened cement paste and the original water/cement ratio (after Christensen *et al.* 1979).

Water/cement ratio by mass	Capillary porosity of cement paste % by volume
0.40	8
0.45	14
0.50	19
0.55	24
0.60	28
0.65	32
0.70	35
0.75	38
0.80	41

Measurement of capillary porosity is a crucial element in the physico-chemical determination of original water/cement ratio (see Section 4.4.3), but, until recently, this important indicator of water/cement ratio could not be assessed by petrography because the capillary pores at around 1 μm in size are too small for observation by optical microscopy. Mielenz (1962), for example, one of the founders of modern concrete petrography, stated that, 'The cement paste of hardened concrete includes ... a system of minute capillaries and pores not visible under the optical microscope'. However, the technique described by Christensen *et al.* (1979), involving impregnation of the concrete sample by low-viscosity resin containing a fluorescent dye and subsequent examination in ultraviolet light, has enabled the capillary porosity to be assessed directly by microscopy. This is the basis of a microscopical method for the determination of original water/cement ratio (see Section 4.4.4).

Mineralogical features

Portlandite is a typical product of the hydration of Portland cement (see Section 4.2.3), but the amount, crystallite size and disposition of Portlandite varies with the original water/cement ratio and compaction. In all Portland cement concretes, unless a mineral addition is present which can react with and remove it, the Portlandite forms within the cement paste and at aggregate/cement interfacial zones. When the water/cement ratio is high, these Portlandite crystallites become unusually large and more clearly defined (Jepsen and Christensen 1989, French 1991a), although the size of Portlandite crystallites is also temperature dependent (see Section 4.2.3). French (1991a) has also observed that the coarser Portlandite crystallites formed with high water/cement ratios tend to occur as clusters, whereas the smaller crystallites formed at lower water/cement ratios are more uniformly distributed. French (1991a) further suggested that 'there is probably slightly more [Portlandite] in concretes made with higher water/cement ratios', but this could be an impression created by the coarser and therefore more visually apparent crystallites.

It seems obvious that higher degrees of hydration (see Section 4.2.3) must indicate higher water/cement ratios, but such would be an over-simplification. Although there is a relationship between the proportion of residual unhydrated cement particles and water/cement ratio, the degree of hydration is also strongly dependent upon curing conditions and the properties of the particular cement involved. According to French (1991a), there is a direct correlation between degree of hydration and original water/cement ratio for concretes made at around 20°C and cured for just a few days: he found cement paste to contain 7 or 8 per cent unhydrated particles at 0.4 water/cement ratio, reducing to 3 or 4 per cent at 0.5 water/cement ratio and virtually none (i.e., completely hydrated) at 0.6 water/cement ratio. However, these guidance values change at significantly lower or higher temperatures during setting and hardening and also relict cement clinker grains are able progressively to hydrate over a period after hardening when there is continuous or intermittent exposure to moisture.

The practical usefulness of these mineralogical factors, Portlandite and degree of hydration, for indicating original water/cement ratio is therefore limited, although they might be helpful indicators when taken together with other features, such as the amount of bleeding and the level of capillary porosity. Original water/cement ratio is most frequently estimated for hardened concrete using the physico-chemical method (see Section 4.4.3), but fluorescence microscopy (see Section 4.4.4) has some advantages, as well as some pitfalls, and is a useful additional aid to the concrete petrographer.

4.4.3 DETERMINATION OF WATER/CEMENT RATIO BY THE PHYSICO-CHEMICAL METHOD

The physico-chemical method for determining original water content and water/cement ratio is detailed in BS 1881:Part 124 (1988) and described in Figg (1989). It is mentioned briefly here as the method is frequently carried out to augment the petrographical examination for concrete composition.

To establish the original water/cement ratio it is clearly necessary to determine the respective proportions of cement and water in the original mix. The cement content may be estimated from a partial chemical analysis, particularly using the values obtained for soluble silica and acid soluble calcium, providing the aggregates do not contain acid soluble calcium. The method is detailed in BS 1881: Part 124 (1988). The original water content is estimated from determined values of combined water of hydration and water in excess of that required for hydration. Combined water is determined by measuring the amount of water driven off from a pre-dried powdered sample of the concrete by ignition. Excess water is calculated from the volume of capillary pores, which is in turn determined by the vacuum-saturation of a dried slice specimen with a liquid of known density (1:1:1 trichlorethane).

According to Figg (1989), in the case of sound, uncarbonated portions of concrete, within the normal range of mixes containing 200 to 500 kg/m^3 of cement, with low-porosity aggregates and no cement replacement materials, the accuracy of the result is likely to be within ± 0.05 of the actual water/cement ratio and the reproducibility (R) is of the order of ± 0.05 for concretes with water/cement ratios in the range 0.4 to 0.8. However, these criteria may not always apply and, in practice, the determined values of original water/cement ratio by this method should be treated with considerable circumspection, even when obtained by experienced analysts.

Moreover, the method produces an averaged finding for a relatively large slice specimen (typically a 20 mm thick slice across a 150 mm diameter core sample) and is thus insensitive to small-scale variations. The determined water/cement ratio will also underestimate the original water content for concretes affected by excessive bleeding or extensive carbonation, but overestimate the original water content for concretes which have been physically damaged or chemically attacked.

4.4.4   DETERMINATION OF WATER/CEMENT RATIO BY FLUORESCENCE MICROSCOPY

Principle of the method

There is a relationship between the water/cement ratio of a concrete mix and the capillary porosity of the hardened cement paste in the concrete (see Section 4.4.2). Christensen *et al.* (1979) and Thaulow *et al.* (1982) demonstrated that the capillary porosity of a cement paste could be determined, on a comparative basis, by impregnating a suitable sample with low-viscosity resin containing a fluorescent dye and then examining a thin-section under a microscope using ultraviolet illumination. The greater the capillary porosity, the more fluorescent resin is absorbed and the brighter is the fluorescence in ultraviolet light. If reference samples of similar concrete and known water/cement ratio are prepared or available, it is possible, by visual comparison, to allocate an equivalent water/cement ratio to the sample in question.

Transmitted light procedure for determination of equivalent water/cement ratio

Although the technique has become widely used in Denmark, Christensen *et al.* (1979) did not provide any methodological details. The method has been commercially applied in the UK by a number of laboratories and the following description is based upon the Sandberg technical procedure (Sandberg 1987). More recently, St John (1994a) has critically reviewed the method.

A slice specimen of the concrete sample in question is impregnated with a low-viscosity epoxy resin containing either Fluorescein or Hudson Yellow) dye and then used to prepare a conventional petrographical thin-section. The thin-section is then examined under a dedicated fluorescence microscope or, more practicably, under a petrological microscope modified by filters to provide ultraviolet illumination. The intensity of fluorescence of the concrete matrix under the microscope is proportional to the quantity of fluorescent resin that has permeated into the matrix, which in turn is dependent upon the microporosity. As the microporosity of the matrix is associated with the original water/cement ratio of the concrete, in the absence of other significant factors, the intensity of fluorescence is an indirect measure of the original water/cement ratio. By comparison of the test thin-sections with suitable reference specimens of known water/cement ratio, the *equivalent* water/cement ratio of the concrete sample may be estimated to the nearest 0.05 in the range 0.3 to 0.8.

The selection and preparation of the concrete specimens is arguably the most important part of the procedure. If the main uncertainty concerns the *overall* water/cement ratio, to ensure representativeness it will be necessary to examine a range of specimens, taken from different parts of the concrete sample or unit under investigation and with various orientations to the original direction of placement. Otherwise, the technique is particularly useful for studying local variations in water/cement ratio and, for example, often the differences between the near-surface zones and the interior of the concrete will be especially relevant to durability. However, the method will yield misleading equivalent water/cement ratio values for concrete areas which have been carbonated or otherwise altered in a manner likely to have affected the microporosity.

It is particularly critical to ensure that the specimen is impregnated by fluorescent resin to the maximum extent permitted by the microporosity. For this reason, Sandberg (1987) found that it was necessary to employ a two-stage impregnation procedure: firstly a thorough impregnation of the slice specimen and secondly a further impregnation of the ground surface prepared for making the thin-section. St John (1994a) also found incomplete impregnation to be a potential limitation of the method, especially in the case of the denser cement pastes which derive from lower water/cement ratios (say less than 0.4) or blends of cement with mineral additions (e.g., ggbs, pfa, microsilica). Mayfield (1990) overcame this difficulty for the reflected light method (see below) by using an alcohol/dye mixture, but this mixture cannot be used in a concrete thin-section.

Concrete thin-sections are usually prepared to a thickness in the range 20 μm to 30 μm, with 25 μm being optimum but geological laboratories are typically standardised on 30 μm. St John (1994a) found that small variations in thin-section thickness had a roughly proportionate influence on the intensity of fluorescence under the microscope (i.e., a 3 μm variation in thickness changed the intensity by around 10 per cent), so that consistency in thin-section thickness was essential, both within and between specimens.

Unless a special microscope is available, most concrete petrographers will find it convenient to make temporary adjustments to the standard petrological microscope. The microscope must be fitted with a 100 watt quartz-halogen or tungsten-halogen light source, set on maximum setting. A blue excitation filter, such as a Leitz Type BG12, is installed to cover the emergent aperture of the light source and a yellow barrier filter, such as a Leitz Type K510, is inserted as an accessory plate, to detect the broad peak of yellow fluorescence of around 550 nm emitted by the dye. Both the polariser and analyser are removed from the light path and the condenser adjusted to give Köhler illumination. It is usually necessary to carry out this microscopy in a semi-darkened room. The setting of the aperture diaphragm is critical and should not be changed during measurements.

On examining the thin-sections using this system, all regions invaded by fluorescent resin will show a yellow colour, the brightness or intensity being controlled by the quantity of resin absorbed by that part of the specimen. Non-opaque materials which have low microporosity will show as blue in colour. Cracks, microcracks and voids will exhibit maximum intensity of fluorescence. By concentrating attention on the cement paste regions, it will be possible to identify areas relatively exhibiting the maximum, minimum and modal average (i.e. most frequent) degrees of fluorescence. If suitable reference sections are available (see below), it will be possible, on a comparative basis, to allocate maximum, mimimum and overall *equivalent* water/cement ratios to these areas.

The estimation of *equivalent* water/cement ratios should only be attempted when a full range of relevant reference thin-sections is available for the concrete under investigation. Such reference sections should be prepared from laboratory-prepared concrete castings, wherein the type of cement, any mineral addition, aggregates and any admixtures are essentially similar to the concrete in question. A sequential range of mixes should be prepared in which the water/cement ratios are strictly controlled in the range from 0.3 to 0.8. Visual comparisons of these reference thin-sections with those prepared from the concrete sample, using the same system at the same time, enables the closest similarities of fluorescence to be established, including intermediate values, and yielding *equivalent* water/cement ratios to the nearest 0.05.

Photomicrographs illustrating some typical appearances of concrete matrices under fluorescence microscopy are shown in Plate 18. Theoretically, in well-mixed concrete, the degree of fluorescence should be reasonably uniform within the cement paste regions, but a patchy appearance is quite common, indicative of localised uneven distribution of cement and water in the mix. However, if the specimen is not fully impregnated and of consistent thickness this can give rise to a similar patchiness. Local areas of apparently higher water/cement may be associated with bleeding and near-surface areas will frequently exhibit enhanced microporosity when not carbonated.

St John (1994a) confirmed the usual presence of a gradation of fluorescence that could be related to water/cement ratio, but considered that estimates outside the range 0.4 to 0.6 were less reliable. Even within the 0.4 to 0.6 range he thought that values were more realistically gauged to the nearest 0.1, rather than 0.05. At water/cement ratios exceeding 0.6, St John found that the fluorescence emission reached saturation unless the concentrations of dye in the resin had been reduced accordingly (which would also be required in the reference specimens), whereas at ratios below 0.4 it was difficult to be confident that complete impregnation had been achieved. It was also noticed that the proportions of Portlandite and unhydrated clinker grains influenced the appearance of the fluorescence.

Notwithstanding the difficulties involved in obtaining reliably quantified *equivalent* water/cement ratios by this method, the procedure is a valuable addition to the concrete petrographer's armoury, providing the potential limitations are recognised and especially in view of the present lack of an entirely satisfactory alternative technique. The fluorescence microscopy method is particularly useful for detecting small-scale variations in water/cement ratio and has the advantage of being completely complementary to the routine petrographical examination of concrete (using the same thin-section if prepared appropriately and the same microscope assembly). Elsen *et al.* (1995) have started to develop a procedure for conducting the examination phase of this method using automated image analysis techniques.

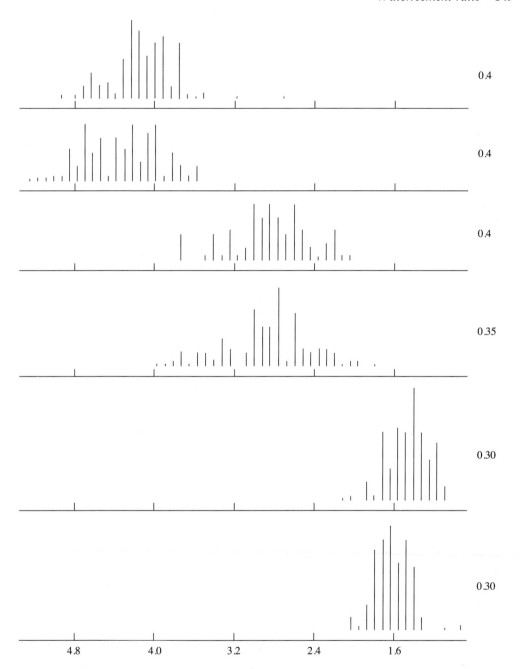

Fig. 4.25 Histograms of emitted fluorescent light in reflected light for cement pastes of various water/cement ratios (from Mayfield 1990).

Reflected light procedure for determination of equivalent water/cement ratio

Mayfield (1990) and his research team have partially developed an alternative fluorescence microscopical method using reflected light with ground and polished specimens. In this

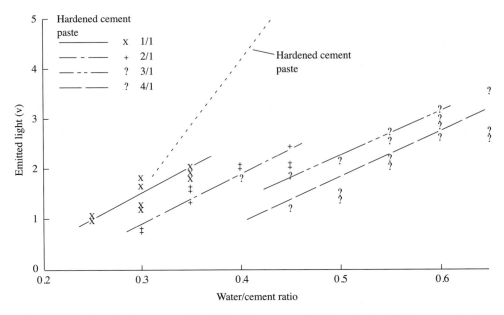

Fig. 4.26 Relationship between emitted light fluorescent light in reflected light, aggregate/cement ratio and water/cement ratio for mortars (from Mayfield 1990).

technique, the concrete specimen, after grinding and polishing, is vacuum impregnated with 10^{-4} molar Fluorescein sodium in alcohol, which permeates the micropore system more easily and thus more reliably than low-viscosity resin. The specimen is then examined using a reflected light (incident-light) microscope, modified in a similar way to the petrological microscope used in the thin-section technique, and estimates of equivalent water/cement ratio are similarly determined by comparison with reference specimens.

Instead of depending upon subjective visual appraisal, Mayfield (1990) demonstrated that it was feasible to quantify the emitted fluorescent light using a photodiode attached to the microscope, an A-D converter and a suitably programmed computer. In this way, he was able to show clear separations in the histograms produced for three different water/cement ratios (0.30, 0.35 and 0.40) when using hardened cement paste specimens (Fig. 4.25).

Mayfield encountered some problems with mortar specimens, because of the dark areas caused by sand grains and unhydrated cement clinker particles: there was a need to correct for different aggregate/cement ratios. Once the sand aggregate content was taken into account, he recorded a relationship between the voltage of emitted light and water/cement ratio (Fig. 4.26). It seems quite likely that the technique could be further refined and perhaps successfully applied to concretes. Wirgot and Van Cauwelaert (1994) have developed a method for measuring fluorescence in the reflected light procedure using image analysis coupled with automatic statistical interpretation.

4.5   Air-void content and air-entrainment

4.5.1   TYPES OF VOIDS IN CONCRETE

A fresh concrete mix is typically compacted, by vibration or ramming, in order to improve the packing of the aggregate and to minimise the residual content of *entrapped* air. Yet, however efficiently the material has been compacted, most concrete inevitably retains some entrapped air. In fresh concrete, such entrapped air forms bubbles which range in size from microscopic spheres the size of cement clinker grains to irregularly shaped gas pockets the size of coarse aggregate particles or larger. Surface-active admixtures may be included in the mix to facilitate the formation of *entrained* air bubbles. After hardening of the concrete, these entrapped and entrained bubbles are fixed in place and become air voids. If a concrete has been under-compacted, or the proportion of matrix is low relative to that of the coarse aggregate, the larger entrapped air voids may be abundant and interconnected, when the concrete is said to be *honeycombed*; this tends to occur in localised areas. Air-void definitions are given in Table 4.15.

Table 4.15 Air-void system definitions (after Figg 1989).

Term	Definition
Air void	A small space enclosed by the cement paste in concrete and occupied by air. This term does not refer to capillary or other openings of submicroscopical dimensions or to voids within particles of aggregate.
Entrained air voids	Air voids characteristically spherical in shape between 10 μm and 1 mm in diameter and should have a regular distribution in the concrete.
Entrapped air voids	Air voids mostly over 1 mm in diameter, typically irregular in shape and usually having an irregular distribution in the concrete, often increasing in size and number towards the surface. Sometimes referred to as compaction voids.
Water voids	Filled with water at the time of setting of the concrete, water voids are irregular in shape, generally elongated and usually very large (several mm in size). Typically found beneath particles of coarse aggregate or reinforcing bars and their interior surface has a granular appearance instead of the usually glazed lustre of air voids.
Gas voids	Formed by the liberation of gas from the reaction between special admixtures and the cement. Their size and shape are variable.
Honeycombing	Interconnecting large entrapped air voids arising from inadequate compaction or lack of mortar.
Air-void content (A)	The proportional volume of air voids in concrete expressed as a volume percentage of hardened concrete.
Specific surface (α)	The surface area of the air voids expressed as mm^2 per mm^3 of air void volume.
Spacing factor (L)	An index related to the maximum distance of any point in the cement paste from the periphery of an air void, expressed in μm or mm. In practice, this is approximately equivalent to half the mean air void spacing within the cement paste.
Air/paste ratio (A/p)	The ratio of the volume of air voids to the volume of the cement paste in the hardened concrete.
Paste content (p)	The proportional volume of the cement paste in concrete, expressed as a volume percentage of the hardened concrete.

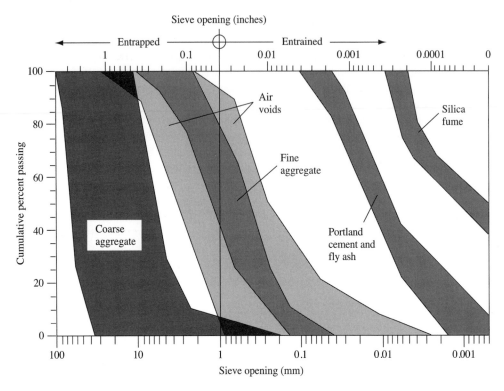

Fig. 4.27 Air-void gradation, compared with aggregates, cement and some additions (from Hover 1993a).

Hover (1993a) has shown that the air void system in a concrete can be represented as a void size distribution in a manner analogous to aggregate gradings (Fig. 4.27). In this way, entrapped air voids might exhibit a grading intermediate to the coarse and fine aggregates, whereas entrained air voids will typically produce a grading equivalent to very fine sand.

Regardless of size or the presence of admixtures, all air bubbles get into fresh concrete by the tumbling and folding actions of mixing and placing (Hover 1993a). The surface-active admixtures serve to stabilise and retain more of the smaller bubbles trapped during mixing and agitating. Thus the so-called *entrained* air voids are actually only an enhanced variety of *entrapped* air voids. Nevertheless, it is convenient, from a practical performance standpoint, to distinguish between entrapped and entrained air voids in concrete. Aerated concrete is a special case in which abundant gas voids are generated either by chemical reactions between additions and the cement or by the use of foaming agents.

Entrapped air voids

Entrapped air voids are typically irregular in shape and, as shown in Fig. 4.27, mainly greater than 1 mm in diameter, although any irregularly shaped smaller air voids might also be taken to be entrapped rather than entrained. The size distribution and disposition of entrapped air voids is also frequently markedly irregular within the concrete, commonly with a tendency for increases in both size and number towards surfaces (especially horizontal surfaces, as-cast).

The content of entrapped air voids should be reduced as the degree of compaction improves for any concrete mix, but virtually *all* normal concretes can be expected to contain at least a small amount of entrapped air. In a non-air-entrained concrete which is *very* well compacted under ideal conditions (as in a laboratory-prepared test cube for example), the entrapped air void content will typically be of the order of 0.5 per cent. On this basis, any air content additional to 0.5 per cent has been termed 'excess voidage' in the UK Concrete Society technical report on core testing (Dewar 1976) and may be used as a comparative measure of concrete quality.

The amount of entrapped air which might be expected in a well compacted *site* concrete depends on a number of factors, but in general contents up to around 3.0 per cent are not unusual. Contents of entrapped air voids substantially in excess of 3.0 per cent, in all or parts of a concrete, might well be symptomatic of important mixing and/or placing deficiencies. The US Bureau of Reclamation (1955) has shown that the content of entrapped air voids in a well compacted concrete varies with maximum aggregate size, being approximately 1.0 per cent, 2.0 per cent and 3.0 per cent air for 40 mm ($1\frac{1}{2}$ inch), 20 mm ($\frac{3}{4}$ inch) and 10 mm ($\frac{3}{8}$ inch) nominal maximum aggregate sizes respectively. In honeycombed areas of concrete, the air void content will probably range between 10 per cent and 30 per cent.

Entrained air voids

Entrained air voids are characteristically spherical in shape and, as shown in Fig. 4.27, mainly range in diameter between 10 μm and 1 mm. Also, in contrast to the entrapped air voids, the entrained air voids usually exhibit a sensibly uniform distribution throughout the concrete matrix, except near to concrete surfaces (see below) and for poorly mixed concrete.

Air-entrainment in concrete is achieved by the addition of organic surfactants, which, according to Usher (1980), 'lower the surface tension of water and facilitate bubble formation ... uniform dispersion and stability are achieved by the mutual repulsion of the negatively charged air-entrainer molecules and the attraction of the air-entrainer molecules for the positive charges on the cement particles'. These surfactants include neutralised wood resins, such as vinsol resin, animal and vegetable fats and their salts, and wetting agents such as alkyl aryl sulfonates, alkyl sulfates or phenol ethoxylates.

The factors affecting the success of air-entrainment are examined in Brown (1983) and Neville and Brooks (1987). Air content can be increased or decreased by variations in *inter alia* temperature during mixing, mix workability, sand content and/or grading, cement fineness and/or alkali content, the presence of other organic materials and/or mineral additions (e.g., pfa), and even 'hardness' of the mix water (see Table 4.16). The influence of these factors varies for different types of air-entraining admixture, for example, the air void system created by vinsol resin is disrupted by high alkali contents, whereas that deriving from sulfonated hydrocarbon is unaffected (Pistilli 1983).

Controlled air-entrainment of concrete has a number of potential benefits. One anonymous article in a concrete magazine perceived air-entrainment as a 'short cut' to durable concrete (Anon 1976), although that author was careful to add that it could not replace correct procedures in mix design or concreting. The beneficial and detrimental effects of entrained air, variously on fresh and hardened concrete, is thoroughly considered by Sutherland (1974) and reviewed in summary form by Usher (1980), Brown (1982) and Malisch (1990). Broadly, other things being unchanged, in fresh concrete entrained air increases workability, improves cohesion and reduces bleeding. In hardened concrete, the

Table 4.16 Factors influencing the entrainment of air in concrete (modified from Brown 1983 and Neville and Brooks 1987).

Type of factor	Increasing air content	Decreasing air content	Example change	Estimate of effect (for 5% air content)[1]
Cement	Decrease in cement content[2]	Increase in cement content[2]	+ 50 kg/m³	0.5% reduction
	Decreased cement fineness	Increased cement fineness	—	—
	Increased alkali content (0.8%)	Decreased alkali content	—	—
Aggregate	Sand grading coarser	Sand grading finer	Zone 3 to 2 (see Table 12)	<0.5% increase
	Sand content increased	Sand content decreased	35 to 45%	1% to 1.5% increase
	Increased 600 μm to 300 μm fraction	Decreased 600 μm to 300 μm fraction	—	—
	Decreased −150 μm fraction	Increased −150 μm fraction	+50 kg/m³	0.5% reduction
	Inclusion of organic impurity		Inclusion	Positive and negative effects
Additions and admixtures	—	Inclusion of pfa	—	—
	—	Greater than usual pfa carbon content	Inclusion	Significant
	Water-reducing admixture usage[3]	—	—	—
	Positive dispensing tolerance	Negative dispensing tolerance	±5%	±0.25%
Mix water	—	Increased hardness	—	—
Mix and mixing	More workable (i.e., higher slump)	Less workable (i.e., lower slump)	50 to 100 mm	1% increase
	Increased mixing efficiency[4]	Decreased mixing efficiency[4]	—	—
	Faster mixer rotation	Slower mixer rotation	—	—
	—	Prolonged mixing during transportation	1 hour	Up to 0.25% reduction
			2 hours	1% reduction
Environment	Lower temperature	Higher temperature	10°C to 20°C	1% to 1.25% reduction
	—	Steam-curing	—	Incipient cracking caused by expansion of bubbles

Notes
1   These effects are only indicative and are not necessarily cumulative.
2   Inclusive of sand content adjustment.
3   Influence of superplasticisers is less clear.
4   Optimum mixing time is required: too short a time causes a non-uniform distribution of the bubbles, whilst over-mixing gradually expels the bubbles.

prospects for durability are generally improved by a reduction in permeability, with resistance to freezing and thawing cycles and de-icing chemicals being notably enhanced.

The principal disadvantage of air-entrainment is the reduction in density and consequent proportional lowering of strength if there are no other changes to the mix design. However, in practice within the normal range of air contents used, the increased workability brought about by air-entrainment usually allows the water/cement ratio to be reduced, which largely compensates for the potential loss of strength. The addition of a plasticising admixture to the concrete mix, in combination with the air-entrainer, is also used to reduce further the water/cement ratio in order to combat strength loss (Hodgkinson and Rostam 1991).

Air-entrainment adequacy for various exposure conditions and with different coarse aggregate sizes is usually specified by placing limits on the *total* air content of the concrete (i.e., the sum of entrapped and entrained air voids). In the UK, total air content limits are recommended for concrete exposure to freezing and thawing and de-icing salts in BS 5328 (1990) and limits are specified by the Department of Transport in respect of highway pavements (Department of Transport 1991). Hover (1993d) has recently reviewed the specifications for concrete air content of a number of authorities in the USA. A summary of some standardised air content limits is given in Table 4.17. As a guide, these various limits indicate that, in a concrete made with nominal 20 mm aggregate, a total air content of around 5.5 per cent or 6 per cent should provide an appropriate level of protection.

Table 4.17 Some British and American recommended air contents for freeze-thaw resistance (modified from BS 5328: Part 1 1990 and Hover 1993d).

Authorities:	ACI 201.2R[1], ACI 211.1[1]., ACI 318[1] ACI 345R[1], ASTM C94[1,2]		ACI 301	BS 5328: Part 1
Conditions	Severe exposure	Moderate exposure	Destructive exposure	Freezing and thawing actions under wet conditions, when concrete grade is lower than C50
Nominal max. aggretate size mm	Mean total air content by volume %			
75	4.5	3.5	1.5–4.5	—
50	5	4	2.5–5.5	—
40	—	—	—	4.5
37.5	5.5	4.5	3–6	—
25	6	4.5	3.5–6.5	—
20	—	—	—	5.5
19	6	5	4–8	—
14	—	—	—	6.5
12.5	7	5.5	5–9	—
10	—	—	—	7.5
9.5	7.5	6	6–10	—

Notes
1   A tolerance of ± 1.5 per cent air content is recommended or specified.
2   Also gives recommendations for mild exposure: 1.5 per cent less than those indicated for moderate exposure with aggregates up to 25 mm, and 2 per cent less with aggregates >25 mm.

Although the addition of an air-entraining admixture will obviously produce air bubbles throughout the mix, in actuality it is only the zone proximate to the exposed concrete surface where the air void system is critical. Sandberg and Collis (1982) found this was the very zone in which air was most likely to be lost during compaction and finishing, in addition to any air lost during transportation of the concrete. In investigating some airfield and road pavement failures in which surface damage had occurred despite the apparent adequacy of total air contents measured on the fresh concrete, Sandberg and Collis found that up to 2 per cent air could be lost in transit and up to a further 2 per cent air was sometimes lost during compaction and finishing. Thus the final air content in the crucial uppermost zone of concrete might be inadequate for the prevention of freeze-thaw damage.

Yingling *et al.* (1992) have investigated the loss of air in pumped concrete. It was found to be common to lose up to about 1 per cent air and 25 mm of slump during pumping, but certain types of pump arrangement could occasionally lead to the loss of half the air content. The loss of air during handling and placing of concrete has recently been reviewed by Hover (1993b), who makes a distinction between the relatively unimportant loss of larger voids, which might significantly affect the total air content, and the much more serious loss of the effective finer voids.

The use of some superplasticisers may destabilise the entrained air void system. MacInnis and Racic (1986) determined that a nominal 5 per cent air content was reduced on average to 3 per cent, after the addition of a superplasticiser, and a 7 per cent air content was reduced to 4 per cent. Saucier *et al.* (1991), in summarising an extensive programme of investigation into air void stability, confirmed *inter alia* the occasional incompatibility between superplasticising and air-entraining admixtures. Also, according to Gutmann (1988), some types of air-entraining agents are more prone to the coalescence of bubbles, creating larger air voids and poorer performance.

In addition to the proportional content of air voids by volume, it has long been recognised that the air void *system*, especially the *spacing factor*, is also of critical importance to the effectiveness of air-entrainment in ensuring frost resistance. Powers (1949b) first proposed the significance of the average inter-bubble distance, or 'spacing factor', on the basis of his saturated flow hydraulic pressure hypothesis to explain the empirical observation that air-entrainment of concrete improved the resistance to freezing and thawing. The spacing factor is defined in ASTM C457 (1990) as being an index related to the maximum distance of any point within the cement paste from the periphery of an air void, which is approximately half the average air void spacing. A maximum spacing factor of 0.2 mm (0.008 inch) is often specified (Neville and Brooks 1987, Hover 1993d).

Powers's spacing factor has been frequently criticised, but continues to be accepted as a useful parameter in assessing the likely effectiveness of an air void system. The Powers spacing factor is calculated from the total air void content, but Walker (1980) proposed the differentiation of entrapped and entrained voids and the calculation of a spacing factor for the small entrained air voids only. Philleo (1983) suggested a complicated alternative form of spacing factor, in an attempt to overcome variations in the air void system within a concrete which he suggested might be more important in determining performance. Chatterji (1984) has observed that, whilst Powers's theory was based upon a model in which all the air voids are equal-sized spheres arranged in a simple cubic lattice, this is 'seldom satisfied in practice'. However, Chatterji also refuted the saturated flow hydraulic pressure hypothesis and challenged the 'physical and scientific relevance' of the spacing factor.

Routine monitoring of air content during construction is usually carried out on samples of fresh concrete, commonly using a pressure meter of the type described in BS 1881:Part 106 (1983) or ASTM C231 (1991b). However, such a method is incapable of distinguishing between entrapped and entrained air and obviously gives no information on either the size or spatial distributions of air voids in a concrete after placing and finishing. Moreover, it is quite likely that the air content of concrete will be significantly, perhaps critically, modified between sampling from the mixer or delivery vehicle and its final placement and consolidation (Sandberg and Collis 1982, Hover 1983c). For these reasons, the assessment of air content for suitable samples taken from the hardened concrete is essential to quantify the air void system.

4.5.2 QUANTIFICATION OF AIR-VOID CONTENT IN HARDENED CONCRETE

A number of methods have been devised for evaluating the air void content of hardened concrete, but the direct microscopical measurement techniques, such as those described in ASTM C457 (1990), are the most precise and, given appropriate samples arguably provide the most useful information. French (1991a) considered the ASTM C457 method to be 'far more reliable' than the other options. In the following account, these other options are briefly considered first, then the microscopical technique is covered more fully in Section 4.5.3.

Visual assessment of excess voidage

Air-voids are clearly visible in the sides of a concrete core sample and any routine description should include reference to the apparent abundance and size of such voidage, as an indication of the success or otherwise of the compaction. In air-entrained concretes, the presence of microscopic bubbles in the concrete matrix might also be apparent when they are relatively abundant.

The Concrete Society working party convened by Dewar (1976) developed standardised comparative techniques for evaluating the visible voids in concrete core samples, based on the notion that total air void contents (for non-air-entrained concretes) greater than 0.5 per cent were excessive. This 'excess voidage' is determined either by visual means or from the density test results. In the visual assessment, standard areas of core side surface (125 × 80 mm) are subjectively compared with actual-size photographs provided for five different levels of excess voidage (in the range 0 per cent to 13 per cent). Two of these are reproduced in Fig. 4.28 for illustrative purposes.

In the density method, the actual density of the concrete, measured from core samples, is compared with a higher target density, with the difference being taken to indicate void content. In the Dewar (1976) procedure, the target 'potential density' represents that of a properly compacted and cured test cube at 28-days when measured by water displacement; the 'actual density' is based upon water-soaked cores. The method currently favoured by the UK Department of Transport (1991) for pavement works, is instead based upon a theoretical maximum *dry* density as the target, calculated from the mix proportions and constituent relative densities, and determining the dry density of cores. A worked example of the Department of Transport approach is given in Table 4.18.

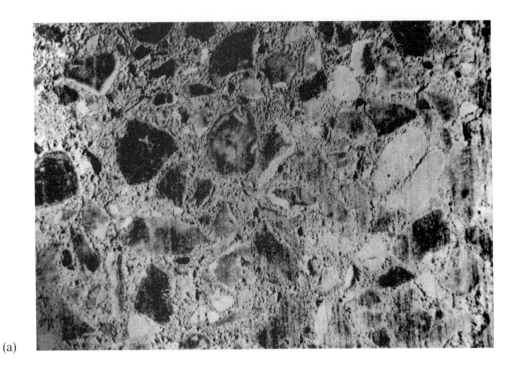

(a)

(b)

Fig. 4.28 Two photographs used for estimating 'excess voidage': (a) 0 per cent (b) 3 per cent (from Dewar 1976).

Table 4.18 Void content estimation from density measurements (worked example of the method given in UK Department of Transport 1991).

The formula given for the Theoretical Maximum Dry Density (TMDD) in DoT (1991) is:

$$TMDD = \frac{[(F \times W_1) + W_3 + W_4] \times 1000}{\dfrac{W_1}{P_1} + \dfrac{W_4}{P_4} + \dfrac{W_3}{P_3} + W_2}$$

where F = time fctor for hydration of cement (given in table in DoT (1991)
  W_1 = mass of cement, kg
  W_2 = mass of total water (in aggregate + added), kg
  W_3 = mass of oven dry fine aggregate, kg
  W_4 = mass of oven dry coarse aggregate, kg
  P_1 = relative density of cement
  P_3 = apparent relative density of fine aggregate
  P_4 = apparent relative density of coarse aggregate

Using data obtained for the concrete in question, the TMDD was calculated as follows:

$$TMDD = \frac{[(1.19 \times 350) + 702 + 1162] \times 1000}{\dfrac{350}{3.12} + \dfrac{1162}{2.52} + \dfrac{702}{2.56} + 150.5} \qquad \text{I}$$

$$= \frac{[416.5 + 702 + 1162] \times 1000}{112.2 + 461.1 + 274.2 + 150.5} \qquad \text{II}$$

$$= \frac{2280.5 \times 1000}{998} \qquad \text{III}$$

$$= 2285 \text{ kg/m}^3$$

This TMDD is then compared against the measured dry density for samples of hardened concrete and the difference taken to be the air void content, as follows:

Calculated theoretical maximum dry density:	2285 kg/m^3
Less dry density measured from samples:	2184 kg/m^3
Difference in kg/m^3:	101 kg/m^3
Difference as % air voids:	4.4 %

Although the density difference methods might appear to produce precise and objective results, this is illusory even for a representative set of core samples, because of the likely errors in establishing the target density. Apart from the relative density values to be chosen for each of the various constituents, it is necessary to assume that the specified mix design has been precisely complied with throughout the concrete being evaluated. In Figg (1989), it is suggested that the overall error may be unreasonably large for non-trial concrete wherein there is uncertainty over composition.

Other methods for assessing air void content

Erlin (1962) proposed a high-pressure method for the determination of total air content in hardened concrete, but, despite being a direct method, this does not seem to have been adopted in practice. The procedure is particularly sensitive to changes in aggregate porosity and, in common with the pressure meter for fresh concrete, cannot differentiate between entrapped and entrained air.

The nature and objectives of an air void system analysis of concrete seem, at first sight, to be ideally suited for modern computerised image analysis. In practice, it has proved difficult to devise an entirely satisfactory specimen preparation technique. If a slice of concrete can be treated to create a sufficiently clear and invariable colour contrast between the air voids and all of the surrounding material, it is possible for a suitable image to be fully processed by a computer, effortlessly yielding all of the required parameters. Procedures have been described by Chatterji and Gudmundsson (1977) and MacInnis and Racic (1986); both consist of blackening the concrete and infilling the voids with bright white material. If satisfactory specimen preparation can be achieved, it is claimed that results can be obtained by image analysis which compare favourably with those found using the ASTM C457 (1990) microscopical method.

4.5.3   MICROSCOPICAL MEASUREMENT OF THE AIR-VOID SYSTEM

The procedures detailed in ASTM C457 (1990) are adopted internationally and form the basis of the following discussion. Specimens are prepared from the concrete sample by cutting a slice using a diamond saw and then progressively grinding one face of the slice to a smooth and flat surface. The finely ground specimen surface is then scanned under a good quality optical travelling microscope, using one of the two techniques described, linear-traverse or modified point-count. The observations collected during these scans are then used to calculate various parameters of the air void system, including total air content and spacing factor.

Adequate preparation of the ground surface is critical. The precise details for grinding with progressively finer abrasives are given in ASTM C457, but the finished product must show 'excellent reflection' of a remote light source at low-angle, be free of scratches and maintain sharp edges around the air voids. Consolidation might be necessary for weaker concretes, to avoid the plucking out of aggregate particles (creating false voids) and/or the crumbling of void edges, but this must not be of a sort that will infill the voids: ASTM C457 describes a molten wax technique. Figure 4.29 illustrates a suitable surface when viewed through the microscope.

Both scanning methods envisage a series of parallel traverse lines across the specimen surface. In the linear-traverse option, the distance travelled across each component encountered as these lines are traversed is measured and cumulatively totalled, also the number of air voids intersected by the traverse line is recorded. The data collected are total length traversed (T_t), total length traversed across cement paste (T_p), total length traversed across air voids (T_a) and total number of air voids intersected by the traverse line (N). Aggregates are not measured as such, but obviously could be obtained by deducting T_p plus T_a from T_t.

In the 'modified' point-count option, the traverses proceed step-by-step and the identity of the component under each stop or point reached is identified and added to the tally for that component. Again, the total number of air voids intersected by the traverse line is also recorded and this represents the modification from conventional point-counting. The linear distance between stops along the traverse (I) has to be determined (it is required to be between 0.64 and 5.1 mm). The data collected are total number of stops or points (S_t), total number of points over cement paste (S_p), total number of points over air voids (S_a) and total number of air voids intersected by the traverse line (N). The separate counting of other components, such as aggregates, or their differentiation, for example between coarse and fine aggregates, can be achieved by adding to the range of tally counters.

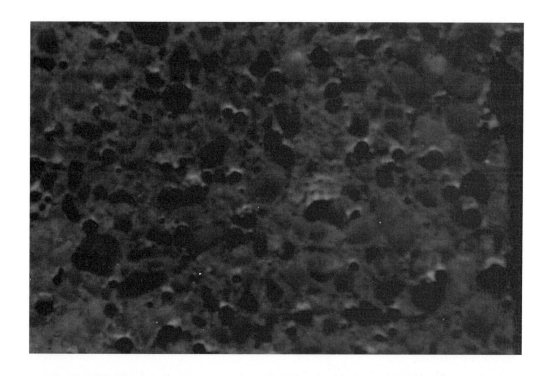

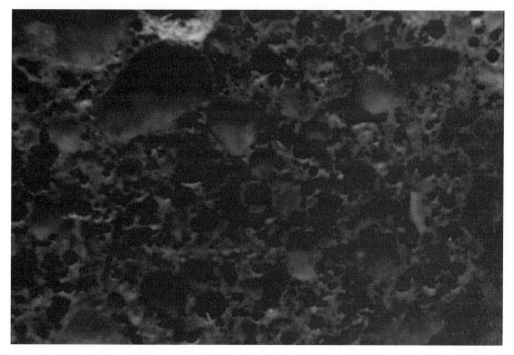

Fig. 4.29 Two examples of suitable concrete surface preparation for air-void point-counting to ASTM C457.

Research has suggested that the two methods produce generally comparable results, but the modified point-counting system is preferred by many laboratories. Computer processors have been successfully linked to the counting devices, thus automating the various calculations given in ASTM C457 (1990) to establish the air void system parameters.

The stipulated minimum area of specimen surface varies with the maximum aggregate size and the mode of calculation. For example, for a nominal 20 mm (19.0) aggregate and employing the standard method of calculation, a minimum area of 7100 mm² is cited; the area of a diametral slice from a 150 mm diameter concrete core is thus more than sufficient. In the case of concretes containing an aggregate with particles in excess of 50 mm in size, when minimum specimen sizes can get impracticable, an alternative calculation procedure based upon the paste air ratio is available, providing the volumetric aggregate/paste ratio is known or has been determined. Table 4.19 gives the minimum lengths of traverse and minimum number of points specified in ASTM C457, which again vary according to maximum aggregate size. In the authors' experience, larger sample areas than those stipulated in ASTM C457 need to be examined when the concrete is unevenly mixed.

Table 4.19 Minimum lengths of traverse and number of points for the ASTM C457 'modified' point-count option (after ASTM C457 1990).

Nominal or observed maximum size of aggregate in the concrete mm	Minimum length of traverse for determination of A, α or $\bar{L}$ mm	Minimum number of points for determination of A, α or $\bar{L}$
150	4064	2400
75	3048	1800
37.5	2540	1500
25	2413	1425
19	2286	1350
12.5	2032	1200
9.5	1905	1125
4.75	1397	1000

Note
According to ASTM C457, these limits should produce a standard deviation not greater than 0.5 per cent for a 3 per cent air content (a coefficient of variation of 17 per cent); the coefficient of variation is reduced for air contents >3 per cent and number of counts >1000.

Formulae are provided in ASTM C457 for calculating air content (A), void frequency (n), paste/air ratio (p/A), specific surface (α) and spacing factor ($\bar{L}$). It has already been mentioned in Section 4.5.1 that Walker (1980) suggested an alternative formula for spacing factor, relating only to the smaller voids. Attiogbe (1993) has devised another formula for calculating 'mean spacing', which he suggests yields a better estimate of the actual spacing of the air voids in concrete than the standard spacing factor.

The statistical reliability of the results of the ASTM C457 test depends on many factors. Hover (1993c) contrasted the 1000 or so air voids evaluated in an average test, with the 10 to 15 billion air voids in a cubic yard of air-entrained concrete and concluded, 'uncertainty or error is expected in such estimates'. The procedure may be expected to be subject to the reasonably predictable statistical errors associated with any counting technique, but in addition will be affected *inter alia* by the quality of the specimen prepared, the apparatus, the magnification selected and the performance of the operator. A major interlaboratory trial, involving twenty laboratories in ten countries, was reported by Sommer (1979) and the precision findings are summarised in Table 4.20.

Table 4.20 Precision of determinations of air void contents and spacing factors (after Sommer 1979) and Figg 1989).

Air-void content, A, %

Concrete:	I		II	
	Prepared and measured in Vienna	Prepared and measured by participant	Prepared and measured in Vienna	Prepared and measured by participant
Number of results	20	18	20	18
Average	2.92	2.73	6.48	5.87
Standard deviation, s	0.34	0.75	0.61	0.89
Lowest	2.52	1.77	5.44	4.39
Highest	3.58	3.90	7.71	7.74
Range	1.06	2.13	2.27	3.35
Precision, 2.8s	r = 0.95	R = 2.10	r = 1.71	R = 2.49
95% confidence limits (±1.96s)	—	±1.47	—	±1.74

Spacing factor, $\bar{L}(\mu m)$

Concrete:	I		II	
	Prepared and measured in Vienna	Prepared and measured by participant[1]	Prepared and measured in Vienna	Prepared and measured by participant[1]
Number of results	20	15	20	15
Average	168	172	87	98
Standard deviation, s	12	26	8	18
Lowest	148	136	75	66
Highest	192	215	99	132
Range	44	79	24	132
Precision, 2.8s	r = 34	R = 73	r = 22	R = 50
95% confidence limits (±1.96s)	—	±51	—	±35

Note
1   All results from two laboratories excluded as 'outliers'.

The precision found in Sommer's trial was poor. Overall, the concrete prepared with 3.0 per cent air void content produced 95 per cent confidence limits of ±1.5 per cent for total air void content, when the specimens were prepared and measured by the participating rather than organising laboratory. The average spacing factor was 170 μm, with 95 per cent confidence limits of ±50 μm. Figg (1989) considered these values could be improved upon and recommended scanning the maximum possible surface area, using at least the minimum number of points recommended in ASTM C457, carrying out specimen preparation to the highest possible standards and using only a ×50 magnification.

In ASTM C457, it is specifically stated that, 'no provision is made for distinguishing among entrapped air voids, entrained air voids and water voids' and that, 'any such distinction is arbitrary'. Nevertheless, the technique involves direct visual observation and differences between void size and morphology are clearly distinguishable. It is therefore common, and here recommended, for entrapped and entrained air voids to be separately counted, so that the approximate relative proportions of those forms of air void, together comprising the total air content (A), can be calculated. Convenient identifying criteria may be derived from ASTM C125 (1996), wherein the following definitions are given: 'an entrapped air void is characteristically 1 mm or more in width and irregular in shape; an entrained air void is typically between 10 and 1000 μm in diameter and spherical or nearly so'. Some example ASTM C457 (1990) results, including this differentiation of air void types, are given in Table 4.21.

Table 4.21 Microscopical determination of air void content and parameters of the air void system with example data, including differentiation of entrapped and entrained air.[1]

Depth from surface, mm:	20	40	60	80	100	Total/overall
Data determined – ASTM C457						
No. points – Aggregates	325	331	307	301	316	1580
Cement paste	135	123	124	149	126	657
Entrapped air	8	7	18	18	24	75
Entrained air	4	9	16	6	9	44
Total voids	12	16	34	24	33	119
Total points	472	470	465	474	475	2356
No. air voids intersected	15	42	35	24	28	144
Calculations – ASTM C457						
Voids traversed, N	27	58	69	48	61	263
Traverse length, l, mm	1.3	1.3	1.3	1.3	1.3	1.3
Total traverse, Tt, mm	597	594	588	599	600	2978
Entrapped air content, %	**1.7**	**1.5**	**3.9**	**3.8**	**5.0**	**3.2**
Entrained air content, %	**0.8**	**1.9**	**3.4**	**1.3**	**1.9**	**1.9**
Total air content, A, %	**2.5**	**3.4**	**7.3**	**5.1**	**6.9**	**5.1**
Void frequency, n	0.05	0.10	0.12	0.08	0.10	0.09
Paste content, p, %	28.6	26.2	26.7	31.4	26.5	27.9
Paste-air ratio, p/A	11.2	7.7	36	6.2	3.8	5.5
Av. chord length, I	0.56	0.35	0.62	0.63	0.68	0.57
Specific surface, α	7.1	11.5	6.4	6.3	5.8	7.0
Spacing factor, $\bar{L}$, mm	**0.94**	**0.49**	**0.57**	**0.81**	**0.65**	**0.69**

Note
1   The total air content is 5 per cent but the entrained air content is only 2 per cent.

In considering the interpretation of ASTM C457 results, it is important to recognise the objectives of the testing in question, particularly whether the issue is compliance with a specified requirement, or performance of a given concrete surface, or conceivably both. Most, but not all, specification limits refer to pressure meter testing of fresh concrete prior to placement, so that the relevance in that respect of air contents determined later from samples of hardened concrete using a quite different technique is obviously questionable, particularly in marginal cases. It has already been explained in Section 4.5.1 that significant losses of air content can occur between original mixing and final placement.

Orientation of the specimen surface is a vital consideration. The standard requirement of ASTM C457 (1990) is for the slice to be sawn 'approximately perpendicular to the layers in which the concrete was placed' and this will clearly produce an overall average, which is best suited to any investigation which is aimed at checking the likely compliance with the original air content requirements. However, ASTM C457 (1990) is frequently used in practice for investigating failures, when the *overall* air content might be of limited relevance.

Following their experiences with some UK airfield pavements where surfaces had spalled in winter conditions despite compliant site test results for air content and satisfactory overall findings in the ASTM C457 (1990) tests, Sandberg and Collis (1982) observed that, 'the obvious and paramount requirement is that the entrained air is actually in that part of the hardened concrete which has to resist frost action'. They proposed that, 'contractual acceptance should, ideally, be based on the entrapped and entrained air contents in the as-placed and hardened concrete'. This advice has been followed for some contracts, when periodic coring and application of ASTM C457 has been part of the specification requirements.

In such cases, when the air content and the parameters of the air void system in the near-surface zone of concrete are under investigation, it will be more appropriate to prepare slice specimens which are orientated *parallel* to the exposed surface (i.e., parallel to the placement layering in a pavement concrete). Often it will be appropriate to prepare several slices, providing specimen surfaces at various depths beneath the exposed concrete face. It is suggested that the 20 mm thick zone of concrete beneath the exposed surface is the most critical for freeze-thaw resistance, so that separate air void measurements at levels of 5 mm, 15 mm and 25 mm beneath the surface would provide suitable information about the adequacy of frost protection. Although this might be achieved, for indicative purposes, by separately counting traverses orientated parallel to the exposed surface on a slice sawn perpendicular to that surface, the limited number of counts at each depth level is unlikely to meet the statistical minimum requirement for reliability.

Of course, in many investigations, it might be appropriate to prepare specimens variously perpendicular and parallel to the exposed concrete surface, enabling thorough studies of both likely compliance and actual performance potential to be completed.

Sources of apparent contradiction between air content findings and actual performance include the type of air voids and the spacing factor. The pressure meter and density difference test methods only measure total air content, rather than entrained air, and the ASTM C457 procedure also determines only total air void content in its unmodified form. Yet frost protection is only afforded by entrained air, so that an apparently suitably high total air content result that, in reality, comprises mainly entrapped air, would provide misleading reassurance. It is primarily for this reason that the differention into entrapped and entrained air voids in the ASTM C457 test is being recommended, although the abitrary and approximate nature of such a sub-division must not be overlooked.

As well as the presence of adequate entrained air voids, the spacing factor has been generally recognised as having a major influence on frost resistance (see Section 4.5.1), although it is less frequently included in specifications. Examples might therefore be encountered where freeze-thaw damage has occurred with concrete exhibiting an apparently adequate air content, even entrained air content, but an excessive spacing factor. A threshold spacing factor of 0.20 mm is commonly cited (Hover 1993d), below which concrete might be expected to exhibit satisfactory durability. Although this is lower than the 0.25mm value originally envisaged by Powers (1949b), some research has indicated that even lower values might be required to ensure protection where de-icing chemicals are used (Klieger 1980).

Aarre (1995) has compared the use of thin-section specimens with the more conventional finely ground specimens described above in the ASTM C457 method. She found that the 'apparent' air content measured from thin-sections was systematically lower than both that determined from comparable ground surfaces and the true air content. Air entrainment in concrete is further discussed in Chapter 5.

4.6 Mineral additions and pigments

The use of mineral additions as cement replacement or filler materials has already been discussed in Section 4.2.4 in the context of blended cements. In terms of identification and quantification, the mode of incorporation of the addition into the concrete, whether as an ingredient of the ex-factory cement or instead added as a separate constituent, is immaterial. The following sections will concentrate on the methods for determining the presence, type (or types) and relative proportions of mineral additions in hardened concrete.

4.6.1 *FLY ASH AND PULVERISED-FUEL ASH*

The terms 'fly ash' and 'pulverised-fuel ash' are commonly regarded as synonyms, but this is not strictly true. According to Cripwell (1992), 'fly ash is the generic term for all finely divided residues collected or precipitated from the exhaust gases of any industrial furnace', whereas pulverised-fuel ash (pfa) is, 'that class of fly ash that is produced as a by-product specifically from the pulverisation of higher ranking coals'. This distinction can be important when considering the extensive international literature on 'fly ash' in concrete. Only 'Type F' fly ash material, as defined in ASTM C618 (1996), is used in the UK and would comply with the above definition for pfa, whereas in some parts of the world 'Type C' fly ash, produced from low-ranking coals, are prevalent. A petrographer should thus take care in using the term 'pfa' unless the source and/or type of any fly ash present is known.

A number of reviews provide accounts of the properties and potentially beneficial influences of fly ash as a constituent within concrete (Owens 1980, 1982, Berry and Malhotra 1987, Dewar and Anderson 1992). One of the most recent and comprehensive reviews was compiled by a UK Concrete Society working party (Concrete Society 1991). There are rather fewer published studies of the *actual* performance of fly ash in structures. However, Cabrera and Woolley (1985) reported an appraisal of some power station foundations cast in 1957, in which most of the concretes contained a 20 per cent cement replacement by pfa, but some sections lacked pfa but were otherwise similar.

They found the pfa concretes had developed twice the strength of the plain concretes between 1 and 25 years, also that the porosity was much lower for the pfa concrete, which exhibited almost no carbonation. Thomas and Matthews (1993) investigated a number of structures and found similarly improved strength and reduced permeability for pfa concrete. They also identified significantly increased resistance to chloride penetration and alkali-silica reactivity (ASR).

Although fly ash is recognised to be a variably pozzolanic material, the precise mechanisms by which the material achieves its beneficial effects in concrete continue to be the subject of research. It is a matter of simple observation that, even in comparatively old concretes, much of the visible fly ash appears to be unreacted (French 1991a). When the concrete is young and still in its main strength gain phase, the fly ash behaves as an inert material, its principal effect being to provide nucleation centres for cement hydration (Montgomery *et al.* 1981, Fraay *et al.* 1989). At later periods, the fly ash participates in pozzolanic reactions, which account for the higher long-term strengths and improved impermeability. However, these reactions are complex and pozzolanicity appears to depend on a range of factors, including mineralogical composition and particle size range (Halse *et al.* 1984, Jun-yuan *et al.* 1984, Mehta 1985).

Modern concretes frequently contain 10–35 per cent fly ash by weight of the binder, although lesser and greater proportions may be encountered. Increasingly, blends of different additions, such as fly ash with silica fume (Ozyildirim and Halstead 1994), are being used. One recent major crossing structure in Europe included concrete designed to include proportions of pfa, ground granulated blastfurnace slag (ggbs) and microsilica.

Identifying fly ash or pfa

As fly ash, when used in normal amounts, does not substantially modify the colour of concrete, its presence will not usually be apparent to the unaided eye. However, microscopical examination of thin-sections ground to 25 μm or less should enable the presence of any fly ash to be established with reasonable certainty, although quantification is much more difficult. The examination of concrete fracture surfaces by scanning electron microscopy (SEM) is also effective for confirming the presence of fly ash (Thomas and Matthews 1991).

Fly ashes used for concrete additions are by-products of burning carbonaceous fuels (e.g., coal). Although it is common for fly ash to contain a few per cent of residual unburnt fuel, which is a useful characteristic component when identifying the presence of fly ash, the majority of the material comprises the fused product of clay impurities in the original fuel. These clays, such as kaolinite and illite, give rise to the aluminosilicate glass and subordinate crystalline phases which dominate the mineral composition of fly ash. It is probable that the type and proportions of these clays contained as impurities in the original fuel dictate the final pozzolanicity of the fly ash (Hubbard *et al.* 1985).

Fly ash particles range from about 1 μm to 150 μm or so, with up to around 50 per cent of those particles being smaller than the 20–30 μm thickness of the thin-section, so that only the larger and largely unreacted particles are readily observable by optical microscopy. French (1991a) commented that, 'in the case of pfa the iron-rich particles are the most conspicuous but the least representative of the globules present'.

Pfa mainly comprises solid and sometimes vesicular spherical glassy particles, although some other types of fly ash can include higher proportions of more irregularly shaped grains. A minority of these spherical particles consist of hollow 'cenospheres' (Raask 1968) which, because of their hollow lightweight nature, typically form a significant *volumetric* part of the material and represent a valuable characteristic indicator of the presence of pfa. The presence of *both* residual fuel particles and glassy cenospheres is virtually conclusive proof of the presence of some amount of fly ash. The cenospheres and other spherical grains can be discrete or agglomerated in clusters. Particle colour is very diverse, ranging from colourless to black and frequently being variously yellow, brown, red or grey. The typical appearance of these spherical particles, in thin-section and under the SEM, is illustrated in Plate 7 and Fig. 4.12.

The mineralogical composition of fly ash is typically about two-thirds aluminosilicate glass. In pfa, the next most abundant constituent is usually crystalline mullite (aluminium silicate, $Al_6Si_2O_{13}$), with some proportions of quartz and iron-bearing phases such as magnetite and hematite. Hubbard *et al.* (1985) obtained modal compositions for a range of UK pfa materials and these are reproduced in Table 4.22. Hubbard *et al.* (1984) found the cenopheres to be compounds of mainly glass with inclusions of mullite and quartz, frequently coated with the iron minerals derived from the furnace oxidation of pyrite. It is sometimes possible to observe mullite crystals in pfa in concrete thin-sections and, using SEM examination, Thomas and Matthews (1991) found that relicts of mullite crystals could be identified (see Fig. 4.12).

Table 4.22 Mineralogical composition of some UK pulverised-fuel ashes (after Hubbard *et al.* 1985).

UK region	Source	Alumino-silicate glass	Mullite $Al_6Si_2O_{13}$	Hematite Fe_2O_3	Magnetite Fe_3O_4	Quartz SiO_2	Coal
				% by mass (for each value $n = 15$)			
SW	P13	62	14	6	4	2	12
	P14	65	13	7	6	5	6
SE	P15	50	19	9	9	4	9
	P16	43	30	9	6	8	3
	P17	52	22	10	5	5	6
Midlands	P18	67	11	6	8	5	3
	P19	55	16	11	8	5	4
	P20	63	18	7	8	3	1
	P21	66	13	7	6	3	5
	P22	53	12	10	7	8	10
	P23	63	12	9	8	5	2
	P24	62	16	6	7	5	4
NW	P25	70	12	5	6	2	4
NE	P26	45	28	9	7	7	4
	P27	70	9	7	7	2	3
	P28	70	12	6	6	3	2
	P29	64	11	11	5	2	6
	P30	62	11	11	7	6	3
Scotland	P31	40	41	6	3	9	5
	P32	29	43	10	4	8	7
P-Lytag	P33	48	37	3	2	7	3
	P28/P	72	12	5	6	3	3
	P25/P	73	9	6	6	2	3
	P20/P	74	10	7	7	1	1
	P31/P	45	35	7	4	5	3
	P24/P	67	13	6	7	4	3
Median values:		62	13	7	6	5	4

The 'Type C' fly ashes appear to exhibit higher proportions of crystalline material. McCarthy *et al.* (1984), for example, reported the presence of quartz, lime, periclase, anhydrite, spinel, tricalcium aluminate (C_3A), merwinite and melilite, plus occasionally alkali sulfates, sodalite and hematite, in a range of such ashes from the western USA. Mullite, a characteristic crystalline phase in 'Type F' fly ash, was found to be only a minor component and was not detected in half of their samples of 'Type C' material. Mehta (1985) similarly found no mullite in high-calcium fly ash, in which C_3A and anhydrite were identified. Harrison and Munday (1975) have reported finding some anhydrite in UK pfa. In the authors' experience, very high-calcium types of fly ash can also contain some C_2S.

Quantifying the content of fly ash or pfa

Although it is relatively straightforward to identify the presence of fly ash in concrete, by recognising under a microscope the characteristic composition and particle shapes described in the foregoing section, it is a more formidable task to quantify the amount present. Since compositional analysis by petrography is more and more frequently being carried out to check compliance with a specified mix design, such quantification is increasingly required. This is particularly the case since the conventional chemical analytical techniques, such as those described in BS 1881 Part 124 (1988), are unreliable for concretes containing fly ash.

According to Figg (1989), pfa with a relatively high calcium content has a particularly significant effect on the BS 1881 analysis for cement content; a pfa containing 12 per cent CaO and present as 30 per cent of the binder caused an 8 per cent overestimate of cement content. French (1991a) claimed that the methods given for quantifying pfa and ggbs in BS 1881, 'are prone to some serious and often unknowable errors'. Such uncertainties worsen as the concrete ages, because variably greater amounts of pozzolanic activity will have taken place, thus varying the proportions of soluble silica in a largely unpredictable manner. Gomà (1989) has claimed success for a complicated chemical technique based upon differential solubility rates and Figg (1989) maintained that an earlier technique (Figg and Bowden 1971) could be used if reference samples of *all* the concrete constituents were available.

Just as examination of concrete in thin-section readily permits the presence of fly ash to be identified, so it is possible visually to estimate the relative proportions of fly ash and cement. However, this technique can only provide an extremely rough guide, because a majority of the fly ash particles will not be visible and the extent to which the fly ash and cement constituents have entered into pozzolanic or hydration reactions will not be known. The former objection is overcome in SEM examinations, where image analysis of highly-polished surfaces is possible (French 1991a), but even this cannot take into account the uncertainties arising from the variable degrees of fly ash reaction.

Figg (1989) mentioned the possible use of X-ray diffraction to quantify the content of crystalline mullite, which is usually present in pfa, but this technique would only be feasible if the mullite content of the pfa used was known, was consistent and, when diluted within all the other materials present in the concrete, was satisfactorily detectable to be quantified.

In view of these difficulties with the various chemical and microscopical approaches to the quantification of fly ash content, it would appear that the only reasonably dependable technique involves the use of an SEM and electron probe microanalysis. One procedure has been developed by French (1991a, 1992), but it is important to recognise that techniques involving electron microscopy typically relate to very small specimens of concrete, so that care must be exercised to ensure that any findings can be taken as truly representative of the concrete in question.

French's approach involves collecting microanalyses from isolated particles of unreacted pfa and unhydrated cement under the SEM (at least twenty of each) and also carrying out area microanalyses of regions of cement paste. Then, using averaged compositions for the pfa, the cement and the hydrated paste, the ratio pfa/(pfa + cement) can be calculated from the triangular relationship of the three oxides which show the maximum contrast between pfa and Portland cement: CaO, Al_2O_3 and SiO_2. A precision for the instrumental analysis of ± 0.03 at a ratio of about 0.50 is claimed, but the overall precision is likely to be worsened by a variety of other potential sources of error.

4.6.2   *BLASTFURNACE SLAG MATERIALS*

Blastfurnace slag is a by-product of the iron-making process and, in its air-cooled form, it is a dull grey, largely crystalline, rock-like material. This air-cooled slag, formed from the fusion of fluxing limestone or dolomite with siliceous and aluminous 'gangue' residues derived from the iron ore and the fuel, exhibits a chemical composition dominated by CaO, SiO_2 and Al_2O_3 and a mineralogical composition typically in the melilite field (Smolczyk 1980). Table 4.23 lists the main crystalline components of an air-cooled blastfurnace slag and Plate 19 depicts the coarsely crystalline melilitic structure.

Table 4.23 Mineralogical composition of an air-cooled blastfurnace slag (after Smolczyk 1980).

Main components	Melilite	Solid solution of
	Gehlenite &	$2CaO.Al_2O_3.SiO_2$ &
	Akermanite	$2CaO.MgO.2SiO_2$
	Merwinite (in basic slags)	$3CaO.MgO.2SiO_2$
	Diopside &	$CaO.MgO.2SiO_2$
	other pyroxenes (in acid slags)	
Minor components	Dicalcium silicate, α, α', β, γ (in basic slags)	$2CaO.SiO_2$
	Monticellite	$CaO.MgO.SiO_2$
	Rankinite	$3CaO.2SiO_2$
	Pseudo-wollastonite	$CaO.SiO_2$
	Oldhamite	CaS
Seldom observed minor components	Anorthite (in acid slags)	$CaO.Al_2O_3.2SiO_2$
	Forsterite	$2MgO.SiO_2$
	Enstatite (in acid slags)	$MgO.SiO_2$
	Perovskite	$CaO.TiO_2$
	Spinel	$MgO.Al_2O_3$

The large production of air-cooled blastfurnace slag in the UK is mainly used (>80 per cent) for road aggregate (Gutt *et al.* 1974), with just a few per cent being employed as an aggregate or an inert filler for concrete. However, some blastfurnace slag is rapidly cooled in a granulation or pelletisation process and the glassy product, which has latent hydraulic properties, is ground to cement fineness or finer for use as an addition or cement replacement material in concrete. This 'ground granulated blastfurnace slag' (ggbs) now enjoys widespread and growing usage and will be frequently encountered by petrographers examining hardened concrete.

A number of reviews provide accounts of the properties and potentially beneficial influences of ggbs as a constituent within concrete (Smolczyk 1980, Werner 1987, Hooton 1987, Dewar and Anderson 1992). One of the most recent and comprehensive reviews was compiled by a UK Concrete Society working party (Concrete Society 1991). There is now appreciable practical experience with concrete containing ggbs, spanning more than 50 years worldwide, and it is evident that ggbs concretes are often notably durable. Fig. 4.30 shows part of a wartime concrete submarine installation in Norway, wherein the sections made without ggbs are deteriorated by alkali-silica reactivity (ASR) and other factors, while the adjacent sections made with ggbs are in comparatively good condition (Jensen 1995).

Fig. 4.30 Wartime submarine dock in Trondheim, Norway, showing cracked and uncracked portions for concrete respectively containing no ggbs (right) and ggbs cement replacement (left).

Identifying ggbs

It is common for ggbs to form up to 50 per cent or more of the binder, so that the recognition of its presence should not pose a problem to the petrographer. Concretes made with ggbs exhibit a distinctive dark greenish-blue matrix coloration, although this changes to a creamy colour reminiscent of that displayed by non-ggbs concrete with time and exposure, apparently because of the oxidation of sulfides present in the ggbs. According to French (1991b), this colour change is not accompanied by any visually apparent features, even when examined under the scanning electron microscope (SEM). Oxidation can penetrate deeply into ggbs concrete and the absence of the dark greenish-blue colour from a sample should never be taken as indicating that ggbs is not present. It is possible to confirm that ggbs is present using either thin-section or reflected-light optical microscopy, or SEM, but, as with pfa discussed in Section 4.6.1, quantification is much more difficult.

The composition of ggbs is dominated by a calcium-aluminium-magnesium-silicate glass, which typically forms more than 70 per cent of the material and frequently more than 90 per cent. Crystalline phases are also found as inclusions within the glass and these have been studied by Scott *et al.* (1986), who found that relatively small changes in the chemical composition of the slag melt dictated the principal crystalline material formed. The sections through the $CaO-SiO_2-Al_2O_3-MgO$ quaternary system (at 10 per cent MgO and 10 per cent Al_2O_3) shown in Fig. 4.31 illustrate how slag melts will crystallise either in the melilite or merwinite fields when quenched. Scott *et al.* (1986) reported that oldhamite (CaS) and native iron were the other main crystalline phases in ggbs.

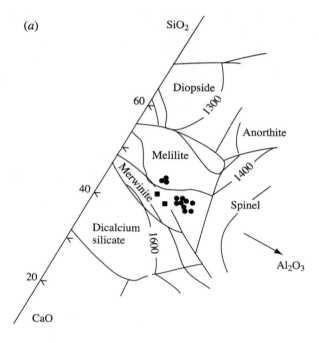

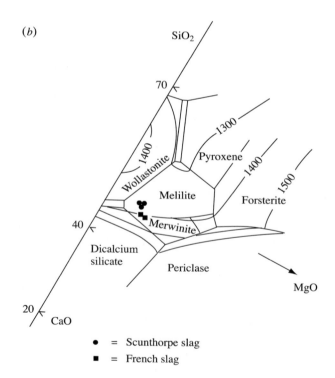

● = Scunthorpe slag

■ = French slag

Fig. 4.31 Sections through the CaO-SiO$_2$-Al$_2$O$_3$-MgO system at 10 per cent MgO (a) and 10 per cent Al$_2$O$_3$ (b), showing the fields for melilite and merwinite (from Scott *et al.* 1986).

The importance of glass content in the hydraulicity of ggbs has been the subject of considerable debate. It was initially assumed that reactivity would increase with glass content, but Demoulian *et al.* (1980) cast doubt on this notion, stating that, 'for the same [chemical] composition, a perfect vitrification is not the criterion of an optimal reactivity'. Hooton and Emery (1983) similarly concluded, 'while an amorphous structure is fundamental to slag reactivity, the highest glass contents were not necessarily indicative of the highest strength development'. They suggested that the crystalline inclusions could act as hydration nuclei or could be associated with more disordered, and hence more reactive, glass than that present in more highly vitrified ggbs.

Frearson and Sims (1991) describe the investigations carried out during the preparation of BS 6699 (1986) and in particular the development of a test for checking that ggbs was adequately glassy. Whilst melilitic ggbs formed mainly particles of transparent glass with occasional cloudy particles containing relatively large well-formed melilite crystallites, it was found that the merwinitic ggbs formed many more cloudy particles containing tiny rod-like or dendritic formations of merwinite crystallites (Fig. 4.32). Despite the apparently lower glass content of the merwinitic ggbs, its actual performance was typically superior to that of the melilitic ggbs in their investigation. A test was therefore devised for BS 6699 (1986), based upon using reflected-light microscopy to quantify the proportions of glass, glassy and crystalline particles, the 'glassy' particles being recognised as being at least as reactive as the pure glass particles and possibly more so. The latest edition of BS 6699 (1992) has instead opted for an X-ray diffraction technique to quantify the overall glass content of ggbs (Gutt 1992), which is required to exceed 67 per cent.

Drissen (1995) found that the estimation of glass content using transmitted light has advantages over other methods and described the techniques required. Taylor (1990) states that both microscopy and X-ray diffraction methods will fail to predict reactivity if microcrystallinity is present. Another optical technique, which has been proposed for determining the hydraulic reactivity of granulated slags (Grade 1968), is based upon the fluorescence colours produced by ultraviolet light. The technique appears to be suitable for ranking slags from the same works, but it is uncertain whether it is applicable to slags from different sources.

French (1991b) suggested, from his observations of completely unreacted ggbs particles in concrete, even at sizes only resolvable using SEM (see earlier Fig. 4.11), that ggbs particles are variously either inert or slowly reactive. If this is true, it is possible that the mineralogical character of the particle is the critical factor, perhaps with the 'glassy' particles and only some of the pure glass particles being the slowly reactive components.

Whatever the composition and microtexture of the particles, it is their generally glass-like nature which characterises ggbs in thin or polished sections of concrete (see earlier Fig. 4.10, Plate 6). The remnant particles are typically abundant, angular even shard-like and glassy in both fracture habit and isotropic properties. Ground particles of pelletised slag will be similar but with an additional tendency to contain gas vesicles. Usually ggbs particles are transparent or translucent and colourless or light brown, but occasionally may exhibit white, green or darker brown coloration. A few opaque discrete iron grains (or inclusions) might be apparent. Ggbs particles also exhibit higher relief than the enveloping cement paste and any fine quartz aggregate particles in the vicinity: the refractive index of ggbs is higher at ~ 1.63–1.64 than quartz at ~ 1.54–1.55 (McIver and Davis 1985).

In reflected light microscopy of concrete, any ggbs particles present usually yield a distinctive white to blue coloration after the polished specimen has been etched with HF vapour. SEM is also effective at confirming the presence of ggbs particles.

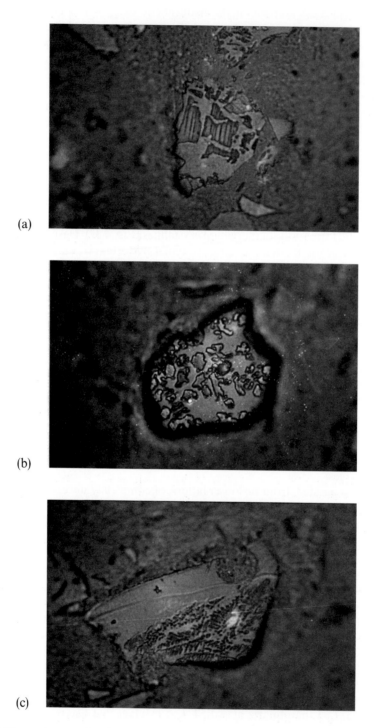

Fig. 4.32 Photomicrographs of ggbs particles in reflected light after etching with nitric acid: (a) melilite crystals within a glassy matrix, (b) rod-like merwinite crystals within a glassy matrix, (c) dendritic merwinite formations within a glassy matrix. (Photographs courtesy of Sandberg.)

Although French (1991b) has mentioned ggbs particles which exhibit no signs of reaction, it is common in older concretes to observe hydration zones surrounding kernels of many of the residual ggbs particles. Sometimes these hydration zones are layered (Plate 20) and it has been suggested by Thaulow (1984) that, in some cases, these might be seasonal in a way analogous to annular rings in tree growth, perhaps offering a means of estimating the approximate age of such a concrete. It is interesting that the boundary between the latest hydration layer and the ggbs kernel is usually rather sharp rather than diffuse as might have been expected. Concrete matrices containing ggbs are also sometimes depleted in Portlandite.

Quantifying the content of ggbs

Like pfa (see Section 4.6.1), whilst the presence of ggbs in a concrete is straightforward to detect, it is much more difficult to quantify. The chemical composition of ggbs is quite close to that of Portland cement and *routine* chemical analysis, to BS 1881: Part 124 (1988) for example, will not necessarily even indicate the presence of ggbs, let alone allow its quantification. Figg (1989) suggests that ggbs can be determined from the 'considerably higher' contents of sulfide and manganese, compared with Portland cement, although the contents of those constituents in the ggbs are required to be known. BS 1881: Part 124 (1988) provides a special chemical method for determining the percentage replacement of Portland cement by ggbs in concrete, based upon the sulfide content, but French (1991) considers this method questionable because of the need to check for other sources of sulfide and also in view of the likelihood that an unknown proportion of the sulfide in the original ggbs will have oxidised by the time of the analysis.

Figg (1989) states unequivocally that, 'determination of the relative proportions of ggbs and OPC in a hardened concrete is not possible by conventional microscopical procedures'. This is true, except that rough estimates are achievable from thin-sections, for example to assess whether or not the content of ggbs, estimated from the usually abundant remnant particles, is consistent with the specification. McIver and Davis (1985) reported a technique in which the thin-section under consideration is compared with reference sections made from concretes of known ggbs content. This approach might enable a reasonably reliable semi-quantitative estimate to be made, providing the ggbs in the reference concretes is sufficiently similar to that in the concrete sample being analysed and the reference concretes are also comparable to the sample in terms of age and exposure history. Obviously this is not practicable for concrete samples of unknown composition.

As with pfa (see Section 4.6.1), a quantitative procedure using electron probe microanalysis offers the best chance of reliably determining the content of ggbs although, as with any techniques involving use of an SEM and microanalysis, it is important to ensure that the specimens analysed are adequately representative of the overall concrete concerned. Figg (1989) mentions a procedure in which spot microanalyses are made of remnant ggbs and unhydrated cement particles in the concrete, to establish the compositions of the original ggbs and cement materials, followed by multiple area analyses of uncontaminated hydrated paste; the relative proportions are then calculated from the data obtained. French (1991b) provides greater detail of this technique (which is broadly similar to that described in Section 4.6.1 for pfa) and also explains some of the probable sources of error, including variations in the water/binder ratio and alterations to the concrete such as the leaching of lime.

4.6.3 *MICROSILICA (CONDENSED SILICA FUME)*

Silicon, ferrosilicon and other silicon alloys are produced by reducing quartz, with coal and iron or other ores, at very high temperatures (2000°C) in electric arc furnaces (Hjorth 1982). Some silicon gas (or 'fume') is produced in the process and reaches the top of the furnace with other combustion gases, where it becomes oxidised to silica by the air and then condenses as <0.1 μm to 1 μm spherical particles of amorphous silica. This material is usually known as 'silica fume' or, more properly, 'condensed silica fume' (csf); in the UK the term 'microsilica' has been adopted. Microsilica is an ultrafine powder, with individual particle sizes between 50 and 100 times finer than cement or pfa, comprising solid spherical glassy particles of amorphous silica (85–96 per cent SiO_2) (Parker 1985, 1986). However, the spherical particles are usually agglomerated so that the effective particle size is much coarser.

Microsilica for use in concrete derives from the manufacture of ferrosilicon alloys and is modified, by densification, micropelletisation or slurrification, to facilitate transportation and handling (Male 1989). World supplies of microsilica are limited with total production probably being between 1 and 1.5 million tonnes. The UK currently imports only modest quantities of microsilica for use in concrete, mainly from Iceland, a country which itself routinely uses blended microsilica-cement as a precaution against alkali-silica reactivity (Asgeirsson 1986). The potential benefits of using microsilica either as a cement replacement material or as an addition to improve concrete properties have been well reviewed by Malhotra and Carette (1983) and Sellevold and Nilsen (1987). Disadvantages would include the health hazards associated with fine silica materials and the increased cost of the concrete.

Although the use of microsilica in concrete has been limited to date, the petrographer may encounter an increasing occurrence in the future and determining its presence or otherwise in a concrete sample is occasionally requested. Detecting microsilica in concrete is complicated by its usually small proportion (between 5 per cent and 30 per cent but typically around 10 per cent) and its extreme fineness, with virtually all of the particles being substantially smaller than the 25 μm to 30 μm thickness of a conventional thin-section. Moreover, because microsilica is an ultrafine pozzolana, it is rapidly consumed in the initial hydration reactions and, according to Figg (1989), after a month it is virtually undetectable even using scanning electron microscopy (SEM). However, the presence of microsilica is frequently betrayed to the petrographer by either or both of two factors: the occurrence of agglomerations and the modifying influence of microsilica on the cement paste.

The microsilica supplied for use in concrete has either been agglomerated by densification or micropelletisation, or is suspended in a slurry with water (Male 1989). These agglomerations are supposedly re-dispersed during concrete mixing and the slurries are often considered to re-disperse most easily; superplasticisers are claimed to aid dispersion (Male 1989). However, St John (1994b) and St John et al. (1996) have challenged this supposition, finding that microsilica agglomerations typically survive and even that some represent fused microspheres, present in the microsilica before densification, that are incapable of dispersion.

St John (1994b) found that this agglomeration caused the mean particle size of microsilica to rise to between 1 μm and 50 μm, rather than the 0.1 μm to 0.2 μm frequently cited. The significance of this finding for the petrographer is that most microsilica concrete can be expected to contain some agglomerations which are coarse enough to survive pozzolanic reactivity and be visually detectable under the SEM and even in thin-section. Also, conclusions that the presence of microsilica agglomerations indicate poor dispersion and

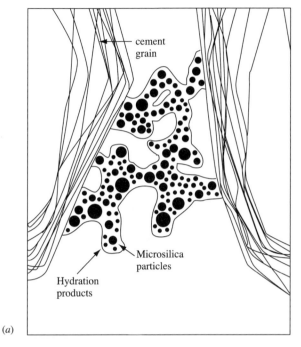

cement
grain

Microsilica
particles

Hydration
products

(a)

Microsilica concrete   Conventional concrete

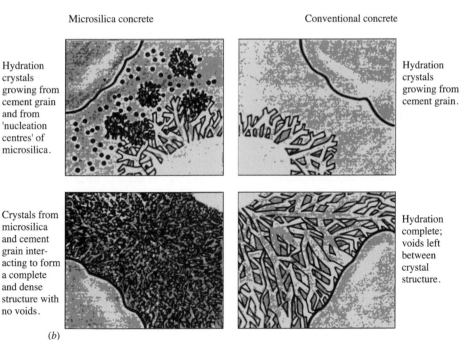

Hydration
crystals
growing from
cement grain
and from
'nucleation
centres' of
microsilica.

Hydration
crystals
growing from
cement grain.

Crystals from
microsilica
and cement
grain inter-
acting to form
a complete
and dense
structure with
no voids.

Hydration
complete;
voids left
between
crystal
structure.

(b)

Fig. 4.33 Sketches to illustrate the theoretical effect of microsilica in densifying the concrete matrix:
(a) microsilica sub-dividing the space between cement grains, (b) comparison between conventional
and microsilica concretes. (From Hjorth (1982) and Male (1989).)

thus a concrete mixing deficiency must be treated with caution, although an *unusual* abundance of coarse agglomerations might justify such an interpretation.

The matrix of microsilica concrete is notably dense to thin-section examination and this optical denseness will frequently suggest the presence of microsilica, even if agglomerations are not apparent. This densifying effect has been attributed to the extreme fineness of microsilica, whereby 50 000 to 100 000 microspheres exist for every cement grain (Male 1989), allowing microsilica hydration products to infill the water spaces usually left within the cement hydrates (Fig. 4.33). Overall coloration of microsilica concrete is often also slightly darker than a comparable material lacking microsilica, but this might not be apparent, especially for lower microsilica contents.

It has been found that calcium hydroxide (Portlandite) is consumed by the pozzolanic reactions involving microsilica (Sellevold *et al.* 1982), the extent of Portlandite depletion increasing linearly with microsilica content. Durning and Hicks (1991) considered this reduction in Portlandite content to be a major factor in the effectiveness of microsilica concrete in resisting chemical attack and stated that the Portlandite content is reduced to zero by a 30 per cent microsilica replacement of ordinary Portland cement. In thin-section examination, the combination of a notably dense paste and a complete or near absence of Portlandite is a strong indication that microsilica might be present. More exhaustive inspection of the thin-section specimen might then reveal the sporadic and confirmatory occurrence of relatively coarse microsilica agglomerations.

Quantification of microsilica in concrete cannot be undertaken by any form of microscopy. Figg (1989) has suggested the use of electron probe microanalysis to establish the composition of the hydrated paste, presumably on the basis of an enhanced silica content relative to that to be expected from Portland cement alone. If the composition of the cement component is known, it might then be possible to calculate an approximate proportion for the microsilica.

4.6.4   NATURAL POZZOLANAS AND OTHER REACTIVE ADDITIONS

Although fly ash, ggbs and, to a lesser extent, microsilica form the majority of the mineral additions used in concrete, natural pozzolanas have been used historically and continue to be added to concrete in some areas of the world. These natural pozzolanic materials sometimes exist in a natural form which facilitates their direct use in concrete after grinding, such as the volcanic or diatomaceous earths, whilst others need to be treated, for example by heating or 'calcining', prior to grinding. Such materials include a range of silica-rich glassy volcanic rocks, such as pumicite, ignimbrites and other tuffs, opaline shales, other shales or clays and opaline cherts. A good review of the natural pozzolanas available in the USA is included in Mielenz (1983).

Generally, natural pozzolanic additions are used to replace between 10 per cent and 30 per cent of the cement in the binder and, for an effective material, the beneficial effects appear broadly comparable with those of fly ash. In the USA, natural pozzolanas are specified, along with fly ash, in ASTM C618 (1992b). In some cases, the presence of a finely ground natural pozzolanic addition could pose a problem for the unsuspecting petrographer and care should be exercised when examining concrete from any areas in which such pozzolanas are traditionally used.

Some discussion of the problems in identifying pozzolanas in concrete will also be found in Chapter 5. The chemical method for identifying pozzolanas involves dissolution of the hardened cement paste and examination of the fine residue. In most cases sufficient of the pozzolanic material is found in the fine residue to allow identification by optical means.

Generally, identification in thin-section will depend on some unique characteristic of the pozzolana being visible either by optical or scanning electron microscopy. For instance sponge spicules, often present in diatomite, may persist in concrete long after the diatom fragments have reacted. Where the presence of a pozzolana is known or suspected, the petrographer should examine samples of the likely pozzolanas for any unique characteristics before examining the concrete in thin-section. This increases the chance of identification.

It must be accepted that in old concretes where the pozzolanic replacement is below 15 per cent, identification may be difficult unless some unique characteristic of the pozzolana is present. Changes in the texture of the hardened cement paste and depletion of Portlandite may be limited, subtle and insufficient to suggest the presence of a pozzolana. In such cases, determination of the Ca/Si ratio by analysis may suggest the presence of a pozzolana, but even this is not possible for wet chemical methods where calcareous aggregates are present.

The effectiveness of mixing mortars using calcined clay and slaked lime was known in prehistoric times. In recent years, considerable claims have been made for the product of calcining a pure clay mineral (kaolinite or 'china clay') ground to an average particle size of 1.5 μm. The properties and effects of this 'metakaolin' have been reviewed by Kostuch et al. (1993) in the UK and by Caldarone et al. (1994) in the USA. Metakaolin appears to exhibit effects similar to those of microsilica for a cement replacement of 5 per cent to 20 per cent. In fact, Kostuch et al. (1993) reported higher pozzolanic activity for metakaolin than for pfa or microsilica. It is thus possible that petrographers in the future will encounter an increasing number of concretes containing metakaolin, when particle fineness will limit the usefulness of microscopy for its detection.

Non-ferrous slags have not generally been considered suitable for use in concrete, either as aggregate or as addition in powdered form. However, some non-ferrous slags might be suitable as a pozzolana. Douglas et al. (1985) have suggested the potential usability of some Canadian non-ferrous slags. In a worldwide survey, a RILEM committee identified 'rice husk ash' as one of the four principal pozzolanic/cementitious siliceous by-products used for concrete, alongside fly ash, ggbs and microsilica (RILEM 1988).

4.6.5 PIGMENTS

The typically white to dark grey colour of concrete is ordinarily determined mainly by the binder, with the fine and coarse aggregates having respectively much less influence. The shade of concrete surface coloration is also affected by other factors, including the type of formwork and the mode of curing. To obtain colours other than white to dark grey, the addition of a pigment is necessary. The use of such pigments is common in precast concrete, such as paving, masonry, cladding units and 'cast stone'. Chemical stains and dyes are infrequently used, because in the longer term they fade in the alkaline environment within concrete.

Generally the use of a pigment will be evident from the concrete colour, but a petrographer may be called upon to confirm the type of pigment used, or even the presence of pigment in cases where only a tinting has been achieved. Most mineral pigments are ultrafine powders, much finer than cement (mainly <1 μm particles), so that optical microscopy is of limited usefulness in identifying pigments and scanning electron microscopy (SEM) will be required. However, the colour imparted to hydrated cement paste by the included pigment particles will be clearly apparent in thin-section, as will its uniformity of dispersion.

Under the SEM, microanalysis of pigment particles can be used to determine the basic type in the case of inorganic materials, although the various colours achieved using iron oxides might not be capable of differentiation. If the composition of the cement is known, or can be established from microanalysis of unhydrated particles, area microanalyses of the pigmented cement paste might then enable the pigment content to be quantified. SEM is unhelpful in the case of carbon black, which has a particle size of ~ 0.01 μm, which is much finer than most inorganic pigments and, as a result, tends to create murky thin-sections.

Table 4.24 Colour and other properties of some types of pigment for concrete (based upon Arnold 1980 and Levitt 1985).

Pigment type	Colour	Example (Arnold)	Particle size[1] μm	Bulk density[1] kg/m^3	Water absorption[1] ml/100g
Synthetic red iron oxide	Red	Pfizer	0.1	900	35
		R-1599	0.7	1500	20
Synthetic yellow iron oxide	Yellow	Mapico Yw	0.2×0.3[2]	800	50
		1000	0.2×0.8[2]	500	90
Synthetic black iron oxide	Black	Pfizer	0.3	1100	33
		BK-5599			
Synthetic brown iron oxide	Brown	R-Coulston	0.3–0.6[3]	900	50
		Brown 537	0.1–0.2[3]	1000	30
Natural brown iron oxide	Brown	R-C Burnt Umber 15			
Chromium oxide	Green	Pfizer G-4099	0.3	1200	19
Carbon black (general purpose)	Black	Mapico Raven 410			
Carbon black (concrete grade)	Black	Mapico Raven 1040	0.01	500	100
Cobalt blue	Blue	Harshaw Blue 7540			
Ultramarine blue	Blue	Frank Davis Blue 410B			
Phthalocyanine	Green or Blue	Minl Pigts Green 5069	0.01	500	0[4]
Toluidine red	Red	Du Pont RT-386-D			
Dalamar (hansa) yellow	Yellow	Du Pont YT-808-D			
Watchung (BON) red	Red	Du Pont RT-761-D			
Chrome yellow	Yellow	Du Pont Y-469-D			
Iron blue	Blue	Milori Blue 50-1750			
Titanium oxide	White		0.2	700	24

Notes
1   The physical property data, after Levitt (1985), do *not* relate to the examples cited by Arnold (1980).
2   Needle-shaped crystals, hence the two particle dimensions.
3   Brown iron oxides have a wider range of particle sizes than the others.
4   Hydrophobic in their undiluted form and will not absorb water.

Concrete pigment types, properties and effects have been reviewed several times in recent literature (Coles 1978, Arnold 1980, Levitt 1985, Spence 1988). A good concise review was presented by Lynsdale and Cabrera (1989). A summary of the main types of pigment likely to be found in concrete is given in Table 4.24, together with their colouring effect and some other characteristics. Some examples of pigmented concrete and cast stone products are given in Table 4.25.

Table 4.25 Some pigmented concrete and cast stone products (based upon Levitt 1985 and Spence 1988).

Effect/colour required	Cement type	Aggregate examples	Admixture(s)	Pigment(s)
Brilliant white	White PC	Calcined flint, Carrara marble		None
	White PC	Calcined flint, quartz sand	Stearic acid	Titanium oxide
Light cream	White PC	Spanish dolomite		Yellow
Mid cream	White PC	Derbyshire limestone		Yellow
Buff	White PC	Grey or brown, with sand	Stearic acid	Yellow
Simulated Bath Stone	White PC	Crushed Bath or Clipsham Stone	Aluminium stearate	Yellow or brown
Simulated York Stone	White PC or OPC	Crushed sandstone, crushed limestone and crushed granite		Brown/ Yellow
Yellow	White PC	Crushed Cotswold limestone		Yellow
Green brick	White PC or WPC & OPC	Light buff limestone	Aluminium stearate	Phthalocya-nine green[1]
Light pink	White PC	Derbyshire limestone		Light red
Deep pink	White PC	Mountsorrel granite		Light red
Red	White PC or OPC	Mountsorrel granite		Red
Red mortar	OPC & lime	Red, brown or buff sand	Stearic acid	Red
Brown	OPC	Mountsorrel granite, or quartzite gravel, or Cotswold limestone		Brown
Bronze	OPC	Criggian Green granite		Dark brown
Green	OPC	Criggian Green granite or green marble		Chromium oxide
Slate grey	OPC	Ingleton granite		Green/blue/ black
Purple slate	OPC	Mountsorrel granite black marble		Red/black
Black	OPC	Clee Hill granite or Belgian black marble		Black

Note
1   10/90 w/w fine silica dilution.

Most of the pigments in common usage are inorganic metallic oxides; the main exception is carbon black. Red, yellow, black and brown colorations are achieved with iron oxides, whilst green is usually derived from chromic oxide and blue from a copper complex with phthalocyanine. Bright white concrete can be achieved by using titanium oxide together with white cement. The most effective inorganic pigments are synthetic rather than natural, because the natural materials are typically less pure and hence exhibit a lower tinting strength. These inorganic pigments are essentially inert and stable, although it has been reported that, in certain conditions, the non-hematite iron oxides can alter to hematite (Koxholt 1985).

As pigment is an inert fine powder, excessive proportions have detrimental effects on the hydrated cement microstructure and on concrete strength. In both BS 1014 (1992) and ASTM C979 (1986) the content of pigment is limited to a maximum of 10 per cent by mass of cement. According to Lynsdale and Cabrera (1989), a dosage in the range 3–6 per cent achieves a 'definite colour'. 'Tinting' only requires ∼1 per cent, but a 'deeper shade' requires up to 10 per cent. Each type of pigment has a colouring saturation level and higher dosages do not necessarily produce even deeper colours.

Various disadvantages have been perceived for pigmented concrete, although in reality these are often features common to all concrete which are just more noticeable in a coloured material (Spence 1988). Alleged 'fading' is frequently caused by a surficial overlay of lime bloom or efflorescence or simply grime from the atmosphere. Black or dark grey concretes pigmented with carbon black are said to be particularly vulnerable to loss of pigment from water-eroded surfaces (Lynsdale and Cabrera 1989). Colouring inconsistency can be overcome by using synthetic inorganic pigments, which achieve their colour saturation at relatively low dosage levels, and by ensuring quality control over concrete mix proportioning and the other variables such as curing.

4.7   Chemical admixtures

Increasingly concrete mixes are designed to include one or more chemical admixtures, which are usually added at low dosage levels (mostly ∼0.1 per cent by weight of cement), to modify the properties of the fresh and/or eventually hardened concrete. Such modifications include workability improvement and water content reduction for a given workability, by plasticisers or superplasticisers, retardation or acceleration of set, air entrainment and waterproofing. Some admixtures achieve combinations of these modifications (e.g., retarder/ plasticiser). Detailed accounts of chemical admixtures for concrete are to be found elsewhere (Rixom 1978, Ramachandran 1984, Neville and Brooks 1987). A list of the main admixture types and their chemical identities is given in Table 4.26.

These chemical admixtures cannot be directly detected by petrography, although in some cases the effects created by an admixture might be observable; the most obvious example being air entrainment (see Section 4.5). Chemical analytical techniques are required for detecting and quantifying the presence of an admixture and various procedures are described by Figg (1989). However, except for chloride-based admixtures, which are now rarely used but will be encountered in many older concretes, most of these materials are organic in nature and present many difficulties to the analyst, particularly in view of the concentrations present as low as 0.02 per cent by weight of cement and as many admixtures are not stable in the long term in the alkaline pore solution. St John (1992) presented chemical data on the analysis and stability of water-reducing agents in concrete; it was found that only the naphthalene sulfonates remained intact.

In some cases, it is possible to extract the organic admixture from a powdered concrete sample with solvents and then identify the nature of the substance using Fourier transform infrared spectrophotometry or other specialised analytical techniques, but the procedures are difficult and may be expensive. Simple qualitative tests for organic matter have sometimes been applied for indicative purposes, but the findings should be treated with circumspection as traces of organic material may well be derived from other concrete constituents (such as the aggregates) or as a minor contamination of the concrete in service.

Table 4.26 Main types of chemical admixture for concrete and some of their varieties (after Figg 1989).

Admixture type	Chemical type	Typical concentration mg/100g concrete
Plasticisers	Lignosulfonates Hydroxycarboxcylic acid salts (e.g., gluconates, leptonates)	20
Pumpability aids (retarders)	Polyhydroxy compounds Tartrates/citrates	20
Superplasticisers	Naphthalene formaldehyde sulfonates Melamine formaldehyde sulfonates	80
Air entrainers	Neutralised wood resins (vinsol) Soaps of fatty acid-resin acid mix (tall oil soaps) Surfactants (alkyl aryl sulfonates)	3
Accelerators	Calcium chloride Calcium formate Triethanolamine Triethanolamine formate Morpholine derivatives	300 300 50
Integral waterproofers	Fatty acids and salts (e.g., calcium stearate, oleic acid)	300
Mortar plasticisers	Neutralised wood resins Surfactants	50

4.8 Modal analysis of concrete

Hardened concrete samples may be subjected to analysis to estimate the original cement content and/or the aggregate/cement ratio and/or the coarse aggregate/fine aggregate ratio and/or the aggregate grading. Standardised chemical analytical methods are used, such as BS 1881:Part 124 (1988), but there are circumstances in which the chemical procedures are inappropriate or unacceptable.

Chemical determination of cement content in concrete is based on the acid soluble silica or the acid soluble calcium values, or preferably both, determined for the concrete. However, this approach may be rendered unreliable when acid soluble constituents are also present in the aggregates or the original concentrations have been modified during the life of the concrete. Aggregates containing acid soluble calcium compounds are common, such as limestone, and some rock materials will release silica during the extraction techniques employed in the chemical analysis of concrete. Chemical alteration of concrete in service is commonplace, ranging from mild water-leaching of exposed surfaces to severe chemical attack. In the case of alkali-silica reaction (ASR), the formation of soluble silica from reactive aggregates within the concrete occurs (see Chapter 6).

In cases where the concrete composition or its condition suggest that chemical analysis might be unreliable, or where in legal disputes an essentially non-destructive technique is required, a 'modal' or 'micrometric' analysis, involving quantitative determination by optical microscopy, can be justified, despite its involved, expensive and arduous procedures. Modal analysis is discussed in Chapter 2 and the micrometric procedure outlined below is taken from that described by Figg (1989). It is possible that, in due course, satisfactory computerised image analysis techniques will be developed for conducting micrometric analyses more objectively and with less operator effort. Deelman (1984) has reported some success for an automated system using a diamond-tipped stylus profilometer to quantify components from the microrelief of a polished concrete surface.

Micrometric analysis is carried out using the same linear traverse or point-counting techniques described for analysing the air void system in Section 4.5.3. Usually point-counting is used for establishing the volume proportions of the cement paste, coarse aggregate, fine aggregate and air void components, from which the original mix proportions by mass are calculated. The cement content is calculated from the cement paste proportion, using the water/cement ratio calculated from capillary porosity, water of hydration and bulk density. Various calculation procedures have been proposed, but, in the interests of standardisation of approach, it is suggested that the formulae given by Figg (1989) are adopted and hence these are reproduced in Appendix 2A. Methods for estimating the original water/cement ratio were discussed in Sections 4.4.3 and 4.4.4.

It is very important to ensure that the ground surfaces used for this exhaustive procedure for determining mix proportions are adequately representative of the concrete under investigation. The slices of concrete must be derived from a location appropriate to the enquiry concerned, be correctly orientated to the placement layering of the concrete and must be large enough in area to overcome the inherent inhomogeneity of concrete. Slice orientation will usually be perpendicular to the placement layering, but other orientations might occasionally be appropriate. For example, a set of slices cut at selected levels parallel to the placement layering might be needed if possible segregation is being investigated. Figg (1989) suggests that the *minimum* surface area should be at least five times the nominal maximum aggregate size, which equates to 100 mm × 100 mm for 20 mm aggregate.

Preparation of the concrete slice specimen surface is broadly similar to that for an air void system analysis (see Section 4.5.3), except that the face must be very finely ground to enable the various concrete constituents to be clearly distinguished (see Fig. 4.34). A more cost-effective alternative to very fine grinding is the painting (or spraying) of a ground surface with varnish or acrylic to simulate a permanently wetted appearance, but care must then be exercised in recording the air voids, some of which might be infilled by the varnish or acrylic.

Aggregate particle size will be under or over-estimated by this technique, because of oblique or non-central sectioning and depending upon particle shape (as explained in Section 4.3.2), so that the uncorrected content of fine aggregate will be under or over-estimated relative to the coarse aggregate content when the differentiation is based only on particle size (say taking 4 mm or 5 mm as the coarse/fine threshold). When the coarse and fine aggregates are petrographically distinctive, for example in the case of a crushed limestone coarse aggregate and a natural quartz sand fine aggregate, this problem can be avoided.

No interlaboratory trials have been published for the micrometric method of determining concrete mix proportions. Axon (1962) reported a trial within one laboratory in the USA, wherein the actual and determined cement contents were remarkably close. In the best of

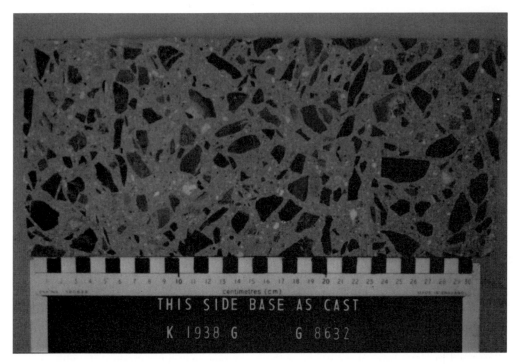

Fig. 4.34 Large finely ground and polished slice of concrete suitable for micrometric analysis.

seven mixes, the actual and determined values were the same and, in the worst, the determined value was no more than 29 kg/m³ different from the actual value of 280 kg/m³, which is a maximum error of about 10 per cent relative. Such accuracy might be difficult to achieve for concrete samples lacking supporting information about their constituents. However, Figg (1989) states that 'research does suggest that the precision of micrometric analysis may be at least as good as, or possibly even better than, that attainable by chemical analysis'. This endorsement of the technique has not yet been tested by a statistically valid precision trial, but is only thought likely to hold for analyses carried out by a skilled and experienced concrete petrographer.

5

The Appearance and Textures of Concrete

5.1 Introduction

It is now accepted that problems of durability are of major concern in modern concreting practice (Mehta 1991). Yet the microscopical examination of concrete in thin-section, which is a powerful investigative technique for this purpose, is not widely used in laboratory investigation and engineering practice. In the following sections some of the more important microtextures of concrete are described, and the changes which occur with time are illustrated to show how observation of these changes may have diagnostic value. As yet there is no systematic description of concrete microtextures similar to that attempted by soil scientists (Bullock *et al.* 1985).

The observation of the microtextures of concrete using the polarising microscope not only requires a knowledge of mineralogy to interpret aggregate textures but also an understanding of the chemistry and physics of Portland cement to interpret the texture of the hardened cement paste. Thus the subject must be approached both from the viewpoints of chemical microscopy as well as classical mineralogy. Even where this type of systematic approach is used many observations of concrete microtexture remain qualitative.

The petrographer is frequently required to examine the microtexture of concrete to explain why deterioration has occurred and relate this microtexture to the materials used, methods of placement and the environmental history of the structure. Microscopical examination of concrete in thin-section can reveal considerable information on deteriorative processes but certain limitations must be kept in mind. Two of these limitations are that thin-sections are essentially two dimensional so the third dimension must be inferred (see discussion by Elias (1971)), and the effective resolution of the polarising microscope is limited to about 1 μm. Since many of the reactions that occur in concrete result in products that are submicroscopic (i.e., less than 1 μm) their observation requires the use of electron microscopy. However, there are problems in relating observations made by the SEM to those made using the polarising microscope because of differences in the types of samples and techniques used which have not been satisfactorily resolved.

Petrographic examination of concrete using thin-section techniques can be a lengthy process. For this reason concrete petrography tends to be limited to structures and works where high standards of safety and durability are of concern. These works are usually constructed of concretes, made with cement contents of 300 kg/m^3 or greater, which have been mixed and placed under supervision. In practice, petrographers do not get many opportunities to examine concretes with lower cement contents that may have been both poorly placed and inadequately cured. It is for this reason that the concrete textures discussed and illustrated are largely confined to the better quality of structural concretes and descriptions of textures are limited to those observed using the polarising light microscope.

5.2   Hardened cement paste

Initially, the immediate products of adding water to Portland cement are the formation of calcium hydroxide (see Section 5.3) and ettringite (Section 5.2.5) which then converts to mono-calcium alumino-sulfate hydrate within the first day. Concurrently, but more slowly the principal cementing phase, an essentially amorphous calcium silicate hydrate is formed together with the calcium aluminate hydrates including those substituted with iron. This formation of calcium silicate hydrate is the principal source of the calcium hydroxide. Apart from calcium hydroxide which is crystalline, these hydration products are beyond the resolution of the light microscope and are best observed on fracture surfaces by scanning electron microscopy.

However, when a cement paste is autoclaved at 170–180°C the principal phase formed is crystalline α-dicalcium silicate hydrate. If the temperature is maintained for a sufficient length of time, such as occurs in geothermal grouts, the crystals of this hydrate may grow large enough to become visible under the microscope as shown in Plate 21. Crystalline α-dicalcium silicate hydrate (α-$2CaO.Si_2O_3.H_2O$) is not considered desirable in concrete products because of its low strength. Silica flour is added to the mix to prevent its formation in favour of a higher strength tobermorite (calcium silicate hydrate) type phase which is submicroscopic (St John and Abbott, 1987).

The textural appearance of hardened cement paste is affected considerably by the thickness of the thin-section. If the thin-section is thicker than 30 μm, the paste has a dark, milky appearance with many of the calcium hydroxide and remnant cement grains either partly obscured or hidden in the thickness of the section. Contrary to Section 11.1.1 of ASTM C856 – 95 Standard Practice for Petrographic Examination of Concrete there appears to be no justification for using section thicknesses greater than 30 μm as little useful detail can be observed in the hardened paste. When the thickness is reduced to 25 μm the hardened cement paste becomes lighter in colour with an ill-defined background texture. Most calcium hydroxide crystals are now reasonably defined and the remnant and partially hydrated cement grains are easily seen. This is the preferred thickness for routine examination of thin-sections of concrete and represents a compromise that yields the greatest information. Below 25 μm the definition of remnant cement grains improves but calcium hydroxide becomes skeletal unless considerable care is taken in grinding the section. A thin-section thickness of 20 μm is preferred by some petrographers especially when they are using a smaller sized polished thin-section as this provides optimum resolution of the hardened cement paste and still allows satisfactory observation of the aggregate. Walker (1979) has shown that if a section is reduced by polishing to 8–14 μm in thickness, remnants of hydration shells become visible. These hydration shells, also known as Hadley shells (Barnes *et al.* 1978), are difficult to observe in 25–30 μm thick thin-sections.

Routine optical examination of hardened cement paste in concrete is restricted to observation of porosity, voids, cracks and fissures, calcium hydroxide, remnant and partially hydrated cement grains, reprecipitated crystals of ettringite and alteration products. While this may appear limited a surprising amount of diagnostic information is possible from this type of optical examination. There are claims by Randolph (1991) that the presence of chemical and mineral admixtures in concrete can be identified by changes in texture on the hardened cement paste including the distribution and morphology of the calcium hydroxide. The effect of chemical admixtures on the morphology of calcium hydroxide has been investigated by Berger and McGregor (1972) and Brandt (1993), but the effective use of these textural changes for the identification of admixtures still requires further investigation.

5.2.1   TYPICAL COMPOSITION OF A HARDENED CEMENT PASTE

The composition of a hardened cement paste at a water/cement ratio of 0.5 is shown in Table 5.1 (Taylor 1984). Volumetric values given in the last column of the table for calcium hydroxide, alite, belite and ferrite phases will not be completely visible because some small calcium hydroxide crystals and fragments of residual clinker may be obscured in the thickness of the thin-section. Obviously the volumes of these phases observable in field concretes will also depend on the cement content, water/cement ratio and curing. However, the results given in Table 5.1 make a useful starting point for discussion.

Table 5.1 Calculated weight and volume percentages for a Portland cement paste hydrated for 91 days. Water/cement = 0.5

Phase	Saturated		Dried at 110°C	
	Weight %	Volume %	Weight %	Volume %
Alite	6	4	8	4
Belite	4	3	6	3
Aluminate	0	0	0	0
Ferrite	2	1	3	1
Calcium hydroxide	14	12	19	12
Calcium carbonate	1	1	1	1
Calcium silicate hydrate	42	40	45	25
Hydrated aluminate and ferrite	18	16	19	13
Pores	12	24	0	42

5.2.2   REMNANT OR OVERSIZED GRAINS OF CEMENT CLINKER

The remnant grains of cement clinker observed in the hardened paste of a mature concrete are those particles which are retained when a cement is passed through a 90 μm sieve. Most of these oversize cement grains are surprisingly resistant to hydration and often will have barely observable hydration rims even where a concrete has been exposed to the weather for many years (Idorn and Thaulow, 1983).

Both the quantities and particles sizes of remnant clinker grains in cements will vary according to their period of manufacture. Lea (1970) quotes averaged results of residues recorded by Redgrave and Spackman as 10 per cent on the 600 μm sieve and 45 per cent on the 150 μm sieve as being typical for the year 1879. Lea then states that by 1910 the residue on the 90 μm sieve for many British and Continental cements had dropped below 10 per cent and by 1970, most ordinary Portland cements were averaging sieve residues around 5 per cent and rapid hardening cements from 1–3 per cent. Currently in those cement works with modern grinding circuits it is the practice to exercise close control over the particle size distribution and to try and limit the residue on the 90 μm sieve to as little as 2 per cent.

From the residues quoted above we would expect that older concretes would have both a larger and a greater number of remnant grains of cement clinker embedded in their hardened pastes. This is illustrated in Plate 22. Further examples of remnant grains in more recent cements may be found in Plates 23 and 24 and many of the other micrographs. A noticeable characteristic of remnant grains of cement clinker is the predominance of belite embedded in brown interstitial phase. In many cases these grains have been derived from nests of belite due to oversize particles of quartz in the raw feed and they may not represent the overall composition of the cement.

A convenient and arbitrary distinction can be made between those residual grains of cement clinker which will eventually hydrate provided they are in contact with sufficient water and space to accommodate their hydration products, and remnant grains which are resistant to hydration because of their large size. This distinction means that the quantity and particle size range of partially hydrated cement grains depends solely on the cement content and the degree of hydration of the concrete.

Sumner, Hepher and Moir (1989), give typical particle size ranges for cements ground in modern conventional open-circuit mills as shown in Table 5.2. They claim that closed-circuit mills do not differ significantly in the particle size range produced so the figures quoted form a useful guide.

Table 5.2 Typical particle sizes and range of cement ground in open-circuit mills.

μm	Typical %	Range %
<90	98	96–100
<63	95	90–98
<45	88	81–93
<30	75	69–81
<20	60	53–65
<15	48	44–52
<10	36	32–39
<5	21	18–23
<2	8	7–9

Lea (1970) quotes depths of hydration for Portland cement particles as being about 7 μm at 1 day and 20 μm at 7 days, and other data to show that particles of cement less than 5–7 μm in diameter mainly contribute to 1–2-day strength, particles less than 20–25 μm to 7-day strength and particles less than 25–30 μm to 28-day strength. It is difficult to compare this data but it suggests that in a concrete of moderate cement content, which has been adequately cured for 28 days, few partially hydrated grains of less than 30 μm should be present. This suggestion is in reasonable agreement with microscopic observation on adequately cured laboratory concretes.

However, where concrete is inadequately moist cured more partially hydrated grains may be present in the outer shell than the inner core because premature drying of the outer shell has halted hydration. Where a concrete contains a high cement content and is mixed at a low water/cement ratio it may become almost impermeable to curing water and there may be insufficient water for hydration leading to a condition called self desiccation. In addition, at very low water/cement ratios pore space to accommodate only a limited volume of hydration products may be present. An illustration of the effect of high cement content and low water/cement ratio is shown in Plate 23.

The effect of extremes of hydration on the amounts and particles sizes of partially hydrated grains of cement is shown in Plate 24. A cement grout, mixed at a water/cement ratio of 0.62 was exposed down a geothermal steam bore at 160°C for 3 months. Because of the prolonged exposure only vestigial remnants of cement grains are present. This is an extreme that will be rarely met in normal concreting practice. By contrast, the super high strength (DSP) mortar was mixed at a water/cement ratio of 0.23, moist cured for 28 days and then exposed for 3 months to daily cycles of wetting at 21°C and drying at 60°C. In spite of the area photographed being only a few millimetres below the surface, small unhydrated grains showing sharp angular outlines are still present and there is sufficient partially hydrated cement remaining to effectively act as a filler. This is the other extreme which only applies to special materials such as super high strength DSP mortar that are highly impermeable and hydration is limited by self desiccation.

Estimates for the amounts of residual grains of cement present in concrete cured at 20°C have been given by French (1991a). He claims that above a water/cement ratio of 0.61 hydration is virtually complete with 3–4 per cent of unhydrated material remaining at water/cement = 0.5 and 7 to 8 per cent at water/cement = 0.4. If the concrete is made at a temperature of close to 0°C the amount of unhydrated material increases while in concrete cured above 40–50°C hydration is generally complete. Addis and Alexander (1994) put forward a concept of cement saturation for concrete which concludes that hydration above water/cement ratios of 0.35 to 4.0 hydration should be complete because hydration is not limited by lack of space or water. However, it should be noted that under normal conditions of exposure, experience shows that concrete in most structures may only be 80 to 90 per cent hydrated irrespective of the water/cement ratio.

Lea (1970) quotes analyses of various size fractions of ground cement clinker which shows that the alite content tends to be higher in the fine fractions and the belite content higher in the coarse fractions. The rate of hydration of the phases is generally stated by Mindess and Young (1981) as being aluminate >alite >ferrite >belite. Combining this data and applying it to a well cured concrete of moderate cement content suggests that the aluminate phase should be rare or absent and belite and ferrite phases the most common phases to be observed in remnant and partially hydrated grains. The amount of alite present should be less than the belite and ferrite but this may be affected by some of the factors discussed above. This is in agreement with observation of well cured concretes of moderate cement content where it is found that belite and ferrite are the most frequently observed residual clinker phases.

5.2.3   INFORMATION FROM REMNANT AND PARTIALLY HYDRATED GRAINS OF CEMENT

If a sample of concrete is polished and then etched either with HF vapour or with 1 per cent nitric acid in ethanol, textural details of the clinker grains remaining in the hardened paste may be observed by examination in reflected light. Application of this technique to distinguish between sulfate resisting and ordinary Portland cements was developed by Grove (1968) and is now incorporated in BS 1881: Pt 124: 1988. The practical considerations of using the technique are discussed in Chapter 4. It is claimed the technique may also be used to determine the water/cement ratio by point-counting residual grains of clinker as described by Ravenscroft (1982).

A method for determining the apparent water/cement ratio of a concrete based on counting in thin-section the number of residual clinker particles per millimetre traverse of paste has been proposed by French (1991). He considers that although it may appear crude, in fact the method is very robust and gives satisfactory results provided a wide range of standard materials is used for calibration.

In ASTM C856-85 (Table 4) under the heading 'Relict cement grains and hydration products' it states 'Cements from different sources have different colors of alumino ferrite and the calcium silicates have pale green or yellow or white shades. It should be possible to match cements from one source', presumably using these colours. There appears to be little published information on the use of this technique.

5.2.4 THE COLOUR OF HARDENED CEMENT PASTE

The normal, greenish-black colour of cement clinker is mainly due to the ferric ion Fe^{3+} in the ferrite phase which in thin-section is a deep brown colour as illustrated by Campbell (1986). Lea (1970), discusses the constituents, principally manganese and magnesium, that darken the colour of the ferrite phase and produce darker coloured cements. More recently Ichikawa and Komukai (1993) reported that substitution of Al^{3+} and Fe^{3+} by Mg and Zn changes the ferrite colour to grey while an increase of the ratio of Si and Ti to the divalent ions prevents this colour change. The effect by various metallic ions on white cement is discussed by Kupper and Schmid-Meil (1988) and methods to deliberately colour cement clinker have been discussed by Laxmi *et al.* (1984).

Overall, the colour of fresh concrete is controlled by both the colour of the cement and the aggregate. Sand particles less than 0.5 mm have an important effect while the colour of the cement is controlled by the raw materials used in its manufacture. Mineral admixtures may also affect colour. Silica fume and fly ash dependent on its carbon content will darken the hardened cement paste while interground limestone and slag will lighten the colour. The colour of the external surfaces of old concrete change as they weather and become covered with efflorescence, moss and lichens.

Mielenz (1962), describing the appearance of fracture surfaces in a sound, well-cured concrete of low water/cement ratio, states

> the fracture surfaces through the cement paste are grey, and the freshly fractured cement paste is somewhat vitreous and amorphous in appearance. Close inspection reveals black or dark-brown, vitreous, angular coarse particles of the cement embedded in the paste and still largely unhydrated ... with an increase in water/cement ratio ... the cement paste is weaker, softer, lighter in colour, and less vitreous in lustre and possibly somewhat granular in appearance, in contrast to equivalent concrete of high quality.

The differences in the texture of fracture surfaces as described above are visible under the low-power stereomicroscope. They are a useful rough guide to concrete quality and their evaluation forms part of the initial examination of a concrete.

The colour of the hardened cement paste in thin-section is primarily a function of light absorption which is affected by both thickness and the composition of the hardened paste. Typically in a 25 μm thick section the paste is light brown in colour but may often have a greenish tint or more rarely a distinct yellow cast. Two dark coloured concretes, sampled from bridges, when examined in thin-section were yellow and contained distinctive small fragments of a reddish interstitial phase attached to the belite. The belite was generally of low birefringence. The dark colour suggested the cement used was high in manganese. French (1991a) reports that where moisture passes through concrete or the water/cement ratio is high the calcium silicate hydrates tend to be lighter in colour, more transparent and of a coarser texture than paste of a low hydration state.

There are difficulties in accurately reproducing the subtle colours observed in thin-section as colour prints have a limited contrast ratio which is often exceeded in thin-section. This results in 'green disease' and other colour distortions as discussed by McCrone and Delly (1973).

5.2.5   ETTRINGITE

Ettringite ($3CaSO_4.3CaO.Al_2O_3.26\text{-}32H_2O$) is produced by a reaction that requires an excess of the sulfate ion SO_4^{2+} over the aluminate phase in the pore solution. Its formation during the setting of concrete is expansive but as the concrete is still in a plastic state the volume increase is easily accommodated. As hardening proceeds more of the aluminate phase moves into the pore solution converting the ettringite to monocalcium sulfo-aluminate ($CaSO_4.3CaO.Al_2O_3.12H_2O$). This conversion results in a small volume decrease and thus may occur in the hardened paste without damage. Both ettringite and monocalcium sulfo-aluminate hydrate formed during setting and hardening are usually submicroscopic and require the use of the SEM for their observation. Only where ettringite is secondary and reformed from reaction of the SO_4^{2+} ion with monocalcium sulfo-aluminate do its crystals become large enough to be observed in thin-section. A general review on the role of ettringite in concrete has been made by Negro (1985).

Where ettringite is observed in cracks and voids it appears as needles (Plate 25(a) which have a distinctive low birefringence and low refractive index, it is often found in conjunction with calcium hydroxide. Under high magnification well crystallised needles of ettringite are revealed as rods, Kennerley (1965) and not needles as shown in Plate 25(b). Because the ettringite is the stable phase with respect to monocalcium sulfo-aluminate below 60°C deposits of ettringite are more common in concrete than would be expected. Situations such as movement of water through a concrete or the formation of alkali-silica gel appear to be able to cause conditions that allow it to reform. In old concretes these deposits can be massive (Plate 26). However, as the ettringite is reformed in void space its expansion can be accommodated.

This should be clearly distinguished from the formation of ettringite by sulfate attack, or the situation where a concrete has been steam cured and later reaction of carbon dioxide with monocalcium sulfo-aluminate causes its formation, Kuzel (1990). Ettringite formation in the fabric of the hardened paste due to the ingress of the sulfate ion, as opposed to redeposition in pores, cracks and fissures as described above, is always potentially damaging (Deng Min and Tang Mingshu 1994). French (1991a) claims that significant formation of ettringite due to seepage of water through a concrete can, in some instances, lead to expansion and cracking.

Ettringite is subject to carbonation, decomposing to calcite, gypsum and alumina gel (Grounds *et al.* 1988, Nishikawa *et al.* 1992) when in contact with water containing carbon dioxide. Under dry conditions the fibrous habit of ettringite persists as its decomposition is retarded by these conditions.

5.2.6 HIGH-ALUMINA CEMENT

High-alumina cement, also known as calcium aluminate cement or 'Ciment Fondu' is commonly manufactured from bauxite and limestone. However, its composition is more variable than Portland cement because of the wider range of raw materials and sintering processes used in its manufacture (Robson 1964). The principal cementing compound, monocalcium aluminate ($CaO.Al_2O_3$), forms about 60 per cent of the composition with up to 10 per cent of di-calcium silicate ($2CaO.SiO_2$), 5–20 per cent of gehlenite ($2CaO.Al_2O_3.SiO_2$) and 10–25 per cent of minor constituents. The latter are a mixture of $12CaO.7Al_2O_3$, ferrous oxide (FeO), the ferrite phase and pleochroite. The important properties of high-alumina cement are its ability to produce strengths in 1 day which are equivalent to the 7-day strength of Portland cement and its sulfate resistance. It can also be fired to produce a refractory material.

Concrete made with high-alumina cement has a serious disadvantage in that over time it undergoes a significant loss of strength when exposed to warm moist conditions (Neville 1975) especially if it is mixed at a water/cement ratio exceeding 0.4. This loss of strength is primarily associated with a slow conversion of the two main hydrates $CaO.Al_2O_3.10H_2O$ and $2CaO.Al_2O_3.8H_2O$ to $3CaO.Al_2O_3.6H_2O$ which increases the porosity. This increase in porosity can affect both strength and durability. To detect and measure the extent of the conversion it is necessary to analyse the concrete using X-ray diffraction and thermal analysis techniques (Midgley and Midgley 1975 and Redler 1991). Recent work indicates that blending high-alumina cement with ground granulated blastfurnace slag may avoid this problem of slow conversion of the phases (Majumdar et al. 1990a,b) and an overview of the recent use of high-alumina cement has been given by Bensted (1993).

The hydration reactions of high-alumina cement are too complex to discuss here. The detailed reactions have been given by Midgley and Midgley (1975). An important point is that no calcium hydroxide is produced as any di-calcium silicate present reacts with alumina formed during the hydration. No calcium hydroxide is formed by hydrolysis. However, concretes made with high-alumina cements are still subject to carbonation as outlined in the reactions given in Section 5.4.

The microscopical examination of high-alumina cement is more complex than Portland cement because of its greater variability in composition, solid solution between minerals and extensive formation of glasses. Examination of the mineral phases in high-alumina cement is best carried out on polished sections with selective etching (Parker, 1952). Lea, (1970), gives some optical data on the mineral phases and Junior et al. (1990) and Venkateswaran et al. (1991) illustrate some textures in polished section. The appearance of a high-alumina cement mounted in immersion oil is illustrated in Plate 27.

Plate 28 shows details of the texture of a grout made from high-alumina cement which after normal curing has been exposed to steam at 150°C. The texture consists of remnant and partially hydrated grains of the cement embedded in an undifferentiated matrix that appears isotropic at the magnification used. There does not appear to be any description of the textures of high-alumina concretes as observed in thin-section and the effect, if any, of conversion of the phases on these textures.

5.2.7 PORTLAND LIMESTONE CEMENT

Between 10–30 per cent of limestone may be interground with Portland cement clinker. It primarily acts as a diluent allowing the production of a cement requiring less energy input for its manufacture. As the limestone is softer than the cement clinker, intergrinding produces a finer particle size range which has a plasticising effect in concrete. Any loss in strength due to dilution of the cement with the limestone is reduced by lower water demand, less bleeding, and increased workability. While some claim the durability of Portland limestone cements do not appear to differ significantly from Portland cements in equivalent concretes, problems of increased plastic shrinkage and cracking have been reported (Baron and Douvre 1987, Moir and Kelham 1989).

Finely ground powders can have a catalysing effect on the hydration of the silicate phases (Ramachandran 1988). The limestone may also react with the aluminate phases to form carbo-aluminates in a similar manner to gypsum so the limestone acts as more than an inert filler in a number of ways. Surprisingly, many standards for ordinary Portland cement permit the addition of up to 5 per cent fillers without restriction but any limestone addition exceeding 5 per cent classes the cement as a Portland/limestone cement. Typically, limestone addition to cement is in the range 10–15 per cent although European specifications permit additions of up to 30 per cent. While standards may not require the limestone to contain more than 75–80 per cent calcium carbonate in practice a minimum of 90 per cent is recommended and limits on clay and dolomite contents may be imposed.

Data on the lowest level of limestone addition that can be observed in a thin-section of concrete does not appear to be available. It is clearly visible in the texture of a concrete made from a Portland/10 per cent limestone cement given in Plate 29. Apart from the presence of the high-order birefringence of the fine particles of the limestone, the texture of the hardened paste appears little different from that of an OPC. Where doubt exists examination of a fracture surface of the concrete by scanning electron microscopy should reveal the presence of small rhombs of calcium carbonate.

5.2.8   PORTLAND POZZOLANA CEMENTS

A wide range of materials, under the generic name of pozzolana, can be used to partially replace Portland cement. Common materials that have been used as pozzolana are the industrial waste products fly ash (Berry and Malhotra (1987), Concrete Society (1991)); silica fume (ACI Committee 226, (1987), Concrete Society (1993)); natural materials such as diatomite, pumicite, tuff and trass (Mehta (1987)) and calcined shale and calcined clays. Swamy (1992), Massazza (1993) and ACI Committee 232.1R, 232.2R and 234R (1994) have reviewed the properties and uses of pozzolana cements and German experience with blended cements is discussed by Schmidt *et al.* (1993).

The simplified reaction of pozzolanas such as silica fume and diatomite with Portland cement is as follows:

$$SiO_2 + Ca(OH)_2 = CaO.SiO_2.nH_2O$$

The end product is a calcium silicate hydrate which may be indistinguishable from similar hydrates formed from the hydration of the Portland cement. This still remains the dominant reaction for those pozzolanas with an alumino-silicate type composition but calcium alumino-silicate and calcium aluminate hydrates are also formed which once again may be indistinguishable from similar cement hydrates.

While pozzolanas cover a considerable range of materials the reactive constituents they contain are restricted to three mineral groups. These are species of amorphous silica, as found in silica fume and diatomite, glassy and amorphous alumino-silicates present in fly ash, pumicite, calcined shale and clay, and altered alumino-silicates of a zeolitic nature which may be present in tuff and trass. The difference between the last two groups is not well defined.

In examining the textural effects of incorporating pozzolana into concrete the petrographer is faced with a complex situation. The two observable effects on texture are the removal of calcium hydroxide due to reaction of the pozzolana and the presence of pozzolanic residues. Observation of both of these largely depends on the reactivity of the pozzolana and the amount of residue remaining when a pozzolana is passed wet through a 45 μm sieve. As a rough indication natural pozzolanas, which have been ground to give minimum water demand, should have a mean particle diameter not exceeding 10 μm and residues of not more than 12 per cent.

The most recent pozzolana to come into general use is silica fume produced as a by-product of the silicon and ferrosilicon industries (Aitcin 1983). It differs from other pozzolanas in consisting of spherical particles of reactive, amorphous silica which range from 20 nm to 1 μm in size. In theory this approximates to a mean particle size of about 0.1 μm which is about 100 times finer than any other pozzolana in use (ACI Committee 226 1987). Up to 10 per cent of other material may be present mainly as coarser material which Bonen and Diamond (1992) report consists mainly of agglomerates of silica fume, quartz, cristobalite, silicon carbide, metallic silicon and carbon particles.

Silica fume is collected as a loose powder with a bulk density of about 130 kg/m^3 which is 'densified' into a more manageable form by blowing air through the storage silo. After one day's aeration the bulk density increases to about 350 kg/m^3 which is marketed as 'undensified' silica fume and after a week's aeration the bulk density increases to about 700 kg/m^3 which is marketed as 'densified' silica fume (Plate 29(b)). It is also densified by spraying water onto fume in a drum pelletiser (Malhotra and Hemmings 1995). Silica fume is available for use in concrete in both powdered form and as a 50 per cent stabilised slurry. The densification of silica fume is caused by agglomeration of particles which has a fundamental effect on its properties. It has been shown that the ability to disperse agglomerates of silica fume by ultrasonic dispersion varies with source and time of storage (Asakura *et al.* 1993 St John 1994) and that once some silica fumes are agglomerated they cannot be redispersed. It has been suggested by St John (1994) that the tangling of the spherical particles fused into chains observed by transmission electron microscopy is primarily responsible for agglomerates that cannot be redispersed. He also proposed that the ultrasonic dispersivity of a silica fume as measured by laser diffractometry is a good indicator of whether a silica fume will disperse well in concrete.

Effectively, the mean particle size of silica fume varies between about 1 and 50 μm dependent on the source of the fume and time of storage. This applies both to powders and slurries. Sufficient individual particles remain and some agglomerates are so small that the special effects deriving from silica fume packing between cement grains is maintained but this is highly variable and is the probable explanation for the different efficiencies of silica fume in practice. When mixed into concrete and super-high-strength DSP mortars agglomerates larger than about 30 μm persist both in concrete (Sveinsdottir and Gudmundsson 1993, Shayan *et al.* 1993) and can, under appropriate conditions of alkali and moisture content, lead to alkali-aggregate reaction (St John *et al.* 1994).

The texture of concrete containing silica fume will vary according to the amount and dispersivity of the silica fume and the water/cement ratio. For ultra-high-strength DSP materials, which contain up to 25 per cent of silica fume by weight of cement and are mixed at very low water/cement ratios, agglomerates will always be present irrespective of the method of curing and mixing (Plate 24(b)). In addition no calcium hydroxide will be

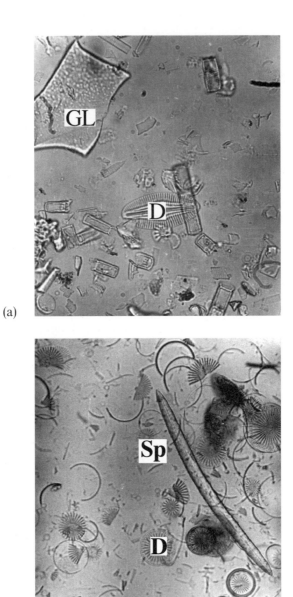

(a)

(b)

Fig. 5.1 The appearance of some diatoms and glass shards found in diatomite and pumicite. (a) This diatomite (Whirinakei, New Zealand) is a mixture of diatoms (D) rhyolitic glass (Gl) and some clay. The glass shards have distinctive conchoidal fractures as shown in the large particle seen here. It is the unreacted ground-up remnants of the structures shown here that are used for their identification in concrete. Width of field 0.25 mm. (b) This is a cleaner diatomite (D), (Kaharoa, New Zealand), which also contains sponge spicules (Sp) which are less reactive than the diatoms and thus persist in concrete. They are often the only remnant that can be identified. A large, intact sponge spicule is clearly visible. Width of field 0.3 mm.

observed. This is confirmed by scanning electron microscope observation of fracture surfaces of DSP which show only a featureless hardened cement paste with no transition zone between the paste and the aggregates.

For high-strength concrete which will normally contain only about 10 per cent of silica fume by weight of cement, some calcium hydroxide may be present dependent on age and curing. Agglomerates of silica fume will be present but may require a careful search of a large area of paste to be observed. These agglomerates are all greater than about 25 μm in diameter, often have a light-brown interior and may show signs of scattered points of birefringence indicating partial reaction has occurred. Even in typical structural concretes, once the agglomerates have formed they are persistent as shown by the report by Sveinsdottir and Gudmundsson (1993) on Icelandic concretes. It is also of interest to note that the silica fume used in these Icelandic concretes was interground with the cement used and was originally present as 7 per cent by weight of cement.

Diatomite, which consists of the opaline silica shells of minute marine and freshwater plants, requires grinding to destroy the complex structures of the diatoms. Otherwise its use as a pozzolana will cause excessive water demand in concrete. To achieve an acceptable water demand the grinding must reduce the mean particle size to about 5 μm and residues must not exceed 5 per cent on the 45 μm sieve. The residue may contain fragments of sponge spicules which are somewhat more resistant to grinding than the softer diatoms Fig. 5.1(b). Diatomite is a less reactive material than silica fume, mainly because of its larger particle size. In concretes older than a few weeks not many diatoms will be visible under the optical microscope but careful searching by scanning electron microscopy may reveal fragments. The only fragments likely to be visible under the optical microscope are the remains of sponge spicules.

Pulverised fuel ash (PFA), commonly referred to as fly ash, is currently the pozzolana that finds greatest use. It varies considerably in composition dependent on the type of coal from which the ash is derived and this is the basis of its specification in ASTM C618. Class F fly ash is produced from burning anthracite or bituminous coal and Class C from lignite or sub-bituminous coals. In broad terms Class F is a low-lime ash and Class C a high-lime ash whose calcium oxide content varies from about 10–30 per cent which gives it some weak cementitious properties. Class C ashes also contain a higher proportion of non-reactive material such as quartz and magnetite (Fig. 5.2). The reactive materials in fly ash are glassy spheres consisting of an alumino-silicate glass in which incipient needles of mullite are present. The size range of the spheres varies from about 10–100 μm in diameter but the actual particle size range present is dependent on the zones of the electrostatic precipitator from which the material is collected. Swamy (1993) has reviewed the requirements of European standards for the use of fly ash.

Even in old concrete, fly ash is easily recognisable under the microscope. Some of the larger spheres persist although they may be coated with calcium hydroxide and it may be possible to observe particles of magnetite. Detection is easier as fly ash is commonly used in the replacement range 20–40 per cent unlike silica fume and diatomite which are used in the 10–15 per cent range. Fly-ash darkens the overall colour of concrete and this is an additional clue. The textural effects in the hardened cement paste are not as clear as with silica fume and diatomite. The amount of calcium hydroxide remaining at six months, for replacements varying from 0–60 per cent fly ash have been measured by Taylor *et al.* (1985). These results show that even at 60 per cent replacement 25 per cent of the calcium hydroxide still remains while at 25 per cent replacement some 80 per cent remains. Thus the effects of fly ash on concrete texture may not be apparent at lower replacement levels when

(a)

(b)

Fig. 5.2 The appearance of fly ash and its cenospheres. Width of fields 1 mm (a) This high lime fly ash (Meremere, New Zealand), contains typical glassy spheres, lumps of lime-rich material, opaque fragments which are probably magnetite and other debris. It is remnants of the glassy spheres and particles of magnetite that are mainly used for identification in concrete. (b) If fly ash is separated in water many of the glass spheres float. These are known as cenospheres and consist of hollow spheres of glass material containing incipient crystals of mullite. Both these photomicrographs were taken with combined transmitted and oblique lighting.

attempting to observe the amount of calcium hydroxide and densification present in the texture.

Pozzolana, based on volcanic alumino-silicate glasses such as pumicite, are slow-reacting materials and residues can be observed in concrete under the microscope as isotropic fragments with conchoidal fracture (Fig. 5.1(a)). Depletion of calcium hydroxide will be less than that occurring with the more reactive silica fume and diatomites. However, it is impossible to detect a pumicite where the concrete sand is derived from acid and intermediate volcanics as the glassy material present is similar in both materials. The effects on texture are similar to those found with fly ash. Pumicite and similar materials may be used in concrete in the 15–30 per cent range so there is a reasonable chance of their detection.

The remaining types of pozzolanas require the same approach for their detection. In each case, detection will be dependent on being able to recognise some distinctive residue and this requires a knowledge of the microscopic appearance of the pozzolana both as raw material and in its ground state. The problem reduces itself to looking for 'needles in a haystack' and requires careful searching in the thin-section.

In many cases it is far more difficult to positively identify whether a pozzolana has been used than is commonly assumed. This is especially the case at lower cement contents and replacements as the following case illustrates. A laboratory concrete, containing 190 kg/m^3 of cement and 27 kg/m^3 (12.5 per cent replacement) of pumicite ground to optimum fineness, was examined after 25 years of exposure to the weather. The texture of the hardened cement paste was considerably depleted in calcium hydroxide and the remaining calcium hydroxide crystals were small in size. No observable fragments of pumicite were visible in thin-section and after careful, prolonged searching of fracture surfaces by scanning electron microscopy one fragment of volcanic glass was observed. It is most unlikely that this one fragment would have been located in a routine examination.

5.2.9 INFORMATION FROM THIN-SECTIONS CONTAINING POZZOLANAS

Pozzolanas used in concrete have been described in Sections 4.6.4 and 5.2.8, but unless the pozzolanic residues can be positively identified by microscopy or scanning electron microscopy it is only possible to infer the presence of a pozzolana in a concrete from the change in texture caused by a reduction in the amount of calcium hydroxide present. Fortunately, in the case of fly ash and silica fume, which are the most commonly used pozzolanas, visible residues are usually present which allow unequivocal identification. In old concretes many residues may be difficult to locate and identify without extracting them by dissolution of the hardened paste and this difficulty increases with the age of the concrete. These difficulties are compounded when the original materials are no longer available for comparison purposes.

Most pozzolanic residues can be identified in concretes which are less than one month old because the reaction has not had sufficient time to consume too much of the pozzolana. In these circumstances, quantitative estimates are possible either by chemical methods (Concrete Society 1989) or electron microprobe analyses (Barker 1990 and French 1991a) but the reliability of these estimates is variable.

5.2.10 PORTLAND BLASTFURNACE SLAG CEMENTS

Blastfurnace slag is produced as a waste product from the smelting of iron ores. The compositions and properties of blast-furnace slags are discussed by Lea (1970) and more recently have been reviewed by Hooton (1987), Concrete Society (1991) and Swamy (1993). Non-ferrous slags are also available but as yet have found little use (Douglas and Malhotra 1987).

If the slag is crystalline, a mineral of variable composition called melilite is formed together with a range of other minor minerals (Nurse and Midgley 1951 and Minato 1968). Crystalline blastfurnace slag has little hydraulic activity and is principally used as an aggregate or for road fill. To give the slag hydraulic properties it is granulated by rapid cooling of the molten slag with water to produce a glassy structure and then ground to a specific surface of approximately 4000 cm^2/g. With the addition of activators setting properties similar to those of Portland cement can be achieved.

Crystalline melilite forms a solid solution series with the end members akermanite ($2CaO.MgO.SiO_2$) and gehlenite ($2CaO.Al_2O_3.SiO_2$) the percentages of which can be estimated by measuring the refractive indices of the melilite (Winchell and Winchell 1964). The melilite glass that forms during granulation of a slag approximates to a portion of the melilite solid solution series and it is claimed the composition can be estimated from the refractive index (Battigin 1986). However, the refractive indices of slag glasses are not the same as those of the crystalline melilite and they should not be confused with each other. As many specifications set limits on the composition of slag cements, estimating their composition from the refractive indices of the glasses present would provide a rapid method of analysis.

The percentage of glass in a granulated slag is considered to be an important indicator of its reactivity. Hooton and Emery (1983) have developed a microscopic method for estimating the glass content of granulated slags by crushing to about 50 μm and examining the grains in immersion oil and Dengler and Montgomery (1984) report some results using immersion oil techniques. The technique is described in the Canadian Standard (CSA 1983). The British Standard methods and the difficulties of quantifying the proportion of slag present have been discussed in Section 4.6.2. However, Drissen (1995) has found that the estimation of glass content using transmitted light has advantages over other methods and describes the techniques required.

In thin-section, melilite crystals have a distinctive appearance (Plate 30(a)), with rectangular or square cross-sections often zoned and show an anomalous blue interference colour under crossed polars (Insley and Fréchette 1955). Granulated slag is usually examined in reflected light where it is necessary to etch the polished section to reveal the details of structure. The truly glassy fragments do not etch but it will be found that many fragments are partially or completely crystalline dependent on the degree of vitrification of the slag. Many distinctive and rather beautiful patterns are formed by etching these areas of crystallinity. Plate 30(b). It is difficult to identify these areas of crystallinity by optical means without the aid of X-ray analysis. The general appearance in transmitted light of ground granulated blastfurnace slag in concrete is shown in Plate 31.

Another optical technique has been proposed for determining the hydraulic reactivity of granulated slags that is based on the flourescence colours produced by UV light (Grade 1968). The technique appears to be suitable for ranking slags from the same works but there is some question whether it is applicable to slags from different sources. It should be noted that the optical methods of estimating the reactivity of granulated slag are mainly applicable to the unground material. If the activity of the ground slag is required the alkali activity test as specified in ASTM C1073 – 91 or the measurement of compressive strength as specified in BS 6699 (1992) may be used.

The commonest use of granulated slag is to replace between 25–70 per cent of the Portland cement either by intergrinding or preferably as an addition of the separately ground material. The lime and alkalis in the Portland cement are sufficient to activate the granulated slag. Portland blastfurnace slag cements are covered in specifications such as BS 146 (1991), BS 4246 (1991), BS 6699 (1992), ASTM C595 (1995), ASTM C989 (1995) and those from many other countries. A specific code of practice for its use in construction, Japan Society of Civil Engineers (1988), has been published in Japan.

Since ground granulated slag is whiter than Portland cement it imparts a lighter colour to the concrete and the presence of calcium sulfide interacting with the ferrite phase may impart a bluish colour (Powers-Couche 1992) which will fade with oxidation. Detection of the presence of granulated slag in concrete presents few problems either in thin- or polished section because granulated slag hydrates more slowly compared with Portland cement and the larger grains tend to remain relatively intact and are often coated with calcium hydroxide. The effect on the texture of the hardened paste is to produce less calcium hydroxide than would be found in Portland cement alone and this effect reaches a maximum at about 60 per cent slag replacement (Taylor 1990). As the principal hydration products are similar to those of Portland cement the appearance of the hardened paste texture will not be significantly different because of the presence of slag apart from the presence of residual grains.

Supersulfated cement consists of granulated blastfurnace slag with activators. The minimum slag content allowed is 75 per cent but typically it contains 10–15 per cent of anhydrite and about 5 per cent of Portland cement as activators. Its principal use is for concrete which is required to withstand severe sulfate and chemical attack. The composition and other properties required are specified in BS 4248: 1974. The principal phases formed on hydration are laths of ettringite, up to 120 μm long and 0.5 μm thick infilled by sub-microscopic calcium silicate hydrate. No calcium hydroxide is formed (Midgley and Pettifer 1971). Thus the texture of supersulfated cement in hardened concrete will be distinctive and consist of interlocked laths of ettringite set in a matrix of unresolvable calcium silicate hydrate with a complete absence of calcium hydroxide. Supersulfated cement is more sensitive to carbonation during storage than ordinary Portland cements (Lea 1970). Concretes made from supersulfated cement lose strength due to carbonation (Manns and Wesche 1968) but as this is usually restricted to the outer surface of a structural member it is not structurally significant.

The use of granulated slag activated by alkalis in the form of potassium and sodium hydroxide or carbonates or soluble glasses is finding increasing use in eastern Europe (Talling and Brandstetr 1989) where chemical resistance and lower costs are required. The product, known as F-cement, produces calcium silicate hydrates without any calcium hydroxide and while it has good chemical resistance is more prone to carbonation than ordinary Portland cement. The carbonation products tend to line microcracking in the hardened paste (Byfors *et al.* 1989) and the final product of carbonation is silica gel which can be reconverted to calcium silica hydrates by realkalisation (Bier *et al.* 1989).

The quantitative estimation of slag content in concrete presents considerable problems which increase with the age of the concrete. Point counting grains of slag in thin- or polished-section is questionable even in young concretes and the best methods available utilise determination of the chemical composition of the hardened paste the interpretation of which requires some knowledge of the compositions of the cement and slag used (French 1991a, Concrete Society 1988).

5.2.11   THE MICROSCOPICAL EXAMINATION OF BLASTFURNACE SLAGS

The microscopical examination of granulated blastfurnace slags can be used to determine both the composition and amount of glass present which are useful parameters for estimating the potential reactivity of the slag. Microscopic examination of concrete in thin-section can identify the presence of slag but quantitative estimation requires the use of chemical or electron probe microanalyses.

5.3   Calcium hydroxide

Calcium hydroxide, also known as Portlandite, is produced by the hydration of the silicate phases, tricalcium and dicalcium silicate, and free lime (CaO) as shown in the approximate equations given below. No calcium hydroxide is produced from the hydration of the aluminate and ferrite phases.

$$2(3CaO.SiO_2) + 6H_2O \rightarrow 3CaO.2SiO_2.3H_2O + 3Ca\,(OH)_2$$
$$2(2CaO.SiO_2) + 4H_2O \rightarrow 3CaO.2SiO_2.3H_2O + Ca(OH)_2$$
$$CaO + H_2O \rightarrow Ca(OH)_2$$

From the first two equations it can be seen that weight for weight the hydration of tricalcium silicate ($3CaO.SiO_2$) produces three times the amount of calcium hydroxide as does the hydration of dicalcium silicate ($2CaO.SiO_2$). Thus the ratio of these two phases will affect the amount of calcium hydroxide formed. The range of the calcium silicates to be found in cements has been recently reviewed by Concrete Society (1987) and is given in Table 5.3.

Table 5.3 Percentages of tricalcium silicate and dicalcium silicate in cements for the period 1914–1984.

Date	−1914	1914–22	1928–30	−1944	−1960	−1980	−1984
% Tricalcium silicate	25	15–48	19–58	30–50	36–55	45–64	54–63
% Dicalcium silicate	45	81–26	53–14	45–20	37–12	30–11	27–8
Ratio	0.6	0.2–1.8	0.4–4.1	0.7–2.5	1.0–4.6	1.5–5.8	2.0–7.9

The effects of these changes on the amounts of calcium hydroxide produced have been estimated for seven British cement plants (Concrete Society 1987) (see also Table 5.4) and are less than would be expected mainly because the cements high in dicalcium silicate also have higher free lime contents.

The results quoted in Table 5.4 for a cement with the maximum tricalcium silicate are typical of current ordinary Portland cements while the cement composition with the lowest tricalcium silicate approximates to a low-heat cement. As the above results were calculated for a fully hydrated cement paste with no free water present they are somewhat artificial and a more practical idea of the amount of calcium hydroxide likely to be present is given in Table 5.1. The above results indicate that changes in cement composition over the years have not had a marked effect on the amount of calcium hydroxide present in concrete.

Table 5.4 The effect of changes in cement composition on calcium hydroxide and calcium silicate hydrate gel contents of hydrated cement paste.

	Typical composition		Range of annual average compositions in 1988	
Cement minerals (%)	1930	1980	Lowest tricalcium silicate cement	Highest tricalcium silicate cement
Free CaO	2.6	1.7	2.2	1.0
Tricalcium silicate	39	53	45	64
Dicalcium silicate	32	18	21	11
Calcium hydroxide calculated	25	27	25	30
Calcium silicate hydrate gel calculated	52	49	47	50

5.3.1   THE OCCURRENCE OF CALCIUM HYDROXIDE IN HARDENED CEMENT PASTE

Bache *et al.* (1966) have reviewed the occurrence and morphology of calcium hydroxide in cement paste and give a series of micrographs showing its appearance in thin-section. In pores, where sufficient space and fluid is present to allow adequate crystallisation, large laths of calcium hydroxide occur or a fissure may be filled as shown in Plate 32.

Where fluid has not filled the pore or fissure, calcium hydroxide crystals will only form on surfaces. However, the greater portion of calcium hydroxide present is scattered through the hardened cement paste where it is in the form of irregular clumps of variably sized crystals filling void space. The maximum size of capillary void space varies with the water/cement ratio and this will affect the sizes of calcium hydroxide crystals in the paste as shown in Plate 33. Other examples of calcium hydroxide in the paste texture are illustrated in many of the figures in this chapter. Bache *et al.* (1966) point out that because the saturation concentration of calcium increases with decreasing calcium hydroxide crystal size there will be a tendency towards the growth of large crystals at the expense of small ones dependent on the rate of diffusion of calcium hydroxide in the liquid-filled system.

In the above discussion it has been assumed that all the calcium hydroxide present in a hardened paste is crystalline and visible. Midgley (1979) estimated that about 18 per cent of the calcium hydroxide present in a hardened paste is amorphous material by comparing the results of thermal analysis with those from X-ray diffraction. He considered this amorphous calcium hydroxide to be present as small diameter globular particles of about 0.1 μm but Taylor (1990) states that there does not appear to be any convincing evidence for this claim. Groves (1981) and Groves and Richardson (1994) claim to have observed the presence of microcrystalline calcium hydroxide but this is disputed by Rayment and Majumdar (1982). However, it is wiser to assume that the upper limit for observable calcium hydroxide in thin-section may only be 80 per cent of the theoretical and in practice concretes may contain considerably less.

Mather (1952) suggests that the formation of calcium hydroxide crystals at aggregate interfaces is affected by the composition of the aggregate and that smaller crystals are observed on limestone aggregate interfaces than on siliceous aggregates. Bache *et al.* (1966) did not find this effect and considers that the surface texture of the aggregates may be of greater importance. French (1991a) considers that calcium hydroxide is more conspicuous on siliceous fine aggregates such as chert and quartz which may accumulate relatively thick layers whereas on carbonate rocks the layer is thin and irregular. In general, finer sands result in greater development of calcium hydroxide on surfaces. Serpentinites develop thick haloes and added materials such as the larger particles of fly ash and slag are often heavily coated with calcium hydroxide. Brandt (1993) observed the formation of an unusual spherulitic calcium hydroxide in voids and cracks where the concentration of superplasticiser exceeded 25 g/l of hardened paste.

The effects discussed above are localised concentrations controlled by both the amount of water and space at particle surfaces with crystallisation modified by the surface texture and porosity of the aggregate. Crystallisation of calcium hydroxide at surfaces is an important factor in strength and is discussed in Section 5.3.4.

5.3.2   THE EFFECTS OF CURING, LEACHING AND REACTION ON CALCIUM HYDROXIDE

Since calcium hydroxide is ubiquitous and tends to form in any space or on any surface where pore fluid is present, its absence or alteration can be indicative of the past history of the concrete. The most obvious example is the carbonation of calcium hydroxide which is discussed in Section 5.4. Where it appears to be absent from a hardened cement paste this may indicate that either high-alumina or supersulfated cement has been used because these cements do not produce calcium hydroxide during their hydration. Alternatively, the concrete either may contain a siliceous addition such as silica fume (Plate 24(b)) or be a high-pressure, steam-cured product containing a silica flour. Although in these cases most of the calcium hydroxide will have reacted it is not unusual for some residual calcium hydroxide to be observed.

Porous aggregates may have a surrounding zone of paste (Plate 34(a)) in which calcium hydroxide is deficient or absent and in rare cases calcium hydroxide will have crystallised in the porous rims of such aggregates. Where an aggregate has reacted to produce an alkali-silica gel which has bled into the surrounding paste (Plate 35) calcium hydroxide will be absent as the calcium is taken up by the alkali-silica gel.

The absence of calcium hydroxide from pores, cracks and fissures indicates that there has been insufficient fluid in these spaces for its formation. Along the margins of cracks through which water has percolated, (Plate 34(b)) calcium hydroxide will often be depleted by leaching. Careful observation of the way in which calcium hydroxide is present in localised areas of a hardened cement paste can provide useful information on the history of the concrete.

5.3.3   EFFECT OF TEMPERATURE ON CALCIUM HYDROXIDE

The solubility of calcium hydroxide is depressed by increasing temperature. The solubility of calcium hydroxide is given as 1.37 g/l at 0°C, 1.28 g/l at 15°C, 1.04 g/l at 40°C and 0.51 g/l at 100°C (Kaye and Laby 1986). This effect of decreasing solubility with temperature reduces the size of calcium hydroxide crystals formed as noted by Berger and McGregor (1973) in spite of the solubility being considerably reduced by the presence of alkali sulfates in the pore solution. Larger crystals of calcium hydroxide should form in concretes exposed to cooler environments.

5.3.4   EFFECT OF CALCIUM HYDROXIDE ON STRENGTH

Calcium hydroxide is of considerable importance in concrete as its formation both in the fabric of the hardened cement paste and at interfaces affects strength. Bentur and Mindess (1990) have reviewed the effect that duplex layers of calcium hydroxide and calcium silicate hydrate, which form at aggregate and reinforcement faces, have on bond strength. The contribution of calcium hydroxide to the strength of concrete is controversial and has been reviewed by Barker (1984) who also discusses morphology. Yilmaz and Glasser (1991) have reported that the morphology of calcium hydroxide is strongly affected by superplasticisers. Calcium hydroxide is also important in autogenous healing of cracks and porosity in concrete where it is undoubtedly the main agent involved (Lauer and Slate 1956) although Clear (1985) showed that carbonation of calcium hydroxide by bicarbonate in water percolating through cracks is also involved. The relationship of submicroscopic calcium hydroxide at interfaces and in the fabric of the paste to the calcium hydroxide observed under the microscope is poorly defined in structural concretes.

5.3.5   INFORMATION FROM THE OBSERVATION OF CALCIUM HYDROXIDE IN CONCRETE

Since calcium hydroxide is ubiquitous and one of the few visible components of hydration in a cement paste, its quantitative estimation by microscopy would seem obvious. However, by its very nature calcium hydroxide is best observed qualitatively. An attempt to determine the amount of calcium hydroxide in concrete by point-counting has been reported by Larsen (1961).

Because calcium hydroxide forms in void space and fissures it may be thought that it would make a good indicator for defining such space. However, these spaces must be filled or partially filled with sufficient fluid for crystallisation of the calcium hydroxide to occur so its presence or absence in voids or fissures is as much an indication of moisture conditions and degree of hydration as the presence of the space itself. This can be of use if the environmental history of the concrete is known but these observations by their very nature remain qualitative.

The depletion or absence of calcium hydroxide in the cement matrix indicates that either a non-Portland cement or a siliceous admixture has been used. Apart from surface layers and along crack margins, calcium hydroxide leaching from the body of a concrete is usually restricted to porous water-retaining structures. If structural concretes are so porous that leaching is able to occur from normal exposure, carbonation occurs long before leaching of the body of the cement matrix takes place. The phrase 'the concrete is leached' should not be used and more correctly should be stated as 'the surface layers of the concrete are leached of their calcium hydroxide'.

Calcium hydroxide is highly susceptible to carbonation and the observation of the extent of this reaction is of great importance in the petrographic examination of concretes and as an indication of the permeability of the surface layers. When calcium hydroxide occurs abnormally the cause should always be investigated. The ability to interpret such observations will depend on the experience of the petrographer.

5.4   Carbonation

The effects of carbonation are of importance to the durability of reinforced concrete and the subject continues to be widely investigated. An extensive literature review of the carbonation of reinforced concrete has been published by Parrott (1987) and a damage classification by Parrott (1990).

Carbonation of a hardened cement matrix most commonly takes place by the reaction of both calcium hydroxide and cement hydrates with atmospheric carbon dioxide. The reactions are as follows:

$Ca(OH)_2 + CO_2 \rightarrow CaCO_3 + H_2O$
Calcium silicate hydrate $+ CO_2 \rightarrow$ various intermediates $\rightarrow CaCO_3 + SiO_2nH_2O + H_2O$
Aluminate hydrates $+ CO_2 \rightarrow CaCO_3 +$ hydrated alumina
Ferrite hydrates $+ CO_2 \rightarrow CaCO_3 +$ hydrated alumina and iron oxides.

These reactions show that concretes made from high-alumina cements are also subject to carbonation because of the reaction of the aluminate hydrates with carbon dioxide even though no calcium hydroxide is produced in the hydration of this type of cement. The detailed reactions that are possible between the components of concrete and carbon dioxide have been reviewed by Calleja (1980). Pozzolanic materials may increase the permeability of concrete unless it is inadequately cured leading to an increased rate of carbonation.

Carbonation acts through solution and some pore fluid, such as a film on internal surfaces, as a prerequisite for the carbonation reaction to proceed. However, the rate of carbonation is minimal for a saturated concrete, because the permeation of carbon dioxide is hindered by saturation of pores. The rate increases to a maximum where the concrete is allowed to dry to the range 70–50 per cent RH. Below this humidity range the rate reduces rapidly (Verbeck 1958) because there is no longer sufficient water in the pores and capillaries of the paste.

For the practical purpose of explaining the depth of carbonation observed, the major controlling factors are the water/cement ratio, the porosity of the aggregates, curing and the environment to which the structure is exposed. However, Brown (1991) found in an extensive survey of four hundred structures that carbonation depths correlated well with concrete quality but there was no significant relationship with reported exposure conditions. In general, any factor which increases the permeability of a concrete to carbon dioxide will also increase the rate of carbonation providing the necessary internal moisture conditions are present. Thus if a concrete is kept continuously moist the effective permeability to carbon dioxide may be low and the rate of carbonation will be minimal. The same concrete protected from the weather and exposed to relative humidities of 50–70 per cent will carbonate more rapidly while the normal outdoor exposure of wetting and drying will result in intermediate rates of carbonation.

For structural concretes the water/cement ratio has the greatest influence on carbonation. The effect of water/cement ratio on the porosity and permeability of a cement paste and its relationship to durability has been discussed by Hansen (1989) who used the classical work of Powers and co-workers to illustrate the effect of permeability on durability.

$$p = \frac{w/c - 0.36m}{w/c + 0.32}$$

where $p =$ porosity
$w/c =$ water/cement ratio
$m =$ degree of cement hydration, $0 < m < 1$

If we assume $m = 0.75$ for a typically cured concrete, $m = 0.5$ for a poorly cured concrete, and $m = 1.0$ for the theoretical limit of hydration, porosities can be calculated from the above equation as shown in Table 5.5.

Table 5.5 Porosity of cement paste calculated from water/cement ratio and degree of cement hydration (*m*).

W/C	Percentage porosity		
	m = 0.5	m = 0.75	m = 1.0
0.4	30	18	6
0.45	35	23	12
0.5	39	28	17
0.55	42	32	22
0.6	46	36	26
0.65	48	39	30
0.7	51	42	33

Powers *et al.* (1958) postulated that a paste with a porosity of less than 30 per cent would be watertight because the capillary pores in the paste become discontinuous. They also determined the time required to moist cure a concrete so that $p = 30$ per cent for a range of water/cement ratios as shown in Table 5.6. The data in Table 5.6 also show that at low water/cement ratios carbonation will be largely controlled by a degree of curing readily achievable in practice. As the water/cement ratio exceeds 0.5 the time required to reduce the paste porosity to 30 per cent or less exceeds the practical time of curing and the depth of carbonation will be controlled by the water/cement ratio.

Table 5.6 Times required to cure cement paste to obtain a porosity of 30 per cent (after Powers *et al.* 1958).

W/C	Time	Degree of hydration
0.4	3 days	0.5
0.45	7 days	0.6
0.5	14 days	0.7
0.6	6 months	0.9
0.7	1 year	1.0

Parrott (1987) quotes the results of numerous investigations which obtained good correlations between carbonation depth and the strength of the concrete. In most respects this is merely a restatement of the principles stated above. Higher strength concretes use more cement and require lower water/cement ratios than lower-strength concretes. For practical engineering purposes this method is useful as the strength of the concrete is often known. However, it can be in error if the quality of the cover concrete differs markedly from the concrete in the core of the structure because inadequate compaction can lead to an increased water/cement ratio in the outer layers. This situation is quite common in field concretes and in these cases it will be found that there may be poor correlation between depth of carbonation and overall water/cement ratio (French 1991a). However, Parrott (1992) found a near linear relationship between carbonation depths and water absorption in cover concrete in specimens stored for up to one and half years in indoor and outdoor exposure. Further work by Parrott (1994) shows that the air permeability and water absorption rate are very sensitive to the moisture content of a concrete.

The chemical reactions of carbonation result in volume changes. Calcium carbonate has a molecular volume which is 11 per cent greater than the calcium hydroxide which it replaces, but approximately 2.5 per cent less than the cement hydrates. The newly formed calcium carbonate with its increased volume now fills previously empty pores and capillaries reducing permeability. In spite of these volume changes carbonation results in an overall shrinkage of a concrete indicating the processes are more complicated. The theory of carbonation shrinkage is beyond the scope of this book and reference should be made to a discussion of the various theories by Ramachandran *et al.* (1981).

While it is generally agreed that the overall rate of carbonation is controlled by the permeability of the concrete, little information on the individual rates applying to calcium hydroxide and the cement hydrates appears to be available. French (1991a,b) states that in practice it is usual to find the calcium silicates alter to carbonate as readily as the hydroxide and even that residual unhydrated clinker particles become pseudomorphed to amorphous silica.

5.4.1   EFFECT OF CARBONATION ON POROSITY AND STRENGTH

Carbonation of an ordinary Portland cement hardened paste increases the strength of the paste and decreases its porosity as shown by a number of studies quoted by Parrott (1987). In these respects it improves the properties of the paste in concrete with the disadvantage that it lowers the pH. The decrease in porosity will also decrease the permeability of the paste but where the outer layer of a concrete is highly porous this reduction in permeability is usually insufficient to prevent oxygen and chloride ions penetrating to the reinforcement.

The effect of carbonation on high-alumina cement is claimed to be similar to that on Portland cement (George 1983) although the carbonation products formed are sometimes carbo-aluminates and not calcium carbonate. Cusino and Negro (1980) claim that the use of fine limestone enhances the formation of the carbo-aluminates and overcomes the strength losses accompanying conversion. In the case of blastfurnace slag cements, carbonation may have a deleterious effect on the hardened cement paste as it appears to decrease strength and increase porosity (Meyer 1968, De Ceukelaire and Van Nieuwenberg 1993). Carbonation causes a marked increase in porosity and reduction in strength of supersulfated cement probably due to the carbonation of the ettringite (Manns and Wesche 1968). Further references to the effects of carbonation on hardened pastes are give by Parrott (1987).

5.4.2   TYPICAL TEXTURES OF CARBONATION

The calcium carbonate formed in hardened cement paste is almost invariably present as clumps of crystals. In its initial stages carbonation appears as diffuse birefringent material. This texture occurs along the margins of cracks and at surfaces which have been coated with a sealant soon after hardening (Plate 36).

It is also observed in thin-section where the specimen has not been adequately protected against carbonation during storage and preparation. As carbonation becomes more intense and proceeds inwards from the surface the crystals form a dense texture which darkens the paste. Under crossed polars the carbonation now appears as a birefringent mass of light which eliminates microscopic detail in the hardened matrix. This is the typical carbonation which will be found in the outer layer of concrete Plate 37(a). In its final stage the carbonated paste takes on a fretted appearance and now may appear lighter in colour Plate 37(b). At this stage of carbonation it is probable that the cement hydrates have been largely destroyed. In many cases remnant cement grains and islands of uncarbonated or partially carbonated cement paste may be observed, but overall where the carbonation is intense most observable detail of the hardened cement paste is either obscured or destroyed. A mechanism to explain why islands of non-carbonated material occur in carbonation zones has been proposed by Houst and Wittmann (1994).

5.4.3 CARBONATION OF THE OUTER LAYERS OF CONCRETE

Almost as soon as a cement starts to harden, carbonation commences if superficial drying of the surface is allowed to occur. This is graphically illustrated in Plate 36(b) where a thin layer of diffuse carbonation has taken place when the surface was briefly dried prior to sealing with a polymeric coating. Even in mortar bars mixed to comply with ASTM C227 (1990) and stored under continuously moist conditions carbonation penetrates about 15 μm into cast surfaces and about 0.5 mm into the top surface. In a well compacted, dense concrete the average depth of carbonation, even after years of exposure, will rarely exceed 2 mm and is often as little as 0.5 mm. However, at the margins of aggregates located near the surface, carbonation may extend along fissures caused by plastic settlement and bleeding. In lesser quality concretes carbonation often penetrates more than 2 mm while a carbonation depth greater than 5 mm indicates porous concrete in the outer layer. Some of these effects have been illustrated by Strunge and Chatterji (1989) who concluded that thin-section analysis gives more detailed information on carbonation than that detected by phen-olphthalein.

When carbonation depths exceed 5 mm the nature of the carbonation zone may change. Islands of uncarbonated cement paste are often present in the carbonation zone and isolated areas of carbonation may be seen beyond the carbonation zone in the paste and around some aggregates. These isolated patches of carbonation indicate that a porous pathway was present out of the plane of the section. The margins of the carbonation zone are irregular and carbonation will penetrate deeply into cracks if these are present. This behaviour is an expression of the localised variability which occurs in concrete especially in the porous surface areas of a concrete.

5.4.4 CARBONATION ASSOCIATED WITH CRACK SYSTEMS

A surface crack provides a pathway for external elements to penetrate the body of the concrete. Fine cracks of less than 10 μm in width may carbonate to depths of 1–2 mm even where the average surface depth of carbonation is less than 0.5 mm. These fine cracks are primarily due to shrinkage Plate 38(a) and invariably are filled with carbonation products. As the crack width increases the depth of carbonation also increases and the crack will be open unless fluid has been present to enable crystallisation to occur Plate 38(b).

A crack 60 μm wide may be carbonated along its margins up to a depth of 50–100 mm, as well as allowing carbonation of aggregate margins and areas of porosity close to the crack. Cracks wider than 100 to 150 μm clearly allow the environment to penetrate to considerable depths. The extent of reaction between the environment and the hardened paste in these wide cracks will depend on the crack width, the severity of that environment and the quality of the concrete (Plate 39).

The width of a crack at the surface of concrete may be less than in the interior, because calcium carbonate can partially or completely block the entrance. In thin-section cracks which appear to be closed at the surface will quite often prove to be open for the remainder of their length Plate 38(d). Carbonation of the paste along crack margins may vary from a light dusting of carbonate to irregular bands of carbonated paste extending the full depth of the crack. Crystals of carbonate will often line crack margins, but a crack wider than 20 μm is rarely completely filled with carbonation products unless the concrete has been saturated for a considerable period of its history.

The above observations on carbonation are in general agreement with the recommendations of BS 5337 which limited crack widths to 100 μm in locations alternating between wet and dry. ACI Committee 224 also recommends a similar limit for water retaining structures. However in BS 8007 (1987) which replaces BS 5337 (1986) these recommendations have been omitted.

5.4.5   CARBONATION ASSOCIATED WITH LIGHTWEIGHT AGGREGATES

For normal density concretes the porosity of the aggregates does not significantly contribute to carbonation. However, if vesicular aggregates such as scoria, pumice or manufactured lightweight aggregates (St John and Smith 1976) are used the effect on carbonation can be significant. This effect is illustrated in Fig. 5.3, where the depth of carbonation is clearly related to the bulk density of the concrete as shown in Table 5.7 where the results of exposing 50 $\times$ 50 $\times$ 300 mm beams to the weather for five years are given. Work by Swamy and Jiang (1992) correlated carbonation depth to the pore structure of lightweight concrete.

However, in structural lightweight concrete it is common practice to use coarse lightweight aggregates mixed with normal density fines. It is claimed that provided cement contents of 350 kg/m^3 or more and low water/cement ratios are used the carbonation of structural lightweight concretes is little different from that of normal-weight concretes (Clarke 1993). Mays and Barnes (1991) inspected forty structures containing structural lightweight concrete constructed prior to 1977 and concluded it was not less durable than normal-weight concrete although its durability may be more sensitive to poor workmanship.

Table 5.7 Mix details, length changes and carbonation areas of test beams of concrete containing lightweight aggregate from outdoor exposure.

Specimen	Cement content (kg/m^3)	Water/cement ratio	Unit weight (kg/m^3)	Length change (%)	Carbonated area (%)	Alkali content (kg/m^3)
F8/1/1	290	1.1	1120	+0.014	54	0.92
F8/1/3	350	0.83	1215	+0.044	28	1.12
F8/1/4	435	0.67	1275	+0.088	12	1.39
F8/1/6	450	0.60	1340	+0.071	10	1.44
F8/1/5	530	0.60	1445	+0.060	5	1.69
F8/1/7	560	0.56	1415	+0.126	8	1.79

Figure 5.3 illustrates many aspects of carbonation already discussed. For the lower density samples which are more permeable because of the lower cement content and higher water/cement ratio, the carbonation is not only greater in area but much more intense. As the bulk density increases the carbonation depth decreases. There is also an increase in the amount of cracking because of the alkali-aggregate reaction which is undoubtedly present. This was not known at the time of the original publication and other causes for the expansion were put forward at that time.

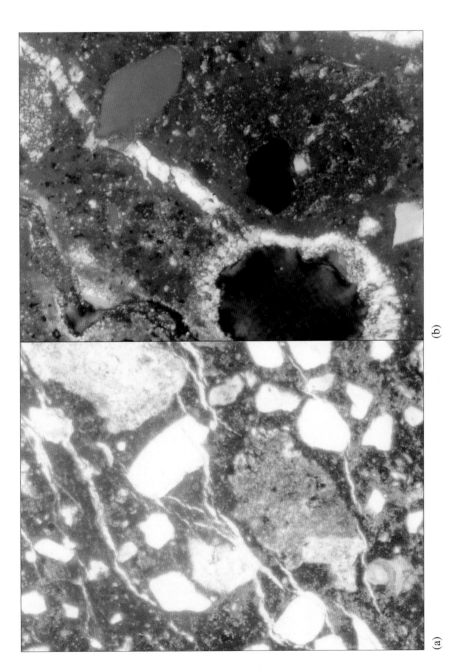

(a)

(b)

Plate 48 The texture of attack by sodium sulfate. (a) The zone of sulfate attack contains an extensive system of fine cracks that form an exfoliation texture. Many of the cracks are filled with crystals of gypsum. Width of field 2.2 mm. (b) At a higher magnification and under crossed polars it is seen that the gypsum tends to align itself across the cracks. Width of field 0.8 mm.

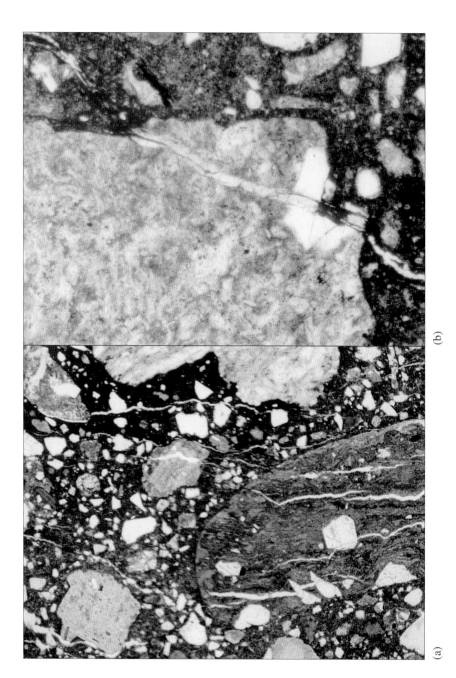

(a)

(b)

Plate 49 Alkali-aggregate reaction associated with attack by sodium sulfate. (a) In the aggregates undergoing alkali-aggregate reaction the cracking was often aligned with aggregate undergoing alkali-aggregate reaction the cracking was often aligned with the exfoliation caused by the sulfate attack. Width of field 6.5 mm. (b) Detail of another aggregate undergoing alkali-aggregate reaction. In this region of the sample sulfate attack was not as apparent. Width of field 0.2 mm.

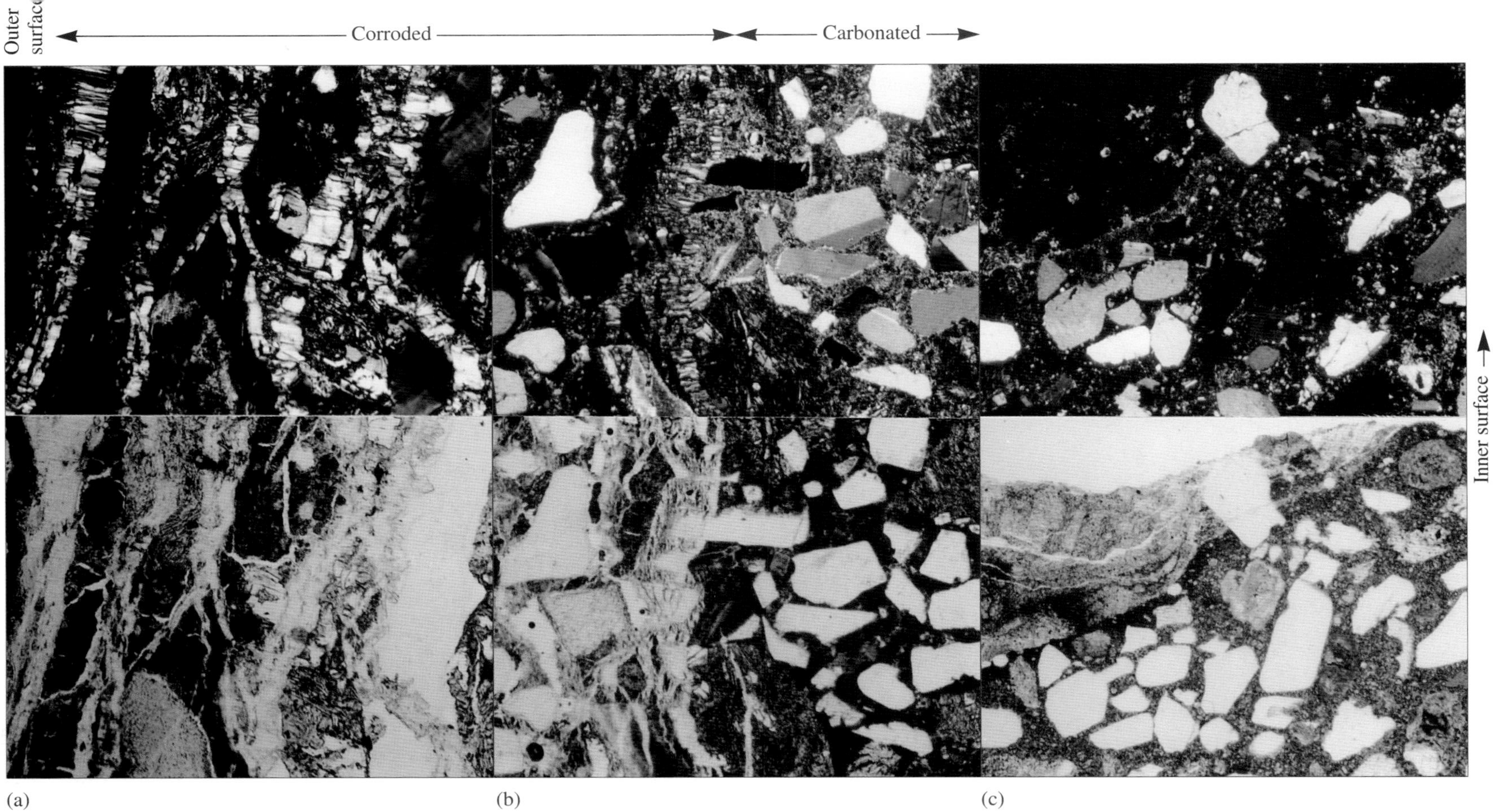

Inner surface →

(c)

(b)

(a)

Outer surface

Corroded ←——→ Carbonated ←——→

Plate 50 Micrographs comparing the textures of the attack on the inside and outside surfaces of pipe A. (a) The hardened paste in the outer zone of the pipe has been completely corroded to an exfoliated amorphous silica with embedded aggregates. The cracks contain aligned gypsum crystals which are clearly visible under crossed polars. Width of fields 0.8 mm. (b) The texture shown in the above micrographs extends about 5 mm into the wall where it grades abruptly into a 5-6 mm wide zone of carbonation.

The interface with this carbonation zone is shown here and stands out clearly under crossed polars. Width of fields 2 mm. (c) On the inside surface of the invert the texture is more typical of acid attack. The texture shown is at the edge of a corrosion pit and consists of a corroded area separated by a thin zone of carbonation from the hardened paste. A leached zone is not apparent. Width of fields 2.8 mm.

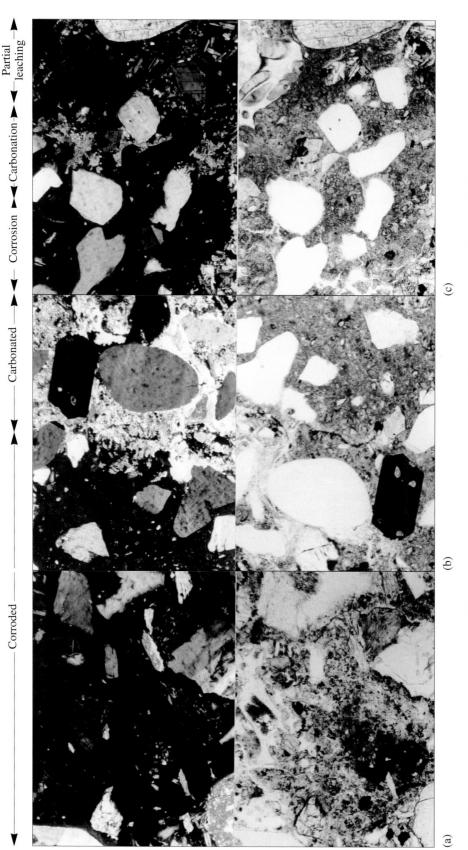

◄— Outer surface ———————————————— Inner surface —►

Corroded ——◄ ►—— Carbonated ——◄ ►—— ◄ Corrosion ►—— ◄ Carbonation ►—— ◄ Partial leaching ►—

(a)

(b)

(c)

Plate 51 Micrographs comparing the textures of the attack on the inside and outside surfaces of pipe B. Width of fields 0.8 mm. (a) The texture shown is typical of the corroded zone. The hardened paste has been corroded to amorphous silica as shown under crossed polars. There is some darkening and yellowing due to the attack but gypsum is absent. (b) About 5 mm into the wall a well defined band of carbonation was present. This texture is typical of acid attack. (c) On the inside of the pipe the corrosion zone consisted of amorphous silica bounded by a narrow irregular zone of yellowing. Under crossed polars it is seen that this yellowing consists of a mixture of carbonation and an amorphous material. The zone of leaching is not well shown.

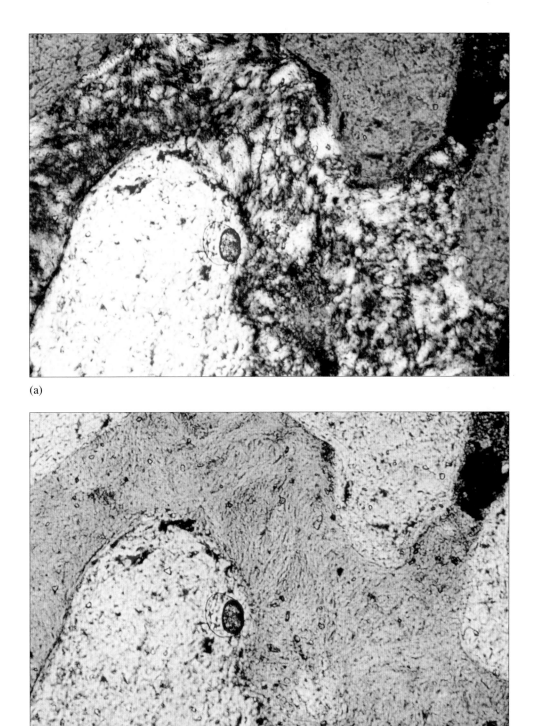

(a)

(b)

Plate 52 The appearance of thaumasite. Width of fields 0.75 mm. (a) The matted felt of fibrous looking crystals is quite similar to massive deposits of ettringite. (b) Under crossed polars it is immediately apparent that the fibrous material has a higher birefringence than ettringite and is showing traces of first-order yellow birefringence which helps to identify it as thaumasite.

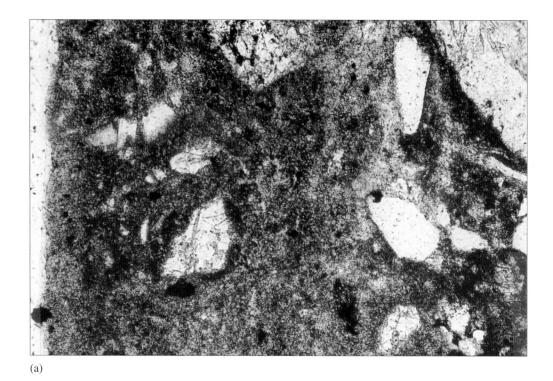

(a)

(b)

Plate 53 The zones of alteration in a concrete exposed to sea water attack. The width of field of each photomicrograph is 1.4 mm. (a) The hardened cement paste in the outer zone is largely altered to fine-grained brucite and carbonation. The exterior surface is at the left-hand edge. (b) Under crossed polars the carbonation is clearly visible but the brucite is too fine grained to be visible.

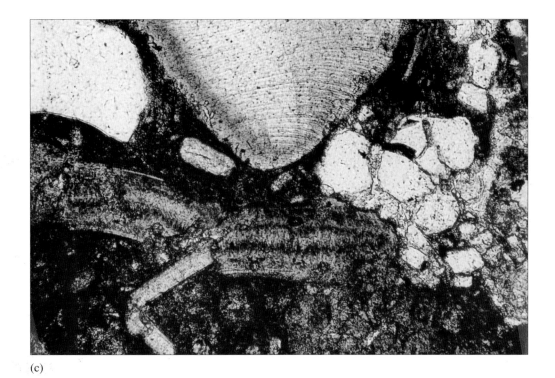

(c)

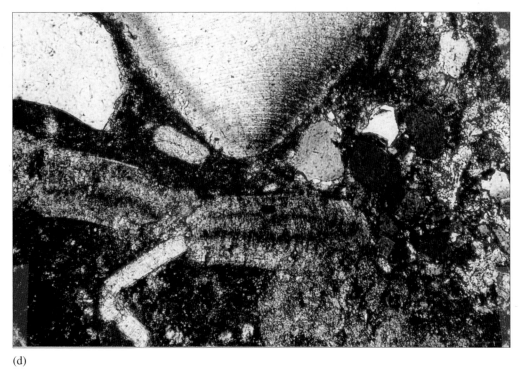

(d)

Plate 53 (*contd*) (c) Part of the zone between 3 and 7 mm from the surface showing patchy darkening of the paste due to leaching of calcium hydroxide and the development of aragonite. (d) Under crossed polars the band of aragonite is visible and calcium hydroxide is largely absent from the hardened cement paste. Shell fragments from part of the aggregate

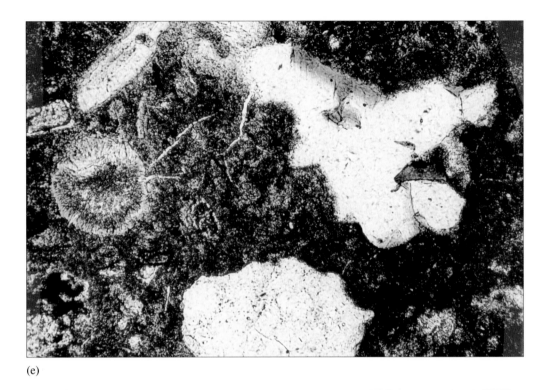

(e)

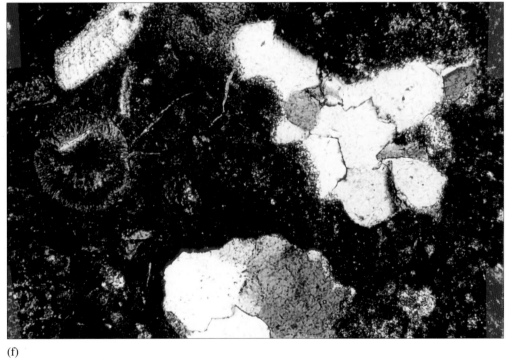

(f)

Plate 53 (*contd*) (e) Innermost zone showing development of acicular ettringite crystals in the matrix and small air voids. The small shrinkage cracks are due to drying the specimen. (f) Under crossed polars low birefringence distinguishes the ettringite in the pore.

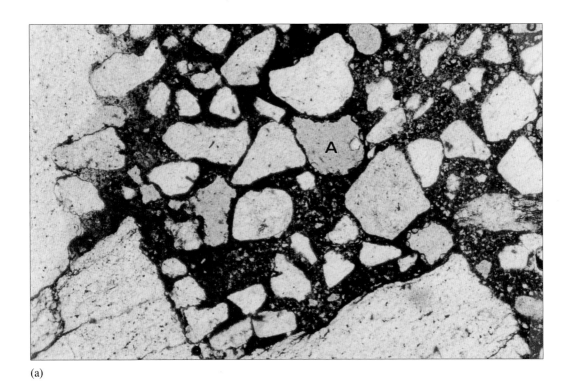

(a)

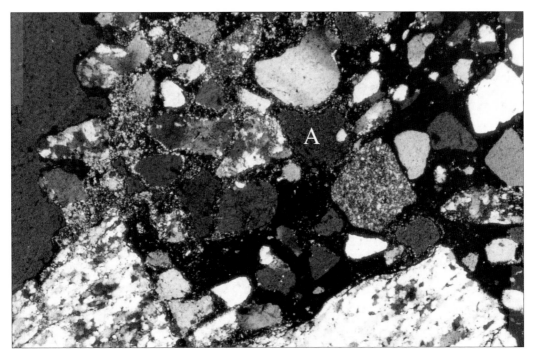

(b)

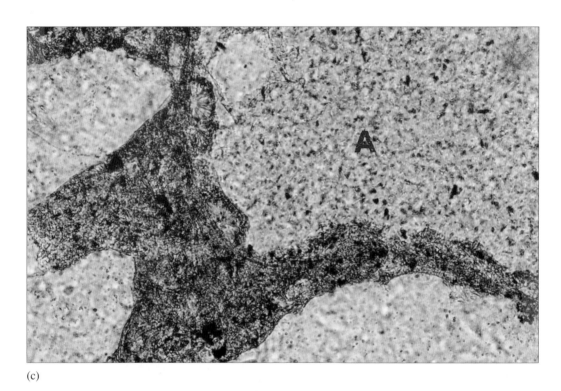

(c)

Plate 54 The texture of a precast concrete pipe exposed to sea water attack. (a) The outer zone (left-hand side) is altered to brucite and calcite which merges into a darkened porous zone leached of calcium hydroxide with development of ettringite in the cement matrix and air voids. Width of field 3.8 mm. (b) Under crossed polars the outer zone is clearly shown by the carbonation. Width of field 3.8 mm. (c) Enlarged view of the area around the air void (A) showing the ettringite development both in the void and the surrounding matrix. Width of field 0.2 mm.

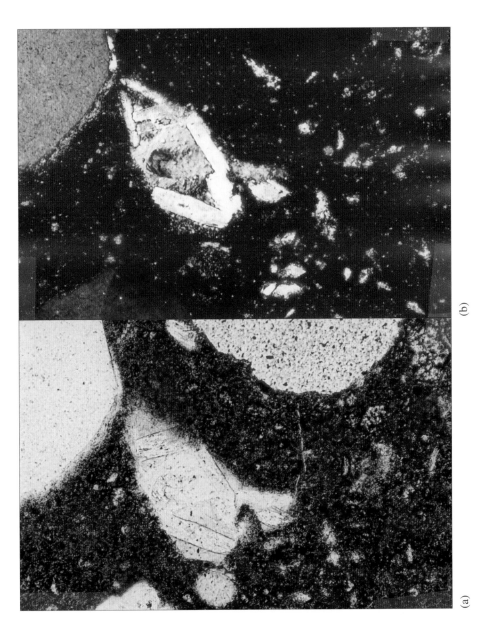

(a)

(b)

Plate 55 The texture of internal sulfate attack. Width of fields 0.8 mm. (a) Laths of calcium hydroxide and gypsum have crystallised in a pore and microcracking is present in the hardened cement paste. (b) Under crossed polars the gypsum is distinguished by its low birefringence and is seen to be scattered throughout the paste.

Plate 56 The typical appearance of pop-outs and rust staining caused by aggregates containing pyrite.

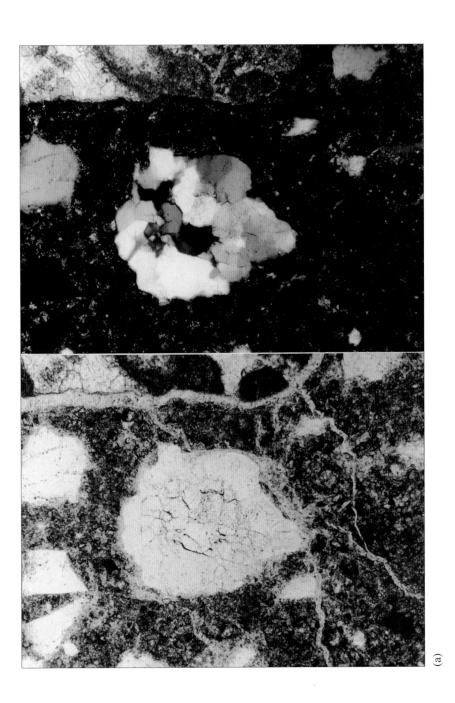

(a)

Plate 57 The texture of delayed ettringite expansion. Width of photomicrographs 0.9 mm. (See also Fig. 4.8). (a) The cracking and fissures have been delineated by the use of a yellow dye. A fissure, filled with aligned ettringite, runs along the margin of the large limestone aggregate to the right of the micrograph and less well defined fissures, also filled with ettringite, surround most of the other aggregates. The cracking in the hardened paste is also filled with ettringite. Most of the ettringite is barely visible here but can be clearly seen when viewed under the microscope.

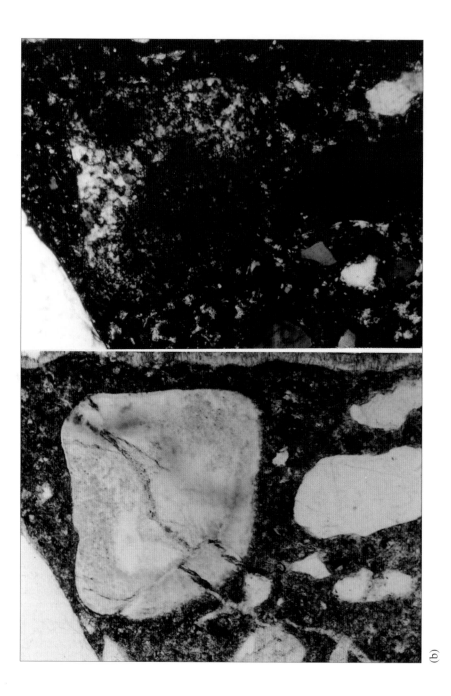

(b)

Plate 57 (*contd*) (b) The chert aggregate in the centre of the micrograph is clearly undergoing alkali-aggregate reaction. There is a fissure along the margin of limestone aggregate on the right hand side of the micrograph filled with clearly visible aligned ettringite. Some of the other aggregate present are also ringed with filled fissures.

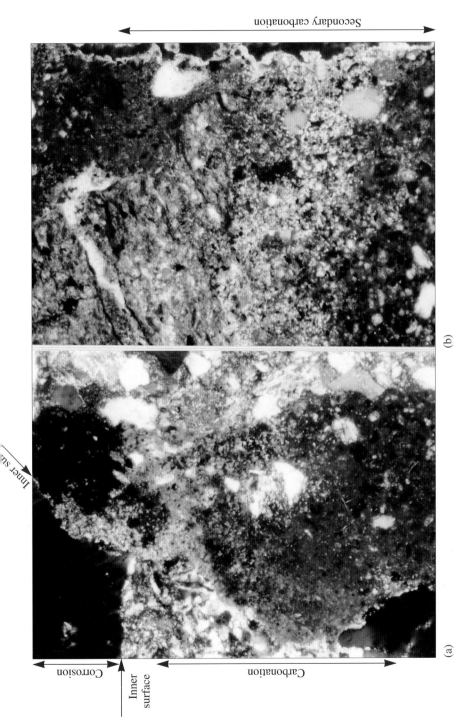

Plate 58 The corrosion texture of attack by the bacterial oxidation of hydrogen sulfide. Width of fields 0.9 mm. (a) Under crossed polars. At the outer surface between two greywacke aggregates, which are standing proud, the hardened cement paste is altered and consists of a complex mixture of carbonation and staining from iron compounds. Behind this is a layer of paste depleted in calcium hydroxide with red specks of iron compounds. Overlaying parts of these zones is an irregular layer of amorphous corroded paste (not shown here) which is often stained with yellow and red material.

(b) Under crossed polars. Varying in depth of 0-2 mm behind the zonation caused by the acid attack is a separate zone of carbonation which merges into sound cement paste. This interior zone of carbonation has probably occurred subsequent to the acid attack and shows that the carbon dioxide present in the sewer has been able to easily penetrate the acid corrosion layer. Note how carbonation has concentrated along a fissure at the aggregate margin at the right side of the micrograph.

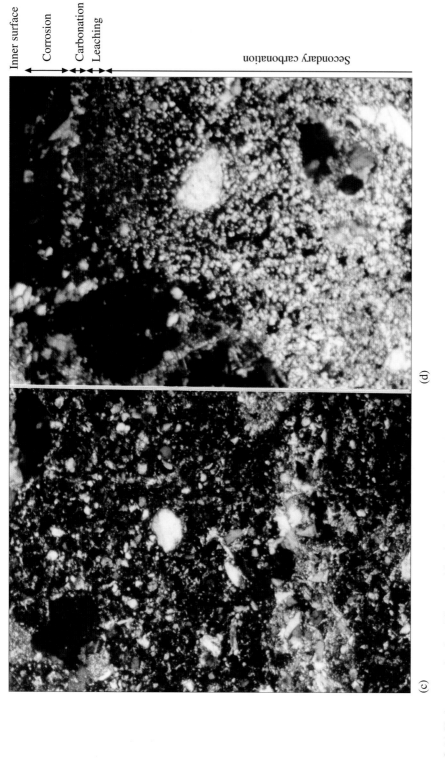

Inner surface

Corrosion

Carbonation

Leaching

Secondary carbonation

(c)

(d)

Plate 58 (*contd*) (c) Under crossed polars. The sound hardened paste in the interior of the pipe wall consists of a rich dense texture with numerous remnant and partially hydrated cement grains in a close packing of aggregates. (d) Under crossed polars. In some areas the subsequent carbonation has merged with the initial acid attack. Unless large sample areas are examined it is these variations in texture which can lead to erroneous conclusions. The textures shown in (a) and (b) dominated the surfaces examined.

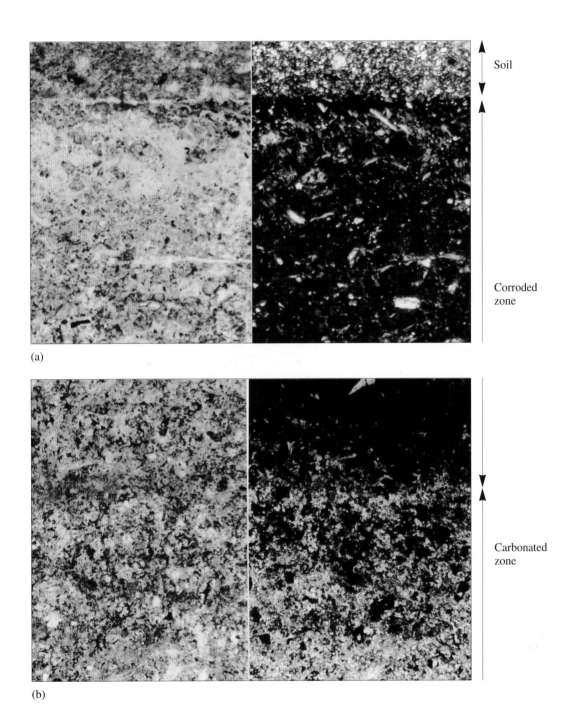

(a)

(b)

Soil

Corroded
zone

Carbonated
zone

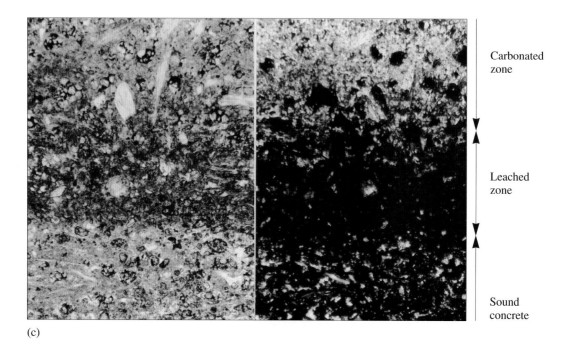

Carbonated
zone

Leached
zone

Sound
concrete

(c)

Plate 59 The zonation of soft water attack on asbestos cement pipe. The fine texture of these pipes, which were not autoclaved, clearly illustrates the zones associated with the attack. Width of fields 0.9 mm. (a) The interface between the soil and the corroded zone often contains deposits of brownish-yellow iron compounds which are visible at the soil interface on this sample. The corroded zone of the pipe consists of amorphous silica, asbestos fibre and some indistinct relics of miscellaneous materials. Under crossed polars the birefringence of the soil minerals and the interface stained with iron compounds stands out clearly. The corroded zone is almost entirely isotropic and only remnants of asbestos fibre are visible. (b) The interface between the corroded and carbonated zones is well defined by a layer of iron staining and the transition to the carbonated zone. Under crossed polars the isotropic nature of the corroded zone, the iron staining and the carbonation are clearly illustrated. The dark spots in the carbonated zone are relics of cement grains. Asbestos fibre is present but is not easy to observe as it tends to merge into the texture at this magpification. (c) The leached zone appears as a darkened layer of cement paste which under crossed polars is deficient in calcium hydroxide. The transitions from carbonated to leached zones and from leached to sound hardened paste are clearly visible under crossed polars. The sound hardened paste is full of partially hydrated and remnant cement grains as well as calcium hydroxide. Asbestos fibre is just visible in all the zones.

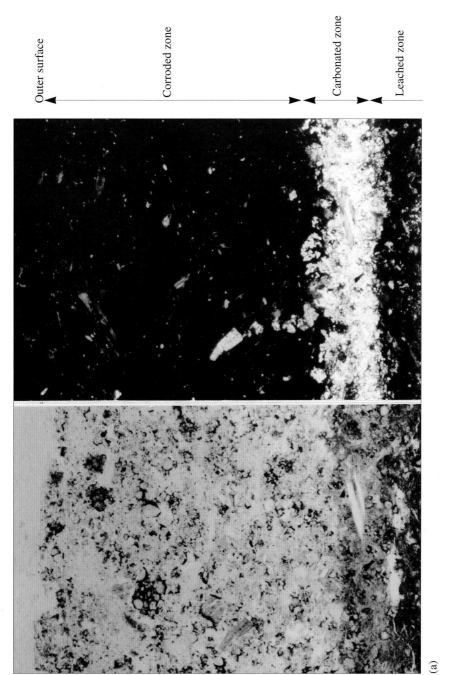

Outer surface

Corroded zone

Carbonated zone

Leached zone

(a)

Plate 60 Variations in the texture of corrosion of a non-autoclaved asbestos cement pipe. Width of fields 0.9 mm. (a) Relics of cement clinker grains are present in this corroded zone which is unusual. Since this pipe contains no silica flour it is rich in remnant grains which may be part of the reason for their presence in the corroded zone. Normally the corroded zone contains little if any part of the cementitious system apart from amorphous silica. Note the absence of iron staining and the way the corroded zone terminates in a ragged manner at the outer edge because adherent earth was cleaned off during sampling. Under crossed polars the corroded, carbonated and leached zones are clearly defined.

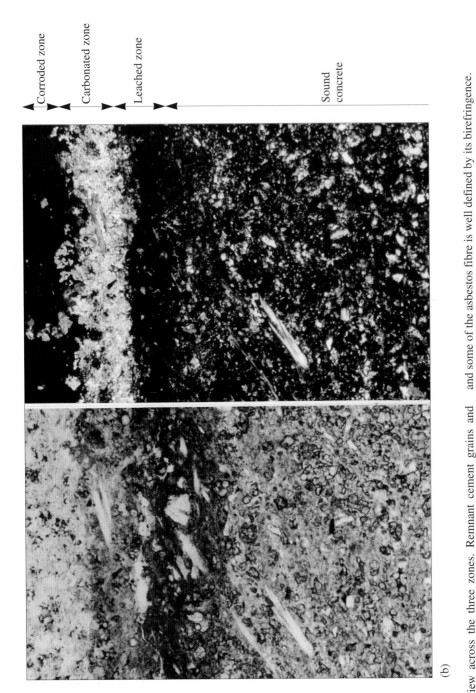

Corroded zone

Carbonated zone

Leached zone

Sound concrete

Plate 60 (*contd*) (b) A view across the three zones. Remnant cement grains and asbestos fibre are visible. and some of the asbestos fibre is well defined by its birefringence. The zones are even more clearly defined under crossed polars

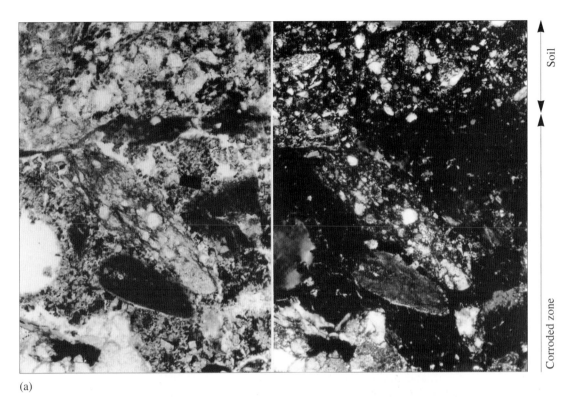

(a)

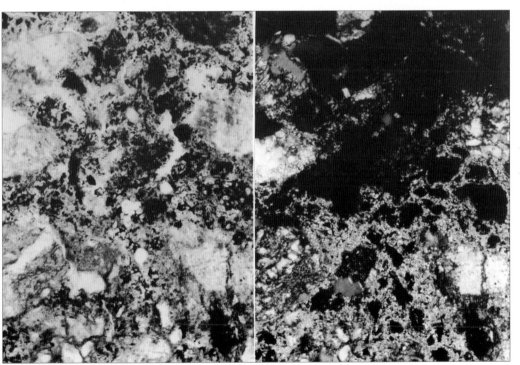

(b)

Soil

Corroded zone

Carbonated zone

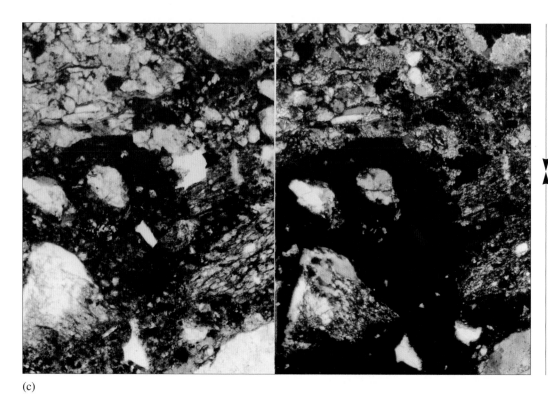

Carbonated zone

Leached zone

(c)

Plate 61 The zonation of soft water attack on centrifugally spun concrete pipe. The close-packed aggregate texture of these pipes makes the zones of attack both difficult to see and illustrate. Widths of fields 0.12 mm. (a) The soil/corrosion interface. A layer of iron staining covers the surface of the pipe at the greywacke soil interface. The texture of the corroded cement paste is stained brown but otherwise is isotropic under crossed polars. The aggregates present are greywacke. (b) The corrosion/carbonation interface. The change in colour from corroded to the carbonated zones is subtle. The portion of the corroded zone is darkened by possible relics of cement grains. Even under crossed polars the two zones are not clearly defined and appear to merge into one another. (c) The carbonation-leached zone interface. Once again these zones are not well differentiated. Even under crossed polars the carbonation is muted. In concrete specimens it is easy to overlook the presence of the leached zone because of the crowded texture.

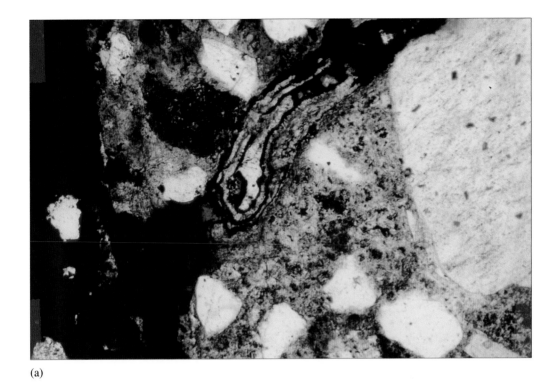

(a)

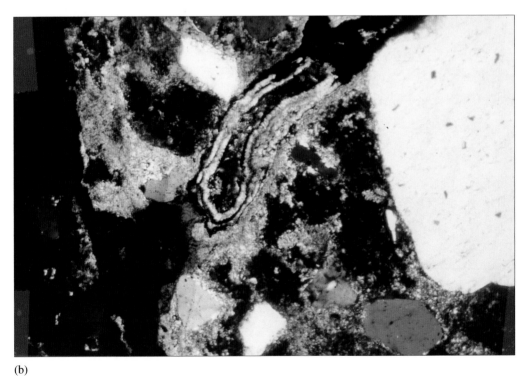

(b)

Plate 62 An example of the effects of corrosion of steel reinforcement on the adjacent concrete. Width of fields 0.8 mm. (a) The reinforcing steel is to the left and is separated from the hardened cement paste by a dark brown band of hydrated iron oxides which extrude into the concrete towards an elongated void. (b) Under crossed polars the relationship of the corrosion products and associated carbonation are visible.

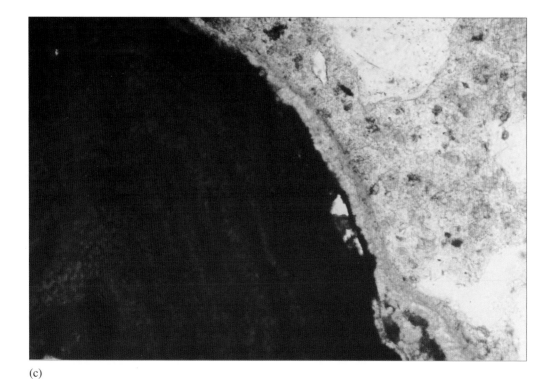

(c)

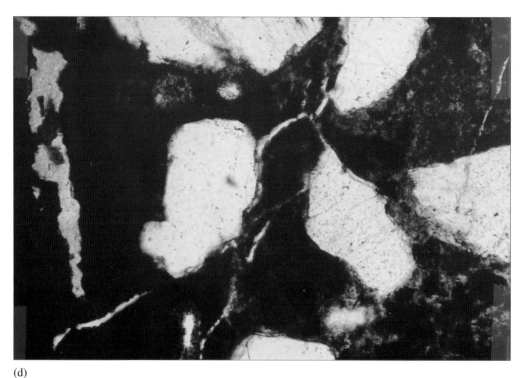

(d)

Plate 62 (*contd*) (c) Detail of the boundary between the hydrated iron oxides and the concrete showing zonation in the corrosion products and an adjacent dark band of carbonation in the hardened cement paste. (d) Hydrated iron oxides have penetrated the hardened cement paste which is cracked because of the volume increases of the corrosion products.

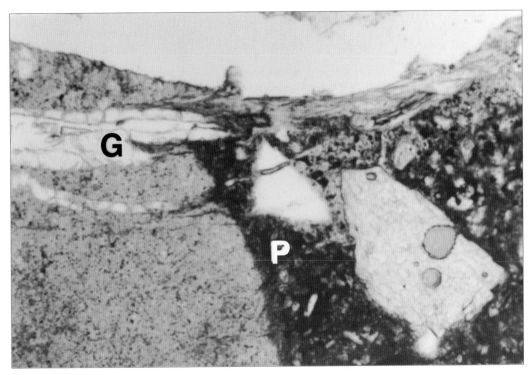

(a)

(b)

Plate 63 Carbonated gel. Width of fields 1.3 mm. (a) Micrograph of a fracture surface at the bottom of a core sample. Gel (G) runs along a crack in the rhyolite aggregate across to the fracture surface where it has carbonated during storage (paste: P). (b) Under crossed polars the gel near the surface can be seen to be carbonated. The carbonation extends into porous hardened paste. Gel extruded onto an outer surface will tend to have the same texture of carbonation as shown here.

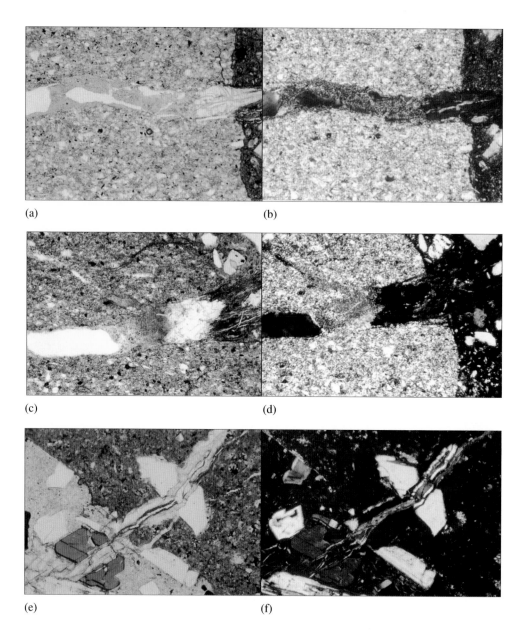

Plate 64 Some textures of typical gel plugs associated with reacting aggregates. Width of fields 1.4 mm. (a) The hornfelsed shale coarse aggregate in this South African hydro structure has a crack which is mainly empty apart from a plug of gel which protrudes into the cement paste. Inside the aggregate there is an area of material that has a granular appearance. Where the gel protrudes into the paste it is clear and then becomes darkened. (b) Under crossed polars the granular material inside the coarse aggregate is crystalline and birefringent and hence is not a true gel. The plug of gel which protrudes is isotropic irrespective of whether it is clear or darkened. This is the appearance of a typical gel plug. (c) The blue-green argillite coarse aggregate in this eighty-year-old Canadian dam has a crack which is empty apart from an area of granular material and a plug of gel which protrudes into the cement paste. In this case darkened gel protrudes into the aggregate as well as where it contacts the paste. (d) Under crossed polars the granular material inside the coarse aggregate is crystalline and birefringent and grades into isotropic gel. (e) The andesite aggregate in this sixty-year-old bridge has a crack filled with clear gel which is difficult to distinguish from the glassy matrix of the rock. Displacement of the crack margins is clearly shown by the pyroxene and feldspar crystals. There is only minimal darkening of the gel in contact with the cement paste. (f) Under crossed polars where the gel runs through the paste it is partially crystallised and birefringent granular material is not present inside the crack. This absence of birefringent gel is fairly typical of reacting volcanic materials.

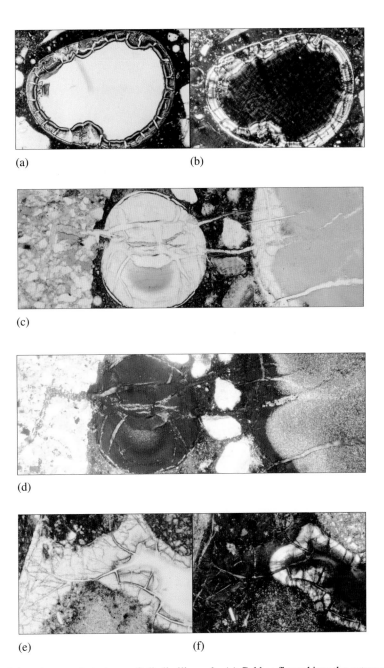

(a)   (b)

(c)

(d)

(e)   (f)

Plate 65 Some of the microscopic textures of alkali-silica gels. (a) Gel has flowed into the entrapped air void and rimmed its surface. It has curved (conchoidal) cracks due to drying and is layered and stained brown. Gel will often be found rimming pores and can be present as a thin layer which can be difficult to distinguish from hardened cement paste when it is thinned to a wedge during sectioning. Width of field 1.4 mm. (b) Under crossed polars the gel is crystallised in spite of the fact that it has been continuously moist cured and the sample is only two years old. This micrograph was taken from a section of an ASTM C227 mortar bar specimen containing basalt coarse aggregate and rhyolitic sand. Width of field 1.44 mm. (c) The hornfelsed shale contains a crack filled with a typical gel plug which has extruded gel into two connected pores. The pores contain clear gel with conchoidal cracks together with some gel which appears to be stained brown. Width of field 2.8 mm. (d) Under crossed polars the brown gel is seen to be crystallised. Inside the crack the granular infilling material is also birefringent. Width of field 2.8 mm. (e) Clear gel with conchoidal cracks fill part of the large entrapped void space in this forty-year old bridge and shows no sign of darkening in contact with the hardened paste. Width of field 1.4 mm. (f) Under crossed polars its outer margin is shown to be crystallised. Width of field 1.4 mm.

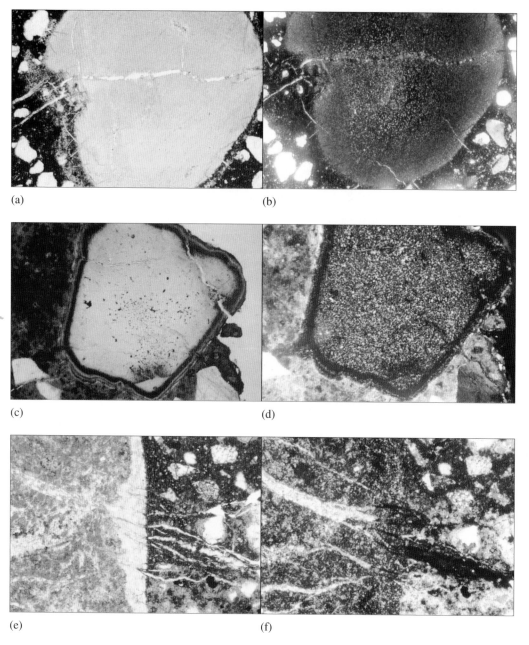

Plate 66 The appearance of reaction and weathering rims in thin-section. (a) This chert aggregate from a thirty-year-old bridge in the Midlands of the UK has a somewhat diffuse brown margin intersected by numerous fine cracks filled with gel. The dark rim could still be due to weathering but the balance of evidence suggested that it is due to reaction as it possibly contains gel. Width of field 3.2 mm. (b) Under crossed polars the centre of the chert is showing some signs of disintegration and the cracking is highlighted. Width of field 3.2 mm. (c) In contrast the dark rim on the chert aggregate from a structure in the south-west of the UK is definitely due to weathering as was apparent when it was compared with other unreacted aggregates present. However, the aggregate is reacting as the weathering rim is penetrated by fine cracks with gel. Width of field 1.4 mm. (d) Under crossed polars note how post-sampling carbonation has obscured the texture of the hardened paste. Width of field 1.4 mm. (e) The rhyolite coarse aggregate in this sixty-year-old dam in the USA appears to have a light-coloured reaction rim. Inside the aggregate granular material and gel fills stranded cracks and extrudes as clear gel into the hardened paste. Width of field 3.2 mm. (f) Under crossed polars the paste texture is partly obscured by post-sampling carbonation. The granular material is birefringent, the gel is isotropic and the assumed reaction rim is obviously altered. Width of field 3.2 mm.

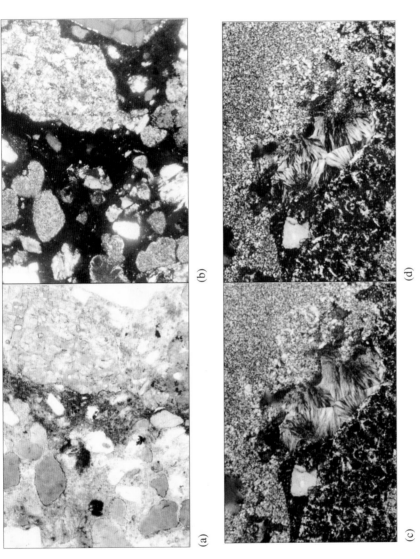

(a)

(b)

(c)

(d)

Plate 67 The appearance of the silica minerals, opal, chalcedony, tridymite and cristobalite. Width of fields: (a), (b), (c), (d), (g), (h) 1.4 mm; (e) and (f) 3.1 mm. (a) Opal. The opal forms the brown matrix to this glauconitic gritstone. A more common occurrence in alkali-aggregate reaction is where the opal forms rims around particles. (b) Under crossed polars the opal is seen to be isotropic. Opal is a rather nondescript material and is mainly distinguished by its low refractive index of 1.4-1.46 and lack of other distinguishing characteristics. (c) Chalcedony. The fibrous texture and low refractive index identifies the chalcedony present at the edge of this English chert aggregate. (d) Under crossed polars the low birefringence of the chalcedony is revealed and the typical cryptocrystalline texture of the chert is also illustrated.

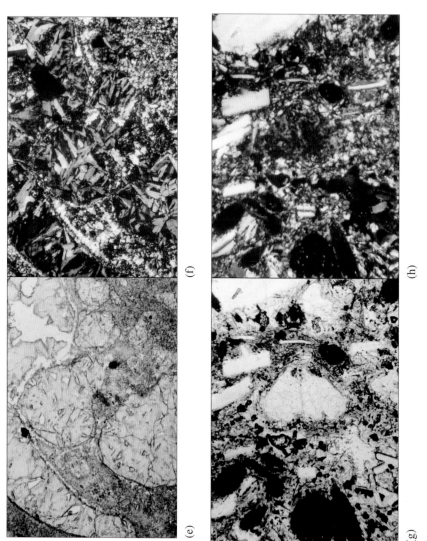

(e)

(f)

(g)

(h)

Plate 67 (*contd*) (e) Tridymite. The low refractive index suggests the presence of a pool of tridymite in this Whakamaru rhyolite. (f) Under crossed polars the wedge-shaped crystals with simple twinning and low birefringence clearly identify the material to be tridymite. The orange-yellow crystals are pyroxene and the white material probably quartz. There is also some fine matrix material present. (g) Cristobalite. Cristobalite tends to form small pools of low refractive index material with a texture that appears to be like overlapping roof tiles. The pools can be quite small and it needs a keen eye to distinguish cristobalite in the matrices of volcanic rocks. This also tends to be the case for the other silica minerals. (h) Under crossed polars the pools of cristobalite have a low birefringence which, combined with the low refractive index and tile-like texture, identify it in this Daisen andesite from Japan.

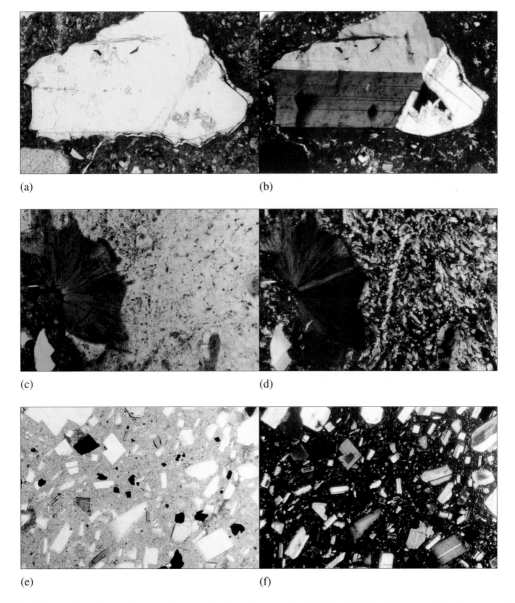

(a)　　　　　　　　　　　　　　　　(b)

(c)　　　　　　　　　　　　　　　　(d)

(e)　　　　　　　　　　　　　　　　(f)

Plate 68 Typical matrices of intermediate and acid volcanic rocks. Width of fields 1.4 mm. (a) The lower margin of the feldspar crystal is attached to an irregular rim of clear, acid volcanic glass, which has reacted and formed a crack, and some associated gel which is not visible. The feldspar crystal is set in a poorly differentiated matrix of this particle which is probably derived from rhyolite. (b) Under crossed polars the glass rim is completely isotropic and the matrix consists of a mixture of isotropic material in which are embedded finely crystalline material. The fissure on the upper right of the feldspar crystal is filled with calcium hydroxide. Whether either the glass rim or the matrix of the rock has reacted, or perhaps both, is impossible to distinguish. (c) The matrices of many fresh rhyolites exhibit wide ranges of texture; the most spectacular being spherulitic texture which consists of radiating areas of feather-like material. The remaining matrix is also often partially devitrified as shown here and filled with minute crystals of feldspar, quartz and ferromagnesians. It is claimed that one of the devitrification products of spherulitic texture is cristobalite but in practice it is the brown undifferentiated matrices of rhyolite that are most commonly associated with akali-aggregate reaction as shown in Plate 69(a). (d) Under crossed polars details of the crystallisation of the matrix are revealed. (e) The matrix of the fresh Egmont andesite is a relatively clear brown glass in which are set crystals of feldspar, magnetite and pyroxenes. Both cristobalite and tridymite are usually absent from this rock which is the main reason why the aggregate has no pessimum proportion. (f) Under crossed polars the glass matrix has partially devitrified as minute crystals of material scattered throughout in spite of the freshness of the rock. The glassy matrices of many dacites, andesites and basalts are often filled with devitrification products, often to such an extent that it is difficult to distinguish the glass. This is one of the reasons why in these cases petrography is unable to do more than indicate the need for further testing.

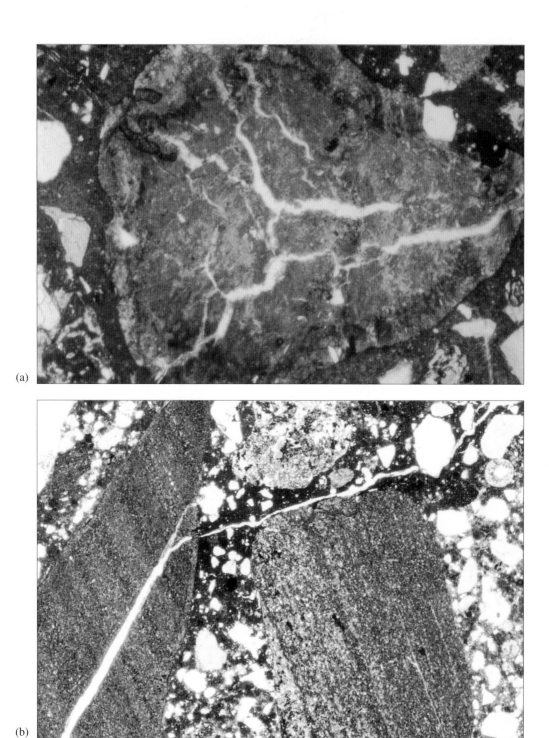

Plate 69 Two types of cracking in reacting aggregates. (a) The texture of the rhyolite particle is disintegrating into a complex of internal cracks and fissures with minute tongues of gel protruding into the paste. Width of field 2.3 mm. (b) The blue-green argillite coarse aggregate has one main crack with gel plugs at the crack entrances. While the cracking tended to run along bedding planes of the aggregate, in this case it is clearly crossing the bedding. Width of field 2.4 mm.

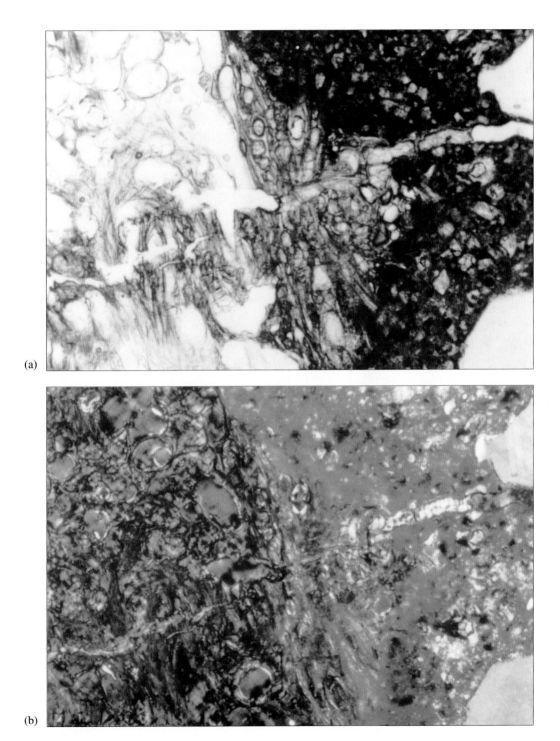

(a)

(b)

Plate 70 The texture of gel associated with a pumice aggregate. Width of fields 1.4 mm. (a) The glassy vesicular structure of the pumice aggregate has an irregular crack which emerges at its apex. The gel plug is difficult to distinguish until it protrudes into the hardened cement paste. (b) Under crossed polars the gel is even more difficult to distinguish and is seen to be crystallised in contact with the hardened paste.

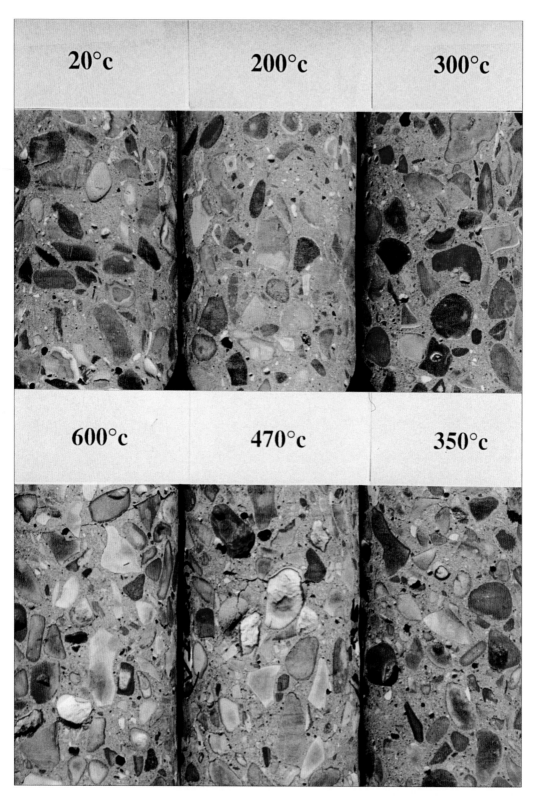

Plate 71 Flint aggregate concrete cores which have been heated for two hours at the temperatures indicated.

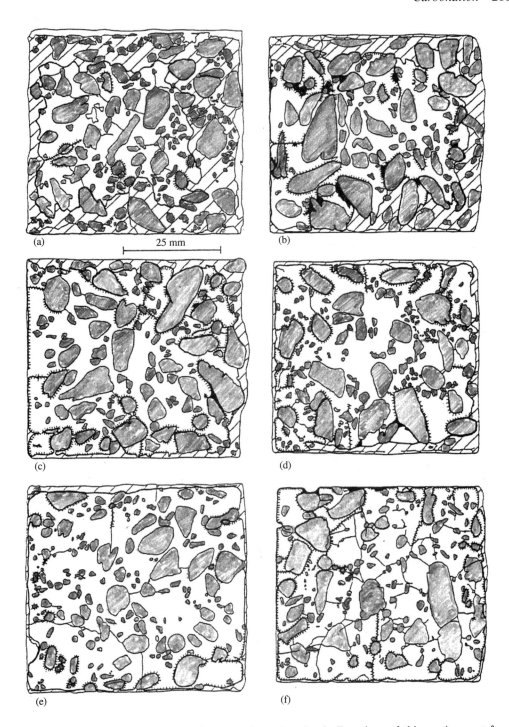

Fig. 5.3 The effect of concrete density on carbonation depth. Drawings of thin-sections cut from 50 × 50 × 300 mm beams containing expanded lightweight argillite aggregate illustrate the effect of exposure to the weather for five years. The hatched areas delineate carbonation in the paste and at aggregate margins. Most of the cracks also contain carbonation (redrawn from St John and Smith (1976), sample numbers as Table 5.7).

The problem of increased carbonation with decreasing bulk density of the concrete is recognised in construction codes for the use of lower bulk density lightweight aggregates. They specify additional protection against corrosion by requiring the reinforcement either to be coated with a cement slurry or some other protective material. However, for structural lightweight concretes apart from some codes requiring extra cover there are no special requirements for protection against carbonation. The petrographer needs to take into account the variables that can affect carbonation depth when assessing the quality of lightweight concretes.

5.4.6   CARBONATION ASSOCIATED WITH LIMESTONE AGGREGATES

There is one form of carbonation that is intrinsic to concrete which is not dependent on atmospheric CO_2. This is the formation of basic calcium carbonate rims around some carbonate rocks which is discussed in Section 6.7.12.

5.4.7   CARBONATION SHRINKAGE

When hardened cement paste carbonates it undergoes irreversible shrinkage. Verbeck (1958) showed that if the paste dries and then carbonates the overall shrinkage is greater than if the two processes occur simultaneously. For normal exposure, simultaneous drying and carbonation of concrete is probably the norm but inside commercial and covered car parks where carbon dioxide levels may be increased, damage due to carbonation shrinkage can occur. The concrete dries because it is protected and elevated levels of carbon dioxide then cause further shrinkage to occur which results in fine map cracking which in turn is outlined by entrapped dirt.

However, the severity of carbonation shrinkage is largely a function of the quality of the concrete as a substantial shell of carbonation has to form in the surface layers for sufficient strain to occur before visible cracking occurs. Since dense concrete only carbonates to depths of about 0.5 mm or less, any shrinkage cracking will be superficial. It is only in more permeable concretes of lesser quality that the carbonation shell becomes deep enough to allow serious carbonation cracking to occur.

5.4.8   CARBONATION CURING OF CONCRETE PRODUCTS

The irreversible nature of carbonation shrinkage can be used to advantage in precast products such as concrete block where it is necessary to limit drying shrinkage. After low-pressure steam curing the precast products are cured at 65°C for up to 18 hours in flue gas containing 5–10 per cent carbon dioxide. This treatment reduces subsequent drying shrinkage by about 30 per cent. The products are dry enough after steam curing to allow sufficient carbonation to occur in the short time of exposure (Toennies 1960). A microstructural study of vacuum curing of mortars with carbon dioxide has been described by Hannawayya (1984).

5.4.9   WELL-CRYSTALLISED CARBONATION

Occasionally well-formed rhombs of calcium carbonate will be observed in cracks, indicating that the carbonate has had space to crystallise freely from solution. However, in the majority of cases the carbonate tends to be irregular in shape. On the surface of concretes exposed to wet conditions, deposits of calcium carbonate may sometimes be observed with both rhombs and irregular crystals present (Plate 40). It should be noted that in wet conditions at temperatures of 60°C or more, needles of aragonite rather than calcite tend to form on surfaces.

Where water is percolating through a crack reaction can occur between calcium hydroxide leached from the concrete with bicarbonate present in the percolating water to form calcium carbonate. Deposition of this carbonate on the walls reduces the crack width and in the case of a narrow crack can completely block it. A combination of calcite deposition and mechanical blocking has been shown to be the principal mechanism of healing cracks in water retaining structures (Clear 1985).

5.4.10 *CARBONATION AND THE pH OF HARDENED CEMENT PASTE*

A saturated solution of calcium hydroxide has a pH of approximately 12.6 at 20°C. In sound concrete the pore solution is not only saturated with calcium hydroxide but also contains sodium and potassium ions which raises the pH into the range of 13–14. The relationship between pH and the constituents present in the pore solution and paste of a concrete and has been reviewed by Diamond (1975). When concrete carbonates the reactions given in Section 5.4 occur. In addition the sodium and potassium ions in the pore solution react as follows:

$$Na^+ \text{ or } K^+ + CO_2 + H_2O \rightarrow NaHCO_3 \text{ or } KHCO_3$$

Thus the end products of carbonation will consist of calcium and alkali carbonates and bicarbonates, and hydrated silica and alumina. The pH of the pore solution in contact with fully carbonated cement paste will depend on the HCO_3/CO_3 equilibria which in turn will be controlled by both the amounts of carbonation and alkalis present and can be assumed to be about pH 10.4. Phenolphthalein changes from colourless to red over the pH range 8.3 to 10 but in practice the eye may just detect the red colour on a concrete surface painted with the indicator at about pH 8.6. To lower the pH to 8.6 it is necessary for the concrete to lose alkalis and it has been suggested that this occurs either by leaching or the alkalis being taken up by the hydrated silica formed from the carbonation of the calcium silicate hydrate (Andrade *et al.* 1986). Leaching of alkalis is possible for concrete exposed to the environment but in protected situations leaching is unlikely and the alkalis must be removed from solution by interaction with some component.

Between the pH of 8.4 of fully carbonated paste and the pH of uncarbonated paste there is a gradient. The question is, where does this gradient lie in the carbonation zone as observed by the microscopist compared with the carbonation zone measured by the colour change of phenolphthalein solution applied to a concrete surface? Work cited by Parrott (1987) indicates that phenolphthalein solution under-estimates the depth of the carbonation zone and that thermal analysis or other methods are required to establish the width of the zone separating fully and non-carbonated concrete. Strunge (1988) concluded from condition surveys that a combination of phenolphthalein staining and petrographic examination in thin-section gives the best evaluation of carbonation in concrete.

The pH gradient has been measured by Parrott (1991) using indicators and Ohgishi and Ono (1983), who related measurement of carbonate concentrations by X-ray diffraction (XRD) to pH of granulated samples in solution. The only reference to comparative measurement of carbonation by microscopy and phenolphthalein solution is by Meyer (1968) who states that they are equivalent. It is always assumed that the carbonation zone observed by examination of thin-sections represents the full depth of reaction but this is probably a simplistic assumption. In high temperature carbonation studies up to 250°C, St John and Abbott (1987) observed a zone of leaching or depletion behind the carbonation zone Plate 21(a). Careful observation of concretes carbonated at ambient temperatures indicates that a narrow, ill-defined zone of depletion of calcium hydroxide is often present. Whether this depletion zone has a lowered pH has not been determined. Microscopical observation of the carbonation zone in concrete provides the most detailed picture but its exact relationship to the pH gradient is an area that still requires investigation.

5.4.11   INFORMATION FROM OBSERVATION OF CARBONATION IN CONCRETE THIN-SECTION

The knowledge of the depth and complexity of carbonation in a reinforced concrete is fundamental to any investigation of its durability. Carbonation essentially reflects the permeability of the outer layers of the concrete and its interaction with the environment. Where cover concrete is readily carbonated it will also be permeable to other external agents such as oxygen, chlorides and sulfates. Thus observation of the carbonation zone in conjunction with other data is used by the petrographer to assess the quality of the concrete in the thin-section (Nepper-Christensen *et al.* 1994). Microscopical observation of carbonation zones in durability investigations is a fundamental technique of measurement and its exclusion from such studies cannot be justified.

5.5   Surface layer of concrete

The development of a surface concrete with different properties from those of the core concrete cannot be entirely avoided during placing, compaction and curing as water and fine materials such as cement and fine aggregate will accumulate near formwork and finished surfaces. Consequently, the surface concrete is richer and its water/cement ratio will differ from that of the interior. The depth of the surface concrete affected varies from a few millimetres to as much as 20 mm under unfavourable conditions. If the surface layers lose too much water after setting, insufficient hydration occurs and a softer surface may result. Concurrently enhanced drying shrinkage will lead to surface cracking (Meyer 1987).

There is now wide agreement that the quality of the surface layer controls many aspects of the durability of concrete. Attempts are now being made to predict durability by measuring the absorption of this layer (Dhir *et al.* 1994). Thus it is surprising that there appear to be few reports of studies of the surface layers of concrete employing petrographic examination of thin-sections. The critical parameter for the durability of the surface layer is its permeability to the ingress of water, carbon dioxide, oxygen, chlorides, sulfates and other aggressive agents. Permeability of surface layers can be qualitatively estimated by impregnating the concrete with a fluorescent dye dissolved in epoxy resin and observing the amount of fluorescence in thin-section (Christensen *et al.* 1984). However, reports of investigations of textural changes in surface areas as related to permeability seem to be a neglected area in concrete petrography.

5.5.1   OBSERVATION OF THE VISIBLE SURFACE

Discoloration of a concrete surface is rarely observable in thin-section because as a defect it has no distinguishable cross-sectional dimension. Often the visibility of staining and colour variation on structures is due to the absorption and/or interference of a very thin layer of material with the incident light. Defects such as pits and shallow irregularities on surfaces stand out when they are illuminated by the contrasting effect of oblique lighting. Again they can be difficult to observe in thin-section as their irregular profiles become smoothed out when magnified and the probability of the section plane intersecting sufficient defects for adequate examination is limited.

The use of the unaided eye supplemented, where necessary by the low-power, stereo-zoom microscope is the most effective method for examining the surface of a concrete which is suffering from discoloration or pitting. These methods are also applicable to most incrustations which can then be removed and either subjected to optical examination in liquid immersion or to chemical analysis. Staining and discoloration are difficult to investigate as there are few useful methods available. Even examination by scanning electron microscopy may fail to detect the nature of some discolorations although Marusin (1995) reports some successful cases using this technique.

However, where corrosion, weathering or abrasion has occurred, that is, the whole nature of the surface has been changed, examination in thin-section yields useful information. Almost invariably the concrete will also be sectioned for other reasons and observation of the surface is only one portion of the overall petrographic examination.

5.5.2   MICROSCOPICAL APPEARANCE OF A WEATHERED CONCRETE SURFACE

Where a concrete has been adequately cast against a surface it will replicate the detail of that surface. This replication is due to crystallisation of calcium hydroxide at the surface and setting expansion (Lea 1970). As weathering proceeds the sharp detail of the surface is removed and the edge of the hardened cement paste takes on a ragged appearance. As more paste is removed aggregates close to the surface start to protrude. During this process dirt deposits and lichens may grow in the irregularities formed (Deruelle 1991).

If the water which contacts the surface is acidic a thin layer, usually less than 100 μm in thickness, of corroded cement paste will form. This layer, which is mainly an isotropic gelatinous silica is fragile, and may be missed if sampling and examination are not carried out carefully. A similar layer of gelatinous silica is formed as the final product of carbonation. In many cases it is not clear whether this isotropic layer is due to corrosion, attack by lichens or to carbonation and it is probable that a combination of these causes may operate for typical temperate climate weathering. An example of a surface exhibiting this type of weathering is given in Plate 41.

5.5.3   EFFLORESCENCES AND EXUDATIONS

Efflorescence is a white incrustation on the surface of concrete consisting of salts derived from the pore solution. Obviously, for visible efflorescence to occur, pore solution must be able to move to the surface to allow evaporation and crystallisation of the salts which are formed from the dissolved calcium, sodium, potassium and sulfate ions. However, the salts containing sodium, potassium, and sulfate ions are readily soluble in water and will be rapidly rain-washed from concrete surfaces unless they are protected. The quantities of these salts are also limited by their concentration in the concrete.

Calcium sulfate is rarely found as efflorescence because any excess sulfate in the pore fluid will preferentially form alkali sulfates. By far the commonest efflorescent salt is calcium

Fig. 5.4 The white efflorescence shown is due to movement of water through cracks and joints dissolving calcium hydroxide and depositing it at the surface as calcium carbonate. Cracks and joints have opened up because of stresses from filling and emptying the reservoir and possibly because of thermal stresses. The margins of cracks are corroded from the soft water present. The darkening is caused by moss and lichens. No rust is present as this sixty-year-old, thrust arch dam is unreinforced.

carbonate derived from atmospheric carbonation of the calcium hydroxide carried to the surface by the pore fluid. It persists because calcium carbonate is almost insoluble.

Where massive exudation occurs it is usually due to the leaching action of water moving through joints of a structure and then depositing leachates as it evaporates. Although water other than the pore solution is also involved, the main source of the efflorescence is still from pore solution. Good examples of massive efflorescence will be found on many old dams, reservoirs and wet tunnels, which may be streaked with spectacular deposits composed of calcium carbonate (Fig. 5.4). The white deposits are often discoloured by dirt, algae and rust.

The soluble efflorescent salts, such as sodium sulfate, potassium sulfate, sodium carbonate, potassium carbonate, trona (Figg *et al.* 1976) syngenite and calcium langbeinite, are found only in protected situations such as under soffits and in tunnels (Fig. 5.5).

In wet tunnels with considerable air movement through them the drying effect of the air can draw groundwater up the walls of porous concrete and bands of efflorescence may occur due to evaporation, particularly around joints (Fig. 5.6). This is a case where efflorescence is not wholly derived from the pore solution and strictly speaking should be regarded as a deposit. Salts such as thenardite and mirabilite (sodium sulfate hydrates) and possibly gypsum are likely to be found in these cases as well as calcium carbonate (St John 1982).

Alkali-silica gel is found on the surface of concrete undergoing alkali aggregate reaction (AAR) although much less commonly than supposed. This is not efflorescence but an exudation which will often be intermixed with calcium carbonate. Most of the gel observed in these cases will be carbonated. This is discussed in Section 6.7.2.

Fig. 5.5 Under the protection of the soffit at the crest of this irrigation dam this white, finely crystalline efflorescence has formed. The surrounding concrete is in good condition and no signs of cracking or deterioration is evident in this sixty-year-old structure. The material was identified as trona, which because it is soluble in water would not remain in a more exposed position.

Examination of efflorescence is practicable in thin-section provided that it forms a reasonably thick layer and care is taken not to destroy it during sectioning. Careful sampling of efflorescence, incrustations and exudations from concrete surfaces for chemical analysis before cutting up concrete samples is adequate for most purposes.

5.5.4 DISCOLORATION DUE TO EFFLORESCENCE

Efflorescent salts crystallise at the interface between saturated concrete and the atmosphere. Because it is dependent on the degree of surface drying, this interface can be either at or below the surface. Since the predominant colour of concrete is grey, and both calcium hydroxide and calcium carbonate are white, crystallisation occurring at the surface, even as very small crystals, will lighten its colour. As the outer layer of the concrete dries, the calcium hydroxide crystallises at a saturation interface which is now below the surface and the grey colour of the surface will be less affected. If these two processes occur on adjacent areas the contrasts between the lighter and darker areas appear as staining or darkening of the surface. It is not practicable to remove these areas of staining or darkening once they have formed. This type of discoloration is one of the more serious problems in the production of cast concrete products (Dutruel and Guyader 1977, Russell 1983) unless care is exercised in their curing and storage.

Often the efflorescent staining discussed above will only be a thin layer over the normal carbonation of the hardened cement paste. As noted in Section 5.5.1 this layer may be very difficult to distinguish in thin-section.

Fig. 5.6 The complex halo visible surrounding the vertical joint in the sixty-year-old tunnel wall has been caused by water wicking up the joint from the saturated rail ballast and then evaporating leaving white crystalline material at the margins of the halo. This was identified as thenardite and is derived from ground water in which sulphate levels were quite low. The concrete was found to be reasonably dense and thus has been able to resist the build up of sulfate, unlike the porous concrete shown in Fig. 5.14.

5.5.5   CORRODED AND ABRADED SURFACES

Abrasion is the mechanical and corrosion the chemical removal of material from the surface of concrete. One of the problems in the petrographic examination of abraded and corroded concretes is that it is not possible to estimate the amount of material that has been removed unless the original position of the surface can be estimated. Thus the degree of attack may be difficult to estimate although in some cases the surface can be inferred by extrapolating a line across the tops of protruding aggregates. The difference between abraded and corroded surfaces is usually obvious in thin-section. A corroded surface should have residues of isotropic silica present as an outer layer backed by a band of carbonation (Fig. 5.4). Where corrosion has been severe there will also be a zone behind the carbonation layer which is wholly or partially leached of calcium hydroxide. This leached zone may be undetectable in superficial cases of corrosion. In contrast, abraded surfaces have a rough, somewhat torn appearance and may show striations. No corrosion will be present and the carbonation layer may be quite thin. In thin-section, abraded surfaces tend to have a 'fresh' appearance provided post-sampling carbonation is prevented.

5.6   Interfaces

It is convenient to differentiate the intrinsic interfaces that are formed between paste, aggregate and reinforcing steel, from interfaces such as those formed by joints, mortars

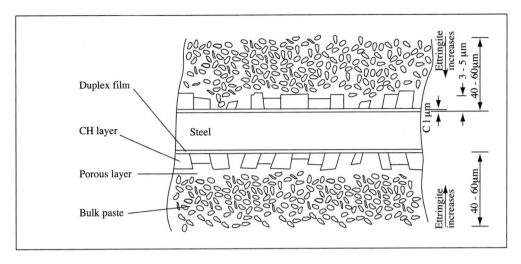

Fig. 5.7 Schematic illustration of steel/paste interface (from Massazza and Costa 1986).

and coatings. The question of the nature of the bond between paste and aggregates, reinforcing and other materials is the subject of considerable investigation (Massazza and Costa 1986) as it is considered to have a fundamental effect on the strength of concrete.

5.6.1 NATURE OF THE AGGREGATE/CEMENT PASTE INTERFACE

The nature of the interface is controlled by the properties of both the aggregate and the cement paste. Aggregates used in concrete have varying degrees of porosity, shape and surface roughness as well as the possibility of adherent dust and dirt. Even in one batch of aggregate a considerable variation will be found. For instance porosity may vary unless the lithology of the parent rock is uniform and many aggregates are a mixture of rounded and crushed materials.

The cement paste in contact with the aggregate is a surface layer and as discussed in Section 5.6.2, water and the finer portions of the paste will tend to concentrate in this layer. Physical contact and/or degree of separation of the paste from the aggregate will be largely dependent on the amount of water concentrated at the interface and the amount of plastic settlement that takes place before setting. Both these factors, which are closely interrelated, create space at the interface in which crystallisation from the pore solution can occur.

In Fig. 5.7 the duplex film, which is about 1 μm thick, consists of calcium hydroxide and calcium silicate hydrate, coats much of the surface of the steel. On this duplex film lies another semi-continuous layer about 3 μm thick of stacked calcium hydroxide crystals with their c axes semi-perpendicular to the steel interface. The zone of calcium hydroxide lies against a weak, porous layer of cement paste enriched in ettringite which changes gradually into normal paste at a distance that appears to vary from about 20 to 40 μm in width. This description is probably applicable to most non-porous surfaces that are reasonably smooth and unreactive. For aggregates that can be both porous and rough surfaced, much greater variation in the details of the aggregate/paste interface would be expected. The only details observable optically are the calcium hydroxide layer and any voids or fissures that may be present.

The interfacial regions of paste surrounding aggregates have been shown to be different from those of the bulk paste and have been called an 'aureole of transition'. Details of this transition aureole are shown in Fig. 5.7 and it should be noted that some of the details are still controversial (Diamond 1986). Most of the finer detail is beyond the resolving power of the light microscope but the larger crystals of calcium hydroxide including their orientation, and possibly increases in paste porosity, are details that should be observable. Carles-Gibergues *et al.* (1993) found that differences between the types of sulfate present in the cement affect the bonding of new-to-old concrete causing variation in both the nature of the calcium hydroxide crystallisation and the amount of ettringite present.

5.6.2   CEMENT PASTE SURROUNDING AGGREGATES IN THIN-SECTION

In thin-section the aggregate/paste interface tends to be more variable than suggested by many of the models proposed. An important variable is the absence or presence of meniscus-like voids underneath coarse aggregates due to plastic settlement and bleeding. This is why it is preferable to cut a thin-section from the vertical plane in the core sample to reveal the presence of this settlement and to use thin-sections of large surface area to show the true extent of this type of variation.

Other voids due to air trapped at the aggregate surface and ragged fissures caused by bleeding may also be present. Whether any of these types of void spaces are filled or lined with calcium hydroxide will be dependent on the amount and time that pore fluid has been present in the void spaces. If a void space has been filled with pore fluid for a considerable period during the life of the concrete it may contain well crystallised calcium hydroxide. In other cases they may be lined with calcium hydroxide and it is not uncommon to find plastic settlement fissures and air voids with only the merest hint, or even apparently devoid, of any visible calcium hydroxide. The dimensions of these voids and fissures usually exceed 20 μm and it could be argued that they are faults and do not form part of the aggregate/paste interface. In concretes with a high water/cement ratio these faults are common and must affect strength. There appears to be no information on systematic investigation by microscopy to relate the incidence of settlement and bleeding faults around aggregates to factors such as water/cement ratio and sand properties except for a report by Hoshino (1988).

Where the paste is in contact with the aggregate surface and not disrupted by bleeding and settlement most aggregate surfaces will be covered with a semi-continuous layer of calcium hydroxide providing that the aggregate is not porous (see Plate 43(b)). This layer can vary from a few microns up to about 20 μm thick even around the same aggregate and is usually in intimate contact, forming a bridge between the aggregate surface and the paste. There appears to be little information on systematic investigation by light microscopy of the orientation of the calcium hydroxide layers. A study of the orientation of calcium hydroxide layers surrounding aggregates in thin-section should be possible but the small size of calcium hydroxide crystals involved could make the use of interference figures difficult.

Many aggregates have sufficient water absorption to require this factor to be taken into account in the concrete mix design. The level of aggregate water absorption at which the following effect occurs is not known but the phenomenon is not uncommon in concrete. The aggregate appears to be surrounded by an aureole of darkened paste and no calcium hydroxide is observable at the aggregate interface. The darkening of the paste is probably illusory and due to a lack of visible calcium hydroxide in the aureole the formation of which appears to have been prevented by suction from the porous aggregate. Sarkar *et al.*

(1992) have reported that paste–aggregate bonding is dependent on the nature of the external shell of lightweight aggregate and that the absence of oriented calcium hydroxide crystals at the interface is related to the water absorption of the aggregate.

Some aggregates, such as limestone (Grandet and Ollivier 1980), interact with the pore solution of the concrete and it is claimed that the sizes of calcium hydroxide crystallising at aggregate interfaces are also affected by the composition of the aggregate (French 1991). At a practical level the main determinants of the incidence of faults and the amount of calcium hydroxide crystallising at aggregate interfacings are undoubtedly the water/cement ratio and the bleeding capacity of the concrete. In typical structural concretes localised variations of mixing and compaction can result in a range of aggregate/paste interfaces being present in the same large-area thin-section. This variability decreases with lower water/cement ratios and adequate compaction of the concrete.

5.6.3   CEMENT PASTE TO STEEL REINFORCING INTERFACES

The interface between steel and paste (see Fig. 5.7) is similar in many respects to that of aggregates except that steel effectively has zero porosity apart from light rust on its surfaces. While steel surfaces are more uniform than aggregate surfaces and proposed models for the interface are more likely to be observed, the paste/steel interface is still subject to the effects of bleeding and plastic settlement. It is rarely possible to observe paste/steel interfaces by transmitted light microscopy because of the difficulties involved in the preparation of thin-sections containing steel. One technique has been proposed by Garrett and Beaman (1985) for retaining diametral cross sections but ideally adequate examination of the steel/paste interface requires longitudinal sectioning of the steel reinforcing. This is one of the neglected areas of concrete petrography and will remain so until adequate sample preparation methods are available.

5.6.4   CEMENT PASTE/FIBRE REINFORCING INTERFACES

Steel, carbon, glass, asbestos, kraft pulp and polymer fibres are also used to reinforce cement products. The examination of their interfaces in thin-section can provide useful information but on the whole these interfaces are better observed by scanning electron microscopy. Reviews covering these materials will be found in references, for example, Massazza and Costa (1986), Diamond (1986) and Brandt (1995).

Some observations with the optical microscope are possible in the cases of glass, asbestos and kraft pulp or other vegetable fibres. As reported from scanning electron microscope studies (Barnes *et al.* 1978, Diamond 1986) unless the individual fibres of a glass strand debond, little if any paste or calcium hydroxide penetrates between the fibres as shown in Plate 42. Along the length of the strands or debonded fibres copious calcium hydroxide is crystallised. In the case of asbestos fibre the high wetting capacity of this fibre combined with the small cross-section of individual fibres appears to prevent the crystallisation of any but large crystals of calcium hydroxide at the fibre interface. It is not possible to make statements about the interface of kraft pulp fibre as the products in which the fibre is incorporated are usually steam cured at high pressure.

5.6.5   EFFECT OF MINERAL ADDITIONS AND CURING ON INTERFACES

Many products containing fibrous reinforcement are steam cured at temperatures of around 180°C and also contain silica flour as a mineral addition. Effectively this removes much of the calcium hydroxide from the cement by reaction with the silica flour and results in an interface deficient in calcium hydroxide. As shown in Plate 43(a) the cement paste appears to be in contact with the kraft pulp fibres with little calcium hydroxide present at the interface.

At ambient temperatures, siliceous mineral additions can also have similar effects. The effect will be greatest for a highly active addition such as silica fume and will decrease with the coarser and less reactive additions.

5.6.6   RENDER INTERFACES

A traditional external rendering consists of a number of coats of cement plaster laid over substrates of concrete or masonry. A general description of current UK techniques will be found in Robinson *et al.* (1992) and American techniques in ACI Committee 524 (1993). Renders which are applied as a single coat are now common in Europe (Mehlmann 1989). Ignatiev and Chatterji (1992) have discussed the compatibility of mortar and concrete.

These interfaces are always characterised by carbonation of the substrate and some depletion of calcium hydroxide in the plaster layer itself due to suction from the substrate. Contact along the interface is never continuous and voids and intermittent fine fissures are evident. The deposition of calcium hydroxide, so often found associated with aggregate interfaces, is not as common at render interfaces probably because of suction from the substrate.

Rendering failures often occur because of the loss of adhesion between the render and the concrete substrate. Modern concreting frequently uses fair faced formwork coated with a release agent to give a smooth surface to the concrete. These surfaces do not make a good substrate for rendering unless all vestiges of the release agent have been removed by either surface preparation or weathering. Residues of the release agent causing the problem are not observable by microscopy and require solvent extraction and chemical analysis for their detection and identification. Thus petrographic examination is of limited use in the diagnosis of adhesion failures and made more difficult in that invariably only the render and not the concrete is made available for examination. Meaningful observation of an interface requires both surfaces to be available for microscopic examination.

The type of render, known as shotcrete or gunite, is a pneumatically applied mortar, or concrete used as a rendering and also for the repair of damaged concrete surfaces (Schrader and Kaden 1987). If a thick layer is required the mortar is applied over a preceding layer soon after initial set has occurred and these interfaces can be identified in thin-section by a diffuse line of calcium hydroxide (Plate 44) and on occasion by some darkening due to absorption by the substrate. It is now common to include silica fume and a range of fibres in shotcrete to improve its rebound and physical properties and these additions will considerably affect its appearance in thin-section as discussed in Section 5.6.5.

Tiles have traditionally been fixed to surfaces with a bedding mortar, which now often includes a latex polymer to improve adhesion (Robinson *et al.* 1992). To level the tile and ensure it is in good contact with the mortar the tiler drives the tile against the mortar by beating. Thin-section examination shows there is an intimate interface between the tile and hardened paste only separated by a semi-continuous layer of calcium hydroxide which is generally less than 5 μm in width (Plate 45). Along the interface void areas are present due to entrapped air. Where the tile is insufficiently beaten during laying only the tile nibs bite into the mortar leaving recesses with large bladed crystals of calcium hydroxide attempting to bridge the gap.

5.6.7 OTHER INTERFACES

Joints between pours of concrete separated by a time interval are often characterised by a line of carbonation in the concrete substrate and depletion of calcium hydroxide in the contact zone of the overlying concrete due to absorption from the substrate. These joints are usually irregular because the underlying surface is sometimes mechanically roughened prior to concreting. Voids, cracks, fissures, dirt and debris are often observed at the interface. Polymers such as epoxy resin used as a coating or topping are characterised by some penetration of the polymer into the substrate with excellent adhesion resulting Plate 36(b). The interface in these cases is to the carbonated surface and does not involve calcium hydroxide.

5.6.8 OBSERVATION OF INTERFACES WITH THE LIGHT MICROSCOPE

The observation of interfaces in concrete and mortars in thin-section can provide information on degree of contact, absorption into the substrate, calcium hydroxide, void spaces, and dirt and debris. Observation is limited to dimensions exceeding about five microns. Submicroscopic examination of interfaces is best carried out by scanning electron microscopy. The broader details of interfaces observed by microscopy need to be correlated with those made by scanning electron microscopy.

5.7 Void space in concrete

To put the subject in perspective the range of voids and pores found in concrete are listed in Table 5.8 together with some of their effects on the properties of concrete. The gel capillaries and pores are beyond the scope of this discussion but porosity involving capillaries exceeding 10 nm affect durability and are important. The conditions under which porosity in concrete crosses over from connected to discrete capillaries have already been discussed in Section 5.4. Where the capillary pore structure is not interconnected its principal effect is on strength (Mindess and Young, 1981). However, where the capillary pore structure is interconnected it not only affects strength but also the durability of the concrete because of increased permeability. This especially applies to the outer layer of a concrete, where durability will be reduced because the interconnected capillaries may allow vapours to penetrate to a sufficient depth to corrode the reinforcement.

As the microscope tends to be restricted to the lower magnifications when working with thin-sections of concrete, effective observation is restricted to dimensions above greater than 1 μm. Thus only those categories in Table 5.6 whose dimensions exceed 5 μm can be easily examined. In practice, the capillary and coarser gel pores require the use of scanning electron microscopy, usually of fracture surfaces, while the finer gel pores may be measured by mercury porosimetry.

The basic difference between a capillary pore and an air void in plastic concrete is that capillary pores are elongated spaces filled with water while air voids are roughly spherical pores formed by the entrapment or entrainment of air. Hence the difference in the minimum dimensions. The minimum size of an air void is limited by its internal gas pressure which in turn is controlled by its radius and the surface tension of the liquid. Small bubbles have high gas pressures and in practice bubbles of less than 4 μm tend to disappear (Dolch 1984).

Table 5.8 Classification of pore sizes in cement pastes (adapted from Mindess and Young 1981)

Designation	Diameter	Description	Role of water in void	Paste properties affected
Voids	2–0.25 mm	Foamed systems	Behaves as bulk water	Strength, density
Voids	large–0.010 mm	Entrapped air voids	Behaves as bulk water	Strength
Voids	1.0–0.005 mm	Entrained air voids	Behaves as bulk water	Strength
Capillary pores	10–0.05 μm	Large capillaries	Behaves as bulk water	Strength, permeability
Capillary pores	50–10 nm	Medium capillaries	Moderate surface tension forces generated	Strength, permeability, shrinkage at high humidities
Gel pores	10–2.5 nm	Small (gel) capillaries	Strong surface tension forces generated	Shrinkage at 50% RH
Gel pores	2.5–0.5 nm	Micropores	Strongly absorbed water: no menisci formed	Shrinkage; creep
Gel pores	< 0.5 nm	Micropores 'interlayer'	Structural water involved in bonding	Shrinkage; creep

An arbitrary distinction is made between entrapped air voids, that is, those voids which are trapped during mixing, and entrained air voids produced by admixtures which have a foaming or air entraining action in the plastic cement paste.

5.7.1 ENTRAPPED AIR VOIDS

Entrapped air voids are invariably produced in the mixing and placing of concrete as has been described in Section 4.5. In an adequately graded and compacted concrete mix the volume present is determined by the maximum size of the coarse aggregate. For instance where 65 mm aggregate is used entrapped air will be about 0.3–0.5 per cent in volume which increases to about 3 per cent for 10 mm aggregate. The dimensions of entrapped air voids range from about 10 mm down, have smoothish margins, and may be spherical to irregular in shape. Both the larger size and irregular shape become more common the higher the water/cement ratio.

In cement mortars, dependent on the sand grading used, entrapped air voids occupy a volume of about 3–8 per cent. Because of the absence of coarse aggregate with a resultant higher cement to aggregate ratio, the entrapped air system in mortars tends to be more regular and in many cases will appear similar to an entrained air system. However, it is not as stable as an entrained system and is more easily removed during placement of the mortar. Measurement of bubble size distributions will also show that entrapped air is deficient in the smaller bubble sizes.

From examination of the air void system in a thin-section it is possible to characterise the parameters of the system as well as observe whether voids contain calcium hydroxide

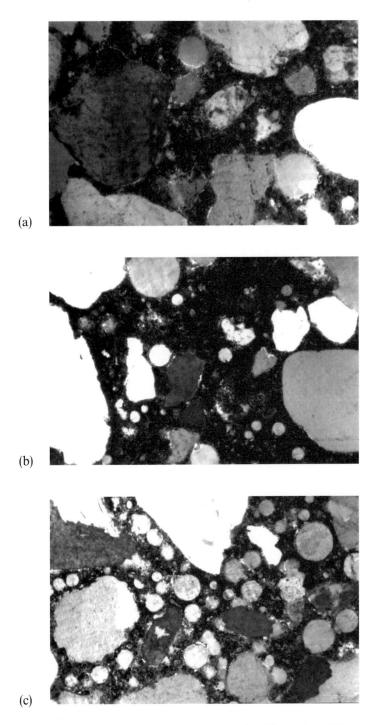

(a)

(b)

(c)

Fig. 5.8 The appearance of entrained air void systems in thin-section. Widths of fields of view 1.4 mm. The appearance of different percentages of air voids in mortars, as measured by an air meter on the plastic mix, and checked by point counting in thin-section are shown above: (a) 6 per cent of entrapped air (b) 10 per cent entrained air (c) 17 per cent entrained air.

and other products. The general appearance in a laboratory mortar of different percentages of air voids (measured by an air meter on the plastic mix) are shown in Fig. 5.8.

5.7.2   ENTRAINED AIR VOID SYSTEMS

The entrainment of a stable, well spaced air void system in plastic concrete and mortar mixes is of considerable importance in those countries subject to severe frost conditions. In addition, entrained air has a lubricating effect on plastic cement paste and produces more workable concretes and mortars. This is especially important for plaster renderings. There is a large literature on the properties of air entrainment in concrete and reference should be made to codes of practice such as ACI Committee 212.3R (American Concrete Institute 1989) and for a general overall discussion of the chemistry and other properties to Dolch (1984).

As described in Sections 2.11 and 4.5 measurement of the air void system is traditionally carried out by observing finely ground sections of concrete with the aid of a low-power stereo-binocular microscope (ASTM C457, 1990). To measure voids down to the smaller diameters so that the system is adequately characterised requires the use of a magnification as high as $\times 120$ (Bruere 1960) and as a result the image of the ground surface tends to be of poor quality. This problem does not exist in transmitted light microscopy where crisp images are readily available at these magnifications. It is not widely recognised that the availability of routine large-area thin-sectioning allows the detailed examination of air void systems and with the use of a suitable point-counting stage accurate quantification is also possible. Measurement of chord lengths as proposed by Lord and Willis (1951), coalescence of bubbles and whether voids have become filled with secondary products can be observed Plate 45. In addition variation of the air void system in the outer layers of the concrete is easily observed.

From examination in thin-section it is possible to measure the air void system and observe its relationship in the texture and whether voids contain calcium hydroxide and other products. The magnification of the photomicrographs is in the range of magnification specified by ASTM C457 (1990) and shows the type of detail that can be observed in thin-section. The laboratory mortars were mixed with a cement/aggregate ratio of 1:2.75 and a water/cement ratio of 0.5 using an OPC, Leighton Buzzard graded quartz sand and vinsol resin as the air entraining agent. The cubes were compacted by rodding in three layers.

5.7.3   CAPILLARY VOIDS

Capillary voids are not easy to observe in thin-section as they are both well dispersed and their dimensions are at or beyond the working limit of observation. In practice this is overcome by incorporating a fluorescent dye in the epoxy resin used for impregnation and visually estimating the amount of fluorescence (Dubrolubov and Romer 1985) and relating this to the water/cement ratio. Strictly speaking this is not observation of capillary voids but an assumption that the coarser capillaries can be satisfactorily impregnated with epoxy resin (Section 2.7).

5.7.4   OTHER VOIDAGE IN CONCRETE

Other voidage, such as cracks (Section 5.8), bleeding channels, voids and fissures due to plastic settlement, excessive air entrainment and inadequate compaction are all faults associated in varying degrees with unsatisfactory mix design, dispensing of constituents, mixing and placement practices.

Bleeding channels and plastic settlement are present in most concretes and become prominent in wet concretes with a high bleeding capacity and increase in frequency in the upper portions of a lift or slab. They tend to be gross faults and do not need highlighting as they are of sufficient dimension to be clearly visible in thin-section. To observe the full extent of plastic settlement the plane of section must be vertical. On the other hand, bleeding channels tend to the vertical and are more likely to be intersected by a horizontal section. In practice the most information appears to be gained from the vertical section.

Fissures due to plastic settlement are associated with the bottom margins of aggregates, may have well defined margins and are often meniscus-like in shape. If pore fluid has been present in them for any length of time they will be lined or filled with calcium hydroxide. Bleeding channels are irregular, tortuous and because a section represents only one plane, will appear to be discontinuous. The channel margins are diffuse and slowly grade into the cement paste suggesting that some of the cement paste at the margins has been partially dissolved during the bleeding process. It is unusual to observe much calcium hydroxide associated with bleeding channels probably because of leaching by the bleeding water. There is a problem in deciding at what point plastic settlement and bleeding are significant from observation made in thin-section. Useful interpretation requires a knowledge of the incidence of these faults for the typical materials used in any one area and examination of large sample cross-sections from a range of sampling points in a structure.

Severe overdosage with an admixture to produce excessive entrained air has a spectacular effect on a concrete. Visually the concrete appears finely honeycombed with air bubbles. By the time the concrete has reached air contents of around 25 per cent it may have little strength and can be crumbled easily and may superficially appear not to have set. Since each 1 per cent of air reduces strength by roughly 5 per cent the reason for this effect is obvious. In thin-section such concretes appear as a mass of small bubbles, sometimes coalesced together, with only thin septums of paste separating bubbles. Because of the close spacing of the air voids these septums do not have sufficient dimensions to provide any strength.

Excess voidage can also be found in concrete due to incomplete compaction and is largely restricted to dryer concrete mixes that do not flow easily. The extreme condition is where insufficient water is used so that the cement and water is not fully dispersed during mixing giving rise to rich layers of sticky cement surrounding aggregates which are almost impossible to compact. This effect is illustrated in Fig. 5.9, in a sample of concrete from an *in situ* rammed pile which had the appearance of having not set. In thin-section it was observed that the concrete was only held together by narrow porous bridges between the rich layers of undispersed cement paste surrounding the aggregates (St John 1983). All the margins of the cement paste facing into voids were ragged and porous and in places very large crystals of calcium hydroxide were present.

Less spectacular cases of incomplete compaction consist of ragged void space which may contain completely unconsolidated material. The bridging spaces between the ragged void space may be porous so that in effect they act as weakened bridges. Kaplan (1960) investigated the effects of incomplete compaction on concrete and showed that an unconsolidated concrete with 25 per cent void space loses about 90 per cent of its compressive strength.

Segregation of the mortar from the coarse aggregate can occur with harsh mixes, mixes that are too wet or too dry, and in those that are sand deficient. Where the segregation is on a wide scale microscopy is inappropriate and the fault should be observed by comparison of large cut faces. Highly localised patches of segregation are not uncommon in concrete

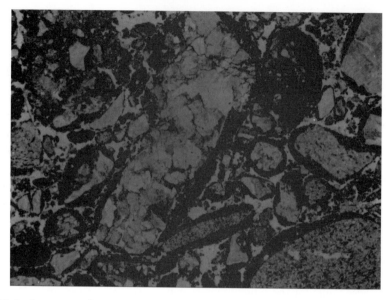

Fig. 5.9 This is the texture of a concrete containing insufficient mixing water. As a result the cement was not dispersed during mixing and adhered to the aggregates as sticky rims. The ramming used during emplacement was insufficient to compact the concrete to more than minimal contact between the cement rims. When the concrete set it was porous, weak and crumbly in texture and superficially appeared not to have set. As the piles were part of a major civic structure the concrete was condemned and replaced. Width of field 12 mm.

and are usually ignored by the petrographer unless their frequency suggests problems in the mix design and placement of the concrete.

5.7.5   *AERATED AND NO FINES CONCRETES*

Aerated concretes, also known as gas, foamed or cellular concretes and mortars are used as semi-structural lightweight and insulating materials (Short and Kinniburgh 1978, Taylor 1974). The aeration, which consists of discrete bubbles which are one to two magnitudes larger in size than air-entrained bubbles, is emplaced by mechanical or chemical foaming. Mechanical foaming may be carried out by adding a preformed foam or foaming *in situ*. Chemical foaming, usually based on the addition of aluminium powder to the mix, generates the foam by the evolution of gas bubbles *in situ*. Many of the aerated products are not concretes but mortars and include additions of silica or blastfurnace slag. Precast products are often autoclaved. The texture illustrated (Plate 46) is of a commercial sample of autoclaved lime silica block. Its aerated structure consists predominantly of pores with diameters of 1.5–2 mm with some interspersion of smaller pores. The smallest pore observed was 0.25 mm in diameter.

No fines concrete is made with 10–20 mm aggregate and a cement:aggregate volume of about 1:8 (Malhotra 1976). Its texture, illustrated in Plate 47, consists of a packing of aggregate coated with cement paste which acts as a bridge to cement the aggregates together. Between the individual aggregates there are large irregular voids with smooth margins.

5.7.6 INFORMATION FROM MICROSCOPIC OBSERVATION OF VOID SPACE IN CONCRETE

In thin-section the light microscope can easily observe details in concrete larger than a few microns. Thus accurate and detailed observation of air void systems, void space due to bleeding and plastic settlement, and incomplete compaction are all possible provided that sufficient sample area is examined so that reasonable estimates of their occurrence can be made. For entrained air void systems microscopic examination of thin-sections is a superior method for detailed examination. It is unlikely to supplant standard ASTM methods because of the time and cost involved and the underestimation of the total air content that occurs when the total air content is estimated from thin-sections.

The examination of capillary void space is not possible by light microscopy and remains the province of scanning electron microscopy. At the other end of the size scale broad faults, such as gross segregation, are better observed by examination of cut surfaces.

5.8 Cracking in concrete

In the Concrete Society Technical Report No. 22 (1992) the following highly pertinent statement is made.

> It is fundamental that reinforced concrete cracks in the hardened state in the tensile zone when subjected to externally imposed structural loads. By means of appropriate design and detailing techniques these cracks can be limited to acceptable levels in terms of structural integrity and aesthetics. Concrete is also liable to crack both in the plastic and hardened states due to stresses which it intrinsically sustains by the nature of its constituent materials. The construction industry does not readily accept that these intrinsic cracks are almost as inevitable as structural cracks, and concrete containing them may often be summarily and unjustifiably condemned.

Unexplained cracking is a common reason for engineering inspection and core sampling of concrete structures. The causes of cracking can be complex and their examination forms an important part of a petrographic examination. Because of this complexity of causes it is vital that the petrographer either inspects the structure or is given an accurate and detailed description of all cracking observed. Too often the inspecting engineer makes sampling decisions without reference to the petrographer. This may result in either the need for further sampling or an unreliable outcome from the petrographic examinations because of insufficient information.

5.8.1 CRACKING AND TENSILE STRAIN

Hardened concrete cracks when its tensile strain capacity is exceeded by stresses applied from any source. The tensile strain capacity of hardened concrete varies from about 0.008–0.015 per cent but whether the concrete actually cracks for any given set of stresses is complex and depends on rate of strain, restraint by aggregates and reinforcing, the age of the concrete, creep and other factors. The relationship of many of these factors to strain capacity of a concrete is beyond the scope of this chapter and reference should be made to Concrete Society Technical Report No. 22 (1992) for further information. Goltermann (1995) has proposed a prediction method for the classification of crack patterns related to strain in the hardened cement paste and aggregates.

Because the surface of concrete is effectively an anisotropic interface for strain it is subject to superficial cracking. This is especially the case for localised drying shrinkage, which may

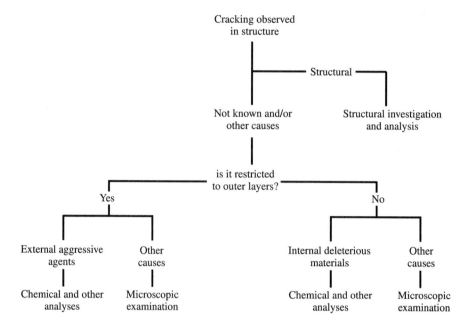

Fig. 5.10 Outline of method required for investigating cracking in structures.

penetrate the concrete to a depth of only a few millimetres. It is an important part of petrographic examination to differentiate superficial surface cracking from other types of cracking which may have more serious consequences.

5.8.2   *DIFFERENTIATING BETWEEN STRUCTURAL AND NON-STRUCTURAL CRACKING*

To distinguish between structural and non-structural cracks requires a systematic approach to diagnosis. Where the cracking is obviously structural in nature its diagnosis lies within the realm of structural engineering and is beyond the scope of this book. For further information some of the common causes of structural cracking are discussed in a simple manner by Kelly (1981) and Gustaferro and Scott (1990) and more extensive reviews have been given by ACI Committee 224.1R (American Concrete Institute 1984) RILEM Committee 104-DDC (1994) and Campbell-Allen (1979). The cracking of concrete associated with corrosion of reinforcement has been discussed by Pullar-Strecker (1987).

Where the cause is in doubt, which is often the case where the properties of the materials are involved, it is necessary for the engineer to follow the examination and analysis procedures as detailed in Fig. 5.10 to enable structural cracking to be clearly distinguished from non-structural cracking. It is rarely possible to determine the cause of non-structural cracking without carrying out some type of examination of cut surfaces. Where the relationship between cracking and deterioration of a concrete fabric has to be established, thin-section examination is often required. The exceptions are cases of expansion due to hard burnt lime and excess magnesia in the cement, and some cases of attack by external aggressive agents where chemical analysis is obligatory and examination of sections may not be justified.

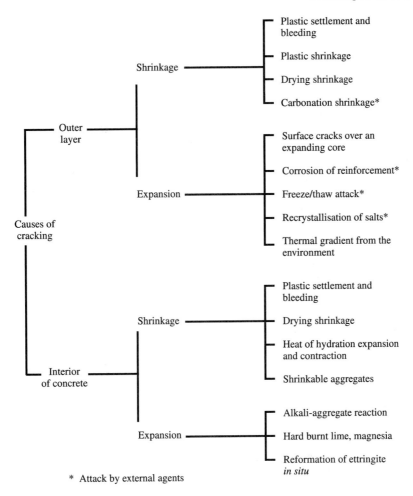

Fig. 5.11 The causes of cracking and their relationship to the outer layer and interior of a concrete.

5.8.3 CLASSIFICATION OF NON-STRUCTURAL CRACKING

There is no entirely satisfactory scheme for classifying the causes of non-structural cracking in concrete. Figure 5.11 is a modification of the classification given in Concrete Society Technical Report No. 22 (1992). It has been modified to emphasise the relationship between the internal and surface cracking which needs to be established in petrographic examination. The outer layer of concrete for the purposes of this table is the thickness of concrete penetrated by the environment. In practice this is a layer that can vary in depth from 1–2 mm in high grade concretes to as much as 50 mm or greater in lower quality concretes.

5.8.4 CRACKING IN THIN-SECTION, AT THE SURFACE, AND THE INTERIOR OF A CONCRETE

The following is an attempt to describe the nature of cracking as commonly observed in thin-section. Since cracking in concrete may be due to a number of causes the following

Fig. 5.12 Scaling of a concrete surface due to frost attack. Some of the surfaces near the base of this sixty-year-old irrigation dam are scaling due to frost attack. Apart from the surface scaling shown this dam is in excellent condition although the concrete does not contain entrained air. The frost attack has occurred because the base of the dam is protected and more subject to frost and some of the surface areas are porous. This is a typical example of mild frost attack.

explanations do not cover all circumstances. These can only be evaluated by considering both the visible effects on the structure together with the observations made in thin-section and other available data. It cannot be over emphasised that adequate diagnosis of cracking requires examination of both the external surfaces and internal fabric of a concrete.

Cracking sub-parallel to and restricted to the near surface

1. *Scaling and exfoliation.* Sulfate, freeze/thaw attack or any expansive agent that penetrates the surface layer can cause scaling and exfoliation. Scaling which is primarily associated with mild freeze/thaw attack and fretting with salt recrystallisation, are merely the initial stages of exfoliation (Fig. 5.12). A surface attacked by frost is rough and irregular and the concrete remaining will be cracked, often with a sub-parallel mesh of cracks, to a depth which is dependent on the degree of attack and the permeability of the concrete (Figs. 5.13, 5.15). The sub-parallel cracking usually grades relatively abruptly into sound concrete. Where severe exfoliation occurs due to sulfate attack sheets of concrete can be detached from the surface (Fig. 5.14). In thin-section the exfoliated area consists of a mass of sub-parallel cracks which run through both the hardened paste and aggregates. Apart from being able to estimate the depth to sound concrete there is little else to be observed. Sampling exfoliated concrete for thin-

Fig. 5.13 The effect of frost on porous concrete. The concrete in this seventy-year-old hydro surge tank is discoloured by lichens and efflorescence and the porous top surfaces of the buttress walls have disintegrated due to frost attack. In this case the exfoliation has left a crumbly texture which gives the appearance that the concrete is more seriously attacked than is the case. Petrographic examination showed that the body of the concrete was still in reasonable condition. After replacement of the affected areas of concrete the structure was returned to operation.

section examination presents difficulties. In core samples much of the exfoliated texture is lost unless special procedures are used to consolidate the surfaces before coring Fig. 5.14.

In cases of sulfate attack crystals of gypsum and/or ettringite may be present filling or lining cracks, pores and fissures. It is rarely possible to observe gypsum or ettringite intimately interspersed in the hardened paste so diagnosis should always be confirmed by X-ray diffraction analysis and where necessary supplemented by chemical analyses for the sulfate content of the affected areas.

2. *Pop-outs.* A pop-out is caused by the expansion of an aggregate lying under the surface of the concrete. In the initial stages a pop-out appears as a semi-circular crack with vertical displacement of the semi-circular margin (Fig. 5.16(a)). As the pop-out develops the crack becomes roughly circular and an inverted cone of material protrudes above the surface and may be removed leaving a crater with remnants of an expansive aggregate at its apex.

The dimensions of pop-outs range from a few millimetres to 100 mm in diameter and up to 40 mm in depth. Expansive materials that cause pop-outs are alkali-reactive particles, iron sulfide minerals, and porous aggregates that undergo freeze/thaw attack. Examination of the area of a pop-out in thin-section is rarely justified as visual examination of the apex of the pop-out crater will often reveal the cause. In thin-section the expansive aggregate which is usually partially disintegrated, can be identified. If alkali-silica gel is present it is also readily identifiable but the iron sulfide minerals may be difficult to observe and require identification by X-ray diffraction analysis.

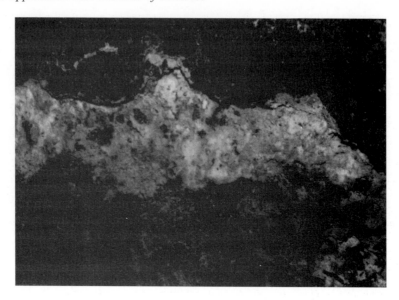

Fig. 5.14 Exfoliation due to sulfate attack. This is the same attack as described in Fig. 5.6, except that in this case the concrete contains a porous limestone aggregate which made it permeable. The area shown is about 0.5 m above the rail ballast and ran the full length of both sides of the tunnel. Exfoliation, which was in a relatively narrow band, was so severe that in many places sheets of concrete fell off when struck with a hammer. The white crystals on the surface of the concrete are thenardite. Petrographic examination showed that exfoliation did not extend deeply below the surface and the concrete was repaired by removal of damaged material and reinstatement with shotcrete.

Cracking approximately perpendicular to, and restricted to, the surface layer

1. *Fine, filled cracking either localised or more general patterns.*   Fine pattern cracking, often widely distributed and attributable to drying and/or carbonation shrinkage, is common on plastered surfaces and may also be present where the finished surface of the concrete has been too wet or too rich in cement (Fig. 5.17).

   Its external appearance is of white lines, often raised because of infilling by carbonation or, where moisture is absent, dark lines because of infilling dirt or simply contrast with a lighter surrounding surface (Fig. 5.17). When very fine the cracking may not be apparent to the naked eye on rough surfaces but can be seen with the aid of a hand lens or stereomicroscope.

   When observed in thin-section this cracking rarely exceeds 10 mm in depth and 10 μm in width and is invariably filled or contains carbonation if the surface has been exposed to the weather for any period. It is not uncommon for the cracks to end at aggregate margins. Where doubt exists about the cause of this type of cracking, thin-section examination is required to show its relationship to the remainder of the concrete. A more severe form of the above cracking is typical of carbonation shrinkage referred to in Section 5.4.6. There are few concretes in which some superficial examples of this type of surface cracking cannot be observed in thin-section as shown in Plate 37(a).

2. *Fine cracking, often open, tends to form patterns in localised areas.*   If the cracking is open at the surface careful examination will often reveal some vertical displacement

Fig. 5.15 This is a negative print of part of a thin-section cut at right-angles to the exfoliated surface shown in Fig. 5.14. Sulfate attack has caused a fine subparallel pattern of cracks to form typical of exfoliation texture which grades relatively abruptly into sound concrete. In this case of sulfate attack there is no softening of texture. Some of the fine cracks bounding the sound concrete are filled with gypsum. Width of field 35 mm.

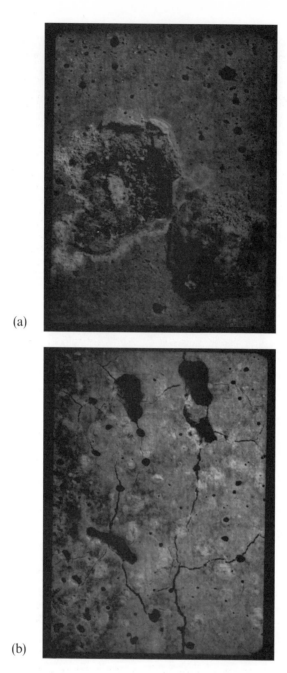

(a)

(b)

Fig. 5.16 A pop-out and an example of open pattern cracking. Width of photographs 25 mm. (a) The pop-out has been removed and lies face down on the surface of the mortar test prism. A reacted aggregate lies at the bottom of the crater and some white alkali-silica gel is visible at the apex of the pop-out. The white halo is probably carbonated alkali-silica gel although analysis would be required to distinguish this from calcium carbonate. (b) Fine, open pattern cracking on the surface of a mortar test prism. The cracking is open and forms a closely spaced pattern which is typical of moist cured concrete. Greater detail is revealed by using a low-power stereomicroscope which should always be used for examination of laboratory specimens. Such examination will often show that crack margins are displaced vertically. Cracking of this type is caused by expansive reactions, the commonest of which is alkali-aggregate reaction.

Fig. 5.17 This is a good example of drying shrinkage in a cement-rich surface outlined by carbonation which is highlighted because it is white and present as a raised line. The paving slabs were laid as site-mixed concrete and before initial set a coloured cement was worked into their surfaces. It is also a good example of a typical deterioration of coloured surfaces by weathering.

at the crack margins. Where the crack is closed because it is filled with carbonation this displacement may not be observable. In thin-section these cracks usually range from about 15–40 μm in width and can penetrate the concrete to depths of about 50 mm. Dependent on exposure conditions they may have carbonated margins or be filled with carbonation for much of their depth. The cracks tend to skirt aggregate margins and take the line of least resistance as shown in Plates 39(a) and (b).

This type of cracking is typical of the first stages of alkali-aggregate reaction in which the core of the concrete has expanded just sufficiently to exceed the tensile strain of the surface in localised areas. In the outer layers of the concrete expansive reaction is absent due to leaching and/or movement of alkalis. If examination of the concrete reveals that potentially reactive aggregates are present it can be assumed that the first stages of alkali-aggregate reaction may be present in the concrete even if no alkali-silica gel or aggregate reaction is observed. Examples of cracking in the early stages of AAR are illustrated in Section 6.6.

3. *Simple cracking, usually open, associated with reinforcement and a few millimetres wide.* The edges of the cracks are often displaced vertically in relation to each other and rust staining may be present but this is not invariable and less common than would be expected. At corners and edges of members a levering action is evident which tends to spall the concrete. This type of cracking is due to corrosion of reinforcement. In its early stages it may be difficult to diagnose but when more advanced is usually obvious. It is rarely possible to successfully make thin-sections of samples containing corroded steel reinforcement. The best method of investigation is to carry out either half-cell potential mapping or resistivity testing or physically remove the affected cover concrete to examine the reinforcement for signs of corrosion, as described by Pullar-Strecker (1987).

Fig. 5.18 Cracking in a plastered surface. The plastered wall of this small landing is cracked for two reasons. The finer dark cracking is due to drying shrinkage which appears dark because the lime bloom on the plaster has weathered away at crack margins. In this case the cause of the cracking is obvious because it is a plastered surface but on concrete surfaces diagnosis is more difficult and may require petrographic examination. The large open crack is due to subsidence.

Surface cracking due to expansive forces in the interior concrete. Best identified in thin-section

1. *Roughly rectangular or triangular patterns, usually open, with widths varying from 60 μm to a few millimetres.* The pattern of cracking may be either restricted to localised areas or be quite widely spread but is never uniform over the structure. A typical example is shown in Fig. 5.19 and many examples of cracking due to alkali-aggregate reaction will be found in the literature, for instance in Hobbs (1988) and Swamy (1992). However, there is a tendency for authors to illustrate the more spectacular cases of cracking. It should be realised that external cracking even in cases of severe alkali-aggregate reaction is often not spectacular (Section 6.7). Vertically displaced margins to the cracks are common and margins may be darkened especially after rain. Where larger cracks are present and water has percolated through the crack system, efflorescence may be present which may be confused with carbonated alkali-silica gel. Clearly visible, external, alkali-silica gel is not common in alkali-aggregate reaction unless the reaction produces large amounts of gel. In thin-section it will be found that the surface cracks transect a layer of concrete up to 75 mm in depth in which alkali-aggregate reaction is visually absent (Fig. 5.20).

    The surface cracks grade into a more closely spaced system of pattern cracks with clear evidence of alkali-aggregate reaction shown by the presence of alkali-silica gel, reacting particles and some traces of ettringite. The crack widths in the interior vary from about 120 μm to as little as 5 μm in width and in severe cases may be transected by wider tension cracks. Crack widths and pattern spacings are variable and dependent on the size of the reacting particles. The smaller the reacting particles the more complex the internal system of pattern cracking as illustrated in Fig. 6.21. A noticeable characteristic of the cracking is that it is often not continuous. Cracks may terminate

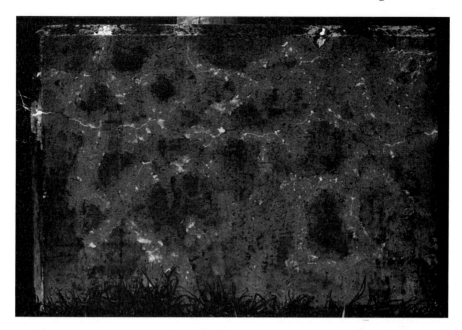

Fig. 5.19 Pattern cracking in advanced alkali-aggregate reaction. The contrast has been deliberately heightened in this photograph of a bridge pier base to outline the cracking caused by an advanced AAR. The darkened margins and open cracks are typical of AAR but the crack pattern is not exclusive to this reaction. Obvious alkali-silica gel was absent even on close inspection but was clearly revealed in petrographic examination of thin-sections. In less severe cases of alkali-aggregate reaction a very much diminished and more localised form of this pattern cracking will be present and can be difficult to interpret without recourse to petrography.

at an aggregate margin or in the paste and restart close by in a displaced position. This is the classical picture of alkali-aggregate reaction where the structure is undergoing normal weathering but may apply to any species of expanding particles where the outer unreacted layer of concrete has cracked in tension over an expanding core of concrete.

A second example of restrained cracking relates to a prestressed beam in a bridge which spanned a narrow marine inlet. The prevailing weather is mild with rain spread throughout most of the year. Cracking and distress was noted within ten years of construction requiring replacement of the deck and cap beams. The deck beams, containing 415 kg/m³ of cement, had a few open cracks which ran parallel to their length. The cap beams, containing 315 kg/m³ of cement, also had a number of cracks which were vertical to their lengths. The deck consisted of precast, hollow cored, prestressed beams while the cap beams on which the deep beams sit were cast *in situ* with conventional reinforcing. In Fig. 5.20 extensive cracking in the sample cores from the deck beams is present because the restraint in the selected plane of sectioning was limited to stirrup reinforcement. The top surface of core A, which was not covered in bitumen, is not deeply carbonated and contains filled tension cracks. The cracking becomes more intense and narrower in width in the interior of the section but is so extensive and interconnected that it appears that only aggregate interlock is holding

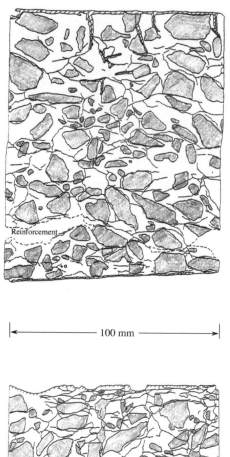

(a) |← ———————— 100 mm ————————→|

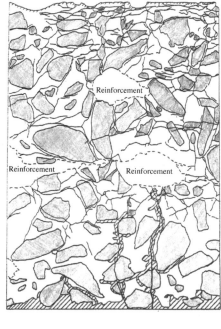

(b)

Fig. 5.20 Restrained cracking due to AAR in twenty-year-old prestressed bridge deck beams. (a) Vertical section from top web of deck beam showing carbonation (crossed hatched), coarse aggregate and cracking greater than 10 μm in width. Extensive finer cracking was present but is not shown (core A). (b) Similar section from bottom web of deck beam (core B).

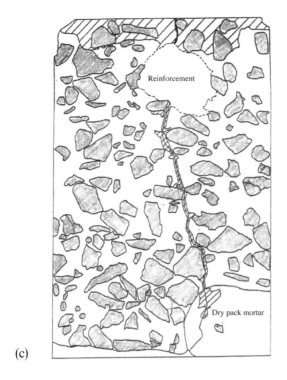

(c)

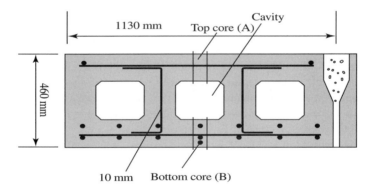

(d)

Fig. 5.20 Restrained cracking due to AAR in twenty-year-old prestressed bridge deck beams. (c) Horizontal section across crack in cap beam. Some tearing of the thin-sections has occurred where reinforcing was present (core C). (d) Dimensions of the beam and locations of cores A and B.

the concrete together. The bottom surface of core A, which is the upper surface inside the hollow core of the beam, is lightly carbonated and contains very fine open cracking. Alkali-aggregate reaction was not detected in the top 40–50 mm of the core but extensive reaction of fine particles of rhyolite together with alkali-silica gel is present throughout the remainder of the core. The same pattern in reverse applies to core B from the bottom web of the beam. Cracking is more severe, wider and because the bottom surface is partially protected from the weather carbonation is deeper. Some signs of delamination from the reinforcement were apparent in these two sections.

In contrast the section of core C from the cap beam shows no signs of alkali-aggregate reaction in spite of similar aggregates being present. Because the cement content was lower and the water/cement ratio higher in the cap beams the depth of carbonation was greater. The simple crack intersected, about 100 μm in width, was carbonated for its full depth. This simple cracking was due to the deck beams expanding and pushing into the cap beams. The textures illustrate the differences in carbonation between grades of concrete and differing exposures. They also illustrate the classical pattern of tensile cracking over expanding inner cores of concrete and how it differs from simple structural cracking. The tensile cracking visible on external surfaces gave little indication of the complex and extensive cracking present in the internal portions of the concrete.

2. *Cracking restricted to one direction, usually open, with widths varying from about* 60 μm *to* 0.5 *mm.*   This is the type of cracking often due to alkali-aggregate reaction because the expansion is restricted in one direction, usually by reinforcement. Because expansion is restricted the widths of cracks may be less. However, in the plane perpendicular to the direction of cracking the same complex pattern of internal cracking may be observed as is shown in Fig. 5.20.

3. *A pattern of fine cracks, often open, sometimes carbonated, with vertically displaced margins.*   In thin-section a relatively closely spaced pattern of cracking is roughly uniform with depth and the range of crack widths is limited as wider tension cracks tend to be absent. This is typical of expansive reactions, including alkali-aggregate reaction, under conditions which have led to reaction throughout the depth of the concrete. In the case of alkali-aggregate reaction cracking requires fairly continuous moist conditions as illustrated in Fig. 5.16(b) and Fig. 5.20(b). Swamy (1992) illustrates this type of cracking for various expansive strains. Alkali-silica gel and evidence of reacting particles should be present before making a diagnosis of alkali-aggregate reaction.

4. *The simple crack, either alone or as a sub-parallel set with variable spacings. Crack widths usually not less than* 100 μm.   In thin-section these cracks usually run completely through the thin-section and in the structure will penetrate either through the member or to considerable depths. The cracks are usually continuous with limited sub-parallel splintering and carbonation will have penetrated to depths controlled by the crack width as shown in Fig. 5.20(c). If water has flowed through the crack the margins may be leached of calcium hydroxide. In many cases it is not possible adequately to determine the cause of these cracks by thin-section examination as they do not embody much information and due to their wide spacing there will be just one crack per thin-section. Typical causes are long-term shrinkage of unreinforced slabs, excessive thermal gradients, settlement of foundations and other structural problems. The purpose of any examination will often be directed to eliminating other possible causes associated with the materials and evaluating the effect of the environment on the concrete surrounding the crack system.

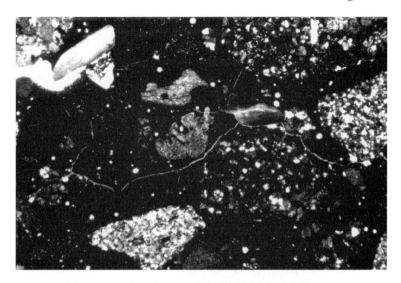

Fig. 5.21 Internal cracking due to localised drying shrinkage. A localised area of paste shrinkage has occurred in this fifty-year-old concrete. The cracks do not exceed 20 μm in width and have a somewhat rectilinear character. Cracked areas of hardened paste, similar to those shown here, are not uncommon in rich undersanded concretes. If this cracking becomes extensive some external cracking must also be present. Width of field 1.4 mm.

Cracking essentially limited to the interior of the concrete. Best identified in thin-section

1. *Fine cracks in the interior of the concrete with few visible effects on the surface.* Because the cracks are largely restricted to the interior of the concrete they are limited to widths of less than about 15 μm and tend to occur only in localised areas. Wider and more general cracking must result in visible surface cracking. Cracking due to internal drying shrinkage tends to be confined to localised areas of the hardened cement paste and often forms a rectilinear pattern. Where cracks run from aggregate to aggregate, or aggregate to void, they are typical of the earliest stage of alkali-aggregate reaction but the diagnosis must be supported by the presence of other evidence of ASR in the concrete (Palmer 1992). Fine cracks restricted to the body of the concrete can also fall into the category of quality faults (Fig. 5.21). There are a number of causes for this type of cracking which have been reviewed by Chatterji (1982).

2. *Cracks around the periphery of aggregates often with radial cracking present.* Shrinking and/or absorptive aggregates can cause localised shrinkage effects in the surrounding cement paste as described in Section 6.6.3. Many commercial aggregates contain isolated particles that fall into this category. If a substantial portion of the aggregate present in a concrete has shrunk extensive surface cracking may occur as reported by Stutterheim *et al.* (1967). The external appearance of this map cracking is shown in Lea (1970) and the internal textures by Roper *et al.* (1964).

5.8.5 *MICROSCOPIC ATTRIBUTES OF CRACKS, CHANNELS, FISSURES AND VOIDS*

At the microscopic level there are a number of attributes of cracks, channels and fissures that can provide information on the time when they formed and also give clues as to their

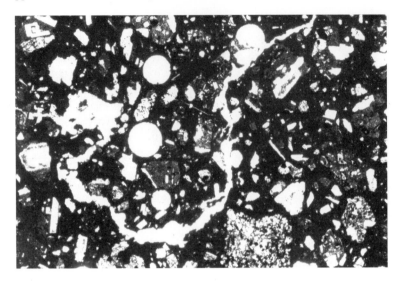

Fig. 5.22 Bleeding channels. A well defined bleeding channel is illustrated in this laboratory sample of concrete. It meanders through the concrete to the outer surface with irregular and poorly defined margins which tend to grade into the fissure. Note how the external mouth (M) of the bleeding channel is quite small compared with the internal width of the channel. It is not common to observe such a well defined bleeding channel. Width of field 1.4 mm.

cause. Concrete cracks in the hardened state, but can only form channels, fissures and voids in the plastic state before setting. This gives a simple method of determining whether formation took place before or after setting of the concrete.

Under the microscope cracking is clearly identified by well defined margins. In higher strength concretes cracking may run through aggregates but in lower strength concrete usually skirts aggregate margins. Voids formed by pockets of air are easily identified by their shape, and generally have smooth margins but a margin can be diffuse if it forms a thin wedge of hardened paste during sectioning. These thin wedges of hardened paste can be mistaken for alkali-silica gel by an inexperienced observer. Channels are irregular tunnels with diffuse margins that carry bleeding water upwards in a meandering fashion. In section they are rarely continuous because they meander in and out of the plane of the section (Fig. 5.22).

Often channels are difficult to observe because they are filled with diffuse laitance like material which requires care to distinguish it from the cement paste. Fissures are cracks in the plastic state that have not been filled with water. They are identified by highly irregular margins, which may vary from sharp to diffuse and do not run through aggregates. Unlike channels they should be continuous in thin-section. Fissures are primarily due to excessive drying of concrete while still plastic and because of this may be concentrated in surface areas. The subject of plastic shrinkage and settlement is well illustrated in the Concrete Society Technical Report No. 22 (1992) and a staining method for their detection in ground surfaces has been given by Ash (1972).

5.8.6 *DIMENSIONS OF CRACKS AND INFILLINGS IN THIN-SECTION*

The minimum crack width in hardened cement paste readily observable by optical microscopy is about 5 μm. Cracks narrower than this tend to merge into the background of the hardened cement paste unless delineated, although they can sometimes be distinguished by darkening of the crack margins. Narrow cracks can be delineated by dyes incorporated into the epoxy resin used to impregnate the concrete. However, in a concrete that is saturated internally it is doubtful whether cracks narrower than 5 μm persist for any length of time as they will be filled by the deposition of calcium hydroxide and merge into the fabric of the paste. In practice most non-structural cracks observed in the body of a concrete will be found to be not less than about 10 μm in width with the majority falling into the 20–60 μm range. Where expansive patterns of cracks have formed those 100 μm in width or wider tend to form a grid of large dimension over the pattern of the finer cracks. Isolated cracks, especially those that run deep into the concrete, will have minimum widths of about 100 μm and maximum widths up to a few millimetres. At the surface of concrete, drying and/or carbonation shrinkage cracking is usually limited to a range of 5–15 μm in width.

Typical ranges for carbon dioxide penetration, plugging and filling of cracks with calcium hydroxide or carbonate are given in Table 5.9 below. This requires some qualification as it only represents data for structures such as bridges and buildings exposed to a temperate climate. The deposition of calcium hydroxide in water-retaining structures will be much greater and is more difficult to categorise.

Table 5.9 The relationship between crack width and depth of penetration of carbonation in field structures

Crack width (μm)	Carbon dioxide penetration (mm)	Surface cracks filled	Cracks in body filled
$\leqslant 20$	< 10	Nearly always	Variable
20–60	< 100	Often plugged	Rare
$\geqslant 100$	> 100	Rarely plugged	Very rare

Dimensions of channels and fissures are more difficult to specify. Since channels have no clearly defined margins and often appear discontinuous in thin-section, dimensions are variable. In practice their widths are observed to be up to a few hundred microns in size. The minimum width of a fissure is not likely to be less than 100 μm because of the way it is formed in the plastic state. This also implies that in practice widths observed will be highly variable.

Jensen *et al.* (1985) have categorised cracking into three groups. Wide cracks over 100 μm in width, fine cracks from 10 to 100 μm in width and microcracks with a width below 10 μm. They use good colour photographs to illustrate these types of cracking.

5.8.7 CRACKING AS AN ARTEFACT OF SAMPLING AND SPECIMEN PREPARATION

Hardened cement paste is brittle and easily cracks during sampling and specimen preparation unless reasonable care is exercised. In those concretes which have undergone normal exposure to climatic conditions most cracks will contain some calcium hydroxide or carbonation. There are usually observable signs to show whether a crack is new or old. If these signs cannot be observed it is then possible that the cracks may have been caused by either sampling or sectioning. A typical example of sampling cracking occurs in the sides of cores due to chattering of the diamond coring bit. These appear as short clean cracks at the edges of the thin-section.

Drying cores at too high a temperature or overheating the thin-section during preparation and some preparation techniques used for SEM examination can cause microcracking of the cement paste. As a general rule concrete used for sectioning should never be heated above 60°C. When this occurs it is usually obvious as the paste has a distinctive shrunk look about it. Another form of cracking that appears to be related to sectioning often occurs at the outer surface of the core. The corroded and carbonated layer undergoes a patchy delamination which in most cases is clearly an artefact.

A question that may be raised with the petrographer is whether the cracking observed in thin-section may have been formed when strains in the concrete were released by core sampling. In making a decision about this the same principle is applied as discussed above. It is the authors' experience that cracking due to release of strain is uncommon although the width of existing cracks may be increased. This is to be expected as releasing strain into existing cracks is more likely to occur than the formation of new cracks.

A more likely scenario is that the sampling misses the cracking present in the concrete altogether. Even in the laboratory apparently badly cracked test samples will often intersect few cracks when thin-sectioned. The probability of intersecting a crack by the plane of the thin-section is somewhat less than would be intuitively expected and some skill is required in selecting the best plane. This is an even more acute problem when sampling a large structure.

5.8.8   *INTERPRETATION OF CRACK SYSTEMS OBSERVED IN THIN-SECTION*

The observation and interpretation of crack systems in thin-section is of fundamental importance in concrete petrography. It is effective only if adequate sampling of the structure is carried out and a sufficient surface area in thin-section is examined. The application of thin-sectioning techniques to fissures formed in the plastic state, widely spaced simple cracks and structural cracking, is rarely justified and examination is better carried out by simpler methods such as observation of the cracks and fissures in cut surfaces. Examination of air void systems in thin-section appears to be a neglected area of investigation as the technique is capable of providing more information than the traditional observation of ground surfaces.

6

Examination of Deteriorated and Damaged Concrete

6.1 Introduction

The durability of concrete involves both intrinsic and extrinsic reactions. Intrinsic attack occurs from deleterious substances or properties incorporated into the concrete mix although an intrinsic component such as alkali may be later augmented from external sources. Extrinsic attack takes place from agents outside the concrete and with the exception of physical agencies usually involves deleterious solutions. This latter type of attack produces a zone of deterioration that moves in from the outer surface of the concrete. From a textural aspect this leads to two important differences for the petrographer. Intrinsic attack will affect a considerable portion of the concrete texture while extrinsic attack is restricted to zones adjacent to outer surfaces. This explains why laboratory testing has on the whole been more successful in predicting durability for an intrinsic reaction such as alkali-aggregate reaction than for an extrinsic problem such as sulfate attack (Cohen and Mather 1991, Mehta 1993).

The concept of an external layer of deterioration is important when investigating the effects of external agents on concrete. In many cases this outer layer consists of three zones. An outer corrosion zone where the attack has gone to completion, an intermediate zone where attack on calcium hydroxide, the hydrates and clinker occurs and salts are deposited, and an inner zone where leaching of calcium hydroxide and possibly other materials from the hardened paste takes place.

Petrographically the outer zone usually consists of amorphous forms of hydrated silica, alumina and iron in which are embedded aggregates. In most cases only hydrated silica and aggregate remain as both the iron and alumina compounds are more readily removed. It is rarely intact as it often suffers damage during sampling and in some cases such as attack by citric acid the zone is of minimal width because silica is dissolved by this acid. The outer zone merges into the intermediate zone. It is the most complex of the zones as recrystallization of salts, removal of calcium hydroxide and decomposition of the cement hydrates all occur together in this zone. It is highly variable in depth depending on the reaction, often contains carbonated material and may be stained by amorphous, hydrated iron compounds. Observable calcium hydroxide is rare or absent indicating that the pH will be below that required for the stability of the cement hydrates. Where salt recrystallisation has occurred the intermediate zone may be extensively exfoliated. The intermediate zone grades into apparently sound hardened cement paste but careful examination will often detect an inner zone where the paste appears to be darkened due to a diminution of calcium hydroxide. This is the chemical gradient which invariably occurs where hardened paste is undergoing dissolution and has been described in detail by Pavlik (1994).

In many cases lack of durability in concrete is a chemical problem which arises during manufacture or later, although to the engineer it is the changes in volume and loss of mechanical integrity that are of importance. The hardened cement paste and many aggregates are stable only under certain conditions of which pH is by far the most important. Mehta *et al.* (1992) in reviewing the 'performance and durability of concrete systems' summarised some relevant data from Reardon (1990) and Gabrisova *et al.* (1991) on the stability of the cement hydrates as follows:

> In the pH range 12.5 to 12.0, calcium hydroxide, calcium monosulfate hydrate and aluminate hydrates, dissolve in sulfate solution from which ettringite precipitates out. Next, gypsum precipitates out in the pH range 11.6 to 10.6; below pH 10.6, ettringite is no longer stable and will start decomposing. Note that the lowering of the pH below 12.5 will also cause the C-S-H phase to be subjected to cycles of dissolution and reprecipitation (Ca/Si ratio is 2.12 at pH 12.5, and 0.5 at pH 8.8) which continue until it is no longer stable at or below pH 8.8. All the foregoing chemical changes in a hardened Portland cement product must be taken into consideration when attempting to understand their physical manifestations, such as expansion, cracking, and loss of strength or mass.

A low permeability in the surface zone is of crucial importance to reducing the rate of attack by corrosive agents. Products formed by the attack may be beneficial or deleterious depending on circumstances. For instance, carbonation may make a concrete more impermeable while reducing the pH but if the concrete still remains sufficiently permeable after the carbonation, oxygen can penetrate and corrode the reinforcing steel. On the other hand the formation of a product such as ettringite can be expansive and disrupt the concrete. When considering whether an increase or decrease in permeability is likely to occur, comparison of molar volumes of reactants and reaction products provides useful information.

Today, intrinsic reactions sometimes occur between aggregates and the pore solution of a concrete because of internal sulfate or salt attack. Unsoundness due to hard burnt lime (CaO) and magnesia (MgO) is rare and as cement manufacturers now control production to avoid these problems they will not be discussed in this chapter. A concrete or mortar made with a cement containing excess hard burnt lime or magnesia can expand up to months or years after setting, leading to the development of cracks. The expansion is due to the slow hydration of the hard burnt lime or magnesia leading to volumetric expansion. Deng Min *et al.* (1995) have proposed a mechanism for the expansion that occurs when hard burnt lime hydrates in hardened cement paste and McKenzie (1994) has described the microscopic methods for studying lime burning and quality. The textures of hard burnt lime illustrated will be applicable to lime in cement.

While the concept of extrinsic and intrinsic reaction is useful to the petrographer classification of concrete deterioration as proposed by Popovics (1987) is of more practical use for engineering purposes. This classification which is a useful summary of the types of attack likely to be encountered, is shown in Table 6.1.

The petrographic investigation of durability requires an assessment of the quality of the *in situ* concrete. If the problem involves the materials it is necessary to attempt identification of cement type, mineral admixtures and mineralogy of the aggregates, estimate the cement content, water/cement ratio and air content, and make a qualitative assessment of aggregate quality and compaction. The data from such an assessment will show that in a surprisingly large number of cases the concrete was not mixed and placed according to specification and accepted practice although the deviations may not be large. Where severe or unusual service conditions apply non-compliance with accepted practice can lead to deterioration of the concrete.

Table 6.1 Classification of concrete deterioration (Popovics 1987).

Class I	Leaching by soft water
Class II	Non-acidic reactions
Class IIA	Base exchange i.e., magnesium salts
Class IIB	Saponification reactions from fats and oils
Class IIC	Reactions with sugars
Class III	Reactions involving excessive expansion
Class IIIA	Sulfate attack
Class IIIB	Alkali-aggregate reaction
Class IV	Reactions with acidic water
Class V	Physical processes other than mechanical
Class VA	Salt solutions causing efflorescence and/or spalling or cracking
Class VB	Freezing and thawing
Class VI	Mechanical deterioration
Class VIA	Abrasion and wear
Class VIB	Excessive shrinkage, uneven thermal expansion, overload, repeated loading etc.

The second part of the petrographic investigation involves identification of the conditions which are causing the materials to deteriorate. In most cases identification of the cause of the aggression and the extent of deterioration is possible *provided that the investigative methods used are adequate*. However, an estimate of the remaining useful life of the materials in a structure requires rates of deterioration to be determined. In many cases it is difficult if not impossible to estimate the rate of attack and any prediction of remaining useful life of the concrete will be based on previous experience.

Given the complexity of investigating problems of durability it is pertinent to define the scope of concrete petrography. Where the service conditions are unusual, petrographic examination combined with chemical investigation of field concretes can provide information that may lead to the better design of concretes. In cases where the mechanism of deterioration is not fully understood, such as in alkali-aggregate reaction and sulfate attack, petrographic field data are vital and take precedence over laboratory testing. The applicability of this field data is largely dependent on being able to characterise the state of the concrete and an important part of this characterisation will be provided by examination of micro textures in thin-section.

The durability of concrete is a broad subject and there is much literature available. Proceedings of international conferences on the durability of concrete and building materials are published by the American Society of Testing Materials, the American Concrete Institute, RILEM and numerous other conferences are held on materials which are related or contain relevant information. Codes of practice to assist engineers in designing durable concrete have been published by the BS5328 (1990), BS8110 (1985), American Concrete Institute (annual publication), Comité Euro-International du Béton (1992) and the Commission Internationale des Grands Barrages (1989). A discussion of the approach used in the USSR to deal with aggressive fluids has been given by Ivanov (1981) and in South Africa by Alexander *et al.* (1994). Monographs by Eglinton (1987) and Mays (1992) intended for the engineer, are useful texts which deal with many aspects of the durability of concrete.

6.2   Sulfate attack

The literature on sulfate attack is complex and confusing. There does not appear to be a consensus on some details of the mechanisms involved in spite of investigation that now extends over sixty years.

The mechanisms by which the various external sulfates attack concrete are still a matter of some controversy (Odler and Gasser 1988, Lawrence 1990, Cohen and Mather 1991, Mehta 1993). However for practical purposes, the extent of attack on any one type of hardened cement paste largely depends on the amount of sulfate in solution which in turn is related to the cations and other anions in solution. Thus the aggressiveness of the sulfates in descending order approximately related to their solubility is ammonium, magnesium, sodium, potassium and calcium. A common form of sulfate attack is from groundwater in gypsum bearing soils which attacks concrete to form ettringite in the hardened cement paste as follows:

$$\text{Ca aluminate hydrates} + CaSO_4.2H_2O \rightarrow 3CaO.Al_2O_3.3CaSO_4.32H_2O$$

The ferrite phase can undergo a similar reaction to form analogous compounds to ettringite. This reaction is the basis for the formulation of sulfate-resisting Portland cement in which the content of the aluminate phase is limited to a maximum of 3 per cent thus restricting the amount of damaging ettringite that can be formed. Other common sulfates are derived from magnesium, sodium and potassium sulfates with the limited possibility of ammonium sulfate as either an effluent or from over-fertilised soil. These sulfates in solution will form gypsum by interaction with the calcium hydroxide in the concrete which may then interact with the hydrated aluminate phases to form ettringite as given above.

$$(NH_4)_2SO_4 + Ca(OH)_2 + 2H_2O \rightarrow CaSO_4.2H_2O + 2NH_3$$

$$MgSO_4 + Ca(OH)_2 + 2H_2O \rightarrow CaSO_4.2H_2O + Mg(OH)_2$$

$$Na_2SO_4 + Ca(OH)_2 + 2H_2O \rightarrow CaSO_4.2H_2O + 2NaOH$$

$$K_2SO_4 + Ca(OH)_2 + 2H_2O \rightarrow CaSO_4.2H_2O + 2KOH$$

Magnesium sulfate also attacks the hydrated silicate phases to form gypsum, brucite and hydrated silica and below a pH of 10.6 it also attacks ettringite.

$$\text{Ca silicate hydrates} + MgSO_4 \rightarrow CaSO_4.2H_2O + Mg(OH)_2 + \text{hydrated silica}$$

$$\text{Ettringite} + MgSO_4 \rightarrow CaSO_4.2H_2O + Mg(OH)_2 + \text{hydrated alumina}$$

The reaction of magnesium sulfate with ettringite tends to be restricted to the outer layer where the removal of calcium hydroxide occurs and the pH is lower. It is also claimed that brucite can very slowly interact with hydrated silica gel to form a magnesium silicate as the final stage in sulfate attack. Lea (1970) states that in practice this step may be reached only after long periods of time.

At temperatures around 6°C under conditions of high humidity, sulfate solutions can attack concrete and mortar to produce thaumasite. This reaction is discussed under the heading 'Formation of thaumasite by sulfate attack' on pp. 258–9.

The physical mechanism by which deterioration occurs from sulfate attack are twofold. These are the expansive formation of either ettringite and/or gypsum in the hardened fabric of the concrete causing cracking and exfoliation, and the softening and dissolution of hydrated cementing compounds to a mush due to direct attack on these compounds by magnesium sulfate or their decomposition by removal of calcium hydroxide reacting with the sulfates. Either or both mechanisms can apply dependent on the temperature, types and concentrations of sulfate in solution and the composition of the concrete. There is still no clear agreement on the exact mechanisms of expansion as referenced in Section 6.2. Also it has been assumed the pH of the groundwater is near neutral. Sulfate attack can be greatly increased by the presence of carbonic, sulfuric, hydrochloric and other acids in water.

Sulfate attack is not restricted to solutions and with increasing air pollution attack from airborne sulfur compounds is becoming an increasing problem. The mechanism of this type of attack has been discussed by Zappia *et al.* (1994).

6.2.1 *GROUNDWATER SULFATES*

The sulfates of greatest concern for the durability of concrete are found in soils and groundwaters. Of the common sulfates found in soils only calcium sulfate dihydrate (gypsum) is effectively limited by its solubility. This is equivalent to about 1200 ppm of SO_3 in groundwater. If greater concentrations of SO_3 are present other cations such as Mg, Na and K must be present to balance the sulfate in solution. However, where the SO_3 content is less than 1200 ppm it must not be assumed that the sulfate has been solely derived from gypsum. Any or all the appropriate cations can also be present in the solution. It should be noted that magnesium and sodium sulfates have solubilities that are 150 to 200 times greater than gypsum.

Biczok (1967) states that the sulfate content (SO_4) of soil is usually in the range of 0.01–0.05 per cent but there are soils in which up to 5 per cent sulfate is present. There are many regions around the world in which soils contain several percent of sulfates. This sulfate is derived from the oxidation of sulfide minerals such as pyrites, the biological decomposition of organic material containing proteins and from agricultural and industrial pollution interacting with minerals in the soil. The oxidation of sulfide minerals and biological decomposition of organic matter require oxygen so that usually much higher sulfate concentrations are present in the groundwater of the upper soil layers than in the deeper artesian waters.

Sampling and analysis of soil and groundwaters for sulfate content and the evaluation of results needs to be carried out with some care as detailed in Building Research Establishment Digest 363 (1991) and BS 1377 (1990). This is because water movements may be vertical or horizontal depending on seasonal variations in rainfall and on the geology of the site and its environs. Also a low sulfate content in the soil does not eliminate the possibility of sulfate attack since groundwater may flow from adjacent areas, particularly if the soil is disturbed by construction. There are also problems in obtaining samples that are not diluted with surface water.

Differing terminologies are used in referring to sulfates. For instance, the term sulfate can refer to either SO_3 or SO_4. Results reported as SO_4 may be converted to SO_3 by multiplying by the factor 0.833. Concentrations may be reported as parts per million, parts per hundred thousand and g/l. The unit g/l may be converted to these other units by multiplying by 1000 and 100 respectively.

Because of the widespread nature of the problem of sulfate attack many countries have specifications and codes of practices for its prevention. Lea (1970), BS 1377 (1990a), Building Research Establishment Digest 363 (1991) and Harrison (1992) are convenient sources for appropriate recommendations. Lawrence (1990) has reviewed current codes in relation to cements together with background data.

Properties of concrete relating to sulfate attack

There is overwhelming evidence to show that concretes of low permeability have good resistance to sulfate attack. This can be achieved by using a low water/cement ratio and high cement content and resistance can be further increased by the use of sulfate resisting cement and cements blended with blastfurnace slag and pozzolanas. However, while sulfate-resisting cements provide good protection against attack, much of this protection can be negated if the concrete is too permeable or additions such as calcium chloride are used. Where pozzolanas are used, highly reactive, low-calcium pozzolanas are required and the concrete must be adequately cured to prevent increased permeability (Harboe 1982, Stark 1982, Dunstan 1982, Mather 1982, Harrison 1992). Al-Amoudi *et al.* (1995) found that concretes made with blended cements, particularly those made with blastfurnace slag gave inferior performance against magnesium sulfate attack when compared to plain cement mortar.

In general, aggregates do not have a decisive effect on sulfate resistance. Porous aggregates which allow hydration and dissolution of salt solutions in their pores are subject to cracking and the development of pop-outs in the surfaces of concrete (Stark 1982). Piasta *et al.* (1987) found that porous limestone aggregates increased sulfate resistance. However, Fiskaa *et al.* (1971) found that replacement of fine aggregates with limestone was unfavourable. Aggregates containing kaolinised feldspars can form ettringite (Coutinho 1979).

Curing affects sulfate resistance principally because well-cured concretes tend to be more impermeable. High pressure steam curing is also effective but where pozzolanas and blastfurnace slag are used autoclaving may reduce durability (Stark 1982). Osborne (1991) found that carbonation is beneficial in reducing the effect of sulfate attack because of the removal of calcium hydroxide from the surface layers of the concrete and ettringite is unstable below pH 10.6. This is why a number of investigations have found increased durability from concretes allowed to dry for some weeks after moist curing. Such drying enhances surficial carbonation.

As would be expected the most common point for sulfate attack is on concrete just above or below ground. Concrete cast in contact with the soil, such as foundation footings, piles, etc., are particularly prone because the permeability of the surfaces cannot be controlled (Harrison and Teychenné 1981). It has been emphasised that where OPC is used in below-ground concrete it is necessary to provide adequate resistance to moisture entry (Hamilton and Handegord 1968, Novokshchenov 1987).

Results of long-term tests reported by Harrison (1992) confirm the importance of using low-permeability concretes especially in cases where the hydrostatic head is applied to the concrete or internal drying of basement floors induces a moisture gradient.

External appearance of concrete attacked by sulfate

The external appearance of sulfate attack appears to be quite variable as would be expected from such a complex reaction. Lea (1970) states that the action of soluble sulfates on cements results in the formation of calcium sulfoaluminate and gypsum accompanied by a marked expansion. In general, concrete attacked by sodium and calcium sulfate eventually becomes reduced to a soft mush but when magnesium sulfate is the main destructive agent the concrete remains hard though it becomes much expanded (Fig. 6.1). This differentiation

Fig. 6.1 Some differences between sodium sulfate and magnesium sulfate attack. Laboratory specimens exposed for 140 days to (a) sodium sulfate solution and (b) magnesium sulfate solution. 1 = OPC, 2 = OPC + 15 per cent silica fume, 3 = sulfate-resisting cement, and 4 = sulfate-resisting cement + 15 per cent silica fume all mixed at a water/cement ratio of 0.3. The specimens attacked by magnesium sulfate have exfoliated and expanded much more than those exposed to sodium sulfate. The results are a good example of the variability that can occur in sulfate attack. From Cohen and Bentur (1988).

of attack by magnesium sulfate is often noted and is particularly applicable to sea water attack on concrete (Fig. 6.1).

The visible effects of sulfate attack are exfoliation of the outer layers of the concrete, corrosion and reduction of the hardened paste to what is commonly described as a 'soft pug' or a 'mush' and the deposition of salts on surfaces and in exfoliation cracks. These effects suggest that essentially two physical processes are occurring in sulfate attack. The expansive recrystallisation of salts in the surface of the concrete and direct corrosive attack on the hardened cement paste. However, which process predominates under any given set of conditions is still an open question.

Descriptions of sulfate attack on field concretes above ground vary across a spectrum of examples where exfoliation of surfaces predominates (Fig. 6.2) to cases where the hardened paste is corroded and softened and exfoliative processes appear absent. However, in many cases illustrations of sulfate attack depict spectacular deterioration and disintegration of concretes which are the terminal results of attack where much of the evidence for the mode of attack has been destroyed. The examination of below ground concretes presents special problems (Fig. 6.3). Examination of concrete surfaces below ground is difficult because

Fig. 6.2 A fence post in sabkah with a saline water table less than 0.5 m from the ground surface. Note the worst damage occurs above ground level.

removal of soil will invariably remove much of the more fragile areas of attack. Thus it is not possible to gain an accurate picture of the sulfate attack unless special care is taken in sampling below ground concrete leaving intact soil in place on the sample and carrying out the examination using a combination of thin-sectioning, scanning electron microscopy and chemical analysis.

Since much of the evidence for relating the nature of visible attack to types of sulfate and other conditions appears to be based on experience from accelerated laboratory testing and thin-section examination is rarely carried out, it is not surprising that controversy exists over the nature of sulfate attack.

In many cases of sulfate attack salts are observed on surfaces. For instance Novak and Colville (1989) found a range of efflorescent sulfates on floors and basements built over sulfate-rich soils. Analysis of these external deposits should always be made but interpretations based on these analyses should be made with caution as they do not necessarily represent the processes of deterioration taking place inside the concrete.

There is another variant of sulfate attack which can cause heaving and cracking in foundations and floors. This is where the fill or soil contains iron sulfides and expansive sulfates are formed *in situ* which cause the subsoil to heave. In many cases direct sulfate attack on the concrete also occurs. Lea (1968) discusses the effect on ground-floor slabs laid over filling containing sulfates and Stark (1987) has reported on a case of heaving of a pavement laid over a Portland cement stabilised slag sub-base.

Fig. 6.3 Destruction of evidence caused by removal of adherent material. This pipe collapsed because of acid sulphate attack. When the pipe was removed from the ground the staff sampling the pipe carefully removed the clay adhering to the outside of the pipe and destroyed vital evidence.

Internal textures of concrete attacked by sulfate

There appear to be only a few published reports of petrographic examinations of laboratory specimens using thin-sections. Nielsen (1966) has presented brief descriptions, together with photomicrographs of thin-sections, showing the effects of eight months of attack on cement pastes (water/cement ratios 0.30, 0.34 and 0.38) by 0.07 molar solutions of sodium, magnesium and ferrous sulfates. Cracking was generally parallel to the surface suggesting the onset of exfoliation. In the case of sodium sulfate attack the one crack present was filled with gypsum and only limited deterioration had taken place. Some ettringite was observed and appeared to have formed after the occurrence of the cracks. In the case of magnesium sulfate attack, gypsum was always precipitated on the surface of the specimen and the magnesium hydroxide (brucite) was precipitated on the gypsum. An air void near the surface was filled with gypsum and ettringite was not observed. In the case of ferrous sulfate attack, gypsum was still the dominant reaction product together with various unidentified iron hydroxides. Ettringite was not observed. The outer layer of the specimen was undergoing exfoliative cracking and the general impression was that deterioration was much greater than in the other solutions. A detailed representation of the microstructure of the zone of deterioration due to magnesium sulfate attack has been given by Bonen and Cohen (1992). Their appears to be general agreement between the two reports given above.

Knofel and Rottger (1985) give a range of micrographs of textures of laboratory concretes which have been attacked by atmospheres enriched with carbon and sulfur dioxides. They found that carbonation penetrated more deeply into the hardened paste than sulfate attack which was no more than a few millimetres deep. Blended blastfurnace slag cements performed slightly less favourably than OPC in their tests.

In most reports examination is limited to visual assessment supplemented by X-ray diffraction identification of the sulfates precipitated in the concrete. For instance, in the long-term investigation of sulfate attack at Northwick Park, Harrison and Teychenné, (1981) of the Building Research Establishment used X-ray diffraction and a staining technique devised by Poole and Thomas (1975) for the identification of sulfate minerals. O'Neill (1992) has illustrated the appearance of some of the common sulfate minerals and concluded that their examination as powder mounts in refractive index oils is a useful method of identification.

Examination of thin-sections cut from a concrete of which one side had been exposed to a mist containing high concentrations of sodium sulfate together with lesser amounts of sodium chloride gave the following textures. The surface zone was undergoing an extensive exfoliation (Plate 48) with many of the cracks filled with gypsum which graded abruptly into apparently sound concrete. Some of the aggregates in this exfoliation zone were exfoliated and also filled with gypsum. No ettringite was observed in cracks or the texture generally. An extensive alkali-aggregate reaction was also occurring in the exfoliation zone involving some of the volcanic aggregates present (Plate 49). However, in spite of the extent of the alkali-aggregate reaction the cracking was dominated by the sulfate attack and the typical pattern cracking associated with expanding aggregates was absent. Since low-alkali cement had been used the alkali required for alkali-aggregate reaction was derived from the release of sodium from the sodium sulfate, and possibly the sodium chloride, reacting with the calcium hydroxide.

A similar texture was found by St John (1982) where concentration of sodium sulfate occurred by wicking and evaporation from porous concrete surfaces in some tunnels, of groundwater containing less than 50 ppm SO_4. In this case non-reactive aggregates were present in the concrete so alkali-aggregate reaction was absent. The same pattern of exfoliation with some cracks filled with gypsum was present and once again the deteriorated exfoliated zone graded rapidly into sound concrete. Serious attack was restricted to porous concrete as some denser concretes in adjacent rail tunnels subject to the same conditions were largely unaffected.

St John (1991) examined the acid sulfate attack on stormwater pipes (Fig. 6.4) which has caused localised failures in portions of the business district of the City of Rotorua, New Zealand. In this case geothermal gases containing hydrogen sulfide and carbon dioxide dissolve in the groundwater at temperatures of about 50°C to form sulfuric and carbonic acids although sulfuric acid probably predominates at the lower pH levels. It should be noted the attack in this case is the reverse of that normally encountered in sewers where the crown is corroded due to oxidised hydrogen sulfides (Section 6.2.4) and the concrete covered by effluent is often relatively sound. The case to be described indicates the complexity

(a)

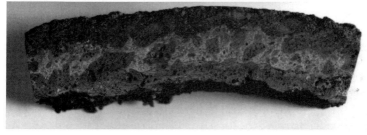

(b)

Fig. 6.4 Acid sulfate attack on stormwater drainage pipes. (a) Because of acid sulfate attack portions of the stormwater drainage system collapsed. The inside surface of the crown of the pipe appears sound but the invert has undergone severe attack. On the piece that has broken away the inside surface is puggy and it can be seen from the fracture surfaces in the main portion of the pipe that both the inside and outside surfaces have been attacked to considerable depths. (b) Some idea of both the internal and external depths of attack can be gauged from the sawn surface of the sample.

of sulfate attack and the type of information that can be gained by examination of the concrete texture.

Two centrifugally spun pipes, corroded to the point of failure were examined. One pipe (A) contained coarse olivine basalt and sand consisting of basalt and acid and intermediate volcanic particles. The other pipe (B) contained a weathered pyroxene andesite coarse aggregate and a sand consisting of acid and intermediate volcanic particles. The texture of both pipes suffered from faults typical of fabrication by centrifugal spinning (Section 7.4) and both were extensively corroded and had collapsed as a result. The outsides of the pipes were extensively corroded even though this was not fully apparent from visual examination. Internally severe corrosion of the inverts had occurred and the texture was soft and mushy. The surfaces of the crowns appeared intact apart from corrosion pits which increased in frequency towards the inverts.

Examination of thin-sections revealed a complex texture with some differences between the two pipes which may have occurred because of differences in corrosive conditions and quality of the pipes (Plate 50). For these reasons the attack on the two pipes may not be directly comparable. Externally pipe A was completely corroded to a depth of more than 5 mm. The corroded layer contained dispersed sulfides and amorphous silica but the aggregates were still intact. Some minor and limited exfoliation and cracking were present. There was a 5–7 mm carbonated zone behind the corroded layer backed by a zone leached of calcium hydroxide. Internally a 10 mm layer of the invert was also corroded in a similar manner to the outside of the pipe with some signs of vestigial gypsum. The transition zone to the hardened paste consisted of a narrow irregular and diffuse zone of carbonation with a distinct yellowing of the texture backed by a narrow zone leached of calcium hydroxide. The remaining intact portion, about one-third of the wall thickness contained veins of ettringite mainly deposited in fissures at aggregate margins.

In contrast to pipe A, the corrosion of the external surfaces of pipe B exhibited an exfoliation texture with aligned gypsum which was present in both paste and aggregates (Plate 51). The hardened cement paste had been completely corroded to amorphous silica and ettringite was not visible. A highly irregular 2–7 mm zone of carbonation was present behind the corrosion layer. Irregular veins of carbonation were also present in the body of the wall. Internally corrosion varied from pits in an otherwise intact surface which was carbonated to a depth of about 0.5 mm to complete corrosion of the surface at the invert. The remains of the corrosion layer were relatively narrow and consisted of a texture of amorphous silica containing exfoliation cracking and some aligned gypsum. The transition to sound paste was discontinuous and consisted of a yellow zone and a thin layer of carbonation. Although the crown was not sectioned there were indications that the body of the concrete would also contain veins of carbonation. Given the moist conditions, high concentration of carbon dioxide in the geothermal gases and the fact that pipe B was fissured along many aggregate margins, this type of deep carbonation was to be expected.

The cases described illustrate the complexity of sulfate attack on concrete. Attack varied from simple exfoliation of concrete to complex corrosion. Where simple exfoliation occurred ettringite was not detected and it is of interest in this respect to note that Ludwig (1980) claims sulfate attack is always initiated by ettringite forming just below the surface causing cracks. The ettringite is then decomposed and the cracks are filled with gypsum. The gypsum formation is a secondary reaction and may possibly contribute to the total expansion.

Formation of thaumasite by sulfate attack

Thaumasite, $CaSiO_3.CaCO_3.CaSO_4.15H_2O$ was identified by Erlin and Stark (1965) in mortar and concrete which had undergone sulfate attack. The occurrence was associated with ettringite either as overgrowths or bands of thaumasite in the ettringite crystals. Van Aardt and Visser (1975) identified rapid thaumasite deterioration of autoclaved mortar bars containing dolomitic aggregate exposed to sulfate solution at 5°C. Lukas (1975) and

Berra and Baronio (1987) identified thaumasite in deteriorated tunnel concretes, Gouda *et al.* (1975) in soil-cements and Crammond (1985) and Moriconi *et al.* (1994) in mortars and renderings from exposed brickwork. Thaumasite has also been identified in hard wall gypsum plasters in contact with Portland cement (Leifeld *et al.* 1970 and Ludwig and Mehr 1986). Deterioration of concrete due to the formation of thaumasite from a sulfide-bearing slate aggregate has been described by Oberholster *et al.* (1984). The occurrence of thaumasite in North America is discussed by Bickley *et al.* (1994).

Crammond (1985) states that thaumasite can form rapidly if the temperature is around 4°C, there is constant high humidity, initial reactive alumina (that is the alumina in the calcium aluminate phase) is between 0.4 and 1.0 per cent and there is an adequate supply of available sulfate and carbonate anions. It is not clear whether thaumasite forms by conversion from ettringite but Taylor (1990) considers that ettringite is only required as a nucleating agent. The microscopic and X-ray diffraction identification of thaumasite is difficult and it can be easily confused with ettringite as it has a similar fibrous character, but in favourable circumstances, as shown in Plate 52, its first-order yellow interference colour in 30 μm thick sections under crossed polars aids identification.

The formation of thaumasite appears to have more serious consequences than the formation of ettringite. Given the right conditions the hardened cement paste can become completely disintegrated by softening. Sulfate-resisting cement may not provide protection against this form of sulfate attack. In tunnels and other underground works where high humidity and low temperatures prevail the possible presence of thaumasite should always be considered.

Petrographic investigation of sulfate attack on concrete

The scarcity of other published reports on the examination of thin-sections of sulfate attack is surprising as this type of examination is essential for providing useful information on the textural aspects of deterioration.

Where it is necessary to determine the depth of deterioration large-area thin-section examination will be required. Considerable care is required to see that the delicate and fragile surface areas are not damaged during sampling and are retained in section for examination. Because of the possibility that ettringite may be present at the submicroscopic level, examination must be supplemented by chemical and X-ray diffraction analyses and if necessary detailed examination of selected areas made by scanning electron microscopy.

6.2.2 SALT AND SEA WATER ATTACK

Sea water attack on concrete has been the subject of investigation since before the turn of the century. The current state of many of these investigations is reported in the international conference on 'Performance of Concrete in Marine Environment' (American Concrete Institute 1980). The problem has become of considerable significance for the construction and durability of off-shore oil installations.

Open ocean water contains approximately 3.5 per cent of soluble salts by weight consisting of approximately 1.1 per cent sodium, 0.04 per cent potassium, 0.1 per cent magnesium, 0.04 per cent calcium, 2 per cent chloride, 0.3 per cent of sulfate (SO_4) and traces of other materials. The pH of sea water varies between 7.8 and 8.3 with about 10 ppm of carbonate, 80 ppm of bicarbonate and small amounts of free carbon dioxide present. In areas of partially enclosed water, for example the Arabian Gulf, salt concentrations can rise to 5 or even 10 per cent in localised areas.

Deterioration due to salt water attack on concrete is most severe in the intertidal and splash zones because wetting and drying can drive salts into the concrete surface and cause salt crystallisation. Wave impact and the physical abrasion from suspended particles and ice can destroy the protective surface layer of brucite and aragonite which forms in contact with sea water. In tropical climates rates of attack are considerably increased and in cold climatic regions the areas subject to wetting and drying are also especially prone to cracking due to freeze-thaw attack. However, in practice the principal cause of deterioration of concrete in sea water is due to the ingress of chloride ions causing corrosion of steel reinforcement. This is discussed in Section 6.5.3.

The principal component in sea water responsible for the chemical deterioration of hardened cement paste is the magnesium ion. This ion attacks the calcium hydroxide and aluminate phases according to the equations given in Section 6.2 but sea water attack in practice does differ in some respects to sulfate attack. Lea (1970) considers the action of sea water is one of several reactions proceeding concurrently;

> Leaching actions remove lime and calcium sulphate while reaction with magnesium sulphate leads to the formation of calcium sulphoaluminate which may cause expansion, rendering the concrete more open to further attack and leaching. The deposition of magnesium hydroxide blocking the pores of the concrete probably tends to slow up the action of dense concretes though on more permeable concretes it may be without much effect.

He also states that unless the pH falls below 7, as can occur in sheltered estuaries where organic matter may cover the sea bed, attack by free carbon dioxide is minimal.

A study of mortar samples immersed in artificial sea water for periods of up to three years (Conjeaud 1980) indicated that SO_4 and Cl^- diffuse rapidly into the mortar. The rate of diffusion is rapidly slowed by the development of an almost impermeable layer of aragonite forming over brucite in the surface of the mortar due to attack by bicarbonate and magnesium ions. This mechanism and the reduction in permeability has been confirmed by Buenfeld and Newman (1984, 1986) who also illustrate some of the textures. Although these layers protect submerged concrete from further deterioration, concrete and mortar surfaces in the intertidal zone are subject to greater damage due to salt crystallisation, impact and abrasion.

Clearly the permeability of the concrete is the most important factor in determining the resistance of concrete to chemical attack by sea water. Dense impermeable concrete is most resistant to damage and Massazza (1985) concludes that pozzolanic or slag cement (>60 per cent slag) make a more durable marine concrete than OPC and noted that lime-pozzolana mortars used by the Romans in sea works are still in good condition. Concrete made with high-alumina cement under conditions which prevent strength regression is highly resistant to sea water and sulfate attack. Although calcium aluminate hydrates are slowly attacked by sulfate solutions with the formation of ettringite there is no calcium hydroxide in the cement paste fabric to react with sulfates and consequently the concrete remains relatively impermeable.

The chemical reactions within the hardened cement paste can produce gypsum, calcium chloride and calcium bicarbonate all of which are soluble in sea water and may be leached. Ettringite and calcium carbonate are also formed. The ettringite may be associated with cracking and expansion of the concrete and the calcium carbonate develops as aragonite rather than calcite in the presence of magnesium ions. The magnesium hydroxide (brucite) formed is almost insoluble in sea water and remains *in situ*.

Brucite is difficult to identify positively in petrographic thin-sections of marine concretes because it is usually present as small crystallites. The birefringence of brucite is moderate ($\eta_\zeta - \eta_\omega = 0.019$) and it may exhibit anomalous reddish-brown colours, but because of the fine-grained scale-like nature of the crystallites in the cement paste these features are often masked. Because the refractive indices and birefringence of calcite and aragonite are similar it is only possible to report the presence of a carbonate. Usually the size of the crystallites in the cement paste prevents other optical properties from being determined so that X-ray diffraction, electron microscopy or infrared spectroscopy are required to confirm the identifications of brucite, aragonite and calcite in marine concretes.

The alteration of concrete in zones adjacent to the outer surface is usually clearly developed in concretes which have been subject to sea water attack and these zones develop and widen with time provided the surface is not removed by abrasion. Plate 53 illustrates these zones which extend up to 10 mm into the mass concrete of a boat slipway which had been exposed to sea water for 60 years. The concrete is of poor quality in spite of a high cement content. The cement paste is completely altered to fine scale-like brucite crystals in a zone extending almost 2 mm inwards from the surface. There is then a zone 5 to 6 mm wide in which large crystallites of aragonite are developed as irregular patches in a microporous paste evidenced by a darkening of the paste because the particle size of the remaining materials is small enough to cause destructive interference of light. In the final innermost zone the cement paste becomes pale coloured because of the extensive development of very small ettringite crystals in the paste itself and in air voids.

Plate 54 illustrates a similar pattern of zonation, but because this concrete is dense and impermeable, and has been exposed to sea water for less than 10 years the zones only extend 3mm into the concrete from its outer surface.

A particularly severe variety of salt damage arises where concrete structures have foundations immersed in saline groundwaters which commonly occur at the coastal margins of desert regions such as the Arabian Gulf region (Fookes and Collis 1977). Saline water is drawn up the surface layers of the concrete by capillary action and evaporation leads to the crystallisation and build up of salts on the concrete surface and within the surface layers. This process causes surface spalling and cracking exposing reinforcement to chloride corrosion. Typically the worst damage occurs a few hundred millimetres from the ground surface. A full treatment of the diagnosis and effects of this type of salt damage is given by Kay, Fookes and Pollock (1982). Sodium chloride salt crystals forming within the concrete can be shown to cause propagation of microcracks within the cement paste as is illustrated in Fig. 6.5.

6.2.3   INTERNAL SULFATE ATTACK

Internal sulfate attack on concrete is due to the inadvertent inclusion of materials containing sulfate, aggregates that can oxidise to produce sulfate and the delayed formation of ettringite.

The inclusion of excess sulfate in concrete occurs most commonly from contaminated aggregates. Where possible, concrete aggregates contaminated with sulfates are avoided and their inadvertent inclusion usually indicates inadequate investigation of aggregate supplies. However, in some areas where aggregates are in short supply contaminated aggregates may have to be used. For instance, in parts of the Arabian peninsular, crushed Sabkha contaminated with calcium and magnesium sulfates is sometimes used for aggregate,

Fig. 6.5 Scanning electron micrograph of a cubic sodium chloride crystal formed within a microfracture in the hardened cement paste of a concrete.

(Fookes and Collis 1975). Samarai (1976) found that where the gypsum content of an aggregate exceeds 4.5 per cent for OPC and 6 per cent SRPC the expansive formation of ettringite can lead to deterioration of concrete. The total SO_3 expressed as the weight of an OPC should not exceed 15.2 per cent. Crammond (1984), from experimental work related to this problem, recommends a maximum limit of 4 per cent SO_3 by weight of ordinary Portland cement and a 2.5 per cent limit for gypsum by weight of aggregates.

The texture of concrete undergoing sulfate attack from contaminated aggregates is shown in Plate 55. Crammond (1984) reports that petrographic evidence shows that coarse crystalline gypsum present as aggregate reacts with the cement alkalis to form calcium hydroxide on the gypsum crystals. The calcium hydroxide crystal then grows inwards towards the centre of the gypsum crystal. This conversion of gypsum to calcium hydroxide releases sulfate to form ettringite with the aluminate hydrates.

A rarer cause of excess sulfate being mixed into concrete, mortars and plasters is when workmen add gypsum during mixing. This is due either to adding the wrong 'grey powder' to the mix or mistakenly adding gypsum plaster in the form of the hemihydrate to accelerate setting and hardening. The resulting expansion occurs within weeks to months because finely ground gypsum reacts quickly. The reaction is spectacular and requires complete removal of the contaminated concrete or mortar. Efflorescence, scaling, spalling and cracking are widespread. Analysis, usually by X-ray diffraction, will show that large amounts of gypsum and ettringite are present and this is usually sufficient to identify the cause. Examination of the texture of this type of failure in thin-section will not provide any additional information to that found from X-ray diffraction and chemical analyses.

Some sulfide minerals such as pyrite (FeS_2), marcasite (FeS_2), pyrrhotite ($Fe_{1-x}S$) and chalcopyrite ($CuFeS_2$) in the presence of oxygen and the alkaline pore solution of concrete can oxidise to produce ferric hydroxide ($Fe(OH)_3$) and a range of sulfates. As the volume of the products is greater than the original mineral, pop-outs occur above aggregates near the surface and in severe cases pattern cracking may also occur as illustrated by Fulton

(1967). The ferric hydroxide, often present as goethite produces a characteristic rust stain asssociated with the pop-outs.

$$2FeS_2 + 2O_2 + 7H_2O \rightarrow 2FeSO_4 + 2H_2SO_4$$

The sulfuric acid and ferrous sulfate then react with the alkaline constituents present in the pore solution of the concrete to give a range of alkaline and calcium sulfates and ferrous hydroxide $(Fe(OH)_2)$ which then oxidises to ferric hydroxide, $Fe(OH)_3$. In some cases ettringite may be formed. These reactions are more complex than given above as among the products that may be found associated with the pop-outs are jarosite $(KFe_3(SO_4)_2(OH)_6)$ and other complex sulfates as reported by Shayan (1988) and De Ceukelaire (1991).

Usually damage due to the presence of sulfide minerals in the aggregates is restricted to pop-outs and iron staining (Plate 56). Fulton (1967) recommends that the total sulfide in aggregates should be limited to 2 per cent for coarse aggregates and less for fine aggregate especially when mine tailings are used as aggregate. However, Midgley (1958) investigated the presence of pyrite in Thames river gravels and found that only some of the pyrite was reactive and that immersion of grains in saturated lime water distinguished reactive pyrite by the formation of a brown precipitate. Cases of severe deterioration of low-quality 'Mundic' concrete blocks have been ascribed in part to decay of sulfides in lime waste aggregate (Stimson 1997).

The petrographic investigation of pop-outs involves examination of the reaction products located around the apex of the pop-out and its socket. Identification is best carried out by microscopic examination of grains in oil immersion or preferably by X-ray diffraction analysis and where necessary, examination and analysis by scanning electron microscopy. Where jarosite is present its yellow colour is distinctive and a good indication that sulfide minerals are involved. It is necessary to ensure that adequate analysis is carried out to determine the cause of pop-outs as some shale aggregates undergo large volume increases on wetting and can also produce pop-outs (Dolar-Mantuani 1983).

Steam curing is often used in precast concrete plants to obtain high early strengths and thus increase production rates. Delayed ettringite formation in concrete can occur when the curing temperature exceeds 75 °C. The conditions and mechanism of delayed ettringite formation depend on a number of factors as shown by Ghorab *et al.* (1980), Heinz and Ludwig (1986), Brown and Bothe (1993) and Glasser *et al.* (1995). Not all cements are affected. It is claimed that molar ratio of SO_3/Al_2O_3 in the cement has to be greater than 0.55 for significant delayed ettringite formation to occur. Thus cements low in alumina, such as sulfate-resisting cements or under-sulfated cements are less likely to be affected. However, Odler and Chen (1995) found that expansion depended on the sulfate and aluminate content but not on their ratio. The extent of expansion due to delayed ettringite formation increases as the temperature is raised to 90 °C which is about the maximum attainable for atmospheric steam curing of products. For higher temperatures, such as those used for autoclaving products the sulfate and aluminate phases are different and as a result delayed ettringite formation has not been reported. Lawrence (1993) concluded from laboratory studies that limiting concrete temperatures to a maximum of 70 °C was a secure way of avoiding the problem. Later work by Lawrence (1995) found a complex relationship between composition and expansion but concluded that involvement of alkali-silica reaction in the expansion is not an essential feature.

A number of cases of damaging expansion resulting from delayed ettringite expansion have been reported (Sylla 1988, Neck 1988, Tepponen and Ericksson 1987) as well as informal reports of widespread damage to railway sleepers and other products. Oberholster *et al.* (1992) investigated cracking in low-pressure steam-cured railway sleepers in South Africa and concluded that in this case alkali-aggregate reaction was the primary cause and not delayed ettringite formation. Similarly, Shayan and Quick (1992) found that railway sleepers in Australia had also cracked due to alkali-aggregate reaction. Stark *et al.* (1992) have found that in concretes with a low water/cement ratio, SO_3 contents as low as 3.3 per cent, and curing temperatures of 60°C, delayed ettringite formation may occur.

This relationship between ettringite and ASR has been recognised for some years (Blackwell and Pettifer 1992). In order for ettringite to form, sulfate ions need to be carried by moisture gradients and concentrated at nucleation sites. The hydrophilic properties of ASR gels provide a suitable mechanism for generating such moisture gradients within the concrete (Poole *et al.* 1996).

Temperatures in the range 75–90°C are typical of low pressure steam curing and also occur in large volumes of freshly placed concrete when temperatures generated by the heat of hydration are not controlled. It is claimed that delayed ettringite formation occurs because the sulfate and aluminate ions become bound in the calcium silicate hydrate structures which under moist conditions are later slowly released to expansively form ettringite (Heinz and Ludwig 1986, Fu *et al.* 1995). A different mechanism has been proposed by Kuzel (1990) who suggests that carbonation of the low calcium sulfoaluminate hydrate to a mono carbonate, $4CaO.Al_2O_3.CO_2.11H_2O$, and gypsum occurs. The gypsum formed by this reaction is then available to react later with the remaining low calcium sulfoaluminate hydrate to form ettringite.

A sample was taken for examination from a cracked structure where it was believed that internal temperatures in the concrete may have exceeded 90°C during early curing. Over a storage period of a year at 20°C and 100 per cent RH a core sample from the structure expanded over 0.5 per cent. In thin-section, the coarse limestone aggregates and many of the quartz sand particles were surrounded by a rim of ettringite up to 100 μm in width. This effect was uniformly spread across the whole of the large-area thin-section. The ettringite needles were usually orientated at right-angles to the smooth margins of the cement paste surrounding the fissures. The needles of ettringite were always on the surface of the paste margins and not the aggregates and gave the appearance of having grown into an existing space (Plate 57(a)). The hardened cement paste contained fine pattern microcracking which was also filled with ettringite and had the appearance of a typical expansion texture for *in situ* formation of ettringite.

A minor alkali-aggregate reaction was occurring in some chert and quartzite grains consistent with the high alkali content of the concrete but was not considered sufficient to cause the amount of cracking and expansion observed (Plate 57(b)). The conclusion from the petrographic examination was that delayed ettringite expansion had clearly occurred in the hardened paste causing cracking and expansion. In addition some mechanism was operating to cause the formation of ring fissures around most of the aggregates which provided space for the subsequent growth of ettringite.

Johansen *et al.* (1994) have reported similar observations in railway sleepers (Fig. 6.6) and put forward a hypothesis that the fissuring around the aggregates is due to the overall expansion of the hardened cement paste. In effect, if each aggregate is considered equivalent to a void and an essentially isotropic expansion of the paste occurs each void must grow in volume proportional to the expansion. It follows that the width of the ring fissure formed around each aggregate will be proportional to the size of the aggregate. Thus the paste gives the appearance of having shrunk away from the aggregates forming ring fissures which then provide space for the growth of the ettringite needles. The texture they illustrate

Fig. 6.6 External cracking of railway sleepers caused by delayed ettringite formation (photograph courtesy of N. Thaulow).

is similar to that described above for the sample from a cracked structure. Scrivener and Taylor (1993) have also concluded delayed ettringite formation is driven by processes occurring within the paste and not by the formation of ettringite at aggregate interfaces.

Although the acicular ettringite crystals are commonly observed in thin-sections to have orientated perpendicular to the walls of a crack or aggregate as described above, scanning electron micrographs show that in three dimensions the orientations are more variable, for example the crystals sometimes form as felted masses aligned parallel to the crack surfaces (Fig. 6.7).

In spite of some controversy there is little doubt that cracking due to delayed ettringite formation does occur. As it is an intrinsic reaction in which the whole paste expands, cracking is typically accompanied by ring fissuring around aggregates in which redeposition of ettringite occurs. Given the appropriate conditions it is possible for alkali-aggregate reactivity to occur concurrently and the mechanism proposed by Brown and Bothe (1993) suggests that alkali-aggregate reaction increases the likelihood of delayed ettringite formation. It is not uncommon in cases of alkali-aggregate reaction to find considerable redeposition of ettringite has occurred in some cracks but ring fissures around aggregates are absent. The formation of ring cracks around aggregates is an important distinction and delayed ettringite formation should not be diagnosed unless this textural aspect of the concrete is present. Marusin (1994) describes a visual method for distinguishing delayed ettringite from alkali-silica gel around aggregate margins.

6.2.4 SULFATE ATTACK IN SEWAGE SYSTEMS

Normal sewage effluents are alkaline and do not attack concrete in contact with them directly. However, anaerobic bacteria present in sewerage pipes can decompose the inorganic

Fig. 6.7 Scanning electron micrograph of a crack surface coated with ettringite crystals. The crack has formed as a 'ring crack' at the surface of a rounded aggregate particle.

and organic sulfur compounds present, releasing hydrogen sulfide to the atmosphere above the effluent. This gas is absorbed by the water film coating the pipe walls which contain aerobic bacteria. The aerobic bacteria oxidise the hydrogen sulfide to sulfurous and sulfuric acids which attack and dissolve the hardened cement paste (Thistlethwayte 1972, McClaren 1984). The result is the development of a concrete surface on the crowns of sewer pipes consisting of aggregates standing proud of severely corroded cement paste.

This acid corrosion of the crown of a sewerage pipe will be most severe if the sulfur content of the effluent is high and the flow rates slow or stagnant so that the sewer becomes septic as shown in Fig. 6.8. Temperature and oxidation-reduction potential will also influence the growth of bacteria and hence the rate of production of sulfuric acid. A detailed study of corrosion of sewerage pipes was undertaken by CSIR South Africa (1967) and these studies were followed up by Barnard (1967), Thistlethwayte (1972) and Pomeroy (1977).

One case examined involved a case of acid attack on the crowns of some 1300 mm diameter main, sewerage pipes made with 50 MPa concrete and greywacke aggregate. In this case the acid attack on the concrete was stopped before the pipes were too severely damaged by running the pipes full of sewage thus preventing any further oxidation of the hydrogen sulfide.

The example shown (Fig. 6.9) has been taken from the crown of a large-diameter concrete sewerage pipe which has undergone acid attack from the oxidisation of hydrogen sulfide by aerobic bacteria. The acid dissolves the hardened cement paste leaving the larger aggregates standing proud of the surface. In severe cases the aggregates fall out as the depth of attack increases.

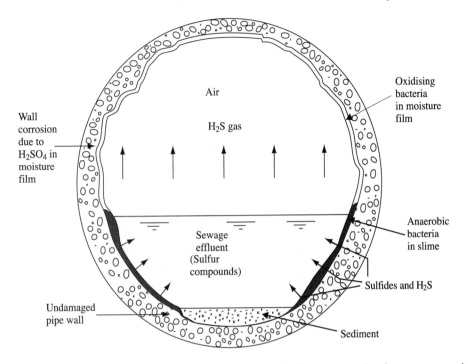

Fig. 6.8 Diagrammatic representation of the mechanisms leading to corrosion in a sewerage pipe.

The surfaces examined had been in contact with sewage for some period and it is likely that some of the outer corroded material may have been removed. The external appearance of the corroded surfaces (Fig. 6.9) was brownish yellow and aggregates were standing proud of the corroded cement paste. In thin-section (Plate 58) the texture of the corroded zone consisted of an approximately 250 μm wide outer layer of amorphous brown material intermixed with carbonation products backed by a 2 mm zone of hardened paste heavily leached of calcium hydroxide. Behind the leached layer there was a zone of intense carbonation varying from 250 μm to 2 mm in width which graded into sound paste.

The texture of the corroded surface of the sewerage pipes is similar to that observed by Chandra (1988) and by De Ceukelaire (1992) and is probably typical of most forms of acid attack where the hardened cement paste is solubilised. A fragile outer zone is formed where the acid attack has gone to completion and only hydrated amorphous silica and alumina remain which may be lightly stained by iron compounds. This outer corroded zone can vary from as little as 20 μm to several millimetres in depth. Inside the corroded zone is a reaction layer consisting of carbonated paste intermixed with brown amorphous material derived from the iron-bearing phases in the hardened paste. Although the corrosion zone may appear as separate layers of brown amorphous material and carbonation, careful examination will usually show that this separation can rarely be justified. Inside the reaction layer is a zone of hardened paste which is depleted in calcium hydroxide which merges in to sound paste. This last zone is often missed in examination but must always be present as acid attack inevitably sets up chemical gradients. The most obvious evidence of the chemical gradient is some depletion of calcium hydroxide.

Fig. 6.9 The effect of bacterial acid attack on the crown of a sewerage pipe.

6.3 Frost and freeze-thaw damage

Hardened concrete may be damaged by frost particularly if it is subjected to freeze-thaw cycles typical of the northern temperate latitudes. In particular, the use of de-icer salts can lead to scaling (Marchand *et al.* 1994). The damage usually starts with a superficial flaking of the surface layers resulting from the development of irregular cracks sub-parallel to the exposed surface and as the damage becomes more severe significant spalling of exposed surfaces and edges occur (Fig. 6.10).

Moisture must be present in the pore structure of the concrete for damage caused by freezing to occur. Thus with well made dense concretes with low water/cement ratios damage is minimal. Air entrained concrete in well made low water/cement ratio mixes, with carefully controlled spacing between air voids, is a widely used method of minimising the effects of freeze-thaw damage.

A variety of explanations for the mechanism of freeze-thaw damage have been proposed. Early theories suggested that the volumetric expansion of pore water on freezing in a confined zone beneath an already frozen surface produced the disruption. This idea was modified by Collins (1944) suggesting that the disruption of the surface layers could result from the growth of ice crystals resulting from the migration of water towards them because of vapour pressure differences. Hydraulic pressure mechanisms caused by water migration through the capillary system resulting in cracking and disruption have been proposed by several authors. Powers (1956) proposed that the expansion due to freezing in the pores within the surface layers forced water inwards through small pores with sufficient hydraulic pressure to disrupt the surrounding material. The effectiveness of air entrainment in overcoming this lends support to the validity of the hypothesis. The theories of frost action have been reviewed by Pigeon and Pleau (1995) who also discuss testing techniques and

Fig. 6.10 Example of frost damage affecting the exposed surfaces on a bridge column. Note cracking parallel to the exposed corner.

methods for producing concrete that is durable in frost conditions. More recent studies by Bager and Sellevold (1986, 1987) using low-temperature calorimetry, indicate that the inward migration of the ice front from the surface is much affected by the previous drying history of the concrete.

It is probable that the observed damage to hardened concrete is the result of the interaction of several of these mechanisms. Petrographically no changes to the cement or aggregate mineralogy can be observed and the damage is restricted to a series of irregular cracks which lie sub-parallel to the exposed surfaces or corners. Such crack systems typically propagate through the cement paste and round the margins of aggregate particles. The distance between the individual sub-parallel cracks depends on the aggregate sizes and the freeze-thaw rates but they are usually equispaced. The cracks furthest in from the surface are impersistent but as the exposed surface is approached the cracking becomes more persistent, more open and bridging cracks approximately perpendicular to the surface become common.

The crack pattern, the details of the local environment, the concrete sample and case history together with the petrographic information relating to the concrete usually provide sufficient evidence for the correct diagnosis of frost damage.

Observations on site, together with a study of the crack pattern developed on resin impregnated slabs cut normal to, and including, the surface are the best aids to diagnosis, though large size petrographic thin-sections can be used in place of the cut slabs.

The effects of cold weather on the setting of concrete are reflected in the microfabric of the cement paste. In petrographic thin-section the cement paste in concrete which has set in cold conditions is lighter in colour than might be expected for a given water/cement ratio. This effect is due to the more open porous nature of the hydrates formed and may be a consequence of slower than normal hydration. Unless care is taken, this effect can introduce errors into the method for estimating water/cement ratio in hardened concrete by calibrating the intensity of fluorescence from a dyed resin impregnating the pore spaces of the cement paste against a series of standards (see Sections 2.7 and 4.4).

In freezing conditions mixing of concrete becomes difficult, the mix water freezes so that the hydration of the cement minerals is partial and uneven with the ice crystals in the mix occupying a larger volume than the water they replace. Thus the concrete does not set properly and remains a friable mass.

6.4   Acid and alkaline attack on concrete

Hardened cement and calcareous types of aggregate are rapidly attacked by acidic solutions with a pH of less than 4.0. Where the hardened matrix contains high-alumina cement it is more resistant above pH 4.0 but below this pH attack is greater than that found with Portland cement (Lea 1970). The composition of ordinary Portland cement, including the use of replacement materials such as fly ash, appears to be less important than the water/binder ratio in determining performance in acid conditions (Matthews 1992). Similarly Ballim and Alexander (1991) found that while ground blastfurnace slag and fly ash reduced leaching from carbonic acid attack the resulting porosity was little different between ordinary Portland and blended cements.

The most important factors in corrosive attack are the amount of fluid flowing over the exposed surface of the concrete and the pH. If a significant flow rate is occurring the attack on the concrete may be considerable even for mildly acidic conditions. This is especially the case where soft acidic waters are permeating through porous concrete and leaking joints in hydraulic structures (Commission Internationale des Grands Barrages 1989). However, if one of the corrosion products formed is insoluble in the attacking fluid it can provide a protective layer on the surface of the concrete. This occurs in marsh water (Lea 1970) where, if humic acid is present it forms an insoluble coating of calcium humate. This may also apply to some extent to many acids which form an insoluble layer of gelatinous silica on the surface of concrete (Grube and Rechenberg 1989).

Between pH 7.0 and 8.5 attack is essentially limited to superficial leaching of calcium hydroxide from the concrete surface. Even at a pH of 6.0 attack is minimal unless the flow rate is significant and is only potentially serious for thin-walled products such as concrete pipe and highly permeable mass concretes. At a pH of 5.0 attack is serious for concrete pipe and dependent on the flow rate of the water will limit the life of such products. BS 8110 (1985) recommends that Portland cement should not be used without protection in conditions where the pH is consistently less than 5.5. In practice it has been found that most calcareous cementing systems will not withstand a pH of less than 4.0 without the provision of an adequate protective coating. However, for massive concrete which is reasonably impermeable, while attack in the pH range 4.0 to 7.0 may corrode concrete surfaces, the depth affected is usually insufficient to be serious unless it leads to corrosion of reinforcement.

Although Portland cement concretes are themselves alkaline they are attacked by strong alkaline solutions where the concentration exceeds 20 per cent but are largely unaffected if the concentration is less than 10 per cent (Biczoc 1967). Concretes containing high-alumina cement are less resistant to alkaline solutions than those made with Portland cements. Alkaline solutions slowly act on calcium aluminate cements by dissolving the alumina gel and attacking the hydrated calcium aluminates.

There are three factors that need to be determined where concrete structures are undergoing corrosion. These are the composition of the attacking fluid determined by chemical analysis, the extent of the corrosive attack on the concrete texture and estimation of the rate of attack. Petrographic examination is able to delineate the nature and depth of the damage to the concrete and in favourable circumstances may allow an approximate rate of attack to be determined.

A collation of abstracts on the subject of chemical attack is given in Appendix 1.

6.4.1   NATURAL ACIDIC WATERS AND THE CALCIUM – BICARBONATE – CARBON DIOXIDE EQUILIBRIA

Most natural waters have acidities that fall into the pH range 4 to 8.5. The pH will rarely fall below pH 5 due to dissolved carbon dioxide unless it is under pressure. Below pH 5 the acidity is usually due to the presence of humic acid which is limited in aggression because calcium humate is almost insoluble in water and forms a protective layer on the concrete (Lea 1970).

Where groundwater percolates through decaying organic matter the carbon dioxide produced is absorbed by the water. The effect of the absorbed carbon dioxide on the pH of the water depends on the amount of bicarbonate already present because only the free carbon dioxide, that is the carbon dioxide not required to stabilise the bicarbonate in solution, affects the pH. The less the water is buffered by bicarbonate the less the amount of carbon dioxide required to lower the pH. A good example is distilled water which has close to zero buffering. Absorption of a few ppm of carbon dioxide from the atmosphere is sufficient to give an equilibrium pH of about 5.7. Distilled water is highly aggressive to concrete as it has both a high solvent capacity because of its low buffering and is also acidic. This situation occurs in the water of some mountain and high moorland areas which are relatively free of dissolved salts (Commission Internationale des Grands Barrages 1989). However, buffering of waters is also affected by other calcium salts such as gypsum and the salts present in sea water as discussed by Terzaghi (1949) so in practice the situation may be more complicated than that described above.

Confusion as to both nomenclature and methods of measurement have occurred in the past when groundwaters were analysed for relative aggressiveness to concrete. Reference should be made to Terzaghi (1949) who first identified some of the problems and Rogers (1973) and Cowie and Glasser (1991/92) who have reviewed the subject. Large errors are possible in the determination of free carbon dioxide and pH and it is essential that all measurements be checked by equilibrium calculations, as described by Stumm and Morgan (1970) and the American Public Health Association (1992). Similarly, groundwaters may be turbid when sampled and measurement of pH can be affected by suspension effects at the liquid junction to the reference electrode of the pH meter (Bates 1964). Many groundwaters, especially those in aquifers, are under pressure so that measurement of carbon dioxide in surface samples may not fully represent the carbon dioxide actually present in the subsurface waters which are in contact with the buried concrete. Special techniques for sampling subsurface waters to avoid the loss of CO_2 have been developed by John *et al.* (1977). Because of the problems involved it is essential that sampling and analysis of groundwaters for dissolved carbon dioxide, alkalinity and pH be carried out by experienced water analysts.

When natural waters containing carbon dioxide attack concrete both the calcium hydroxide and the hardened cement paste are first carbonated and then dissolved. The two attacking species are carbonic acid H_2CO_3 and calcium bicarbonate $Ca(HCO_3)_2$. As calcium bicarbonate is soluble it can be removed by the attacking fluids providing sufficient carbonic acid is present to stabilise the bicarbonate formed.

$$Ca(OH)_2 + H_2CO_3 \rightarrow CaCO_3 + 2H_2O$$

$$Ca(OH)_2 + Ca(HCO_3)_2 \rightarrow 2CaCO_3 + 2H_2O$$

However, calcium carbonate, whether present due to the carbonation of calcium hydroxide or as limestone aggregate, and the hydrated cement phases are also attacked by carbonic acid.

$$CaCO_3 + H_2CO_3 \rightarrow Ca(HCO_3)_2$$

Hydrated cement $+ H_2CO_3 \rightarrow$ intermediates $\rightarrow$ Si hydrates $+$ Al hydrates $+$ Fe hydrates

The end products of the attack are gelatinous forms of hydrated silica and alumina and iron hydroxides. In practice most of the hydrated alumina and iron hydroxides are removed by the attacking fluid so that the only remaining product is gelatinous silica. It is the formation of soluble calcium bicarbonate and its removal by the fluid that is responsible for the actual corrosion of concrete surfaces.

It has been reported by Gerdes and Wittmann (1992) that deterioration of concrete due to similar reactions detailed above can be induced by long lasting electric fields in mortar coatings lining water reservoirs.

Petrographic textures

When a thin-section cut from concrete corroded by soft acidic water is examined it will be observed that the non-calcareous aggregates are left standing proud of the corroded surface of the hardened cement paste. However, many limestone aggregates will be dissolved more readily than the surrounding cement-paste. As the attack proceeds smaller aggregates near the surface may become dislodged as the typical rough corrosion surface develops. The texture of the outer layers of the hardened paste consists of an irregular outer zone consisting primarily of gelatinous silica lying over a zone of carbonation which merges into another zone of hardened paste depleted in calcium hydroxide as illustrated in Fig. 6.11

The zones are not very wide and the gelatinous outer zone may be partially absent as it is easily damaged and lost in sampling and may be difficult to observe because it is isotropic. In addition it is often colourless although it can contain patches of yellow to dark brown occluded material. Occasionally a thin line of reddish-brown material may separate the outer zone from the carbonation zone. While the carbonated layer is obvious in thin-section the zone leached of calcium hydroxide is usually not well defined and is often difficult to observe. A method has been proposed by Koelliker (1985) to overcome the problem of retaining the outer, gelatinous layer during sampling.

This type of zonation is *always present* on the undisturbed surfaces of concretes corroded by soft acidic waters. The outer zone of gelatinous silica represents the completion of the attack. The inner zone depleted of calcium hydroxide is where there is a concentration

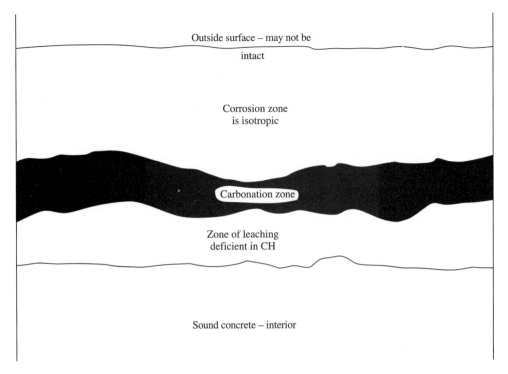

Outside surface – may not be
intact

Corrosion zone
is isotropic

Carbonation zone

Zone of leaching
deficient in CH

Sound concrete – interior

Fig. 6.11 Diagram illustrating the zones of soft water attack.

gradient occurring due to the attack and the zone of carbonation is where the actual dissolution of material occurs.

Estimation of rates of attack

Classification of the aggressiveness of natural waters containing carbon dioxide was made by Werner and Giertz-Hedstrom (1934) based on experience in Scandinavia. Subsequent classifications, such as those given by Lea (1970), are refinements of this basic work and will indicate the degree of protection required by use of concrete in new works. However, where attack has occurred on concrete the situation may arise where the extent of attack needs to be determined so that some estimate of an expected lifetime may be made (Plates 59, 60, 61 and Fig. 6.12).

A number of methods, both theoretical and practical, have been proposed for estimating the rate of attack. All the practical methods depend on estimation of the amount of material removed by the corrosion. The simplest test is to measure weight loss of laboratory specimens (Tremper 1932, Werner and Giertz-Hedstrom 1934) and relate these losses to products and structures. Al-Adeeb and Matti (1984) measured the free lime content and permeability of asbestos-cement pipes to estimate rate of attack. Where water is percolating through a structure the increase in calcium in the leachate has been used to estimate weight loss from the concrete (Commission Internationale des Grands Barrages 1989) but this is restricted to those cases where a leachate can be collected.

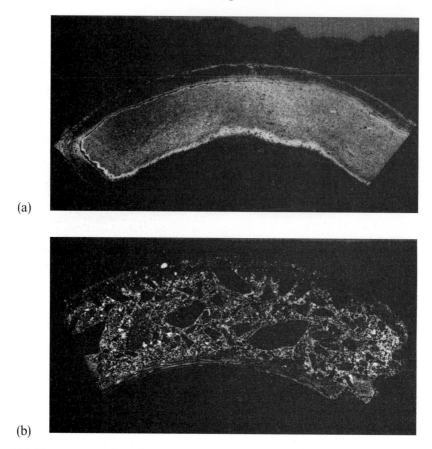

(a)

(b)

Fig. 6.12 The texture of the soft water corrosion of concrete and asbestos pipes from St John and Penhale (1980). (a) Photograph of a negative print of a large-area thin-section cut from an asbestos cement pipe. The pipe has been buried for twenty years on a site with soft acidic water flows (pH ≈ 6). The outer dark zone is where the hardened cement paste has corroded to amorphous silica and the brightish white zone is carbonation. The zone of leaching is not apparent but is clearly visible under the microscope. (b) A large-area thin-section from a concrete pipe exposed as described above. The corrosion and carbonated zones are barely apparent but still clearly visible under the microscope. These two photographs illustrate that zones of deterioration are more apparent in fine-textured fabrics. Hence the need for adequate microscopic examination of concrete textures where alteration may not be apparent in sawn or fracture surfaces because of the presence of aggregate. Width of view in photographs 170 mm.

St John and Penhale (1980) have used another approach. The depth of external corrosion on a three-year-old concrete sewerage system was estimated by systematic core sampling and measuring corrosion depths in thin-section. The corrosion depth was defined as the sum of the corroded, carbonated and leached layers on the assumption that when the leached layer penetrated to the steel reinforcing corrosion would occur. Groundwater analyses were also performed which showed that the water flowing around the pipes had a range of corrosivity. The results were compared, using the same petrographic techniques, with test samples of pipes that had been buried under similar conditions for many years under known conditions.

The rate equations for attack by soft waters are all similar and the simple equation proposed by Valenta and Modry (1969) is adequate for most purposes. It is $t = ax^2$ where t = time in years, a = rate constant and x = depth of corrosion. As the corrosive conditions were variable a range of rate constants were derived which ranged from 0.7 to 1.18. This translates to a minimum life of thirty years for the most aggressive condition found to a maximum life of five hundred years for the least aggressive condition. The results obtained were in reasonable agreement with those of Tremper (1932), and Werner and Giertz-Hedstrom (1934).

6.4.2 CARBONATION AND CORROSION BY GEOTHERMAL FLUIDS

Carbonation and corrosion of concrete by geothermal fluids containing carbon dioxide is much more rapid than at ambient temperatures. If the fluids are insufficiently buffered the carbon dioxide will give rise to acidic conditions and severe attack occurs within months. The cement grouting systems used to anchor geothermal bore casings are one of the products most likely to be severely damaged as the temperatures of fluids surrounding the bore holes can range up to 300°C at pressures of up to 50 bars (Milestone *et al.* (1986). Geothermal cement grouts usually consist of Portland cement mixed at water/cement ratios of about 0.6 or greater and may contain added silica flour in the form of ground quartz to suppress the formation of α-dicalcium silicate hydrate, and bentonite and chemical additions to improve pumpability. Because of their formulation these cement grouts are fairly permeable.

Textures of grouts exposed to geothermal fluids

The textures of grouts exposed to geothermal fluids containing carbon dioxide graphically illustrate many aspects of the corrosion mechanisms at elevated temperatures. In addition, the effect of temperatures ranging from 160 to 260°C on the hydration products formed, with and without the addition of silica flour to the grouts, is of importance for both carbonation and corrosion.

St John and Abbott (1987) examined sections of both autoclaved grout specimens in the laboratory and field samples exposed to downhole geothermal fluids at New Zealand's Broadlands Geothermal Development. The textures are similar to those produced by corrosion from soft acidic water at ambient temperatures and are shown for the autoclaved specimens in Fig. 6.13. The mode of carbonation at these high temperatures is largely controlled by the amount of calcium hydroxide present which in turn is controlled by the amount of silica flour or other material that can react with calcium hydroxide.

When calcium hydroxide and the hydrated cement in the outer margins are carbonated a chemical gradient forms between the carbonated and inner zones to give a clearly defined zone leached of calcium hydroxide as shown in Plate 59(b) and Figs. 6.13(a) and (c). Where silica flour has been added to the grout, most of the calcium hydroxide will have reacted and direct carbonation of the cement hydrates is the dominating process. At 260°C the carbonation zone may sometimes consist of a series of complex carbonation bands rather than a single band, irrespective of the composition of the grout, as shown in Fig. 6.13(e) and (f). The reason why these carbonation bands form, which have been observed both in the laboratory and in the field, is not known.

Where silica flour has been added, the direct carbonation of cement hydrates does not set up the same visible chemical gradient and as a result the leached zone is absent. At

(a)

(b)

Fig. 6.13 The high-temperature carbonation of cement grouts exposed to carbon dioxide and water in an autoclave as observed under crossed polars; neutral pH conditions. (a) After exposure for 14 days at 150 °C the cement grout, containing no added silica, shows clearly defined zones of alteration as described in the text. The outer zone consists of densely packed carbonation and appears darker than the central zone which contains large amounts of calcium hydroxide. The reason for this difference is that the high-order birefringence of the carbonation is not as bright as the lower-order birefringence of the calcium hydroxide. A narrow leached zone is just visible between the outer carbonation zone and the unattacked central zone. (b) Cement grout, containing 40 per cent added silica, exposed as described in (a). Even though the outer zone grades in brightness from the outside surface it is all carbonated material. There is an abrupt transition to the unattacked paste in the centre containing some residual fragments of silica flour which are visible as bright spots. The marked difference in the depths of carbonation is due to the different high-temperature hydrated phases in the hardened paste. No leached zone is able to be present because the calcium hydroxide has reacted with the silica flour. Width of view in photographs 80 mm.

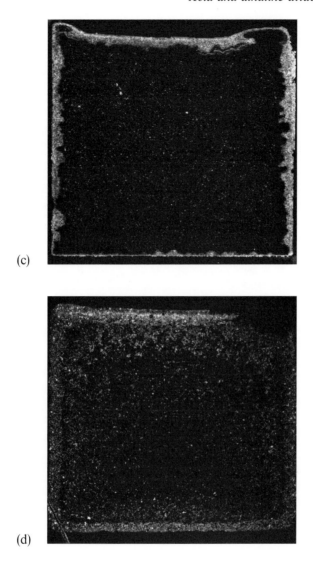

(c)

(d)

Fig. 6.13 The high-temperature carbonation of cement grouts exposed to carbon dioxide and water in an autoclave as observed under crossed polars; neutral pH conditions. (c) Cement grout, containing no added silica flour, after 14 days exposure at 260 °C. The bright outer zone is carbonation separated by a heavily leached zone from the unattacked paste. The white specks in the central zone are calcium hydroxide. (d) Cement grout, containing 40 per cent added silica, exposed as described in (c). The outer zone consists of carbonation which becomes separated into discrete clumps whose frequency lessens with depth. Bright clumps visible throughout the body of the sample are carbonation as calcium hydroxide is absent. Note that this is a large increase in carbonation depth over that observed in (c). Width of view in photographs 80 mm.

160 °C there is an ill-defined zone of alteration which is the interface between the carbonation zone and the uncarbonated grout. At 260 °C this ill-defined alteration zone is obscured by the presence of 'clumped carbonation' but the carbonation zone is still observable because of the deposition of dark material at its inner margin. Apart from this, the

(e)

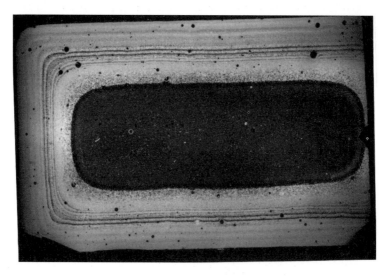

(f)

Fig. 6.13 The high-temperature carbonation of cement grouts exposed to carbon dioxide and water in an autoclave as observed under crossed polars; neutral pH conditions. (e) Cement grout, containing no added silica, exposed for 21 days at 260°C. There is a wider outer carbonated zone followed by narrow bands of carbonation set in a background matrix leached of calcium hydroxide. The conditions which lead to the formation of this banding are not known. (f) Cement grout, containing 1 per cent bentonite, exposed for 21 days at 260°C. Banded carbonation is present. Note how the addition of 1 per cent of bentonite has increased the depth of carbonation. Width of view in photographs 70 mm.

carbonation grades imperceptibly into a texture where clumps of calcium carbonate crystals become less frequent towards the centre of the specimen.

Irrespective of temperature, considerable amounts of silica flour remain unreacted in the outer zones of the grout samples because carbonation removes calcium hydroxide before it can react with the silica. The quantities remaining suggest that initial rates of carbonation are rapid. It is surprising that even after three months exposure at 160°C, modest amounts of quartz as small as 10 μm in size can be observed at the centre of samples. Only at 260°C has the temperature been sufficient to ensure almost complete reaction of the silica flour. Similarly, the presence of remnant cement grains is surprising. Some cement grains appear to be highly resistant to hydration and only at 260°C has essentially complete hydration been observed.

Where the attacking fluid is acidic, even if only mildly acidic, a corroded outer zone of gelatinous silica is formed which appears to be porous. It is very fragile and rarely remains intact at thicknesses of greater than 3–4 mm and appears to offer little barrier to the access of fluids at these temperatures. More rapid corrosion occurs when silica flour is added to the grouts because carbonation of the cement hydrates formed under these conditions leads to increased permeability (Milestone *et al.* 1986). Later laboratory work indicates that any addition that reacts with calcium hydroxide or increases permeability has similar effects. It has been found that data obtained in the laboratory give similar textures and rates of attack to those found with downhole specimens.

The hydrated cement phases present in the grouts can only be identified by X-ray diffraction analysis as apart from α-dicalcium silicate hydrate all the phases are submicroscopic. The types of high temperature phases found have been discussed by Milestone *et al.* (1986) and are briefly described in the Glossary.

6.4.3　ACID TYPE ATTACK FROM SULFATES, IRON SULFIDES, BRINE SOLUTIONS AND MICROBIAL ACTION

Cases of acid attack on concrete are not restricted to naturally occurring carbonic acid. Thornton (1978) has reported on a case of a sulfate rich lake water flowing through concrete outlet tunnels and being converted to sulfuric acid by bacteria. Attack had decomposed surfaces to depths of 30 mm. This is similar to the acid attack that occurs in sewers (see Section 6.2.4). Eglinton (1987) briefly discusses the conditions under which pyrite in rocks soils and wastes can be oxidised to form sulfuric acid, rather than forming sulfates, and mentions cases where the resulting low pH values have caused corrosion.

Concentrated brines containing magnesium chloride associated with some geologic sequences may be encountered when drilling for gas and oil. Oberste-Padtberg (1985) has illustrated the textures of oil well grouts made from either high sulfate resistant or ordinary Portland cement blended with blastfurnace slag and fly ash. These textures show that not only does the typical oriented crystallisation of brucite occur at the surface but magnesium ions can penetrate the interior probably through cleavages and microfissures. This not only allows the expansive formation of brucite internally but also results in acidification of the pore solution by decrease of hydroxyl activity to a level where the calcium silicate hydrate phases may be no longer stable.

The attack by acid slimes on building stone, which will also be applicable to concrete, is reported by Blaschke (1987) and Bock and Sand (1986). The slimes are produced by fungi and algae which collect air pollution leading to sulfate attack and by nitrifying bacteria producing nitric acid.

6.4.4　INDUSTRIAL CHEMICAL ATTACK ON CONCRETE

Industrial chemical attack is here defined as attack by chemicals that either are not found in nature or whose concentration has been increased above natural levels by industrial activity. It covers a wide range of substances and no attempt is made to cover the whole field. For detailed information on the attack by individual chemicals reference should be made to Biczoc (1967). Recommendations for protection against a range of chemicals are made in the report by ACI Committee 515.1R (1985) and a listing of the effect of many chemicals on concrete has been made by the Portland Cement Association (1981). Concrete and mortars are often used in conjunction with masonry to provide protection against chemicals and information on this topic is given by Sheppard (1982).

A fundamental problem in considering the chemical resistance of calcareous hydraulic materials is to know where to set the limits to useful discussion. Because concrete is a relatively cheap building material there is a tendency for the construction industry to extend its use into areas where it is no longer an appropriate material for the exposure conditions. In these cases the problem is not whether the concrete will be attacked but how to determine the rate of attack so that the time to repair or replacement can be estimated.

As discussed in Section 6.4 concrete should not be exposed to fluids or vapours of less than pH 4 either without suitable protection, or the use of more chemically resistant materials such as supersulfated cement, because unacceptably high rates of attack may occur. However, this generalisation covers a complex and variable situation. Kuenning (1966) found that the chemical composition of the aggressive liquid was at least as important as pH and also found no correlation between pH and weight loss from mortar bars. Resistance of mortar was increased by a longer curing times and by a decrease in water/cement ratio.

Kuenning's last observation confirms the point that has been repeatedly observed by investigators. That is, the more impermeable the concrete the lower the rate of attack for any given exposure condition. Over the last decade super-high-strength materials such as densified system with small particles (DSP) have been developed which have increased resistance to aggressive fluids (St John *et al.* 1993, Fattuhi and Hughes 1986) and show promise as protective linings with increased durability. DSP is so impermeable that attack is limited to the surface and the aggressive fluids cannot penetrate any distance into the fabric of the mortar thus limiting the rate of attack as shown in Fig. 6.14.

More recently some other aspects of the chemical resistance of concrete have become important. The ability of concrete structures safely to contain fluids and toxic solids and the solidification of toxic wastes using cementitious materials has become important, as a result of national and international legislation for environmental protection (Morrey 1991, Commission of the European Communities Council 1993).

Macro and micro encapsulation of hazardous and radioactive wastes has led to the development of a wide range of specialist cement-based materials designed to overcome problems of permeability and mechanical and chemical long-term stability. In many cases use is made of cement replacements such as ground granulated blastfurnace slag and fly ash together with hydration modifying admixtures in order to meet the special requirements of such materials. Roy (1988) and Fairhall (1989, 1994) have reviewed the containment of nuclear wastes. Petrographic study of these materials presents no special difficulties and the methods described elsewhere in this text may be applied without modification.

The use of cement to solidify and stabilise hazardous wastes is a well-established practice, though some wastes have been shown to significantly retard or poison the normal hydraulic and pozzolanic reactions (Hills 1993, Jones 1988) and doubt has arisen concerning the long-term stability of some of these materials (ENDS 1992, Webb 1993). Reviews of the methods in current use for the fixation and solidification of hazardous wastes are given by Connor (1990) and Spence (1993). The petrographic techniques applicable to cement

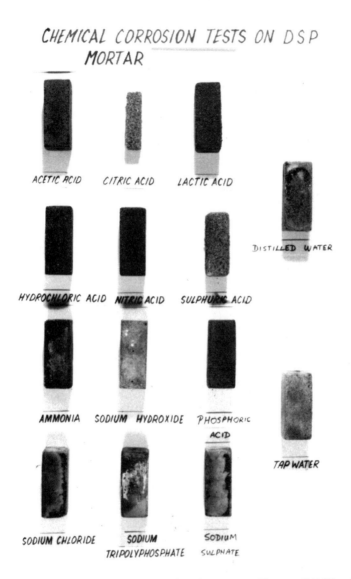

CHEMICAL CORROSION TESTS ON DSP MORTAR

ACETIC ACID   CITRIC ACID   LACTIC ACID

DISTILLED WATER

HYDROCHLORIC ACID   NITRIC ACID   SULPHURIC ACID

AMMONIA   SODIUM HYDROXIDE   PHOSPHORIC ACID

TAP WATER

SODIUM CHLORIDE   SODIUM TRIPOLYPHOSPHATE   SODIUM SULPHATE

Fig. 6.14 Chemical attack on ultra-high-strength DSP mortar. The small DSP prisms have been exposed to 5 per cent concentrations of the chemicals shown, (except for sodium chloride solution which was saturated), for sixteen weeks at 21 °C. The chemical solutions were renewed every four weeks. Significant volume losses due to citric (75 per cent) and sulfuric acids (33 per cent) occurred. Weight losses for the strong acids varied from 77 per cent to 15 per cent indicating that significant leaching occurred even where volume losses of less than 3 per cent were measured.

solidified hazardous wastes have been reviewed by Wakely *et al.* (1992) but petrographic observations are hampered by the common occurrence of iron-rich or carbonaceous particles in the wastes and particularly in incinerated wastes tend to mask the details of the cementitious matrix microstructure. In many examples the solidified wastes are more closely allied to cement stabilised soils than to normal concretes or mortars. Since chemical stabilisation and fixation is as important as mechanical stability, a chemical approach is more often the most useful method of study (Cocke 1990).

Factors to be evaluated

When the petrographer is faced with the evaluation of the chemical attack on concrete, mortars and calcareous materials caution is necessary. Where new construction is involved the experience from the literature may be used. However, where deterioration is occurring in an existing structure due to chemical attack it is necessary not only to evaluate the state of the concrete but also to investigate the chemistry of the aggressive fluids involved. It may be advisable to consult a specialist for advice as the way in which some chemical species attack concrete can be unexpected.

The factors requiring evaluation are the composition of the aggregates, the cement content and type, water/cement ratio, measurement of air void system and the identification of mineral admixtures if present and textural faults in the hardened cement paste. Textural faults such as porous surfaces as shown by excessive depths of carbonation, excessive drying shrinkage cracking of surfaces and plastic settlement around aggregate margins will increase rates of attack.

Identification and concentration of the major ions in solution and the pH must be determined and any temperature fluctuations of the attacking fluid estimated. Where the fluid is in the form of condensation it may be difficult to sample. Condensates may be absorbed into filter paper which is then immediately placed in distilled water to provide a diluted solution for later analysis. Condensates do not necessarily adequately represent the attacking fluids as they will usually be modified by interaction with the surface of the concrete.

Petrographic textures

There are only a limited number of published cases of attack by industrial chemicals where the texture of the concrete has been examined in thin-section.

Chandra (1988) and De Ceukelaire (1992) have described the textures resulting from attack by hydrochloric acid. Chandra used mortars and exposed them to 15 per cent hydrochloric acid for five days and De Ceukelaire studied the attack by 1 per cent hydrochloric acid over a period of two years on ordinary Portland cement, sulfate-resisting cement and blastfurnace slag cement.

Both observed two main zones of attacked concrete on the surfaces of the specimens with some differences in detail. The outer corroded zone consisted of mainly amorphous silica and alumina with varying degrees of coloration. Chandra found a yellow colour which he identified as iron chloride which was absent in De Ceukelaire's samples. The second zone was brown in colour due to the presence of amorphous material which was considered to be ferric hydroxide. De Ceukelaire also observed a thin layer of carbonation between the brown zone and the sound concrete.

A detailed chemical description of zonation due to acid attack has been given by Pavlik (1994) who found that attack by nitric and acetic acids gave similar patterns. In the case of nitric acid attack the outer corroded zone largely consists of silica and grades into a brown area where hydrated iron oxide is present. Below this there is a rapid transition to a layer depleted of calcium hydroxide and enriched in sulfate. Attack by acetic acid differs in that in the outer corroded zone the silica content is lower and the zone is preferentially enriched in hydroxides of alumina and iron. No brown layer is reported but the zone

Fig. 6.15 Spalling of concrete due to rusting reinforcement with no visible rust stains; this is not uncommon.

depleted in calcium hydroxide and enriched in sulfate is still present. Unfortunately no microscopic examination was made of the corrosion zones but the results support the generalised scenario of an outer corroded zone, and intermediate reaction zone where iron is concentrated and an inner zone leached of calcium hydroxide.

Strunge and Chatterji (1989) have described and illustrated the texture of surfaces attacked by acid rain. They identified four zones consisting of an innermost zone of unaltered concrete, bordered by a zone of carbonation followed by a zone free of crystalline material. The outermost zone consisted of porous material.

6.5 Corrosion of steel reinforcement

Corrosion of steel reinforcement is perhaps the most common cause of the deterioration of concrete structures. The expansion resulting from rusting of reinforcement causes the surrounding concrete to crack and spall, the bonding between the steel and the concrete is lost and the steel reinforcement itself is weakened. The first visual indications of rusting of the reinforcement are fine cracks followed by unsightly brown rust stains reflecting the pattern of the underlying reinforcement on the concrete surfaces. However, visible rust staining is not always present at the surface and this can make initial diagnosis difficult in such cases. As the damage progresses the staining becomes more severe, the cracks widen and propagate and segments of concrete overlying the rusting steel are spalled off. The macroscopic appearance of concrete damaged by corrosion of steel reinforcement is illustrated in Fig. 6.15.

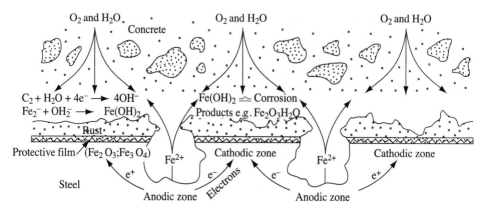

Fig. 6.16 Diagrammatic representation of pitting corrosion of steel in concrete. Oxidation of iron to ferric ions is the principal initial anodic process. At the cathode relatively soluble ferrous hydroxide develops which is then modified and oxidised into the familiar brown rust minerals as hydrated ferric oxides such as goethite and lepidocrocite. (Diagram courtesy of D. Leek.)

6.5.1   STEEL CORROSION MECHANISMS

The corrosion of steel is an electro-chemical process with anodic and cathodic areas developing on the steel as a result of inhomogeneity in the steel itself, for example at locations where the oxide layer is damaged by abrasion with aggregate particles, or as the result of local variations within the adjacent concrete such as permeability or sulfate/chloride concentrations. The electro-chemical cells with the concrete pore fluid as electrolyte may develop on scales ranging from microns to metres in size, depending on the particular circumstances.

Where two dissimilar metals are connected, for example between reinforcement and the tie wires, at welds or where galvanised structural members are in contact with the reinforcement, a corrosion cell can develop with the concrete pore fluid acting as the electrolyte. Corrosion cells on macro- and microscales developing on reinforcing steel need first to break down the protective oxide layer. This layer is generally considered to be a two-tier film with a conduction film of magnetite (Fe_3O_4), in direct contact with the steel, covered in turn by an insulating layer of ferric oxide (Fe_2O_3) (Bloom and Goldberg 1965). This film protects the steel from further corrosion in the alkaline conditions of a sound concrete. Damaged areas of the film may repair themselves, first by the formation of ferrous hydroxide ($Fe(OH)_2$), which will oxidise through intermediate forms, to ferric oxide which is impermeable to ions and so passivates the steel.

If the concrete has carbonated or chloride ions are present, the steel reinforcement will corrode if the oxide layer is damaged and moisture and oxygen are available. The anodic reaction may be expressed by:

$$3Fe + 8OH^- \rightarrow Fe_3O_4 + 4H_2O + 8e^-$$

and the complementary reaction at the cathode by

$$8e^- + 4H_2O + 2O_2 \rightarrow 8OH^-$$

The corrosion cell is illustrated diagrammatically by Fig. 6.16.

6.5.2 CARBONATION AND CORROSION OF REINFORCEMENT

A common cause of corrosion of reinforcement is where carbon dioxide from the atmosphere slowly penetrates the concrete and converts the calcium hydroxide and cementing phases to calcium carbonate. On carbonation reaching the steel reinforcement the calcium hydroxide layer shown in Fig. 5.7 is replaced by calcium carbonate which has a pH of 9.5 which is insufficient to maintain the protective layer on the steel. Provided moisture and oxygen are available corrosion is able to occur (Plate 62). The extent to which carbonation proceeds inwards from the exposed surface depends on the permeability and depth of the concrete cover over the reinforcement, cracking, and the relative humidity of the exposure conditions. The petrographic features of carbonation of concrete have already been discussed in Section 5.4 which is illustrated with micrographs.

6.5.3 CHLORIDES AND CORROSION OF REINFORCEMENT

It is possible to construct a graphical representation of the phase equilibria relationships which exist between metal, the corrosion products and the electrolyte, based on thermodynamic considerations. Such diagrams are referred to as Pourbaix Diagrams, after their originator Pourbaix (1974). They may also be modified to illustrate the corrosion behaviour of iron in solutions containing chloride ions (Cl^-) and can therefore provide a graphical illustration of the influence of chloride ions in steel reinforced concrete. Although such diagrams form a useful means of illustrating the effects of changes in ion type and concentrations, they are based on equilibrium conditions and should not be applied directly to the non-equilibrium dynamic system of reinforcement corrosion in concrete.

The risk of chloride-induced corrosion can arise from a number of possible sources. Salt contamination of the aggregates or reinforcing steel gave rise to numerous corrosion problems in the early developments in the Arabian Gulf region (Kay *et al.* 1982). Alternatively the use of calcium chloride as a setting accelerator in concretes has given rise to corrosion problems and is rarely used in modern concrete for that reason. It also attacks the calcium hydroxide in the concrete and forms calcium oxychloride which can cause disruption of the concrete (Collepardi and Monosi 1992). A third common source of chloride ions is the de-icing salts, either calcium or sodium chloride, used to treat roads in northern latitudes.

The depth of chloride ingress from de-icing salts depends on the ease with which the chloride ions can migrate through the cement paste from the treated surface. Microcracks and permeable concrete assist the diffusion process and the relative concentrations of chloride and hydroxyl ions are important factors. Some of the chloride becomes bound into the cement paste in the form of calcium chloroaluminate hydrate $(3CaO.Al_2O_3.CaCl_2.10H_2O)$ dependent on the amount of aluminate phase present in the cement while the remainder is present as free chloride ions in the pore solution. Chloride ions cause local breakdown of the protective oxide film that passivates the steel even with pH values typical of uncarbonated concrete and these damaged areas then become anodes while the unaffected areas become cathodes. A detailed scanning electron microscope study coupled with microanalysis by Leek (1997) determined elemental concentrations in the regions around the anodic and cathodic areas as illustrated in Fig. 6.17.

Several methods are used to prevent or minimise reinforcement corrosion. These include the use of applied voltage for cathodic protection, the use of sacrificial anodes and corrosion inhibitors which appear to either stabilise the protective film on the steel or absorb oxygen

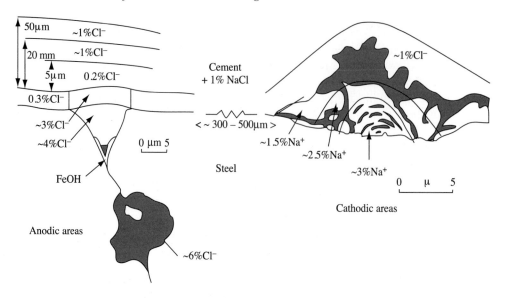

Fig. 6.17 Diagrammatic illustration of chloride and sodium concentrations close to anodic and cathodic areas on a laboratory-prepared specimen treated with a 1 per cent salt solution. The stippled areas represent corrosion products. (Diagram courtesy of D. Leek.)

from the system. Various impermeable organic coatings have been used in attempts to minimise the diffusion of oxygen and moisture into the concrete. Electrolytic methods have been used to realkalise or remove chlorides from concrete to prevent or arrest steel reinforcement corrosion (Vennesland 1995). These methods appear to be successful but may have the disadvantage of activating localised concentrations of sodium ions which may be high enough to activate dormant alkali-aggregate reactions in the concrete (Al-Kadhimi and Banfill 1996).

6.6   Plastic and drying shrinkage

The loss of moisture from a concrete or mortar either during setting or after hardening leads to a volume shrinkage which, if of sufficient magnitude, may lead to the development of cracks. The basic mechanism first involves the loss of free water which causes little or no shrinkage, but as drying continues absorbed water is removed leading to progressive shrinkage. Although the shrinkage is directly proportional to the loss of water for simple laboratory cement paste specimens, aggregate type, grain sizes, mix design and curing conditions all influence the amount and pattern of drying shrinkage.

6.6.1   MACROSCOPIC EFFECTS OF PLASTIC AND DRYING SHRINKAGE

Loss of water due to bleeding and excessive drying during setting may lead to plastic shrinkage cracks which appear within a few hours of the placement of the concrete. Dependent on exposure conditions drying shrinkage develops more slowly over a period from weeks to months once the concrete has set (Fig. 6.18). Commonly drying shrinkage occurs within a few days of casting as a network of fine cracks or crazing restricted to the

Fig. 6.18 Example of drying shrinkage crack at the centre of a wall panel.

cement rich surface layers (Fig. 6.19). Reviews of the effects of shrinkage and the development of cracking in concrete are given by Neville and Brooks (1987), Concrete Society (1982) and Fookes (1977).

6.6.2 PETROGRAPHIC EXAMINATION OF SHRINKAGE CRACKING

Visual examination on site is often sufficient to make a satisfactory diagnosis of the type of cracking and its cause. Petrographic examination of thin-sections cut perpendicular to the surface and including examples of the cracks will allow confirmation of the field diagnosis. Surface crazing does not normally extend more than a millimetre into the surface and is confined to the cement rich surface layer which is usually carbonated. Plastic shrinkage cracks form thin wedges which narrow downward from the surface and may extend from a few millimetres to several centimetres into the concrete. Although they form before the final setting of the concrete the cracking may develop calcium hydroxide crystals on margins. Samples taken from hardened concrete are typically carbonated both on the outer surface and along the margins of the cracks. The depth and distribution of the carbonation will depend on the age of the concrete and its permeability.

The correct diagnosis of the longer-term drying shrinkage cracks will be based on the direct observation made of the structure and on evaluation of the mix design and concrete quality. Petrographic examination of specimens containing the cracking will only confirm that such cracks formed after hardening of the concrete. Calcium hydroxide crystals will not normally be present in cracks unless a crack forms a pathway for moisture which may also allow carbonation of crack margins.

Fig. 6.19 Crazing on the end of a cast *in situ* wall panel.

6.6.3   *SHRINKAGE OF AGGREGATES IN CONCRETE*

Some aggregate types, especially those containing clay and clay-like minerals, are capable of substantial moisture movements themselves, thus contributing to the initial drying shrinkage of concrete and sometimes causing excessive shrinkage cracking. A sandstone aggregate from South Africa was reported to exhibit excessive shrinkage in concrete (Stutterheim 1954) and found by Roper *et al.* (1964) to contain the swelling clay, montmorillonite (smectite). High concrete shrinkage associated with glacial gravel aggregates in Scotland was found to be associated with greywacke, mudstone and sometimes altered basalts and dolerites (Edwards 1970). Cole and Beresford (1980) have described weathered basalt aggregates from Australia that sometimes exhibit considerble moisture instability associated with smectites and chlorite-smectites.

Although potentially shrinkable rock constituents can be identified by routine aggregate petrography, it is common in the UK for a specialised shrinkage test to be carried out. This test, which is a concrete drying shrinkage procedure using standardised mix proportions, was developed by the Building Research Station, following the performance problems encountered in Scotland, and published in their Digest 35 (1963, 1968, 1971). This guidance has been superseded by Building Research Establishment Digest 357 (1991) and the test is now a British Standard method (BS 812 1995).

The pattern of surface cracking of unrestrained concrete caused by aggregate shrinkage is similar to that of conventional drying shrinkage: map cracking. However, the presence of reinforcement will be likely to modify the crack pattern and secondary damage, for example by frost action in Scotland, can greatly worsen and obscure the original shrinkage cracking.

Concrete shrinkage induced or exacerbated by aggregate shrinkage can be diagnosed by petrographical examination of the interior, when aggregate shrinkage creates a fairly distinctive pattern of micro-cracking. Building Research Establishment Digest 357 (1991) suggests that if examination identifies shrinkable rock types as aggregate particles and 'clear evidence' of peripheral cracking (along aggregate particle boundaries) or internal cracking along bedding planes in sedimentary rocks, or both, then 'it is probable that significant shrinkage has occurred'. Care should then be taken to discount other possible causes, for example, delayed ettringite formation (DEF) which can also induce peripheral cracking although caused by cement paste expansion rather than aggregate shrinkage, but when secondary ettringite might also be present and the aggregate types might not be shrinkable in character (see Section 6.2.3.)

6.7   Alkali-aggregate reaction (AAR)

In this section alkali-aggregate reaction (AAR) is used as a general term to cover all the reactions between cement alkalis and aggregates. Where the reacting species is known to be a siliceous component the more restrictive term alkali-silica reaction (ASR) is used.

There is a large literature on the alkali-aggregate reaction which has been referenced by bibliographies published by Figg (1977), British Cement Association (1992), and Diamond (1992). Two books, by Hobbs (1988) and Swamy (1992), also discuss the problem of alkali-aggregate reaction in depth, together with proceedings of ten International Conferences on alkali-aggregate reaction. The following general description of alkali-aggregate reaction has been adapted from two codes of practice published by the Cement and Concrete Association of New Zealand (1991), and the Concrete Society (1995).

The alkali-aggregate reaction is the expansive reaction between the alkalis sodium and potassium in the pore solution of a concrete and minerals in the aggregates. The principal source of alkalis is derived from the cement itself but any source of sodium or potassium can contribute to the reaction provided that the alkali can move into the pore solution of the concrete and create the necessary hydroxyl ion concentrations required.

Three types of reaction are identified in alkali-aggregate reaction:

1. The alkali-silica reaction (ASR): the reaction between alkalis and varieties of silica minerals such as opal, chalcedony, cristobalite and tridymite and volcanic glasses to produce expansive alkali-silica gels.
2. The alkali-silicate reaction: the reaction of alkalis with disordered forms of silica and quartz, often submicroscopic, in sedimentary and metamorphic rocks to produce expansive alkali-silica gels. It was believed that the reaction of this group of rocks was different from (1) above but this is no longer generally accepted and this subdivision appears to be an arbitrary one.
3. The alkali-carbonate reaction (ACR): thin reaction rims at the margins of limestone aggregate particles are not uncommon in concretes but the most common deleterious, expansive reaction involves alkali and argillaceous dolomitic limestone. The expansion is initiated by de-dolomitisation processes in the rock but the mechanism causing expansion has not been clearly identified. The presence of alkali-silica gel is rarely reported and is not necessary for expansion to occur.

The damage resulting from the alkali-silica and alkali-silicate reactions is due to expansion from the formation of alkali-silica gel in the aggregate or at its margins and the subsequent absorption of pore fluid which causes the gel to swell and exert pressure. This internal pressure may be large enough to crack the concrete. The reaction can only proceed if *all* the following three factors are present in the concrete.

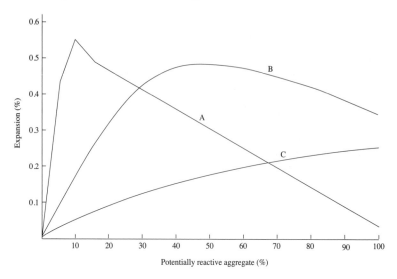

Fig. 6.20 Pessimum proportion curves are shown for three volcanic aggregates. The fresh glassy andesite (C) has no pessimum proportion in contrast to the other andesite (B) which has a broad pessimum. Rhyolite (A) exhibits a more pronounced pessimum. Some aggregates exhibit marked pessimum proportions over a narrow range of concentration. In general most aggregates have pessimum proportion and andesite (C) is not the norm. The curves are the plots of the 12 month expansions obtained from testing some New Zealand volcanic materials by ASTM C227 at 1.5 per cent Na_2O equivalent. (Cement and Concrete Association of New Zealand, 1991.)

1. sufficient moisture, here defined as not less than an 85 per cent relative humidity, in the pore structure of the concrete
2. a sufficiently high alkalinity in the pore fluid surrounding the reacting particle
3. a reactive mineral(s) in the aggregate which often may not react unless it is present in a critical range.

If *any one* of these three factors is absent then the alkali-silica reaction will not proceed. If either (1) or (2) ceases to apply during the course of an existing alkali-silica reaction the reaction will stop but may recommence if the necessary conditions in (1) and/or (2) are again satisfied. Many types of reactive aggregates will only react if they are present in a critical volume proportion range. The proportion of reactive aggregate which produces the maximum expansion for a given alkali content in the concrete is known as the pessimum proportion and typical examples are illustrated in Fig. 6.20.

6.7.1   THE CHEMISTRY OF ALKALI-SILICA REACTIONS

The larger portion of the alkali metal ions in Portland cement are usually present as sodium sulfate (Na_2SO_4), potassium sulfate (K_2SO_4) and as the mixed salts $(Na,K)_2SO_4$ and calcium langbeinite $(K_2Ca_2(SO_4)_3)$ dependent on the amounts and ratio of sodium and potassium present. In this form they are readily soluble. Smaller amounts occur in solid solution in the cement minerals and may be substantially released within twenty-eight days as the cement hydrates. Alkalis may also be introduced as sodium chloride, due to inadequate washing of marine derived aggregates, or less significantly from de-icing salts or deposited salt spray. Admixtures are another potential source of alkalis. Some aggregates themselves also may be potential sources of alkalis.

When water is added to Portland cement the alkali sulfates react with the hydrating tricalcium aluminate and calcium hydroxide to precipitate ettringite, releasing sodium and potassium ions into the pore solution. This pore solution contains considerable concentrations of these ions and very low concentrations of other ions such as calcium, sulfate and chloride. The considerable concentration of alkalis in the pore solution results in a range of hydroxyl concentrations of about pH 13 to 14. It is these hydroxyl concentrations which provide the chemical driving force for alkali-aggregate reaction.

Sodium chloride is readily soluble in the pore solution and reacts with the aluminate phases to form chloroaluminates and the sodium ions are released into the pore solution to further contribute to the overall alkali concentration. The extent to which the chloroaluminates form is dependent on both the concentrations of chloride and the aluminate phases present. At the levels of chloride permitted by many standards in reinforced concrete (0.1 to 0.15 per cent for moist environments), complete conversion to chloroaluminates with release of sodium ions can be assumed. This reaction of sodium chloride can take place in both the plastic and hardened states of concrete.

Alkali-silica reaction is essentially an attack by sodium and potassium ions, with concomitant hydroxyl ions, on varieties of silica to produce alkali-silica gel. The rate of this attack will depend both on the types of silica present and the concentration of the alkalis in the pore solution. Only at the upper end of the pH range does significant attack develop. The gels that form consist of a calcium, sodium, potassium silicate of variable composition. The formation of the swelling form of these gels appears to require a minimum amount of calcium hydroxide to be present in the concrete. Pozzolanas effectively compete for the available calcium hydroxide and this is one of the reasons why they can suppress alkali-silica reaction as without sufficient calcium hydroxide only non-swelling gels are formed. However, Duchesne and Bérubé, (1994) found no correlation between reduction in concrete expansion and calcium hydroxide depletion when testing supplementary cementing materials.

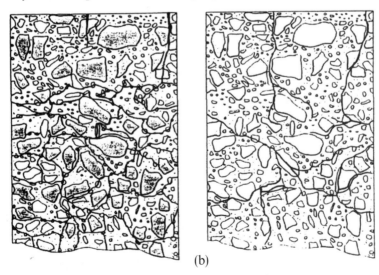

(a) (b)

Fig. 6.21 Internal crack patterns due to caused by alkali-silica reaction. In the earliest stage of reaction the internal pattern cracking is absent and only isolated cracks perpendicular to the external surface will be present terminating at depths of 50 to 75 mm. As the reaction develops the crack patterns shown in (a) and (b) develop. (a) Crack pattern caused by reactive sand particles. (b) Crack pattern caused by reactive minerals in the coarse aggregate (from British Cement Association (1992).

The swelling types of gel are capable of absorbing water into their structure and expanding. It is this expansive pressure which creates tensile strain within the concrete and causes cracking when the local, unrestrained, tensile strength of the concrete is exceeded. The expansive forces are believed rarely to exceed 6 to 7 MPa (Diamond 1989) but the force varies both with the composition of the gel, in a way which is not fully understood, and with the total amount of gel present in the concrete. Spatially, the reaction between aggregate and alkalis to form gel is rarely uniform and nearly always appears as separated point sources of reaction. The point sources are often restricted to a few local areas but in more severe cases will be widespread. It is this tendency to occur as point sources of expansive force that gives the typical, somewhat random to semi-regular pattern of internal cracking seen in concretes affected by alkali-aggregate reaction as shown in Fig. 6.21.

The amount of gel also depends on the amount of available reactive silica and, therefore, up to a point an increase in the amount of reactive silica produces an increase in expansion. Above a certain proportion of reactive silica to alkali so much of the alkali is absorbed that the concomitant concentration of hydroxide in solution is insufficient to maintain the same severity of attack and the expansion decreases again. This is the reason that is given to account for the critical, or 'pessimum', proportion that occurs with so many aggregates (Concrete Society 1995, Cement and Concrete Association of New Zealand 1991).

6.7.2   EXTERNAL APPEARANCE OF CONCRETE AFFECTED BY ALKALI-SILICA REACTION

Cracking caused by the alkali-silica reaction is never distributed uniformly over the surface of a concrete even in test specimens stored under closely controlled laboratory conditions. In the field its appearance can vary from just one or two cracks, to isolated areas of pattern cracking and in severe cases to extensive pattern cracking being evident. It is common to find that only a limited portion of the surfaces appear to be affected while the remainder of the structure shows little sign of the reaction in spite of the same mix design and concreting materials ostensibly being used throughout. It is also common to find the reaction is more severe on the weather side of the structure due to the greater extent of wetting and drying (Wood *et al.* 1996). In severe cases the differential expansion of the concrete is sufficient to cause visible displacement of structural elements (Fig. 6.22). Where concrete is heavily reinforced, cracking may be wholly restrained or only present in the direction of the reinforcement. Examples of various types of external cracking due to alkali-silica reaction are illustrated in British Cement Association (1992) and the assessment of expansion using crack widths has been discussed by Jones and Clark (1996).

Where the alkali-silica reaction is still active cracks will often be open and margins may be vertically displaced. On fine cracks this displacement can be felt by rubbing the finger tips across the crack. Older cracks may be filled with carbonated calcium hydroxide which can be difficult to distinguish from carbonated gel. Exudations of alkali-silica gel are not common and where present indicate that there has been sufficient moisture to carry the gel through to the surface. This exuded gel is usually white in colour because the originally transparent gel carbonates to white crystalline material on contact with air. Removal of exudation followed by microscopic examination in a suitable refractive index oil will sometimes show clear gel is present beneath the carbonated exterior. The texture of carbonated gel in thin-section is shown in Plate 63.

Although gel may not be visible, sufficient may have been absorbed at crack margins to darken them, especially when the concrete surface is wetted. Thus enhanced delineation of cracks may be noted after rain which persists for some period after the remainder of

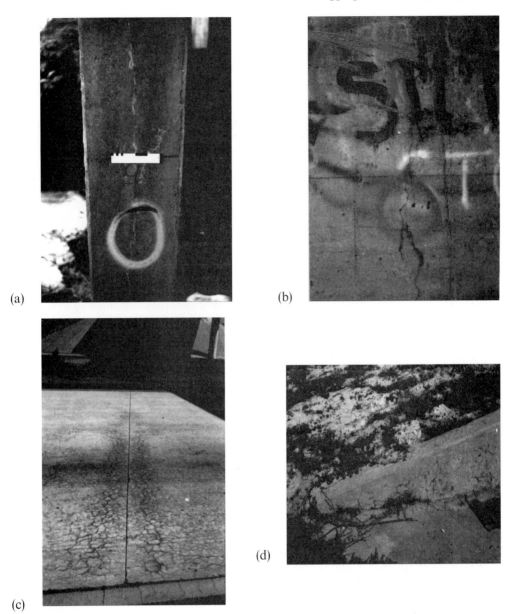

(a)

(b)

(c)

(d)

Fig. 6.22 The appearance of external cracking due to moderate alkali-aggregate reaction in a temperate climate. (a) The slender pier of this twenty-year-old bridge has cracking restricted to a few areas parallel to its length probably due to pre-stressing. The alkali-silica reaction is in its earliest stage and no gel was visible in thin-section. Because the margins of the crack are carbonated they appear white. (b) More typical signs of alkali-silica reaction are the localised somewhat irregular cracks with darkened crack margins. The pier foot of this thirty-year-old bridge sitting in a tidal zone of brackish water was more severely cracked. Internally, clear signs of a minor to moderate alkali-silica reaction were present. (c) The pavement of this South African motorway is extensively pattern microcracked especially around the transverse joints. It is difficult to assign a cause to this cracking from visual examination as the cracking could be due to a number of causes exacerbated by vehicular loading. (d) However, examination of an adjacent drainage culvert, containing similar concrete reveals the typical, somewhat random pattern cracking typical of alkali-aggregate reaction. Note that the crack margins are darkened. Thin-section examination of core samples from the pavement revealed an extensive and severe alkali-aggregate reaction was occurring.

(e)   (f)

Fig. 6.22 The apearance of external cracking due to moderate alkali-aggregate reaction in a temperate climate. (e) Looking across the twenty-five-year old hardstanding area of this airbase displacement of one half of the hardstand relative to the other is clearly apparent along the construction joints. This differential expansion between reacting and non-reacting slabs continues unabated. (f) However, signs of visible cracking are limited and unspectacular. This is in spite of the slab being unreinforced apart from transfer pegs at joints. Thin-section examination of core samples from the hardstand identified the non-reaction and reactive areas of the pavement and showed that a severe alkali-silica reaction was occurring. (Whenuapai airbase hardstanding, constructed 1965.)

the concrete surface has dried. Also there may be a tendency for a pink/brown coloration to form along the margins of cracks, possibly due to alkali bleaching the lichens which may be present on the concrete surfaces.

A number of other surface features can occur with alkali-aggregate reaction. Where sufficient moisture is present and the alkalis are not depleted from the outer layer of the concrete, pop-outs can occur due to the formation and swelling of gel. Pop-outs sometimes occur in laboratory specimens which have been continuously moist cured but are uncommon in structures. In some cases where a structural member is heavily restrained by reinforcing the only evidence of alkali-aggregate reaction will be internal evidence of gel and rarely surface pop-outs. A rare occurrence of staining with pop-outs associated with granite aggregate has been reported as an alkali-aggregate reaction by Mindess and Gilley (1973).

Where unexplained cracking is observed in a structure, especially if some of the features as described above are visible, core sampling and further investigation should be undertaken. The sites for this sampling must be selected with care, as described in Sections 3.5 and 6.7.10, to obtain a reliable diagnosis of the cause(s) of cracking.

6.7.3 MICROSCOPIC TEXTURES OF ALKALI-SILICA REACTION IN CONCRETE

The microscopic textures of concretes affected by alkali-silica reaction are variable and depend on the extent of the reaction, the size and distribution of the reacting aggregates and the environment of the concrete structure. The three types of feature observed in thin-section by the petrographer are cracking, alkali-silica gel and signs of reaction in aggregate particles. These should *all be present* for unequivocal identification of alkali-silica reaction in thin-section. In practice, meaningful observation is possible only if adequate sampling has been performed and sufficient surface areas of the axial planes of the core samples are examined (Sims 1992, Palmer 1992).

6.7.4 INTERNAL TEXTURE OF CRACKING IN CONCRETE AFFECTED BY ALKALI-SILICA REACTION

In the early stages, cracking caused by alkali-silica reaction will be limited to a few isolated cracks which may penetrate up to 50–75 mm into the concrete. Examination in thin-section may fail to observe gel, any cracking apart from the surface cracks and reacting aggregates. This type of cracking occurs because the expansive effects of the alkali-silica reaction have not generally exceeded the tensile strength of the concrete. However the outer surface of a concrete is depleted in alkalis early in its life due to leaching and wetting and drying effects. Thus alkali-silica reaction is absent from the outer 50–75 mm of concrete and this outer shell is in tension with respect to the inner portion of the concrete where the expansive reaction is occurring. This situation typically leads to the localised formation of a few isolated cracks in the outer shell in the early stages of alkali-silica reaction (see pages 234–7). Cases of this type of cracking have been reported by St John (1990) where all the conditions for alkali-silica reaction are present and no other cause for the cracking can be found.

With increasing development of reaction and consequent expansion internal cracking develops and the frequency of surface cracking also increases but this surface cracking has a more widely spaced pattern than that found internally. A detailed description of this type of cracking is given in Section 5.8.4. Finally, in cases of severe expansion, major cracks millimetres in width may develop which penetrate deeply into or through the concrete. The location of a major crack is largely controlled by the configuration, design and loading of the structure and the way in which it interacts to accommodate the large internal expansion caused by the alkali-silica reaction.

A complex pattern of cracking will be found on the surface where environmental conditions have not caused leaching of alkalis from the concrete shell. In these cases, fairly closely spaced, fine pattern cracking will be present on the surface and there will be little gradation in the spacing and frequency of the cracking from surface to the interior. In addition, both gel and reacting aggregates will be found in the outer shell and pop-outs may occur.

The above textural description of the cracking due to alkali-silica reaction is for unrestrained concrete exposed to temperate weather conditions. Where frost attack occurs cracking initiated by alkali-silica reaction is enhanced and may even lead to severe disintegration of the concrete. If restraint is present, most commonly due to the presence of steel reinforcing, cracking of the concrete will be partially or completely suppressed in the direction of the restraint. Internal microcracking is also affected by the size and shape of the reacting aggregate as shown in Fig. 6.21. Where a portion of the fine aggregate has reacted the spacing and widths of the internal microcracking are smaller than where coarse aggregate has reacted (Palmer 1992).

6.7.5 MICROSCOPIC TEXTURE OF ALKALI-SILICA GEL IN CONCRETE AFFECTED BY ALKALI-SILICA REACTION

The alkali-silica gel found in concrete is a material of variable composition consisting of SiO_2, Ca, Na and K together with other minor elements (Poole 1992). In thin-section typical alkali-silica gel is colourless, usually conchoidally cracked due to drying of the gel in sample preparation, and has a mean refractive index of 1.48–1.49 with an upper limit of 1.51 and a lower limit of 1.45 (Idorn 1961). When it is deposited in cracks and pores it is distinctive in texture and immediately recognisable in thin-section. However, in older concretes the appearance of gel is more variable than as described above.

Perhaps the most distinctive variation in the texture of the gel is that present in reacting aggregates. In the interior of reacted aggregates cracks are usually empty apart from crack entrances which are often plugged by layers of granular, birefringent material, clear gel and dark gel as shown in Plate 64. The three layers have been shown to be similar in chemical composition and probably represent stages in the formation of the gel and its interaction with the cement paste as described by Davies and Oberholster (1986), Andersen and Thaulow (1990), and Katayama and Bragg (1996). However, the granular, birefringent layer is not always present and may be difficult to observe in some types of reactive aggregate and where the aggregate particles are small.

In the cracks and voids of the hardened paste alkali-silica gel can also be variable in texture and colour and is prone to crystallisation. This crystallisation has a much lower birefringence than the granular, birefringent material discussed above. The differences are illustrated in Plate 64 and Plate 65. The darkening of gel in contact with cement paste has already been mentioned. This darkening can range from a light tan to a deep brown-black colour. Gel is sensitive to carbonation so that cracks in the outer shell of concrete associated with alkali-silica reaction are often carbonated to depths considerably greater than would be expected. In general, older gels tend to be more crystallised but the circumstances in which gel crystallises are not fully understood and as shown in Plate 65 clear gel and crystallised gel can be in intimate contact. A possible explanation for gel crystallisation has been proposed by Dron and Brivot (1996). It should not be assumed that all gel in contact with cement paste is darkened as clear gel may be observed in cracks running through the hardened paste. Gel can soak into the hardened paste at the margins of aggregates and cracks causing an apparent darkening probably due to the absorption of calcium hydroxide.

In the early stages of alkali-silica reaction the fact that alkali-silica gel is not observed does not necessarily mean that it is absent. Unless gel is present in cracks or voids it can be difficult to observe in thin-section with the optical microscope. This effect can be demonstrated by visual observation of fracture surfaces of core samples where in some cases reaction rims may be clearly visible around some aggregates (Plate 66). However, when the core sample is thin-sectioned these reaction rims often effectively disappear and cannot be observed. For this reason the amount of gel observed in thin-section frequently cannot be correlated with the severity of the alkali-silica reaction and work by Johansen *et al.* (1994) indicates that damage may occur when limited amounts of gel form. Martineau *et al.* (1996) also found that it is not possible systematically to relate expansions to the quantity of gel found by fluorescence.

It is difficult to make meaningful observations on the mobility of gel. There is no doubt that under appropriate conditions gel can move considerable distances but the relationship of gel viscosity varies with aggregate type, point in the life of the structure and the local environment in the concrete. When gel is observed in cracks and voids with no obvious aggregate source it must be remembered that the thin-section is a plane and the actual source may be close by, just below or above the plane of the section. In a well-developed alkali-silica reaction innocuous aggregates may also be cracked because they cannot withstand the expansive forces generated. Mobile gel can flow into these aggregate cracks but should not be taken as evidence that the aggregate has reacted.

Ettringite is commonly found intimately associated with alkali-silica gel but much of this ettringite is observable only with the aid of the scanning electron microscope and cannot be seen in thin-section. There are a range of microproducts associated with gel that have been observed with the scanning electron microscope and while these are of importance in elucidating the mechanism of reaction and gel formation they are generally not observable by the light microscope (Regourd-Moranville 1989). However, bands and accretions of ettringite are also quite commonly found in the cracks and voids of concretes undergoing alkali-silica reaction not necessarily intimately associated with gel and these are observable in thin-section. There is a report by Marfil and Maiza (1993) of the identification of a zeolite of the heulandite-clinoptolite group in microfissures and pores of a concrete highly fissured by alkali-aggregate reaction. No details are given of the aggregates. The micrographs given illustrate material that could easily be confused with ettringite.

The presence of alkali-silica gel is one of the more important observations required for the identification of alkali-silica reaction. Lightweight and cellular concretes are claimed to have sufficient void space to accommodate the gel without expansion but this claim must be treated with caution as there is still no clear understanding of whether the expansive pressures generated are due to the formation of the gel or to its subsequent absorption of water. However, there is experimental evidence to show that expansion decreases with increasing porosity of aggregates and that air entrainment has a similar effect (Collins and Bareham 1987, Jensen *et al.* 1984).

6.7.6 TYPES OF REACTIVE MINERALS AND THEIR MICROSCOPIC IDENTIFICATION

Alkali-silica reaction occurs between the alkaline pore fluid of a concrete and one or more reactive minerals in the aggregate. In many cases the reactive mineral forms a small part of the aggregate or alternatively, only a small portion of the reactive mineral present in the aggregate is available for reaction. Since the reactive mineral will be dispersed in the rock texture, it would be expected that aggregate porosity would be important in that it may effectively control the rate of access of the pore fluid and in addition may affect the magnitude of the expansive forces generated. For example, Rayment (1992) found that the porous cortex was the most reactive component of the flint fragments within UK sand and gravel aggregates. Paradoxically, there is evidence indicating that the denser forms of some aggregates are more reactive (Hobbs 1988).

In alkali-silica reaction there are only two generalised classes of minerals that are known to be expansively reactive with the alkalis in concrete. These are metastable types of silica including some disordered forms of quartz, and alumino-silicate glasses with a silica content usually in excess of 60–65 per cent and not less than 50 per cent (Katayama *et al.* 1989). The natural forms of metastable silica are opal, chalcedony, tridymite and cristobalite (Plate 67) and of the alumino-silicate glasses the matrix of fresh intermediate to acid volcanic rocks (Plate 68). Most commercial glasses, including glass fibres and mineral wools (Keriené and Kaminskas 1993) are also reactive. For instance, the commercial alumino-silicate glass known as 'Pyrex' is so reactive that it is used as the standard reactive aggregate for comparison purposes in ASTM C 441. Synthetic cristobalite can also be used for the same purpose (Lumley 1989, Berra *et al.* 1995).

The identification of opal, tridymite and cristobalite and of the glass matrix of fresh acid/intermediate volcanic rocks, theoretically presents few problems to the petrographer. The practical problem is that small quantities of some of these reactive minerals can be important. It has been shown that as little as 1–2 per cent of opaline material (Stanton *et al.* 1942) and only a few percent of tridymite and cristobalite can cause alkali-silica reaction

(Kennerley and St John 1988). Thus care is required when carrying out petrographic examination of aggregates likely to contain these minerals to see that they are not overlooked. Where doubt exists about the presence of opal, cristobalite and tridymite they can be quantitatively determined as phosphoric acid residues (Talvitie 1964, Cement Association of Japan 1971). A BS gel pat method for the detection of opal in aggregates has recently been developed (BSI 1996).

Determination of the potential reactivity of chalcedony in flints and cherts, disordered forms of quartz, and alteration of the glassy matrix of volcanics present petrographic problems for which there are currently no ready solutions. The petrographer is rarely able to relate the presence of these minerals to potential reactivity. In the case of chalcedony which is often submicroscopic, its presence does not necessarily indicate potential reactivity. Alternatively, the Jones and Tarleton (1958) gel pat test can be used to identify these highly reactive minerals.

Work by Rayment (1992, 1996) has shown that other forms of disordered silica present in English flints and cherts are the more likely cause of reactivity and Groves *et al.* (1987) identified reactive, disordered structures in opal by transmission electron microscopy. Similarly the degree of strain in quartz has been claimed to correlate with reactivity (Dolar-Mantuani 1983) but this is now believed to be incorrect and there are claims that submicroscopic silica in veins and at crystal boundaries is responsible (Grattan-Bellew 1992, Kerrick and Hooton 1992). Thomson *et al.* (1994) investigated the microscopic and X-ray diffraction analysis of potentially reactive rocks and concluded that there is a good relationship between expansivity and a crystallinity index of quartz and that silica dissolution could be related to grain size and crystal defects and deformation. They also claimed that petrographic examination is effective in predicting reactivity. Similarly Wigum (1995, 1996) has correlated microstructural features of cataclastic rocks with their alkali reactivity. The techniques for measuring strain in quartz have been reviewed by Smith *et al.* (1992) and a new method was proposed that eliminates some of the uncertainty in the measurements. The effect on reactivity of the degree of alteration and devitrification of the glassy matrix of volcanic rocks is not certain, however, extensively altered and devitrified volcanic rocks are often non-reactive.

This inability to correlate petrographically chalcedony, disordered silica and altered glasses with potential reactivity has resulted in a range of chemical, mortar and concrete tests being developed to determine the reactivity of aggregates. These have been reviewed by Grattan-Bellew (1989, 1996) and Nixon and Sims (1992). ASTM C289 has often proved unreliable for rocks containing chalcedony and disordered forms of silica but modification of this test by Sorrentino *et al.* (1992) shows promise in overcoming some of the problems. In Denmark a chemical shrinkage test is used to identify reactive flints and cherts (Knudsen 1992), and in Germany a sodium hydroxide dissolution test is used for the detection of opaline materials. Because the mortar bar test, ASTM C227, can give variable results when testing some of these rocks an alternative mortar test carried out at 80°C (Oberholster and Davies 1986) has been developed as ASTM C1260 (1994).

While the tests referred to above are of varying adequacy the petrographer is faced with the problem when examining concretes undergoing alkali-silica reaction of often being unable precisely to identify the reactive mineral(s) in the aggregate. However, while the reactive mineral(s) may not be identifiable, examination of concrete in thin-section does indicate whether aggregates as opposed to minerals have reacted. Such information from structures can be of vital importance in identifying reactive aggregates and indicating the need for laboratory testing. Where satisfactory results from testing aggregates for potential reactivity are not available identification from field concretes may be the only source of reliable information.

6.7.7 TEXTURE OF REACTING AGGREGATES

Because of the problems associated with the identification of reactive minerals the texture of reacting aggregates has assumed some importance in petrographic examination. Idorn (1961, 1967) was one of the first to use thin-sections to investigate systematically the texture of reacting aggregates and has published many excellent micrographs of disintegrating flints and cherts together with some data on phyllites, basalts and shale. Since then there have been numerous petrographic investigations of concrete but illustrations of textures of reacting aggregates tend to be limited. No other systematic study such as that undertaken by Idorn appears to be readily available.

Attack in its simplest form consists of a simple major crack system transecting the aggregate with the crack normally being empty of gel apart from crack entrances which may be plugged by gel layers as described earlier in Plate 64. Reaction rims and disintegration of rock texture are not observed. It might be thought that this simple cracking is typical of the early stages of reaction but this is not the case. For a surprising number of aggregates such as andesites, dacites, greywackes, sandstones, quartzites, granites and coarser grained rocks in general, the visible attack on the rock texture is limited no matter how advanced the alkali-silica reaction (Plate 69(b)).

At the other extreme are those aggregates that are severely disintegrated due to alkali-silica reaction. Much of the internal texture may have been consumed by the reaction which deposits gelatinous material inside a microcracked shell. The finer-grained rocks such as flints, cherts, argillites, shales, phyllites and rhyolites often exhibit this texture which is typical of those rocks where the reactive mineral(s) form a substantial portion of the rock texture (Plate 69(a)). Good examples of this type of disintegration texture in flints are given by Idorn (1967). However, in the earlier stages of alkali-silica reaction these aggregates may exhibit simple cracking with little disintegration so that a range of reaction textures may be found even in the same thin-section.

Reaction rims are rarely observable in thin-section and should not be confused with rims due to weathering (Plate 66). The presence of weathering is easily checked by observing unreacted aggregates present in the section. To observe the presence of reaction rims it is necessary to examine a fracture surface of the concrete preferably supplemented by testing for the fluorescence of gel with uranium salts as described by Natesaiyer and Hover (1989) or to analyse the rims directly by electron probe microanalysis.

In the matrix of glassy volcanic rocks and fine matrix of sedimentary rocks the texture often gives the appearance of having been consumed by the alkali-silica reaction, with an abrupt transition to unaffected matrix. In some cases delineation with fluorescent dyes shows the presence of fine cracks through the unaffected matrix leading to the attacked areas. Where veins of alteration feeding into cracks are observed, which could possibly be due to alkali-silica reaction, it is often the case that these same veins can also be found in unreacted aggregates (Fig. 6.23) and so the relationship of this alteration to the alkali-silica reaction cannot be established.

6.7.8 REACTION OF LIGHTWEIGHT CONCRETES

Concretes containing significant proportions of lightweight aggregates such as pumice, scoria, expanded shales, sintered fly ash and also concretes with substantial air void contents

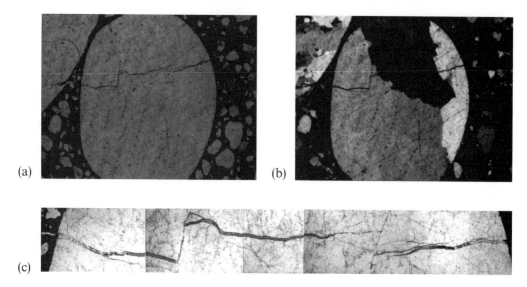

Fig. 6.23 Veins of alteration and reaction in a quartzite aggregate. (a) The round quartzite aggregate is cracked. The entrances to the cracks are plugged with gel as illustrated in Plate 64. The remainder of the crack is filled with brown material which is vaguely birefringent. Width of field 8.5 mm. (b) Under crossed polars the matamorphic character of the quartzite is apparent. Width of field 8.5 mm. (c) A mosaic detail of the crack shows that it is difficult to distinguish reaction from alteration. Similar veins of brown material could be observed in other aggregates where reaction was absent. However, interpretation in these cases is limited by observation being restricted to the plane of the section; length 4.2 mm.

frequently do not undergo expansion even if alkali-silica gel is present. This is based on work by Vivian (1947) as well as field experience over the years. It is possible that the substantial void space in the aggregates or air void system can accommodate the gel and any resulting expansion. However there are a few cases reported where lightweight concretes have expanded and caused problems.

De Ceukelaire (1992) has reported on a bridge in Belgium where chert, present in the sand used in a lightweight concrete bridge, reacted and expanded to such an extent that the bridge had to be replaced. The expanded shale coarse aggregate was not able to accommodate the gel sufficiently to prevent expansion. In Kansas, USA, Crumpton (1988) reported that several bridge decks made with expanded shale lightweight aggregate had to have as much as a third of a metre or more of their length sawn off because of concrete expansion. Initially expansion was slow and alkali-aggregate reaction was not evident but as expansion continued abundant alkali-silica gel was found. It is not stated whether normal weight materials were also present in these concretes. St John and Smith (1976) found that test specimens made with expanded argillite aggregate slowly expanded in spite of low-alkali cement having been used. No alkali-silica gel was observed. Thus expansion in lightweight concretes can occur under appropriate conditions. The problem is that these conditions have as yet not been defined and it is not always clear whether other aggregates may be involved.

The petrographic examination of lightweight concrete is more difficult than normal weight concrete because of the glassy texture of the aggregates used. In the case of expanded shales the aggregate margins are defined by a denser oxidised rim of material which is reasonably smooth and presents a definable interface with the cement paste providing it has not been crushed. Natural pumice and scoria are usually crushed materials and the interface with the hardened paste is highly irregular. The glassy texture in thin-section has a poorly defined texture and it can be difficult to trace cracks and differentiate alkali-silica gel as shown in Plate 70. In addition, both surface carbonation and carbonation down cracks can be more extensive than in normal weight concretes as described in Section 5.4.5. In the more severe cases the carbonation can obscure the concrete texture and make observation difficult.

6.7.9 THE INTERPRETATION OF THE TEXTURES OF ALKALI-SILICA REACTION

If *adequate* sampling and petrographic examination of concrete are carried out the only cases where unequivocal identification of alkali-silica reaction is not possible are those where the reaction is either in its earliest stage or the concrete contains considerable void space. It is the experience of the authors, gained from examination of a wide range of concretes from many countries, that identification of alkali-silica reaction is not a problem providing the procedures in Section 6.7.10 are followed (St John 1985). Difficulties in making unequivocal identification usually indicates faulty petrographic practice and there may be problems in determining the extent to which any damage can be attributed to the alkali-silica reaction identified (Sims 1996).

Identification of any alkali-silica reaction is only one of the aims of petrographic examination and, as discussed in Section 6.1, determination of the quality of the concrete is necessary for interpretation. This is especially the case where the concrete is exposed to severe climatic conditions, aggressive environments and other causes of deterioration. In cases where damage is the result of more than one cause the problem of deciding which is the primary cause can be difficult but is considerably aided by determining the quality of the concrete. If no other cause of damage is identified, the extent and pattern of cracking is a reasonable indicator of the severity of the alkali-silica reaction in the concrete. The minimum extent of localised strain in two dimensions can be crudely estimated for an expanding concrete from summing the crack widths on selected orthogonal axes but this requires a planar area of at least 75×75 mm and is only practicable in thin-section. Such estimates should be treated with caution as many other factors may influence the results.

Since it is only possible to observe gel in open spaces such as voids, cracks and fissures and the conditions under which alkali-silica gel crystallises are not well understood, statements such as 'the gel is old' or 'the reaction has undergone several episodes because of gel layering' should be made with caution. Petrographic examination cannot provide information about whether alkali-silica reaction has ceased, or whether residual expansion is still possible. This requires the expansive movement in the structure and residual expansion on core samples to be determined (Thaulow and Geiker 1992). Nielson (1994) has proposed a method based on dimensions of the structural member for determining if and when repair will be required in cases of alkali-silica reaction and Chrisp *et al.* (1994) a method of non-destructive testing to quantify damage due to alkali-aggregate reaction.

From thin-section examination it can be concluded that some aggregate particles in the concrete have reacted but any statement about aggregates being innocuous needs to be made with caution and requires supplementary data to define the conditions. Supplementary data such as chemical analysis of the concrete to determine alkali content, information on possible pessimum proportion, and data from condition surveys are required to validate statements made about aggregates being innocuous. In this respect the use of fluorescent uranium salt solutions on fracture surfaces and examination by the scanning electron microscope may show gel and attack inside aggregates that are not apparent from microscopic examination and these techniques should always be considered in cases of doubt.

The interpretation of thin-section examination of concretes undergoing alkali-aggregate reaction should not be extended to making statements about structural integrity. While a concrete may be severely micro-cracked internally, aggregate interlock, localisation of cracking, restraint by reinforcement and other factors can reduce losses in strength. It is now generally agreed that the structural effects of alkali-aggregate reaction, apart from movement, have been over-estimated but this does not obviate the need for adequate structural assessment (Institution of Structural Engineers 1992, Walton 1993). Structures undergoing alkali-aggregate reaction do not collapse even though they may require repair and in severe cases replacement. There is only one case of collapse reported and that involved both alkali-aggregate reaction and sulfate attack (Haavik and Mielenz 1991).

6.7.10   EXAMINATION OF CONCRETES SUSPECTED OF ALKALI-SILICA REACTION

Failure to follow necessary procedures when carrying out a petrographic examination of a concrete suspected of undergoing alkali-silica reaction may lead to erroneous conclusions. If a number of cores, all taken from an area of concrete with moderate cracking due to alkali-silica reaction are examined, a marked variability will be found. In some of the cores little if any cracking and gel will be found while in others the reaction is clearly present and commensurate with the external cracking observed. Careful selection of sampling sites is imperative and the petrographer should insist on inspecting the structure and choosing the sample sites. If this is not possible selection must be carried out by personnel who have the skill to make the necessary judgements.

Experience indicates that the outer 150–200 mm of a core sample frequently provides the most information on quality, weathering, external attack and alkali-silica reaction. If the external portions of a core sample are missing the core ideally should be rejected for petrographic examination as the relationship of alkali-silica reaction to external factors cannot be ascertained. If facilities for large-area thin-sectioning (minimum size 75 × 75 mm) of concrete are not available it may be better only to use thin-sections for aggregate identification and to rely on examination of ground surfaces which can be impregnated with fluorescent dyed epoxy resins. While this may give less information than can be obtained from examination of large-area thin-sections it will not lead to erroneous conclusions because too small an area has been examined. Sims *et al.* (1992) have discussed the sampling and methods required for quantifying the microscopic examination of alkali-aggregate reaction.

6.7.11   OTHER FACTORS TO BE CONSIDERED

One of the questions that inevitably arises in the investigation of alkali-silica reaction is the amount of alkali contributed by the cement to the concrete. Most aggregates require $3.0 \, kg/m^3$ of Na_2O equivalent or greater for significant reaction to occur although a number of aggregates have been reported that require as little as $2.0 \, kg/m^3$ (Nixon and Sims 1992,

Woolf 1952). Consider a hypothetical example of a typical bridge concrete supposedly containing 360 kg/m³ of cement which contains an alkali content of 0.55 per cent as Na_2O equivalent. This concrete, if correctly mixed, would have initially contained 2.0 kg/m³ of alkali. However, parts of the bridge concrete are now undergoing significant alkali-silica reaction in spite of laboratory data indicating that significant reaction of the aggregate used requires at least 3.0 kg/m³ of alkali.

There are at least four possibilities to consider in this case. The first and most important possibility is that a high-alkali cement was used instead of the low-alkali cement specified. This occurs more commonly than would be expected. The second is the possibility of the laboratory test data being wrong and in fact the aggregate is significantly reactive at the lower alkali content. Many aggregates are difficult to test in the laboratory because the results are sensitive to pessimum proportion of the reactive component, and to leaching of alkali from the test specimen during storage (Rogers and Hooton 1989). The third possibility is that alkali has been preferentially concentrated in one portion of the concrete although petrography should resolve this question because the signs of reaction will be restricted to those layers. The fourth possibility is that there is another source of alkali apart from the cement. This alkali can be derived externally from salts or be internally derived from aggregates, mineral admixtures or salt contamination of aggregates. These are all possibilities which must be eliminated for the effective analysis of alkali-aggregate reaction in concrete.

One method of determining the alkali content of a concrete is to identify the origin of the cement, determine the cement content by petrography and use alkali results from historical cement analyses (Section 2.17). Where this technique can be applied it may unequivocally resolve the question of the alkali content due to the cement irrespective of whether construction records are available.

The only other alternative is to try to estimate the alkali content by analysing the concrete. Suitable methods of analysis are discussed in Section 2.17. Great care is required in sampling to avoid areas of concrete which are affected by leaching of alkalis. Surface areas which are invariably leached by wetting and drying and also localised areas of concrete undergoing significant alkali-silica reaction should be avoided. On most structures it is possible to find less reacted areas immediately adjacent to severe areas of cracking which are likely to contain levels of alkalis more representative of the concrete as mixed.

Other internal sources of alkali in concrete are salt contamination from marine derived aggregate and the alkali content present in some superplasticisers. Determination of the chloride content from deep inside a concrete will indicate the amount of salt contamination from aggregate because salt spray and de-icing salts are normally limited to penetration depths of about 100 mm in better-quality concretes. Identification of a superplasticiser in an old concrete is very difficult and there is no standard technique available for quantification. This is discussed in Section 2.17. The amounts of alkali that may be contributed from chemical additions have been estimated by the Cement and Concrete Association of New Zealand (1991).

A controversial question is whether the aggregates themselves can contribute alkalis to the concrete because of the alkali-aggregate reaction attacking aggregates. The possibility that volcanic aggregates will release alkali into the pore solution without themselves undergoing expansive alkali-aggregate reaction has been reported by St John and Goguel (1992). In particular, some New Zealand basalts have released sufficient alkali to initiate expansive reaction of other aggregates present in concrete structures. Goguel (1996) and Bérubé, *et al.* (1996) found that a range of aggregates may release alkalis. Gillott and Rogers

(1994) report that dawsonite ($NaAlCO_3(OH)_2$) releases alkali by reaction with calcium hydroxide in the pore solution of concrete although expansion may have also resulted from increases in the solid volume of the dawsonite reaction itself.

There are a number of conditions which may impact on the severity of alkali-aggregate reaction which must be considered in the petrographic examination. These are frost attack leading to freezing and thawing of the concrete, frequent condensation on surfaces typically under desert conditions, salt-laden winds with minimal rainfall and pavements, especially those which are effectively sealed on their underside restricting moisture movement to the top surface.

The relationship of damage due to freezing and thawing of concrete and alkali-aggregate reaction is not well documented. Where the surface of a concrete is disintegrating the primary cause must be frost attack even if alkali-aggregate reaction is present. Alkali-aggregate reaction does not by itself usually cause disintegration of surfaces apart from pop-outs. However, more severe cracking seems to occur in countries with cold winter climates and this suggests that freezing and thawing is opening up the cracks initiated by alkali-aggregate reaction to depths which may be controlled by moisture penetration (Bérubé *et al.* 1996). The contribution to the overall damage in the structure from freezing and thawing can be evaluated only by characterising the air void system and measuring the porosity and permeability of the concrete and the frost susceptibility of the aggregates (Hudec 1991).

Desert conditions can provide a very severe environment for concrete. Hot days followed by cold nights lead to considerable condensation on concrete surfaces. This daily condensation constitutes a pumping effect on the moisture in the surface of the concrete. If salt deposition occurs, as in parts of the Arabian peninsula, these salts are transported into the concrete and both augmentation of alkali-aggregate reaction and sulfate attack may occur (Fookes and Collis 1975, Fookes 1993). Salt-laden winds are common in areas adjacent to the sea and may penetrate considerable distances inland. The salt deposits on surfaces and is available to move into the concrete when dew forms, causing corrosion of reinforcement and augmenting the alkalis in the concrete. In wet temperate climates, such as New Zealand, this does not appear to be a problem as chloride levels measured in exposed concrete indicate that frequent rain is removing the salt before it can penetrate. Unless the salt deposition is severe its effect on alkali-silica reaction is probably not significant.

Pavements, because they are relatively thin and have large horizontal surfaces, are more sensitive to climatic conditions and deposition of salts. If the pavement is effectively sealed on its bottom surface, moisture and salts can only move in and out of the concrete via the exposed upper surfaces. There is little information about what effect this configuration has on the distribution of alkalis and its relationship to alkali-aggregate reaction. Concentration of alkalis causing alkali-silica reaction in a pavement has been reported by Hadley (1968) and in laboratory specimens by Nixon *et al.* (1979). Torii *et al.* (1996) found that cathodic protection significantly increased expansion in concrete containing reactive aggregate. Hobbs (1988) argues that increases in alkali concentration from these causes are likely to be only transitory and will not be significant in alkali-silica reaction. Given the difficulties in making reliable alkali analyses of concrete published data should be treated with caution.

The effect of temperature on alkali-silica reaction has been investigated by Poole (1992). Temperatures of 38°C lead to faster rates of reaction while temperatures of 10–15°C, which are more typical of ground temperatures, give slower reaction as would be expected. However, the total amount of expansion is sometimes reported to be greater at the lower temperature. This suggests that concrete below ground should be more severely cracked than above ground.

Expansion essentially ceases in a concrete that has dried to an internal relative humidity of about 80 per cent or less. Stark (1985) measured relative humidities in some dams in the south-western United States and found that only the concrete within a few inches of exposed surfaces was sufficiently dry to preclude expansion even under desert conditions. It seems to be accepted that the larger structural concrete members do not dry out sufficiently to prevent expansion. Some data on the factors influencing the relative humidity in concrete have been reported by Parrott (1991).

6.7.12 THE ALKALI-CARBONATE REACTION

The alkali-carbonate reaction is principally the interaction of cement alkalis with argillaceous dolomitic limestones. It should be distinguished from the reaction of siliceous limestones where there is observable alkali-silica gel and cracking formed by attack on the siliceous component of the rock. The dividing line between the alkali-silica and alkali-carbonate reactions is considered by some workers to be arbitrary (Katayama 1992) but most workers in this field still consider it to be a separate reaction.

Reactive argillaceous dolomitic rocks in a quarry may be in beds of limited depth and represent only a small proportion of the output of the quarry. Thus the proportion of reactive dolomitic aggregate present can be highly variable which is a complicating factor in petrographic examination. The incidence of the alkali-carbonate reaction is not as widespread as the alkali-silica reaction but can be suspected wherever argillaceous dolomitic rocks are interbedded with other types of dolomite or limestone (Dolar-Mantuani 1983).

There are three reactions associated with carbonate rocks. These are:

1. The expansive reaction by cement alkalis with impure dolomitic limestones. Dolomite contents range from as little as 5 per cent to over 80 per cent. The dolomite crystals, which average about 50 μm in size, are set in a fine-grained matrix of calcite, usually less than 5 μm in size, together with disseminated clay and silica. The clays are usually illite and mixed-layer chlorite and the silica consists of various forms of quartz some of which may be disordered and reactive. These clays and silica comprise the acid insoluble residue of the rock which often tests as reactive by ASTM C289. There is some evidence to suggest that a minimum of 5–10 per cent insoluble residue is required for expansive alkali-carbonate reaction to occur in concrete (Katayama 1992).

    The alkaline pore solution in the concrete reacts with the dolomite ($CaMg(CO_3)_2$) to de-dolomitise the rock, form brucite ($Mg(OH)_2$) and calcite and regenerate the hydroxyl ion (OH^-).

    $$CaMg(CO_3)_2 + 2OH^- \rightarrow Mg(OH)_2 + CaCO_3 + CO_3^{2-}$$
    $$CO_3^{2-} + Ca(OH)_2 \rightarrow CaCO_3 + 2OH^-$$

    It is claimed that this formation of brucite and calcite is not expansive and it is believed that de-dolomitisation opens up the texture and facilitates reaction of the alkalis with the matrix. It has been proposed that the actual expansion is due to the re-wetting of the clays in the matrix of the dolomite (Gillott and Swenson 1969). However, this is disputed by Tang Mingshu (1992), who considers that the formation of the brucite and calcite is expansive because of a topochemical *in situ* reaction. Katayama (1992) considers neither of the above theories adequate and has suggested that reaction of disordered silica in the matrix of the dolomite to produce alkali-silica gels is the cause of the expansion. In later papers Deng Min and Tang Minshu (1993) and Tang Mingshu *et al.* (1994) claim that expansion occurs due to de-dolomitisation and that the rate and amount of expansion increases with increasing pH. Prince *et al.* (1994) showed that the products formed from de-dolomitisation depend on equilibria which is controlled by the calcium:sodium ratio in the pore solution. Higher ratios lead to the formation of calcite and brucite and lower values to the formation of brucite and pirssonite, $CaNa_2(CO_3)_2.2H_2O$.

Alkali-carbonate rocks do not exhibit pessimum proportion and may be reactive with cement alkali contents as low as 0.4 per cent Na_2O equivalent. Marked rimming of aggregates is common. There is one major difference from the alkali-silica reaction. Alkali-silica gel has rarely been observed in expansive concretes containing these impure dolomitic limestones. It has been suggested (Katayama 1992) that the gel is fixed by brucite to form the aggregate rims commonly observed which are enriched in silica (Poole 1981). However, brucite is extremely insoluble and any interaction between hydrous silicate anions and magnesium cations will be very slow. Thus it is surprising that alkali-silica gel is not observed more often.

2. The expansive reaction by cement alkalis with siliceous limestones to give a classical alkali-silica reaction. The siliceous portion of the limestone produces alkali-silica gels. It should be possible for both alkali-silica and alkali-carbonate reactions to be present in the same concrete if both types of aggregate are present. Such a concurrence of ASR has been reported for some Chinese aggregates (Tong Liang and Tang Mingshu, 1996, Tang Mingshu *et al.* 1996).

3. The alkaline dissolution of the surface of a crystalline limestone to form a rim of birefringent material that has the appearance of calcium hydroxide. This reaction is not expansive. The following reaction has been suggested:

$$CaCO_3 + 2OH^- \rightarrow Ca(OH)_2 + CO_3^{2-}$$

However, this reaction is unlikely as calcium hydroxide is more soluble than calcium carbonate. Monteiro and Mehta (1986) identified the rims as being the basic calcium carbonate $2CaCO_3.Ca(OH)_2.1.5H_2O$. Schimmel (1970), who originally identified this basic carbonate, does not give optical or solubility data for this compound but presumably it is less soluble than the calcium carbonate and has similar optical properties to calcium hydroxide.

6.7.13   *PETROGRAPHIC TEXTURE OF CONCRETES CONTAINING EXPANSIVE ARGILLACEOUS DOLOMITIC LIMESTONES*

The external appearance of concrete undergoing an expansive alkali-carbonate reaction is of somewhat finely spaced map cracking as shown in Fig. 6.24. It resembles a concrete undergoing alkali-silica reaction which has been exposed to continuously moist conditions although frost may open up some of the cracks. External deposits of alkali-silica gel are extremely rare and if they are present suggest the possibility that an alkali-silica reaction must be considered.

The internal texture of concrete undergoing alkali-carbonate reaction is not well described in the literature especially with respect to the relationship of the outer shell of the concrete to the interior. It would be expected that the outer shell would be less reacted due to the leaching of alkalis by weathering as for other types of alkali-aggregate reaction but the

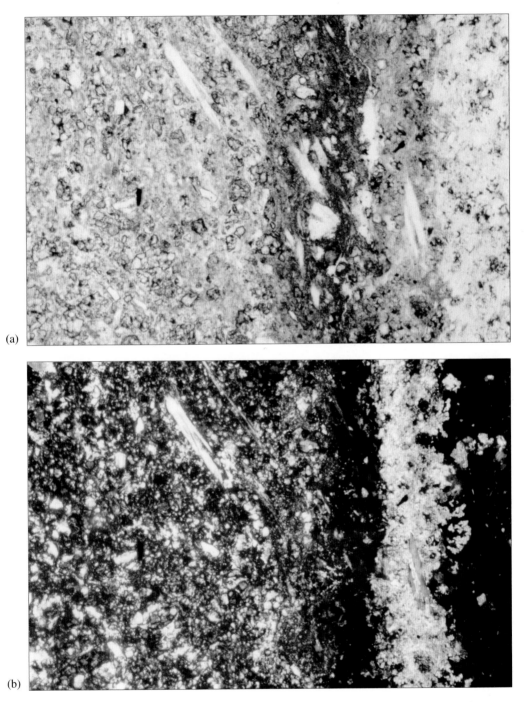

(a)

(b)

Plate 72 The texture of asbestos cement pipe. Width of fields 1.3 mm. (a) This is the region of an asbestos cement pipe which has been attacked by soft water. A few bundles of asbestos fibre are oriented sub parallel to the length of the pipe. The outer surface is to the right. (b) Under crossed polars the texture is seen to consist of quartz flour embedded in a dark matrix containing minimal calcium hydroxide, with the asbestos fibres being outlined by their birefringence. The texture of soft water attack consisting of corroded, carbonated and leached zones is clearly visible.

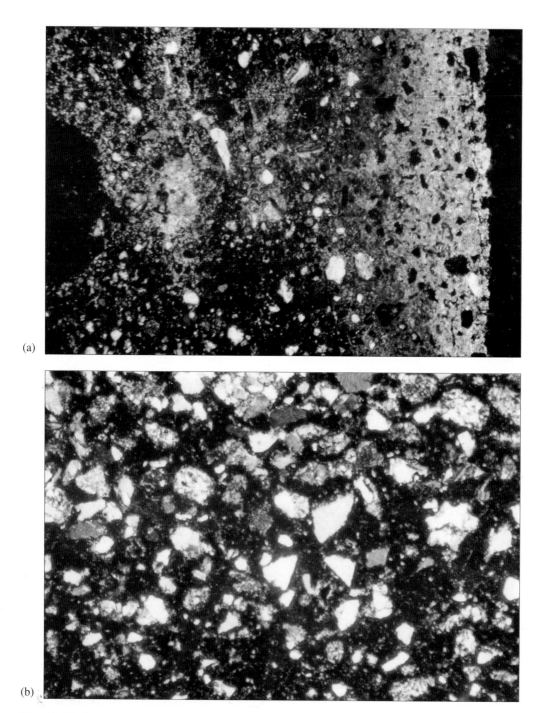

Plate 73 The internal textures of spun pipe. Width of fields 1.15 mm. (a) Crossed polars. The concrete adjacent to the internal surface of the pipe consists of sludge with some remnants of fine sand. This texture extends to a depth of about 15 mm. Carbonation extends to a depth of about 0.5-1.0 mm and a shrinkage crack is carbonated to its full depth with a patch of carbonation at its base. This is an undesirable texture. Inner surface to the right. (b) Crossed polars. In contrast, the texture of the concrete adjacent to the inner surface in this swell-made pipe consists of a rich mortar with closely packed sand. The right-hand margin of the field is about 2 mm from the inner surface. This is the type of texture required for these products.

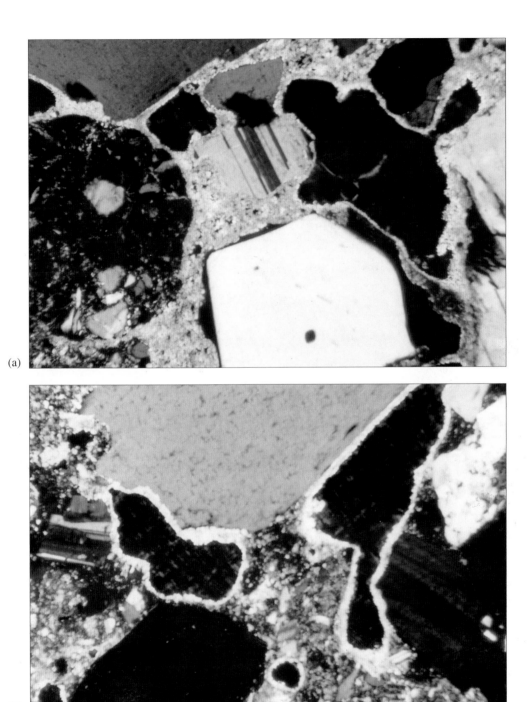

(a)

(b)

Plate 74 The internal texture of concrete brick. Width of fields 1.4 mm. (a) Crossed polars. Several of the typical large pores that are scattered throughout the texture of a concrete brick are outlined by the intense carbonation near the surface. This brick was made by a Columbia machine and cured at 80°C in a carbon dioxide enriched atmosphere. There are remnants of a glass rim around the feldspar aggregate at the bottom of the micrograph. (b) Crossed polars. A few mm inside the brick the carbonation curing has outlined the margins of pores. The hardened paste fills the remaining space between the densely packed aggregate and is only partially carbonated. Carbonation is absent deeper than a few mm from the surface. The texture is of a well-compacted concrete with insufficient cement paste to fill all the space between the aggregates.

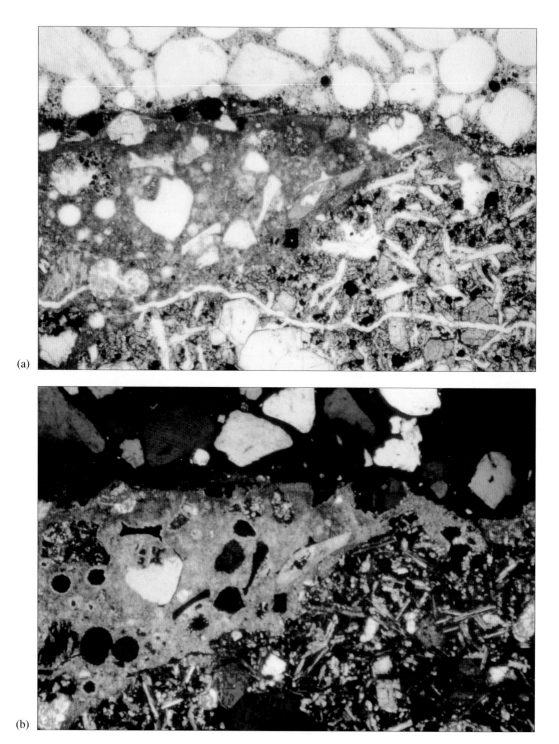

(a)

(b)

Plate 75 The texture of an epoxy grout used for repair. Width of fields 1.35 mm. (a) The epoxy grout filled with quartz and fly-ash is to the top. (b) Under crossed polars the carbonation of the concrete in the top of the crane support wall is evident. In spite of this the adhesion of the epoxy grout is excellent.

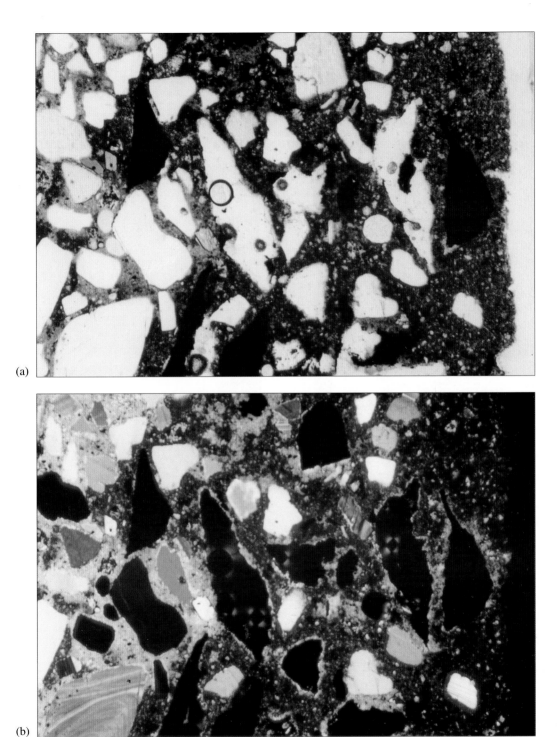

(a)

(b)

Plate 76 The texture of a proprietary floor finish. Width of fields 0.35 mm. (a) A thin layer of floor finish containing fragments of iron and aggregate has been worked into the surface of the concrete. (b) Under crossed polars it is evident that the surface of the original concrete is carbonated before the finish was laid and that the floor finish is partially porous and has allowed some carbonation to penetrate. Top surface to the right.

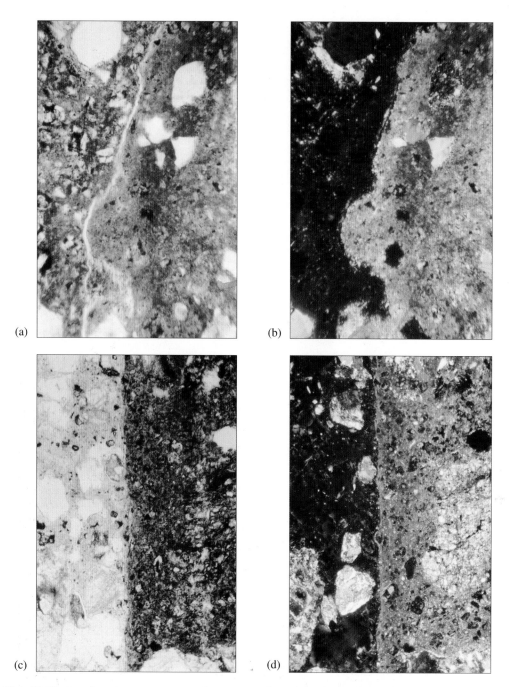

Plate 77 The interfaces between mortar and concrete and mortar and polymer topping. Width of fields 0.9 mm. (a) The mortar to the left is separated from the irregular surface of the concrete by a narrow fissure probably caused by differential movements. This type of fissuring at mortar/concrete interfaces is quite common and in this case extends along a considerable portion of the interface. It is possible that some interlock occurs across the fissure due to its irregular nature. (b) Under crossed polars the surface carbonation of the concrete clearly defines the interface. (c) The polymer topping is to the left and has clearly bonded to the mortar beneath without any fissuring. (d) Under crossed polars the lightly carbonated mortar defines the interface. The polymer topping is isotropic with numerous voids and a quartz filler. Top surfaces are to the left.

(a)

(b)

Plate 78 The textures of a 20-year-old sandlime brick. Width of fields 1.7 mm. (a) Under crossed polars bright calcium hydroxide crystals border the quartz grains and the low birefringent matrix consists of calcium hydroxide and CSH. Only minimal carbonation is present. (b) Under crossed polars the texture is almost wholly carbonated apart from areas marked X between some quartz grains which are assumed to be poorly crystalline CSH(I).

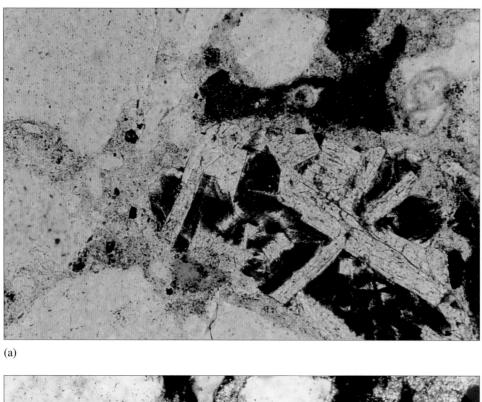

(a)

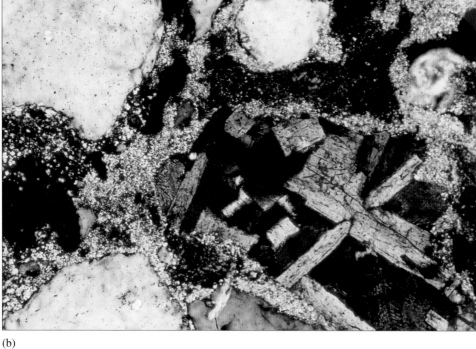

(b)

Plate 79 An unusual crystallisation of products in sandlime brick. Width of fields 0.75 mm. (a) A void is infilled with clear laths and plates and felted greenish brown crystals. (b) Under crossed polars the laths and plates have low birefringence and are possible dicalcium silicate hydrate and the acicular felted mass of green crystals are possibly actinolite. The matrix is partially carbonated.

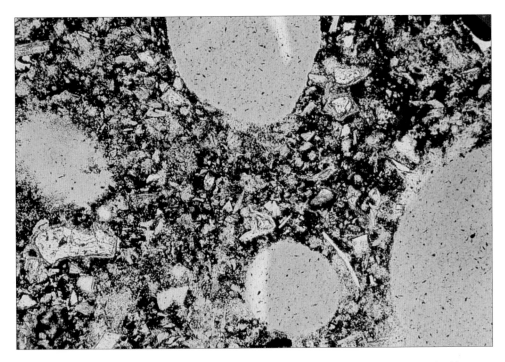

Plate 80 Particles of slag in a foamed concrete block. The isotropic slag particles are rimmed with very small tabular crystals of hydrotalcite [$Mg_6Al_2(OH)_{16}CO_3.4H_2O$]. The air voids are coloured with red dye. Width of field 1.6 mm.

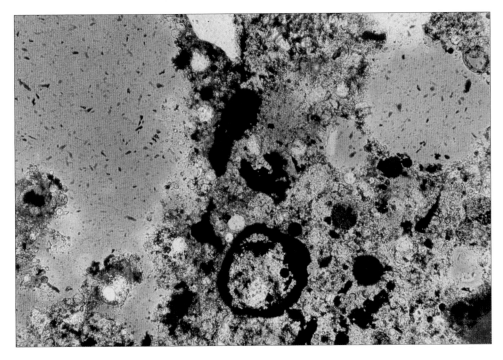

Plate 81 The texture of a lightweight foamed masonry block containing quartz and fly-ash. The transparent hollow spheres and iron-rich particles of the fly-ash stand out from the hardened cement matrix. Pores and porous areas are coloured with red dye. Width of field 0.8 mm.

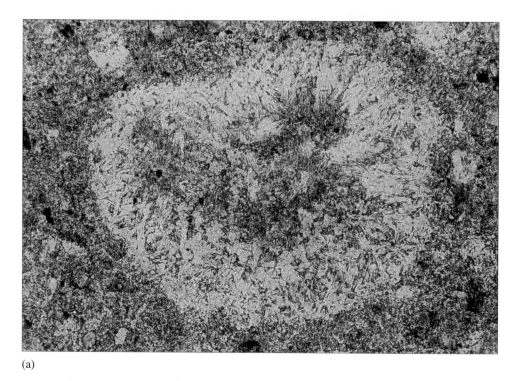

(a)

(b)

Plate 82 Visible gypsum crystals in a void. Width of fields 0.8 mm. (a) This void is large enough to allow the gypsum crystals to grow into visible fibrous laths which is their typical form. Around the void the gypsum crystals are too small to be identifiable. (b) Under crossed polars the low birefringence of the gypsum is shown.

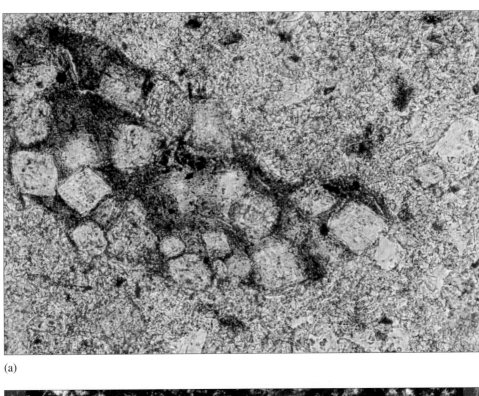

(a)

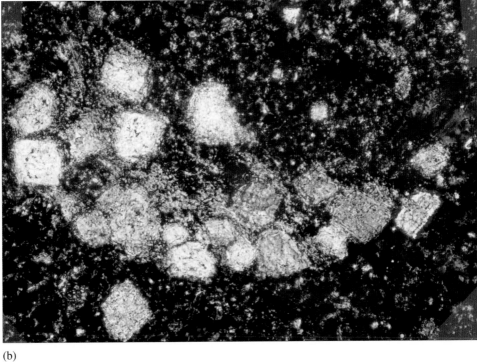

(b)

Plate 83 Anhydrite crystals in hardwall plaster. Width of fields 0.45 mm. (a) The iron-stained particle contains well-formed rectangular-shaped crystals of anhydrite. (b) Under crossed polars the anhydrite crystals show moderate birefringence.

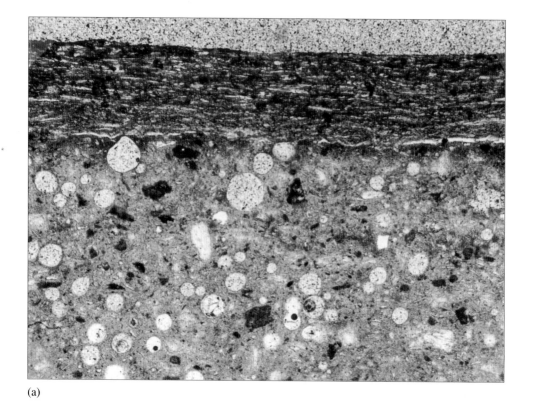

(a)

(b)

Plate 84 Texture of a paper-faced gypsum plaster board. (a) The paper facing stained red by dye and the air void system in this Gyproc plaster board are illustrated. Width of field 2 mm. (b) At higher magnification details of the air voids and the brown aggregate are shown. Crystals of gypsum are not large enough to be visible. Width of field 0.8 mm.

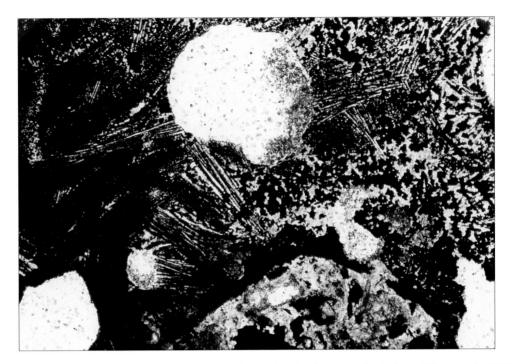

Plate 85 Porous boiler clinker with an opaque matrix showing development of acicular melilite crystals. Width of field 1.5 mm.

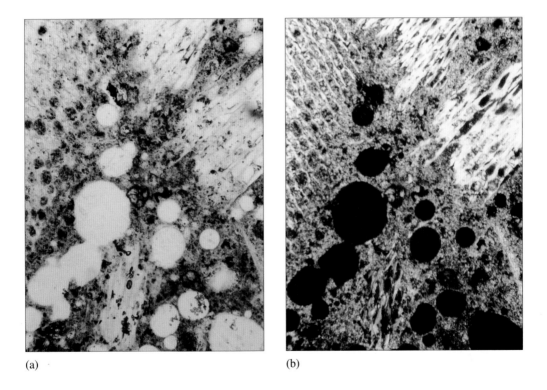

(a)   (b)

Plate 86 The texture of sawdust concrete. Width of fields 0.9 mm. (a) The chips of sawdust are in intimate contact with the hardened cement matrix which contains an extensive air void system. (b) Under crossed polars the cellular structure of the wood and the typical birefringence of the cellulose is shown. Because the product is porous the hardened matrix is well carbonated.

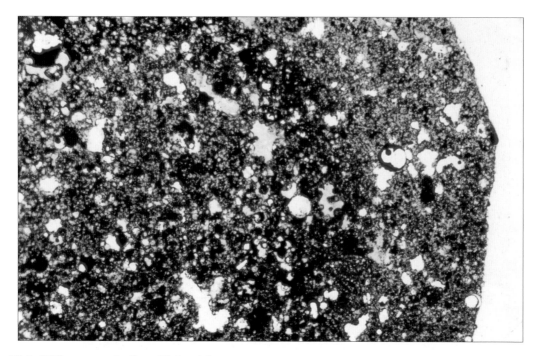

Plate 87 The texture of a 'Lytag' lightweight aggregate made from pulverised fly-ash. Some remnants of fly-ash spheres are still present. Note the zonal structure of the aggregate, the lack of interconnection between the pores and the outer sealed surface. This type of aggregate has only limited vesicularity making it suitable for structural lightweight concrete. Width of field 3.5 mm.

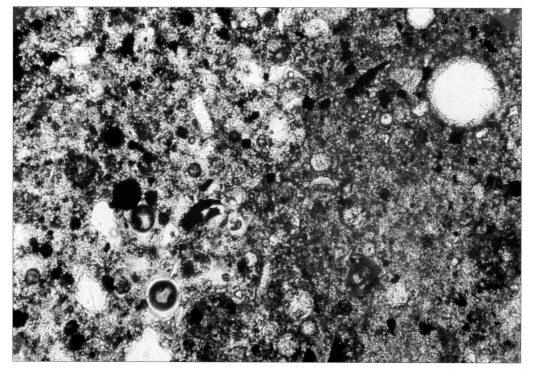

Plate 88 'Lytag' concrete. The 'Lytag' aggregate is to the left. In the concrete to the right some voids contain ettringite and possibly thaumasite crystals. Note the excellent contact between the 'Lytag' and the hardened cement matrix. Width of field 0.8 mm.

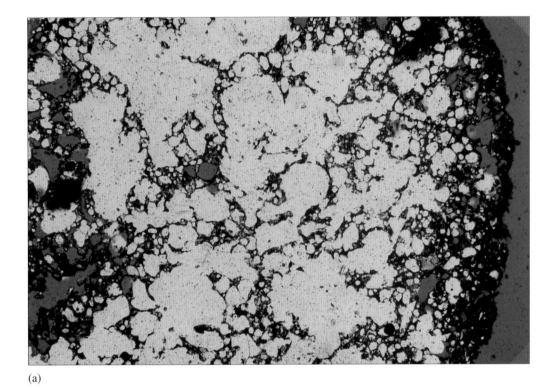

(a)

(b)

Plate 89 The textures of expanded clay aggregates. (a) The brown oxidised rim and the open vesicular structure are typical of expanded clay aggregates that are heavily bloated. Width of field 13 mm. (b) This expanded clay aggregate has been only partially bloated and has much less vesicular structure with some areas of unexpanded clay. Width of field 5 mm.

Fig. 6.24 External pattern cracking due to the alkali-carbonate reaction usually appears as a mesh of closely spaced polygonal cracks somewhat similar to cracking in alkali-silica reaction where the concrete has been continuously moist cured. This Canadian footpath contains Kingston limestone, one of the classical aggregates which undergo alkali-carbonate reaction.

fine mesh of cracks visible externally suggests that this may not be an operative factor. The coarse aggregate produces most of the expansion and cracking which is in contrast to many other reactive aggregates where greater reaction is usually found in the fine aggregate. Cracks originate in the aggregates and extend into the hardened paste and in severe cases form the typical network of expansive cracking. Gel plugs at crack entrances are not observed. Ettringite is occasionally observed but is not common. In some cases cracks may contain a poorly defined reaction product which suggests movement of materials has occurred. Alkali-silica gel is absent although there are reports of gel being present in some cases (Katayama 1992).

The most distinctive features of the alkali-carbonate reaction are the presence of cracking associated with argillaceous dolomitic limestone in the coarse aggregate. The expansive dolomitic limestones are claimed to have a distinctive texture (Dolar-Mantuani 1983). This texture consists of well-developed rhombohedra of dolomite, in the size range 100–25 μm or less, enveloped by a fine-grained matrix of calcite, quartz and clay. In practice reliable recognition of this texture may require previous petrographic experience with these rocks. If in doubt some of the aggregates can be removed from the concrete and analysed for dolomite by X-ray diffraction or chemical analysis of the aggregates carried out. Alternatively, staining with copper nitrate solution may be used to identify the presence of dolomite (Fairbanks 1925).

Another feature of the alkali-carbonate reaction is the formation of rims of carbonation around the margins of aggregates (Mather *et al.* 1964). This comprises a soft, friable, buff-coloured zone in the surrounding concrete paste consisting mostly of fine grained calcite (Sims and Poole 1980). The carbonation rims do not surround all aggregates or necessarily completely ring an aggregate. Examination of a test specimen in thin-section seemed to indicate clearly that these carbonate rims are not due to carbonation by carbon dioxide from the atmosphere. Many rims are isolated by uncarbonated hardened paste suggesting they are an *in situ* formation of calcite from reaction of the alkalis with the dolomitic aggregate.

6.7.14   SUPPLEMENTARY TESTS

Because the mechanism of the alkali-carbonate reaction is still in doubt petrographers have relied heavily on supplementary tests for the identification of reactive dolomitic rocks. Testing dolomite rocks using ASTM C289 rarely gives a positive result although it is claimed that if the rock is first dissolved in acid and the insoluble residue tested this gives more reliable results (Katayama 1992). This assumes that expansion in concrete is essentially caused by alkali-silica reaction. Until recently the most accepted test was to immerse rock cylinders or prisms in a 1 molar solution of sodium hydroxide and measure the volume change as specified in ASTM C586. However, this test gives false results for some types of dolomitic rocks, the so-called late-expanding type (Dolar-Mantuani 1983), which do not expand in concrete. An autoclave test has been proposed by Tang Mingshu *et al.* (1994). Recently ASTM has developed a concrete prism test for alkali carbonate reactive aggregates (ASTM C1105-95).

6.7.15   PETROGRAPHIC TEXTURES OF THE ALKALI-CARBONATE REACTION

Identification of reactive dolomitic rocks in deteriorated concretes requires that other causes of expansion be eliminated. This elimination process can be difficult. It requires a large sampling area to allow careful appraisal of all the aggregates present, changes in texture of the hardened paste from the outside surface inwards, the type, distribution and frequency of aggregate rims, nature of the cracking and the possible presence of gel and other alteration products. Fracture surfaces should be tested for the presence of gel with uranyl salts as proposed by Natesaiyer and Hover (1989).

6.8   Fire-damaged concrete

Normal concretes are incombustible and exhibit good resistance to fire although they are not classed as refractory materials. Nevertheless, heating due to fire or other causes will induce alterations to the mineralogy and fabric of the material. These alterations are dependent on the maximum temperature reached, the length of time at that temperature and the rate of heating and cooling or quenching if water is used to extinguish the fire.

The severity of the deterioration will also be partly dependent on the cross-sections of the concrete elements involved because of the thermal gradients developed and the differential thermal expansions between the various components within the concrete, including steel reinforcement if present. The effects of fire damage can include discoloration, surface cracking, surface spalling and dehydration of calcium hydroxide and other cementitious phases. Internal cracking within the cement paste, cracking around and across aggregate particles and distortion and loss of bonding to reinforcing steel also occur.

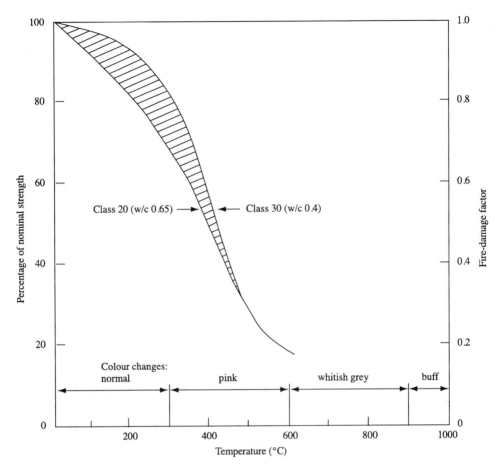

Fig. 6.25 Reduction in strength with temperature of normal weight concrete and relation to colour changes (Concrete Society 1990).

Factors such as these together with melting points of external objects, provide important evidence concerning the maximum temperature that the damaged portions of a structural element have attained in a fire. This allows the engineer to estimate the amount of concrete to be removed and replaced to allow reinstatement of the member since strength loss can be correlated with temperature as illustrated by Fig. 6.25. A review of methods of assessing temperatures reached during fire damage to concrete is given by Kramf and Haksever (1986) and Concrete Society (1990).

6.8.1 *THERMAL EXPANSION OF CONCRETE*

A cement paste sample first expands on heating as a result of normal thermal expansion, but this effect is counteracted by shrinkage of the hydrates as water is driven off with this effect eventually becoming dominant. The temperature at which maximum expansion occurs depends on specimen size and heating conditions but may be as high as 300°C for rapidly heated air-dried specimens. At a higher temperature, shrinkage continues and can produce reductions in dimension of 0.5 per cent or more resulting in severe cracking. Calcium hydroxide loses water above 400–450°C to form calcium oxide. If this calcium oxide rehydrates due to the presence of moisture there is considerable volume expansion which may disrupt the concrete.

The aggregates present in mortars and concretes typically undergo progressive expansion as temperature rises opposing the shrinkage induced in the cement paste. These opposing effects tend to weaken and crack the concrete. There is considerable variation in the thermal expansive behaviour of the aggregate types, with fine-grained basic rocks such as basalt or dolerite showing the best fire-resistant properties. Individual crystal grains in aggregate particles can exhibit anisotropic thermal behaviour. Quartz for example has a coefficient of linear expansion α of 7.5×10^6 parallel to and 13.7×10^6 perpendicular to the crystallographic axis. Such variations can lead to thermal stressing and cracking of aggregate particles particularly if the crystal grains are relatively large.

These effects can be detected in concretes which have been heated to temperatures in excess of about 180°C. There is an additional problem with quartz, which is a common constituent of sand and gravels, in that there is a significant increase in volume associated with the phase transition from α to β quartz at 573°C. This expansion has a disruptive effect on a concrete containing quartz chert or flint aggregates. Sandstone aggregates do not usually cause such severe cracking because the intergranular cement between the quartz grains tends to shrink on heating, thus in part compensating the α to β expansion. In general terms the various expansion and shrinkage effects in a concrete are illustrated in Fig. 6.26.

The differences in the thermal properties of the cement paste and aggregate cause stresses to be developed in the concrete. Normally, for temperatures below about 250°C these stresses are not large enough to cause significant cracking, though on cooling there is a residual contraction which is essentially a drying shrinkage of the cement paste. At higher temperatures cracking becomes increasingly severe so that on cooling from high temperature a residual expansion and irreversible damage remain.

6.8.2   MINERALOGICAL CHANGES ON HEATING

The cement hydrates in the hardened matrix of concrete undergo a progressive loss of water with heating with resultant loss of strength. Initially only the free water is lost with minimal loss in strength but with increasing temperatures over 200°C the cement hydrates start to lose their bound water and by 400°C this loss is serious and calcium hydroxide starts to dehydroxylate. By 500°C the cement hydrates are largely dehydrated and by 650°C over 80 per cent loss in strength may have occurred (Malhotra 1956).

6.8.3   VISUAL EFFECTS OF FIRE DAMAGE

Depending on the severity of the heating, the nature of the aggregates and concrete mix design, a number of fire damage features may be observed.

There is a change in colour from grey to pink or red which occurs at temperatures in the range 250°C–300°C and gives a visual indication of a significant reduction in the concrete's residual strength. The colour change is thought to be dependent on the presence of iron compounds such as limonite, goethite and lepidocrocite in the aggregate which all dehydrate and oxidise in this temperature range and if these minerals are not present the colour change may not occur. The intensity of pink coloration is dependent on aggregate type with flint aggregate concretes perhaps producing the most striking colour change. At

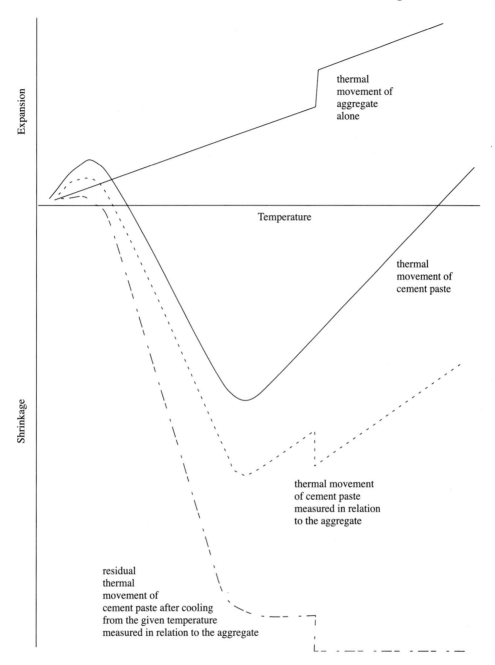

Fig. 6.26 Expansion and shrinkage of cement paste and aggregate when heated (Zoldners 1968).

temperatures in the region of 500–600°C a further colour change from pink/red to purple/grey occurs. Bessey (1950) suggests that this change results from the reaction between ferric oxide and lime forming lighter-coloured calcium ferrites. This suggested mechanism is supported by the observation that concretes which do not contain much lime remain pink up to 1000°C. The colour changes that occur are illustrated by Plate 71 which is of test pieces of concrete heated to a series of different temperatures in a furnace.

Spalling of the surface layers of concrete is a common effect of fire damage and often tends to give an exaggerated impression of its true extent. Surface spalling may be grouped into two types. Explosive spalling is erratic and typically occurs during the first 30 minutes or so of exposure to fire. A slower spalling occurs as cracks form parallel to the fire-affected surfaces leading to a gradual separation of concrete layers and detachment of a section of concrete along some plane of weakness such as a plane of steel reinforcement. According to Castillo and Durrani (1990), spalling of high-strength concrete under load is most likely in the 320–360°C temperature range. The mechanisms of spalling are discussed by Harmathy (1968). He suggests that the thermal gradients will drive water vapour inwards from the surface leading to a moisture saturated layer parallel to the heated surface. If this saturated layer cannot migrate inwards quickly enough it will be overtaken by the advancing heat front vaporising the moisture with consequent pressure build up and spalling. The decomposition of a limestone aggregate to calcium oxide with the liberation of carbon dioxide takes place at about 900°C and this has been cited as the reason for the degradation of concretes containing them. However, the reaction is slow and fire tests indicate that unless the heating is prolonged only the surface material is decomposed.

6.8.4   PETROGRAPHIC STUDY OF FIRE-DAMAGED CONCRETE

The differential expansions and contractions within the fabric of a concrete that has been exposed to fire introduces cracking on various scales, loss of bond to steel reinforcement, and decomposition of the cement matrix. Depending on the scale of the cracking, the nature and pattern of the cracks may be studied using either cut and polished slab specimens or appropriate petrographic thin-sections cut from the affected structure.

The temperature gradient is steepest over the first 20–30 mm from the surface exposed to heating and consequently develops the most severe cracking. These near surface cracks tend to reduce the thermal conductivity of this zone which may then provide partial protection to the inner concrete. Typically in fire-damaged structures, surface spalling and the high thermal conductivity of reinforcing steel ($48/Wm^{-1} K^{-1}$ for carbon steel as against $1.5/Wm^{-1} K^{-1}$ for concrete) reduce the effectiveness of a low conductivity surface zone.

In addition to the development of cracking, changes to the mineralogical composition of the cement paste also occur as the concrete is exposed to heat, as has been noted above. The most important changes are the dehydration and later possible rehydration of the calcium hydroxide phase and cement hydrates. There is also the possible dehydration of iron compounds in aggregate particles and the calcining of limestone aggregate or carbonated concrete if sufficiently high temperatures are maintained for a prolonged period. X-ray diffraction and electron microscopic techniques allow most of these changes to be identified. However, because of the fine grain size of the minerals in a cement paste, only gross changes may be observed with the petrological microscope.

6.8.5   OPTICAL PETROGRAPHIC EXAMINATION

Exposure to temperatures below about 250°C causes little significant change in the general appearance of a concrete in petrographic thin-section, though scanning electron microscopy and observation of the effects of low-stress cycling of samples indicate additional microcracks may develop in concrete heated at temperatures as low as 180°C. Above this temperature

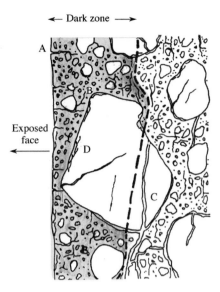

Fig. 6.27 Cracking produced in a concrete with a flint coarse aggregate which had been subjected to a fire with temperatures at the surface estimated to be between 300°C and 600°C. Note the 4 mm wide darkened zone where the cement paste has partly dehydrated. A: Shrinkage cracks perpendicular to the outer surface. B: Irregular cracking within the dark zone. C: Cracking sub-parallel to the surface cutting both coarse aggregate and cement paste. D: Cracking within the coarse aggregate due to thermal mismatch between the expansion of the aggregate and matrix. (Drawn from a thin section provided by courtesy of STATS Ltd.)

crack patterns develop and increase in severity as the temperature of exposure increases. Since the cracks develop principally as a result of differential thermal movements between the components of the concrete, the maximum temperature reached by a portion of concrete is more important than length of exposure. However, prolonged exposure will allow the cracks to extend and widen, and networks to become more extensive.

The severity of cracking will also depend on the thermal conductivities of the aggregate particles and the thermal movement mismatch between them and the paste. Flint aggregates have one of the highest thermal conductivities and as a consequence concretes with this aggregate show severe cracking which cuts across both the paste and aggregate particles and also produces cracking at cement/aggregate interfaces as illustrated in Fig. 6.27. Similar effects are observed with granitic and sandstone aggregates except there is less peripheral cracking at aggregate/cement interfaces possibly because these aggregates can more readily accommodate differential expansion along intergrain boundaries of these poly-crystalline materials. Limestone aggregates similarly appear to accommodate thermal movement well, with the majority of cracking confined to the cement paste and the margins of fine aggregate particles as is shown in Fig. 6.28. A general feature of this microcracking is the tendency to develop as a series of sub-parallel fractures parallel to the original heated surface. Locally, fracture orientations may be controlled by local modifications to the stress field in the concrete, in for example, areas close to air voids.

Although mineralogical changes are difficult to observe directly with the petrological microscope, the normally featureless cement paste begins to show a patchy anisotropy with

(a)

(b)

Fig. 6.28 Photomicrographs of concretes which had been exposed to heating at an exposed surface to temperatures between 360°C and 600°C. Width of field of view for each thin-section photomicrograph is 1.5 mm, plane polarised light. (a) Shrinkage crack perpendicular to the exposed surface (right-hand edge) and irregular cracks within the darkened zone. (b) Cracking through the cement matrix and a quartzite coarse aggregate particle as well as round its margins. The main cracking is sub-parallel to the exposed surface.

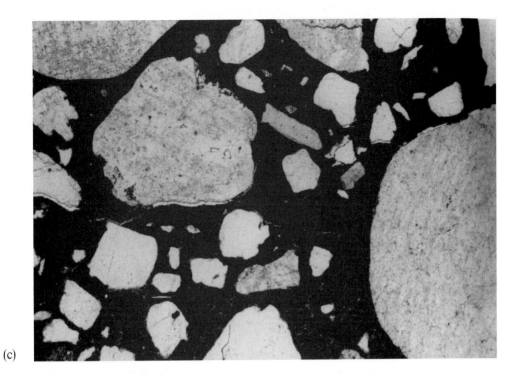

(c)

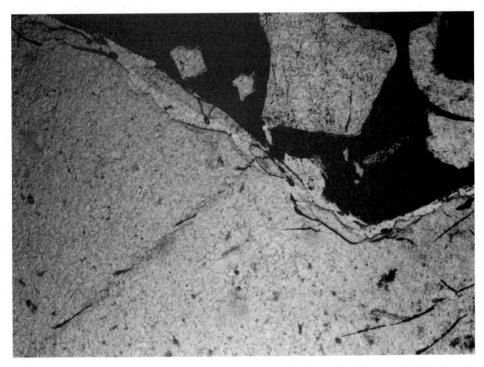

(d)

Fig. 6.28 (contd) (c) Cracking in the cement matrix close to the boundary between the dark outer (top) and light inner zone (bottom). The cracks are sub-parallel to the outer surface of the concrete. (d) Peripheral cracking on a quartzite coarse aggregate particle shows differences between the thermal expansion of the aggregate and cement matrix.

pale yellow-beige birefringent colours at exposure temperatures over about 450°C, Riley (1991). The more or less complete development of this birefringence in the paste has been taken as an indication that the affected concrete has been exposed to temperatures in excess of 500°C. The birefringence indicates that the calcium silicate hydrate gels and the calcium hydroxide have developed significant amounts of new crystalline calcium silicate and calcium oxide phases as they have become dehydrated. The possible reason for this birefringence is the development of a 9 Å calcium silicate hydrate known as riversideite (Heller and Taylor 1956) as the temperature is not sufficient to form β-wollastonite (Urabe 1990). However, the birefringence of riversideite is not large enough to give a yellow-beige colour under crossed polars so that other minerals must be involved. X-ray diffraction analyses confirm the disappearance of calcium hydroxide and the development of calcium silicates at around this temperature.

6.8.6 *OTHER METHODS OF EXAMINATION OF FIRE-DAMAGED CONCRETE*

A thermoluminescence test for fire-damaged concrete has been proposed by Placido (1980) and further developed by Chew (1988). This technique involves determining the residual thermoluminescence of sand extracted by acid from samples of fire-damaged concrete. It is claimed the method is simple and objective with an accuracy of 10 per cent. Harmathy (1968) investigated the use of thermogravimetric analysis and thermal dilatometry to determine the temperature history of concrete following fire exposure. Small powdered samples excluding as much aggregate as possible are used for the thermogravimetric measurements and prisms of concrete cut parallel to the surface are measured for thermal dilation. All samples must be obtained within 1 or 2 days of the fire occurring as rehydration of some of the components in the concrete is possible. It is claimed that fairly accurate temperature estimates can be obtained even without carrying out tests on a reference specimen.

The earliest stages of thermal cracking in concrete may be observed directly using scanning electron microscopy of polished surfaces. Semi-quantitative comparisons with unaffected concretes from the same structural element may be accomplished using a traverse crack counting method (Section 4.8) or by measuring strain energy loss resulting from low stress cycling tests.

6.9 General aspects of deterioration and weathering

The deterioration of concrete with time is dependent on factors ranging from faults at the design and construction stage, through choice of materials to environmental conditions. Typically, several mechanisms operate together contributing towards the overall deterioration.

Concrete structures are rarely built under ideal conditions and defects in the concrete arising from imperfections in mix design and casting caused by plastic shrinkage, plastic settlement cracking, blow holes, honeycombing and scouring of surfaces can provide pathways for ingress of water or other fluids and areas of weakness that can be preferentially exploited by aggressive environmental conditions.

The choice of inappropriate aggregates or cement may also lead to problems of concrete deterioration which develop in the longer term. One example, the alkali-aggregate reaction has been discussed in Section 6.7, but other examples include the use of aggregates containing soft or unsound particles or shrinking aggregates which exhibit excessive moisture movement (see Section 6.6.3).

The use of calcium aluminate cement in conditions where the conversion of the hydrate $CaO.Al_2O_3.10H_2O$ to $3CaO.Al_2O_3.6H_2O$ is likely to lead to an unacceptable loss of strength and possible failure is another example where unsuitable materials have been used in the past. High alumina cements are dealt with in Sections 4.2.4 and 5.2.7 and in detail by Neville (1975).

Another example involves the incorporation of small percentages of 'reactive' pyrites in concrete aggregates which can lead to unsightly, if superficial, brown stains on concrete surfaces. Reactive pyrites may be identified by their reaction with lime water, giving blue-green ferrous hydroxide which turns to brown ferric hydroxide on standing.

Climatic conditions, which involve wetting and drying, freezing and thawing, and thermal movement may have a major effect on rates of deterioration of concrete structures. Consequently, the effects of deterioration are often most severe on the exposed weather sides of structures. In condition surveys plotting the exposure conditions of the structure on a plan using a scale of five levels from most severe to most sheltered, can be a useful aid to the prediction of the future pattern of deterioration of the structure.

Particularly aggressive environments such as sulfate-bearing groundwaters, attack by acids, salt damage from deicing salts and other types of chemical attack have been dealt with in the preceding sections. In most examples the deterioration progresses inwards from the outer surface of the concrete and along existing cracks, joints between lifts, and other pre-existing defects. Usually the deterioration affects the cement matrix leading to complete breakdown of the structure of the surface layers allowing the aggregate particles to fall away.

Cracking, whether induced by overloading, cyclic stressing leading to fatigue, impact or other causes such as freeze-thaw action or alkali-silica reaction is the most commonly observed sign of deterioration. Analysis of the causes of various types of crack pattern have already been covered in Chapter 5 and described and discussed in a number of publications, for example Concrete Society Report No. 22 (1982) and Fookes (1977).

The correct interpretation of crack patterns during the field survey of structures is perhaps the most important aid to the diagnosis of deterioration. Other observations which may assist include vertical displacement of crack margins, discoloration or degradation of surfaces, and the occurrence of pop-outs.

The careful correlation of visible features of deterioration including the relative severity of deterioration with respect to position within the structure, and the correlation of deterioration with the original constructional sequence and the severity of environmental exposure, will provide clear guides towards diagnosis. However, a field examination of structures, even when constructional details are available, is rarely sufficient to allow an unequivocal and clear diagnosis of the cause of a deterioration to be made without a more detailed laboratory investigation of cored samples taken from carefully selected locations on the structure. In many cases this will require that petrographic examination of the samples be performed.

7

Concrete Products and Miscellaneous Materials

7.1 Introduction

A wide range of materials, products and techniques are associated with the use of Portland cement some of which will be discussed in this chapter. The related materials, lime and gypsum, both of which antedate Portland cement, are also included. The choice and grouping of topics chosen illustrate where petrography can be applied and is not intended to be exhaustive. Some overlap with previous chapters has been necessary to ensure that each topic is adequately covered.

The term precast concrete applies to products that are made as modules or units that are later fitted together in construction. To provide economies of scale, organisation and better product control they are usually made in factories rather than on site. The types and methods of manufacture of precast products are discussed by Glanville (1950), Childe (1961), Levitt (1982), Richardson (1991) and proceedings of fourteen triennial International Congresses of the Precast Concrete Industry organised by the Bureau International du Beton Manufacture. The last reported conference was held in Washington in 1993.

Descriptions of techniques and products intended for use by architects and engineers have been outlined in the annual publication 'Specification' (e.g. Wiliams 1994). This has been an excellent source of information on current practices and usage; it is now only available on a CD-ROM basis. General texts on finishes and materials intended mainly for architects have been written by Deane (1989) and Everett (1994).

The petrographer may not only be required to examine laboratory test specimens and recent construction, but also structures that date back to the introduction of Portland cement in the mid-nineteenth century. An attempt has been made to detail some of the older techniques and products. However, pre-nineteenth-century techniques and products have been largely excluded as these are close to archaeology and are beyond the scope of this book.

The key to successful petrographic examination of techniques and products lies in an understanding of the chemistry and physics of the materials and fabrication processes used. When these are understood petrographic examination of texture may confirm whether the claimed materials have been used and it may be possible to identify the changes in texture that have occurred with time.

7.2 Examination of laboratory specimens

The first step in the examination of a laboratory specimen involves a thorough inspection of the sample using the low-power stereo microscope. This particularly applies to test specimens used in alkali-aggregate studies where it is important to determine the incidence of microcracking and gel present on the surfaces of the specimen before carrying out thin-sectioning.

A common mistake is to use too high a magnification on the stereo microscope for scanning surfaces. A ten times magnification is adequate and higher magnifications should only be used when a point of particular interest needs to be examined in detail. It is also necessary to use a micrometer eyepiece so that the dimensions of cracks and other features can be estimated.

Many laboratory specimens are stored under moist conditions so that surfaces are saturated with water. It is often suggested that microcracking and deterioration become more visible if the surfaces are allowed to partially dry and that examination should be carried out in this condition. This is because the edges of cracks dry at a different rate to the surrounding concrete and shade contrasts develop from this differential drying. If the stereo microscope is correctly used to examine saturated surfaces little additional information is gained from examination of partially dried surfaces and there is a danger that the contrast that develops will over accentuate some types of deterioration and bias the examination. The use of partial drying to enhance surface deterioration should be restricted to visual examination.

The question of choice of plane and area of sectioning requires careful consideration and must be related to the problem. In general laboratory specimens will be cast either as bars or cylinders and less commonly as slabs or discs. Where the ratio of the thickness to the length and breadth of the specimen is small, such as in the case of 25 mm mortar bars used for testing aggregates for alkali-aggregate reactivity, surface cracking tends to run across the mortar bar. Longitudinal sections should intersect more cracking and also have a high surface to volume ratio and are more likely to provide useful information.

For cylinders, which usually have a length to diameter aspect ratio of 2:1, axial sections appear to provide more information. However, if surface deterioration is being examined, and the internal texture of the concrete is of secondary importance the plane of sectioning should be chosen to provide the greatest length of surface.

Petrographic examination often shows that the texture of concrete or mortar present in laboratory test specimens is less homogeneous than concrete in structures. This is hardly surprising when it is considered that the volumes of test mixes and cast specimens are small and the probability of localised inhomogeneity is increased. This problem requires that a sufficient number of sections of large enough size be cut to avoid faulty conclusions being drawn from the petrographic examination.

7.3 Fibre-reinforced products

The first widely manufactured fibre/cement composite was developed around 1900 utilising asbestos as the fibre. Since the initial use of asbestos other fibres made from glass, steel, carbon, kevlar, polypropylene, nylon and the natural fibres sisal, jute and kraft pulp have all found use.

The petrographic examination of fibre-reinforced products in thin-section requires some contrast between the fibre and the hardened cement paste if details of fibre distribution and fibre/matrix interface are to be observed. For instance, glass fibres may be difficult to differentiate in a cement matrix and thus are often better examined using the scanning electron microscope.

The properties of fibre-reinforced cementitious composites have been reviewed by Bentur and Mindess (1990) and Brandt (1995) and of natural fibre reinforced cement and concrete by Swamy (1988). ACI Committee 544 (1988) and BS 6432 (1990) specify the various tests applicable.

7.3.1   ASBESTOS-CEMENT PRODUCTS

Asbestos cement is composed of finely divided asbestos fibre dispersed in a matrix of hardened Portland cement. Where the product is high pressure steam cured up to 40 per cent of the cement is replaced by silica flour. Flat and corrugated sheet contains 9–12 per cent, pressure pipe 11–14 per cent and fire resistant board 20–30 per cent asbestos fibre. Three principal methods of manufacture are used. The wet transfer roller or Hatschek process and its variant for making pipes known as the Mazza process, the semi-dry Magnani process principally used for making corrugated roofing sheet and the Manville dry extrusion process. More recently an injection moulding process using hydraulic pressure to de-water the product has come into use for making sheet from 10 to 100 mm thick. General details of these processes and the properties of the products are described by Concrete Society (1973), Ryder (1975), Coutts (1988) and Bentur and Mindess (1990).

The Hatschek and Mazza processes, the two most commonly used methods of manufacture, lead to preferred orientation of the asbestos fibre. In thin-section, more sub-parallel fibres will be present in a section cut along the length of the sheet than across it. To get an overall view of the distribution of fibre it is necessary to cut sections both along and across the sheet. In the case of pipes sub-parallel orientation occurs along the length of the pipe.

There is a range of asbestiform minerals which can be readily identified by infrared and optical techniques, their optical details are given in the glossary of minerals. The predominant form of asbestos used in building products is chrysotile (Middleton 1979). Some amosite (grunerite) is added to improve the filtration properties of the wet slurry used in manufacture especially in the Hatschek process and amosite predominates in fire-resistant boards because of its good high temperature resistance. Crocidolite was also added in the manufacture of pipe in the USA (Rajhans and Sullivan 1981). From the early 1970s partial replacement of asbestos by either kraft pulp or glass fibre has occurred and since the 1980s the use of asbestos fibre, especially crocidolite, has been banned in many countries. Thus it is necessary when examining asbestos-cement products to be aware that other types of fibres may be present even for pipes dating back twenty-five years.

The appearance of the hardened matrix depends on the method of curing. Asbestos cement products are usually steam cured to increase production. Products cured at ambient or with low pressure steam at about 80°C may only contain cement as the matrix. Where a more durable matrix is required the products will be cured at about 170°C in high-pressure steam. This can always be detected as the matrix will contain the residues of the silica flour added to suppress the formation of the low strength αC_2SH in favour of the stronger tobermorite type phase.

The microscopic texture of asbestos-cement consists of a hydrated matrix containing numerous partially hydrated and remnant grains of cement and well dispersed calcium hydroxide in which are embedded fibres of asbestos. Where silica flour has been added the coarser grains of silica, usually quartz, remain unreacted and the amount of calcium hydroxide present is considerably reduced but is rarely completely absent, Plate 72. The asbestos appears as bundles of fibres, usually about 3–10 μm in diameter and its optical characteristics clearly distinguish it from any other likely fibre to be used in these products (see Glossary). Where the product has been in use for a long time considerable depths of carbonation can occur. Thin sheets can be completely carbonated especially in the case of autoclaved products.

Bentur and Mindess (1990) in reviewing the long-term performance state that in natural weathering the composite is excellent. However, there are two problems which the petrographer may have to examine. These are carbonation leading to embrittlement of the product and attack by aggressive water on pipes either due to soft water or sulfates. It is necessary when considering these problems to distinguish between products cured below 100°C and autoclaved products as they are will respond differently to the environment.

Opoczky and Péntek (1975) examined commercial sheet in thin-section which had been weathered between two and fifty-eight years. From about fifteen years the fibres showed signs of corrosion from the alkaline pore solution with the formation of possibly brucite. Magnesium carbonate and low lime calcium silicate hydrates were also identified. Testing showed that even where over 50 per cent of chrysotile fibres were corroded the sheet still had excellent mechanical properties. However, Sarkar *et al.* (1987) failed to detect either corrosion of fibres or brucite and magnesite in sheet attacked by sulfate rich waters.

There is little doubt that embrittlement of asbestos-cement due to carbonation frequently occurs with ageing and the extent of this will depend on the permeability of the product and how it is cured. In this respect while autoclaving gives enhanced mechanical properties its resistance to carbonation is likely to be less than a comparable product cured at 80°C or ambient (Section 6.4.2). Examination in thin-section is an excellent method for determining the fine detail of carbonation texture but there appear to be no petrographic studies reported in the literature.

Asbestos-cement pressure pipe has been used extensively for water circulation and is specified in BS 486 (1981) and BS 4624 (1992). Its ability to withstand soft water and soluble sulfates is of some importance. ASTM C 500 (1991) gives guidelines for internal and external corrosion factors based on an aggressivity index of the water being transported and the acidity of the soil and soluble sulfate present in external water. There are additional recommendations that uncombined calcium hydroxide in the product should be less than 1 per cent, and for aggressive conditions Type II cement should be used.

It is reported by Al-Adeeb and Matti (1984) that asbestos-cement pipes in Kuwait have life expectancies of only nine to thirty years. Tests showed the water to be moderately aggressive according to ASTM C 500 and calcium hydroxide contents averaging 17 per cent suggested that the pipes may not have been autoclaved. There are also reports from some areas of New Zealand that asbestos-cement water pipes are starting to fail after twenty-five years. Once again the water is moderately aggressive but in this case the pipes have been autoclaved which may be contributing to their longer working life.

There is an important difference between internal and external attack by aggressive water on asbestos-cement pipe. For significant external attack to occur the pH must be below 7 as discussed in Section 6.4. Internal attack sufficient to cause significant loss in strength takes place at pHs in the range of 7 to 9 and is dependent on the carbonate bicarbonate equilibria and its ability to dissolve calcium hydroxide. This is the basis of the ASTM aggressivity index and the limit of 1 per cent uncombined calcium hydroxide. The limit on calcium hydroxide is probably more an indicator of pipe quality rather than having much effect on corrosion rate. 1 per cent calcium hydroxide can only be attained in pipes where reactive silica flour and adequate autoclaving has been used so that a dense product results.

There appears to be no reported petrographic studies on the failure of asbestos-cement pipe from internal aggressive water attack. Bursting of the pipes by splitting seems to predominate over beam type failure in these cases suggesting that the hardened cement matrix is weakened. The preferential alignment of fibres along the length of the pipe will also contribute to the splitting failure.

7.3.2   SYNTHETIC-FIBRE REINFORCED PRODUCTS

For convenience glass, steel, carbon and organic polymer fibres are grouped together. The properties and use of these fibres have been reviewed by Bentur and Mindess (1990) and discussion will be mainly restricted to the use of glass fibre. Initially the composition of glass used to make the fibres was an electrical grade of alumino-silicate glass known as E glass.

It was soon found that the alkali resistance of E glass fibre in a Portland cement matrix is limited in moist conditions and that corrosion and brittleness in the fibres over time can seriously weaken the product. The use of a chemical-resistant glass fibre containing zirconia known as CemFIL appeared to offer a solution to this problem but it merely slows the reaction down and does not prevent it. Similarly it was believed that the use of silica fume might have a beneficial effect but as reported by Yilmaz and Glasser (1991) silica fume inhibits notching damage from calcium hydroxide crystals but does not inhibit other attack mechanisms.

Because of potential long-term reduction in strength and increase in brittleness with ageing fibre glass cement products have been limited to non-structural or semi-structural components such as thin sheet and claddings. According to Bentur and Mindess (1990) the principal cause of failure in panels is due to cracking. The shrinkage of the composite can range from 0.05 to 0.25 per cent dependent on the sand content. Differential strains induced by thermal and moisture movements can lead to sufficient stress generation to cause cracking.

Most of the detailed examination of fibres in hardened cement has been carried out by scanning electron microscopy and there appear to be no reported petrographic studies using thin-section examination. Plate 42 illustrates the texture of a laboratory test prism made by mixing chopped strands of alkali-resistant glass with cement and lime. The individual glass fibres are about 15 μm in diameter and in longitudinal section have a thin dark margin which is the sizing. Being a glass they are isotropic apart from some minor strain birefringence. In cross-section the strands of fibres that have not dispersed are partly penetrated by calcium hydroxide crystals but the space between the bunched fibres is otherwise empty. Where strands have opened up both cement paste and calcium hydroxide penetrate. In longitudinal section individual dispersed fibres are in intimate contact with the hardened cement paste and deposition of calcium hydroxide at the interface is not marked. In general the less the dispersion the more the deposition of calcium hydroxide at interfaces probably because of the space created by the strands starting to open up.

The petrographic investigation of failed glass fibre cement composites requires a number of factors to be considered. It may be necessary to determine that alkali resistant glass and not E glass fibre has been used. The presence of zircon in the glass can be detected by X-ray analysis on the scanning electron microscope but not by light microscopy as both E glass and alkali resistant glass have similar refractive indices (see Glossary). It is advisable to use a low alkali cement to minimise alkaline attack on the fibre and chemical analysis should indicate the alkali content to be less than 0.6 per cent sodium oxide equivalent.

Thin-sectioning will indicate the nature and characteristics of the sand and whether Portland cement has been used, the presence of pozzolanas such as silica fume and the extent of carbonation. Some idea of fibre dispersion can be gained but generally this is not uniform on the local scale and requires large-area thin-sectioning cut from orthogonal planes to have any meaning. Where high alkali cement has been used a careful examination should be made for the presence of alkali-aggregate reaction. This may be difficult to observe as the amounts of alkali-silica gel involved may be quite small. Examination of fibre matrix interfaces should be carried out by scanning electron microscopy on fracture surfaces.

7.3.3   NATURAL-FIBRE REINFORCED PRODUCTS

The range and properties of natural fibres used to reinforce cement composites has been reviewed by Bentur and Mindess (1990). Apart from the use of otherwise waste natural fibres to produce low cost cement products the most important natural fibre is kraft pulp currently used as a replacement for asbestos in sheet products manufactured in Australasia and some other countries. The discussion will be restricted to the use of kraft pulp fibre.

There are a number of problems in making a strong and durable sheet using kraft fibre. Because the effective density of natural fibres is about 1.2 compared with 2.6 for glass and chrysotile asbestos fibres, greater volumes of natural fibres have to be used to attain the same weight addition to the composite. This combined with other factors leads to flexural strengths of about 20 MPa as compared with 40 MPa for asbestos sheet. Unless the sugars are completely removed from the kraft pulp and it is adequately beaten the setting of the cement and dispersion of the fibre can be affected (Coutts 1983) which can considerably reduce the flexural strength of the final product. Because of these effects kraft pulp fibre cement products are invariably autoclaved at about 170°C to attain maximum strength.

The durability of kraft pulp fibre cement sheet is affected by moisture conditions and carbonation. Bentur and Mindess (1990) report that shrinkage strains can exceed 0.15 per cent which may be partly due to the presence of the kraft fibres. Sharman and Vautier (1986) found that alkaline corrosion is unlikely in autoclaved sheet because of the removal of lignin and hemicellulose during processing of the fibres. The data discussed by Bentur and Mindess (1990) on carbonation is confusing. However, for autoclaved product it can be assumed that thin sheet will normally be in an advanced state of carbonation and some increased brittleness will have occurred after exposure to the weather for any length of time. There appear to be no petrographic reports describing the texture of weathered sheet.

The texture of autoclaved kraft pulp fibre cement sheet, fresh from the factory, is shown in Plate 43a. Kraft pulp fibres varying from 10 μm to over 1 mm in length are preferentially aligned in a hardened matrix of hydrated cement containing fragments of quartz. Longitudinal kraft pulp fibres have parallel extinction, are length slow and show first-order interference colours which are typical of cellulose fibre. Fibres are of the order of 10 to 20 μm in width but this is variable as many fibres appear partially flattened with a tendency to undulate. They are rarely straight. In cross-section fibres are hollow with walls about 5 μm thick. Considerable amounts of quartz flour remain ranging from over 100 to 10 μm in size. Reaction rims are present on some quartz grains but are not common. The hardened matrix consists of dark essentially isotropic material in which are embedded remnant grains of cement with reaction rims and some very fine low birefringent material which is probably calcium hydroxide. Otherwise calcium hydroxide appears absent. The outer margins of the sheet are carbonated to depths of 0.5 to 1.0 mm. This carbonation is reasonably intense at the surface but grades into scattered clumps of carbonation which give the texture a spotty appearance. Overall the texture is compact with few voids although in some samples there is a tendency for residual fissures to be present parallel to the length of the sheet derived from the original laying up of the product.

7.4   Concrete pipe, block, brick, tile and pavers

These products represent a major section of the concrete industry. The products discussed have been limited to those containing normal weight aggregates and conventional reinforcement. Most of the differences between these products are the result of variation in mix design due to the method of fabrication used. In spite of automated methods being readily available a surprising amount of production is still manufactured by older methods.

7.4.1   CONCRETE PIPES

Concrete pipes were originally made by hand methods or machine tamping to consolidate the concrete into the mould. This latter method is still used today for unusual shapes. The mechanical processes used for making pipe can be divided into horizontal and vertical forming methods (Dutton 1977). Around the 1870s the first pipe-making machine was developed which employed a rotating dolly to simultaneously form the inner wall and compact the concrete in a vertical mould. In 1910 the Hume process of centrifugally spinning a horizontal mould in which concrete is placed to form the pipe wall was introduced. In spite of being labour intensive it still remains one of the principal methods of fabrication (Fig. 7.1) in many parts of the world. This versatile process is used to make pipes and hollow poles and allows the use of steel reinforcing and prestressing. Around 1960 the 'Roller Suspension' process originally known as the Rocla process was introduced whereby concrete placed in a slowly rotating, horizontal mould is compacted by a high pressure roller. In some respects it is variant of the Hume process where the pressure roller replaces the centrifugal forces to compact the concrete.

There is a range of national specifications covering concrete pipes among which the following are applicable. Pipes are generally described by BS5911 (1981), ASTM C76 (1995) and ASTM C497M (1996). Prestressed concrete pipes are described in BS4625 (1970), BS 5178 (1975) and ASTM C1089 (1988). The standards generally require the mix design to be such that specified absorption, hydrostatic head and load tests can be attained. In ASTM C1089 (1988) a minimum concrete cover of 13 mm over reinforcement is required and provision is made for the use of galvanised steel. As with the majority of precast products the quality of the concrete is controlled by performance tests on the product.

Levitt (1982) recommends a typical mix design suitable for spun concrete as three parts by weight of clean 10 mm crushed rock, 1.5 parts of sharp concreting sand to 1 part of a coarsely ground cement and a water/cement ratio of 0.32. Where segregation is a problem he claims that replacement of 20–40 per cent of the cement by fly ash is beneficial. Glanville (1950) states that aggregate/cement ratios vary between 3.5:1 and 2.5:1 depending on the class of pipe indicating that in practice a wide range of cement contents may be found in older pipes. BS 5911 (1984) allows the use of ordinary, rapid-hardening, Portland-blastfurnace and sulfate-resisting cements and requires a minimum cement content of 360 kg/m³. In practice the water/cement ratio appears to be closer to 0.4 than 0.3. It is

Fig. 7.1 Making a centrifugally spun pipe. This 3600 mm diameter pipe is being made in a temporary factory near the project. The inside has been finished and the mould is undergoing its final spinning sequence to ensure compaction of the concrete. The concrete feeder used to place the concrete is in the foreground. In the United States the Packerhead process is widely used and highly automated. The process utilises a vertical outer mould and a rapidly rotating roller head which packs the concrete by flinging it outwards under high centrifugal force. It is possible to include reinforcing using the Packerhead process but the pipes are about 30 per cent heavier than those made using the centrifugal process. Another automated process uses a very dry mix which is fed into the top of a mould with a vibrating core which rises slowly to form the pipe wall.

not clear as to how much excess water is removed during the spinning process but reasonably wet mixes are required to get good compaction. The use of low-frequency high-amplitude vibration enables dryer mixes to be used. Examination of the internal textures of pipe walls suggests that excess water is largely removed from the layers close to the internal surface of the pipe but no quantitative data appears to be available.

Details of the internal texture of concrete pipe cast by the old method of compaction with a rotating dolly in a vertical mould are not available. The texture of pipes made by the 'Roller Suspension' process is that of a well compacted low water/cement ratio concrete rich in cement while the texture of pipes made in vibratory vertical casting machines is similar to modern concrete block. One form of pressure pipe made by the 'Roller Suspension' method consisted of prestressed strand wound round the fully cured pipe which was then protected by a layer of pneumatically applied mortar. Experience indicated that the applied mortar was permeable to chlorides resulting in some failures due to corrosion of the post-stressed reinforcement. There is no information on the internal texture of pipes made by the Packerhead process.

The internal texture of the wall of a centrifugally spun pipe is zoned because the fine material in the concrete mix partially segregates. Some of the excess water is forced to the inner surface as a thin slurry and is removed during the spinning process. Segregation at the inner surface leads to a layer of rich mortar being formed which enables this surface to be smoothed with a forming bar. Where necessary, localised rough patches on the inner

Fig. 7.2 The concrete around the spigot area of this centrifugally spun pipe has not been adequately compacted during spinning. This can be a problem with this method of pipe production unless care is taken in fabrication.

surface are repaired by sprinkling neat cement over the rough area during the final forming process.

Some other faults can occur in centrifugal concrete pipe production many of which are obvious on stripping the mould. Unless care is taken compaction around the spigot area can be a problem (Fig. 7.2). Spun concrete pipe is commonly steam cured at 80 to 90 °C to give adequate early strength for stripping but care in handling is still required to prevent cracking at this stage of production. External surfaces of spun pipe are usually excellent but can be marred by inadequate mould release agents (Fig. 7.3).

The commonest fault in a centrifugally spun pipe is caused by inadequately graded and wrongly shaped aggregate which allows too much segregation to occur. This can result in a 10 to 15 mm deep zone on the inside of the pipe of almost neat cement paste grading into a rich mortar containing only the finest fraction of the sand (Plate 73). This texture is prone to shrinkage cracking. In a well-made pipe the inner zone is still present but the almost neat cement paste is present only as a thin skin and the dense mortar contains closely packed sand which forms an effective barrier to the ingress of fluids to the underlying concrete in the pipe wall. The overall depth of the segregated zone may be as little as 5 mm while the remainder of the wall consists of a dense concrete containing well-packed aggregate. Where the concrete mix is too harsh, excessive fissuring along margins of aggregate and reinforcing occur because of aggregate interlock and the inner surface may require excessive use of neat cement for smoothing.

Many aspects of segregation and zonation can be seen by examining a finely ground section cut from across the wall of a pipe. If further detail is required it is necessary to examine large-area thin-sections. Both the depth and nature of the segregated zone are important. It is desirable but not critical for the segregated zone to be limited in depth as this indicates good mix design. It is more important that the segregated mortar is densely

Fig. 7.3 The outer surface of this spun pipe has been damaged due to inadequate release agent damaging the surface. This damage is superficial and is unlikely to affect the durability of the pipe.

packed with sand to provide good durability. Fissuring along aggregate margins and reinforcing steel caused by lockup and localised inhomogeneity of the hardened cement paste due to the bleeding of water during the spinning of the pipe are features present in most pipes. This is especially the case where double reinforcing cages are used (Fig. 7.4). The petrographer should be cautious in condemning a pipe on these grounds especially where the pipe complies with the requirements of the absorption, hydrostatic and load tests of the appropriate standard. A further indication that most of the channels and fissures in spun concrete pipes are not critical is that, although pipes that fail the hydrostatic test are left full of water for a number of days they then usually pass the hydrostatic test. Thus channels and fissures present are easily blocked off by the formation of calcium hydroxide and autogenous healing.

While concrete pipe rarely fails underground unless it is overloaded or attacked by aggressive fluids the same pipes used above ground may exhibit serious circumferential cracking. A number of cases of cracking in exposed pipe lines and pipes used as permanent formwork have occurred in New Zealand as shown in Fig. 7.5. The reason for this type of cracking has not been investigated but it is thought to be due to insufficient longitudinal rods in the reinforcing cage being present to resist differential temperature stresses.

7.4.2 *CONCRETE BLOCK, BRICK, TILE AND PAVERS*

Concrete masonry units are produced in a variety of sizes and shapes and may be divided into four specified broad groups, solid load bearing blocks, ASTM C145 (1985), hollow load bearing blocks ASTM C90-946 (1992) and solid or hollow non load bearing blocks ASTM C129 (1994). The load bearing blocks are further subdivided according to their intended use and degree of exposure. In the UK precast concrete masonry units are specified in BS6073 (1981).

Fig. 7.4 A core through the wall of this double reinforced pressure pipe illustrates the problem of inhomogeneity that can occur due to aggregate lockup on reinforcing cages. The thickness of the wall is 180 mm and the inner surface is to the left of the photograph.

Concrete blocks, bricks, tiles and pavers are now commonly manufactured using one casting machine for the whole range of products. These casting machines are proprietary equipment which use efficient, high energy vibration combined with pressure to compact semi-dry mixes of concrete. After casting the concrete is stiff enough to be immediately removed from the mould in a process which resembles building sand-castles. Cement contents range from 230 kg/m^3 for blocks and bricks and up to 350 kg/m^3 for tiles and pavers. Water/cement ratios range from about 0.45 to 0.35. The maximum size of sand

Fig. 7.5 Circumferential cracking of spun pipe exposed above ground. This pipe is close to the sea which has accelerated corrosion of the reinforcement. Further inland pipes were showing similar cracking but corrosion of reinforcement was more limited.

used is about 8 mm and the fineness moduli range from 3.5 to 5.0 indicating that very coarse sand gradings are used and this combination of a coarse sand and a low water/cement ratio results in a semi-dry mix. The product may be cured either at ambient, or in low-pressure steam at about 80°C or in high-pressure steam at 180°C. In the last case silica flour will be added to ensure the formation of the higher strength tobermorite type cement hydrates. Partial replacement of the cement by fly ash, slag or pozzolana are sometimes used.

The texture in thin-section of concrete block, brick, tile and pavers results from the semi-dry mixes used. The external surfaces are irregular with many re-entrant pores and aggregates which protrude to the outer surface of the paste. Internally, there are numerous large, irregular pores as well as smaller pores due to entrainment of air, Plate 74. Crystals of calcium hydroxide may line and protrude into the large pores, line aggregate margins and often partially fill the smaller pores. In the hardened paste the crystals of calcium hydroxide tend to be minute. The aggregate appears closely packed so that no large areas of cement paste are observed. This is the texture of harsh mortar with insufficient cement paste to fill all the space between the aggregates. The addition of silica flour, fly ash and slag is unlikely to materially change this texture apart from the presence of their residues and some reduction in the amount of calcium hydroxide observed. In the products examined drying shrinkage cracks have not been observed which is to be expected given the cement contents and water/cement ratios used.

Concrete block masonry made with plastic mixes presents no specific features for petrographic study different from those already considered in Chapter 5 and elsewhere in this text. Similarly, pavers and tiles compacted by hydraulic pressure present textures typical of dense concrete. These products are normally plain but special surface finishes are sometimes used including 'glazes' formed by mixing selected aggregates materials and pigments with a thermosetting binder which may be hardened by a high temperature high-pressure curing process.

The properties of interest in these products are strength, permeability, moisture movement and appearance. Petrographic examination of the concrete is usually not relevant to the investigation of these properties apart from appearance. A common problem with unsightly concrete block is caused by efflorescence. This problem is discussed in Section 5.4.3 and a more unusual case is shown in Fig. 7.6. An investigation of unacceptable colour variation is described in the following case.

Intermittent subtle colour variations were occurring in a concrete brick made with ordinary Portland cement coloured with a titanium pigment. The bricks were initially low-pressure steam cured in an atmosphere enriched in carbon dioxide. It was claimed that the bricks were all fabricated under exactly the same conditions. Usually the subtle colour variation was not detected until the bricks were used in construction when the colour differences became apparent. The differences in shade could be clearly seen when bricks were compared side by side and could also be measured on a colour comparator. It was established that the change in colour affected the whole surface and was not an edge effect.

Comparison of bricks of different colour in thin-section failed to reveal any significant differences. Examination of fracture surfaces by scanning electron microscopy showed that on the surfaces of the bricks which were different in colour there was less calcium than the 'normal' bricks. That is, undetected variations in the production process were intermittently reducing the amount of efflorescence present on some of the bricks and allowing more of the titanium pigment to become visible.

Fig. 7.6 Calcium sulfate and calcium sodium sulfate efflorescence on new load-bearing precast concrete blocks.

Concrete blocks made from no-fines concrete are composed of a nominal single-sized coarse aggregate mixed with cement and water but contain no fine aggregate. Such blocks have a moderate compressive strength but high porosity and consequently have good thermal and sound insulating properties. The water/cement ratio used is critical and adequate compaction is required to avoid the formation of cavities. Reviews of no-fines concrete are given by Malhotra (1976), Spratt (1980) and Neville and Brookes (1987). No-fines concrete, because of its high void content, is typically well carbonated but except for the absence of fine aggregate presents no petrographic differences from normal concrete which has carbonated. The texture of a no-fines concrete is shown in Section 5.3.

7.5   Plaster, mortar and grout based on Portland cement

These materials can be divided into those based on Portland cement, on lime and on gypsum. There are subdivisions within these groups and lime is also used in combination with both cement and gypsum plaster. This section is restricted to materials where Portland cement is the major cementing material. Lime and gypsum plaster are discussed in Sections 7.7 and 7.8.

The term plaster is used to describe smoothing of internal surfaces and the terms rendering or stucco applied to external surfaces although in practice the term plaster is used indiscriminately. The examination of failed interfaces between layers of plaster and plaster and substrate is a difficult area. The problem is not so much the microscope examination which is quite capable of providing good data but the sampling. It is difficult to obtain a satisfactory cross-section of a plaster and substrate without coring and there is a high probability of induced cracking which may make petrographic interpretation difficult. In many cases the interface has already failed and coring is not practicable.

7.5.1 PLASTER

Cement plaster is a mixture of cement and sand which is used to cover, seal and decorate surfaces. It is customary to use either lime or air entraining agents to give water retention and workability. To improve adhesion, additions such as polyvinyl acetate, styrene-butadiene and acrylic copolymers may be added (Ramachandran 1984). The plaster may either be applied traditionally in two or three layers or as a single layer (Mehlmann 1989) dependent on the job and local practice. The range of finished surfaces can vary from smooth to rough cast and from trowelled to thrown. There is a considerable variation from country to country in plastering methods. Current UK practice is described by Wickens (1992), Deane (1989) and specified in BS 5262 (1991), BS 5492 (1990) and American practice in ASTM C 926 (1995) and ACI Committee 524 (1993). Given the range of surfaces and materials that are plastered it is not surprising that the skill and experience of the plasterer are the most important factors in producing durable plastered surfaces.

Plaster mixes vary from proportions of 1:3 to 1:8 of cement and sand and 25 to 200 per cent of lime by weight of cement may be added to the mix. The leaner the cement mix the larger the amount of lime added. Air entrainment is often used in place of lime. One of the key ingredients is the sand. The types and grading of sands previously used in the UK have been discussed by Cowper (1950), and some current gradings are specified in BS 1199 (1976) and ASTM C 897 (1988).

The durability of a plastered surface depends on the ability of the plaster to adhere to surfaces and to provide a barrier against moisture. Some cracking is inevitable due to moisture movement but this can be controlled by construction joints. In general more problems are encountered with external renderings than internal plaster. Plaster that has been applied with a trowel and finished with a smooth surface is usually less durable than thrown plasters with a rough cast or heavily textured finish.

Analysis and petrographic examination of plasters

Petrographic examination may be used where plaster has failed because of lack of adhesion, cracking and surface problems such as efflorescence and discoloration. BS 4551 (1980) describes the sampling and chemical methods required to determine the composition of the hardened matrix and the grading of the sand and these should form part of the petrographic examination. It is important in any examination that the proportions of materials in the plaster be determined as these have a decisive effect on relative moisture movements. In some cases it will be found that where tradesmen hand batch their materials by volume it is possible for serious mistakes to be made in the batching and this may be the cause of the problem.

BS 5262 has a section on durability of plaster and discusses a number of factors which influence it. These are the nature and durability of the background, type of rendering and proportions of the mix, method of application, moisture movement in both background and rendering, and differential shrinkage within the plaster itself. This is a formidable list of factors to be evaluated to which must be added the skill of the tradesmen. It is not surprising that diagnosis of a plastering problem depends more on experience and often it is not possible to establish the cause of the failure beyond doubt. Low-power stereo microscopic examination of the plaster can be helpful but thin-section examination is unlikely to provide useful information that cannot better be gained by chemical analysis unless identification of the components is required.

The modern practice of casting walls in fair face shuttering is common. The dense smooth surfaces which result are often sealed by curing agents which are designed to weather away before further treatment of the surfaces is applied. This weathering effect is dependent on the curing membrane being correctly applied and for sufficient time for weathering to take place which may vary according to the season. The plasterer may try to combat the effect of possible residues on such surfaces by the use of bonding agents rather than laboriously preparing the surface by removing the surface skin from the concrete. Unless the bonding agents are correctly used they can also cause adhesion failure as noted in BS 5262. It is possible to analyse for residues of curing agents and the presence of bonding agents by solvent extraction of samples followed by appropriate methods of organic analysis. The results may indicate possible reasons for the failure but are rarely sufficient proof of cause.

At a fundamental level plaster cracks and fails either because the adhesion is broken by differential movements exceeding the strength of the bond or adhesion has not been adequately developed due to incompatibility of the substrate with the plaster. The two factors are closely interrelated. Ignatiev and Chatterji (1992) identified the high shrinkage strains and elastic moduli of modern mortars as the principle cause of cracking and stress the need for using mortar with a low modulus of elasticity. As already discussed, residues on concrete surfaces can prevent adequate adhesion ensuing failure irrespective of the plaster formulation.

Problems can also arise from efflorescence due to the movement of soluble salts to external surfaces and unsoundness of lime causing cracking and popping of surfaces. Chemical analysis of the efflorescence is usually sufficient to identify the source of the salts. Detection of unsound lime in a hardened plaster is difficult and it is really necessary to obtain a sample of the lime used and test it for unsoundness according to BS 890 (1995) or ASTM C 110 (1995) to clearly identify this problem.

7.5.2   SPRAYED CONCRETE

Sprayed concrete is applied by feeding a specially designed concrete mixture into a stream of high-velocity air so that it is blasted onto the surface. If the maximum aggregate size is less than 10 mm it is referred to as gunite and if over 10 mm as shotcrete. The cement and aggregate may be premixed dry and the water added at the spray nozzle or the wet pre-mix pumped to the spray nozzle and sprayed in a stream of high-velocity air. The dry process gives better compaction and higher strengths than the wet process and according to Tabor (1992) give strengths of 40–50 MPa and 20–30 MPa respectively.

Schrader and Kaden (1987) state that the strength of concrete mixes used for shotcrete vary from 20 to 80 MPa and give examples of water/cement ratios varying from 0.37 to 0.42 and cement contents from 500 to 400 kg/m^3. Set accelerators (Mailvaganam 1984), styrene butadiene latex, polyvinylidene chloride and silica fume may be added and steel, glass and other fibres included in the mix. Schrader and Kaden (1987) stress that the most important factor in the application of sprayed concrete is the skill and experience of the nozzleman and provide data on durability.

Sprayed concrete is used for building up and repairing surfaces where the concrete has had to be removed because of failure, fabricating underground works and for applying a protective layer to large surface areas. It is usually applied by specialist operators. Specifications and codes of practice for sprayed concrete have been published by the Concrete Society (1979, 1980), ACI Committee 506 (1990), American Society of Civil Engineers (1995), ASTM C 1116 (1991) and ASTM C 1141 (1989).

Localised lack of bond and the formation of large voids are common problems and the *in situ* methods for investigating these are given by ACI Committee 506. Adhesion problems can be caused by the relatively impermeable sprayed concrete locking in moisture in the substrate. The high cement content of sprayed concrete results in shrinkage values up to double that of typical concrete although this can be more than halved by the use of additions such as styrene butadiene latex. Because of the strengths of both the substrate and the sprayed concrete shrinkage rarely causes loss of adhesion but manifests itself as closely spaced fine cracking which can allow the penetration of moisture and salts. It is more durable to use a mix with lower strength and higher permeability that does not crack than stronger mixes which are more prone to cracking, Schrader and Kaden (1987).

Petrographic examination of thin-sections cut from cores drilled through the sprayed concrete into the substrate can provide information on hardened paste volume, air content, type, distribution and volume of aggregate, number of applied layers, cracking, carbonation of the surfaces of both the sprayed concrete and the substrate and the presence of alkali-aggregate reaction can identified or measured. The examination of three cases of sprayed concrete illustrate some of the textures which result. The first case dates from the 1950s and is of a gunite used to repair an exfoliating tunnel lining attacked by sulfates. The second is of a new shotcrete used to protect and decorate a new concrete surface and the third of the gunite covering over the wound post stressing wires of a Rocla pipe dating from the 1960s which had failed due to chloride attack.

In the first case the gunite was built up in four layers totalling about 100 mm in depth. The interfaces of the layers were clearly demarcated by a 50–100 μm wide bands of diffuse carbonation and subtle changes in texture across the interfaces indicating some time lapse between applications (Plate 44). The outer surface of the gunite was carbonated to an average depth of 100 μm and up to 1.5 mm in the areas of porosity that were scattered along the outermost zone. The interface with the concrete was excellent in spite of the concrete being carbonated to a depth of 2 mm. The band of gunite adjacent to the concrete was cement rich compared to the remainder of the texture. The hardened cement paste was extremely rich in partially hydrated and remnant cement grains in which clusters of belite predominate. Abundant calcium hydroxide was present as large clusters of crystals and in the smaller voids indicating a reasonably wet mix was used. A limited number of small air voids were present together with larger entrapped voids but overall the void content was low. The bond between interfaces was excellent apart from one interface which was narrowly fissured for much of its length. Isolated drying shrinkage cracks were present in the body of the gunite and one larger shrinkage crack extended from the surface through the two outer layers. The maximum aggregate size was about 3 mm, was well graded and gave a compact texture to the gunite.

In the second case the cores of the shotcrete did not include the concrete substrate. The shotcrete varied between 100 and 140 mm in depth and was built up in three layers which can be seen by examining the section with the naked eye. However, in thin-section these layers were difficult to detect under the microscope apart from bands of poor compaction and very light carbonation. The outer surface was carbonated to about 1 mm and was characterised by poor compaction and numerous large irregular voids. The hardened cement paste was rich in remnant cement grains in which the ferrite phase was predominant and the abundant calcium hydroxide was reasonably well dispersed indicating a lower water/cement ratio was used. Moderate amounts of calcium hydroxide had crystallised in pores and voids. The maximum aggregate size was about 10 mm, gap graded to a well graded sand. There are numerous cracks and fissures associated with the interface zones and the overall impression was of dense shotcrete which had been incorrectly applied in the interfacial zones.

In the third case the gunite had been applied as one 25 mm thick layer over the prestressing strand. The texture was dense apart from the zones at the surface and adjacent to the prestressing strand. These were porous and full of large irregular voids. Carbonation had penetrated from 3 to 6 mm from the outside. The texture of the hardened matrix was similar to the first case in many respects.

The three textures described illustrate that in the second and third cases the operators applying the sprayed concrete were having problems in obtaining adequate compaction at the beginning and end of each layer. While the textures are only described qualitatively, there would be few problems in making quantitative measurements

7.5.3   GROUT

The term grout describes cementing materials that are fluids rather than plastic mixes. They are intended to fill spaces such as cracks, joints, prestressing ducts, to bed machinery and make preplaced concrete. Chemical grouts are used extensively in geotechnology to consolidate ground for large works. Grouts are based on Portland cement with additions and sand, filled polymers such as epoxy resin and proprietary mixtures of chemicals. Apart from simple cement grouts used for masonry and tiling most grouts are proprietary formulations and usually put in place by specialist firms. Chemical and polymer grouts are outside the scope of this book and texts such as those by Karol (1990) should be consulted for further information.

Masonry grout is specified in ASTM C 476 (1995) and consists of Portland cement plus 0–10 per cent of hydrated lime mixed with either 2–3 parts of fine sand or 1–2 parts of coarse aggregate by volume. The minimum compressive strength required is 14 MPa but there is no limit on the drying shrinkage specified. ASTM C 938 (1980) outlines the practice for proportioning grout mixtures of preplaced aggregate concrete which includes admixtures containing cement, pozzolana, fine aggregate, fluidifier and other chemical admixtures. Tests for fluidity, expansion, bleeding, water retentivity and unit weight are specified without performance requirements. Three types of packaged non-shrink hydraulic cement grouts are specified in ASTM C 1107 (1991) intended for use under applied loads such as bedding machinery or structures. These are pre-hardening volume-adjusting, post-hardening volume adjusting and combination volume adjusting types. Compressive strengths of 34.5 MPa are required and limits of 0–4 per cent presetting and 0–0.3 per cent post hardening expansions are specified.

Since the principal purpose of a grout is to fill a space so that loads can be transferred across the space, any shrinkage that reduces the load capacity is undesirable. Grouts based on Portland cement will shrink on drying unless additions are used to counteract this effect. Older antishrink additions are based on the expansion of rusting iron powder or the use of gas grouts based on aluminium powder, azides and other chemicals and fluid coke additions (Shaw 1985). Modern cement grouts often use shrinkage-compensating additions based on either calcium sulfoaluminates or lime or mixtures of Portland cement, high-alumina cement and gypsum (Mailvaganam 1984). In practice proprietary formulations are used on which precise information is difficult to obtain.

There seems to be general agreement that cement-based grouts are satisfactory provided that they do not have to withstand vibrating or rolling loads. In these cases epoxy resin-based grouts are usually recommended. Petrographic examination of grout is limited to

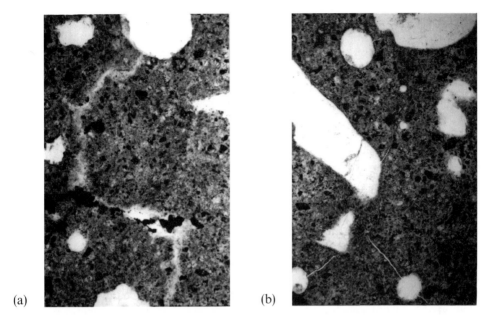

(a)   (b)

Fig. 7.7 The texture of a failed grout containing iron particles. (Widths of fields 0.9 mm). (a) The dark irregular particles of iron are visible in the cement-rich texture of the grout which also contains some ground granulated slag. The sand present is quartz. A bleeding channel meanders through the texture. (b) In spite of the use of iron particles as an anti-shrink additive fine drying shrinkage cracks were present in parts of the texture of this grout.

examination of texture unless a sample can be cut across the interfaces with material being grouted. The textures in thin-section of epoxy and cement materials used to grout a crane rail foundation are described below. The original cement grout had failed and attempts had been made to regrout the rail using an epoxy-based material.

The epoxy grout was heavily filled with rounded quartz sand ranging from 1000 to 50 μm with fly ash as a fine filler. About 5 per cent of a well-dispersed air void system was present with voids ranging from 150 to 50 μm in diameter. Adhesion to the old carbonated concrete foundation surface was good and the integrity of the grout was excellent (Plate 75). In contrast the Portland cement based grout had extensively failed due to either exfoliation or rectangular patterns of cracking and the adhesion to the foundation was poor or absent. The grout consisted of Portland cement, quartz sand 2 to 0.1 mm in size, iron particles 0.5 to 0.1 mm in size, slag particles and unidentified chemical additions. The hardened matrix was very rich in calcium hydroxide and partially hydrated cement grains and pores were filled or lined with calcium hydroxide and in some cases ettringite was present. The top surface was carbonated to a depth of 5 mm and carbonation extended up to 20 mm down cracks. In places an unidentified brown material was present and this combined with other observations suggested inadequate mixing had occurred, Fig. 7.7.

7.6   Floor finishes

The wide range of finishes used on concrete floors are described by Deane (1989). Those of interest are floor screeds, terrazzo, tiles and polymer coatings. BS CP 204 (1970) now partly replaced by BS 8204 (1987–94), specifies the requirements for *in situ* floorings and comprehensive information on the design of concrete floors for commercial and industrial use has been published by the New Zealand Portland Cement Association (1962). Magnesium oxychloride cements for flooring were covered in BS 776 (1972) but this standard has now been withdrawn. Information on this cement is given by Lea (1970).

7.6.1   FLOOR SCREEDS

The simplest method of finishing a floor is to screed it level soon after placing the concrete, to wait until initial set has occurred, and then trowel the surface to form a skin of compacted cement without the formation of laitance. This technique produces a thin skin of cement-enriched concrete with a low water/cement ratio on the surface. If required the surface can be later treated with surface-hardening solutions which further increase hardness and impermeability. If the hand trowelling is replaced by power floating even harder surfaces can be made. Provided that the surface is adequately cured a hard-wearing, non-dusting surface is produced which is adequate for many purposes.

The texture in thin-section of this type of surface finish will consist of a narrow band of cement-enriched concrete but the use of surface hardeners may be difficult to observe. Carbonation should not penetrate much beyond the enriched surface zone and drying shrinkage cracks should be only a few microns wide.

Where the floor is intended for heavy duty industrial use increased abrasion and chemical resistance may be required. In this case a special topping is used which may be laid over the still plastic floor as a thin monolithic screed or be placed as a thicker screed over an already cast floor. The topping is specifically designed for the conditions of use and may include selected aggregate to increase abrasion resistance. Alternatively, a thin layer of an armour-plate finish containing iron particles may be laid on the surface. Because the topping is relatively thin it is possible to use special materials and low water/cement ratios as well as finishing with a power compacter.

The texture of a proprietary industrial finish is shown in Plate 76. The black-coloured topping was variable in thickness and no more than 3–4 mm thick, contained ragged iron particles 25–1000 μm in length and was well bonded to the concrete underneath. Carbonation of the surface varied from almost nil to a depth of 1–2 mm. The concrete underneath was also carbonated to a depth of about 2 mm indicating the topping was laid over a hardened surface. The black pigment was not apparent but scattered throughout the topping were fine particles of moderate birefringent material. Overall the texture is somewhat confused and has the appearance of a powdered material that has been smoothed over a moderately rough concrete surface.

Petrographic examination of floor finishes can provide considerable information on the detailed texture of both the topping and the underlying concrete. Providing that adequate coring and sectioning is carried out it is an excellent method for examining the interfaces and relationship of the various layers making up the floor system.

7.6.2   TERRAZZO

Terrazzo, also known as Venetian mosaic, is a specialised concrete finish consisting of marble chips set in a matrix of white or coloured cement the surface of which has been ground to give a smooth finish. The process is described in Specification 94 (Williams 1994) and the properties of terrazzo tiles are specified in BS 4131 (1973). The marble chips

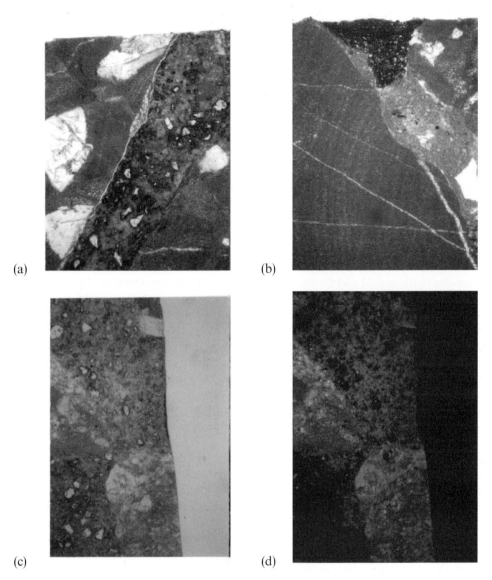

(a)　　　　　　　　　　　　(b)

(c)　　　　　　　　　　　　(d)

Fig. 7.8 The textures of a commercial terrazzo floor. Widths of fields 0.9 mm. (a) A band of rich hardened cement paste runs down between two limestone aggregates and a drying shrinkage crack runs along the margin of the left-hand aggregate. The top surface is relatively smooth from the grinding of the surface. This is a typical texture of a terrazzo floor. (b) The same floor in an area of pitting damage which became apparent soon after the floor was laid. A deep pit between two aggregates is filled with dark material which appears to be a mixture of debris. This type of pit tends to stand out clearly especially when incident light grazes the surface. Pits of this nature are difficult to remove by regrinding. (c) The same floor with a shallow indentation which has trapped the dirt. This is more typical of normal wear and tear and is easily removed when the surface is reground. The small particles in the texture are partially hydrated and remnant cement grains. (d) Under crossed polars the hardened cement paste is seen to be quite deeply carbonated and the dirt appears as a thin isotropic patch. To the lower right is an area of uncarbonated paste which has been shielded by the limestone aggregate at the surface. (Note: The surface in (a) and (b) is at the top, whereas the surface in (c) and (d) is on the right-hand side.)

are graded in size according to the surface appearance required and must be free of dust. A rich cement mix varying between 1:2.5 to 1:3 cement-aggregate ratio is used and up to 10 per cent of pigment may be added to give the depth of colour required. Water/cement ratios are usually about 0.5 so that care is required in mixing and laying to ensure that segregation of the marble chip is avoided.

The thickness of the terrazzo varies according to the job. For floors 20 mm is recommended but thinner dimensions for precast slabs and wall finishes may be used. Terrazzo is laid as a finish either on a backing, such as concrete levelled with a mortar, or using a cheaper backing mix in the case of precast tiles. In the case of floors the terrazzo may be isolated either by a bitumen or sand layer or firmly bonded by the use of grout or by directly laying on plastic concrete.

After placement the terrazzo mix is screeded level without bringing laitance to the surface and wet cured for at least seven days before coarse grinding. This removes the top few millimetres to reveal the aggregate and any unfilled voids. These are grouted with the same cement and pigment used in the original mix and the surface is again wet cured to harden the grout. The surface is then ground with successively finer grades of carborundum wheels until the desired finish is achieved. In the case of precast tiles the terrazzo mix is placed in the bottom of the mould and the backing mix laid over the top. Compaction is achieved by vibration and pressing.

The texture in thin-section of a terrazzo topping is typical of an under-sanded concrete containing a high cement content. If coloured cement has been used much of the fine detail of the hardened cement paste may be obscured by the pigment. Because of the rich wet cement mixes used drying shrinkage cracks, about 10 μm in width will always be present especially along the margins of aggregates (Fig. 7.8). The extent of drying shrinkage present is largely a function of paste content. Surfaces are usually carbonated to a depth of 1 to 2 mm and carbonation may extend up to 5 mm down aggregate margins and fill the entrances to shrinkage cracks (Fig. 7.8). Isolated patches of carbonation may also be present dependent on compaction.

One of the faults encountered with terrazzo is the presence of unsightly and excessive cracking. As with most types of concrete floor finishes it is essential that sufficient joints in the backing are used to prevent movement in the structure from cracking the terrazzo. Additional joints restricting the area of the terrazzo topping to a maximum of 1 m^2 are required to control cracking due to drying shrinkage. Some visible, isolated drying shrinkage cracks are inevitable but if these become extensive they soon fill with dirt and become unsightly. Larger cracks are almost invariably due to structural movement.

Because the surface of terrazzo is smooth and appears polished its appearance is sensitive to surface defects. Some problems are generated during construction. Reducing curing time before grinding can result in weakened surfaces which only become apparent with wear. New terrazzo is particularly prone to damage from other sub-trades and wherever possible should be laid last. There is also a problem of slippery surfaces. More coarsely ground surfaces are safer but lack a polished appearance and may be waxed to improve their appearance which is counter productive. Incorrect cleaning procedures can damage terrazzo and is usually evident as pitting of the marble chips. Kessler (1948) investigated the effect of soda ash, tri-sodium phosphate and a synthetic sulfonate on terrazzo and found that tri-sodium phosphate was less injurious than soda ash and that the synthetic sulfonate was relatively free from injurious action.

Terrazzo is a decorative finish which the trade claims to be a hard and durable finish for walls and floors. These claims are true only if the terrazzo is correctly formulated and laid by skilled contractors. In addition, floors must be adequately maintained to prevent damage.

With time terrazzo floors become scratched and ingrained dirt in surface defects give it an unacceptable appearance (Fig. 7.8(c), (d)). Provided these defects are superficial they can be removed by light regrinding. Where deeper pits and extensive cracking are present due to faulty construction and incorrect cleaning procedures effective regrinding of surfaces may not be possible (Fig. 7.8(b)).

7.6.3 TILED FINISHES

The tiles used on floors and walls are usually set in a cement mortar which now commonly contains bonding agents based on styrene-butadiene latex. The joints are filled with a grout which in many cases contains no Portland cement and may be based on polymers. Successful tiling depends both on the correct use of materials and the skill of the tiler. Current UK practice is described by Robinson *et al.* (1992) and Williams (1994).

The failure of tiled surfaces mainly involves loss of adhesion either between the tile and the bedding mortar or the bedding mortar and the substrate. Tiles and mosaics are dimensionally stable relative to mortar and concrete and unless potential movements of the concrete and mortar are controlled by appropriate joints, lifting of tiles may occur.

Coring of tiled surfaces and examination of thin-sections provides information on composition of mortar and concrete but is justified only if compositions are in question. Chemical determination of cement content may be required where it is suspected that an over-rich mortar has been used. On walls where adhesion between the bedding mortar and the concrete has failed the cause of the failure can be difficult to determine.

The texture in thin-section of a floor tile setting is shown in Plate 45. Between the mortar bed and the tile there is a layer about 0.5 mm thick of neat cement which was known to have been mixed with styrene-butadiene latex although there is no visible evidence for the presence of the latex. The hardened matrix of the bonding layer is very rich in cement grains most of which do not show obvious signs of hydration rims. A limited amount of calcium hydroxide is present in the matrix with larger crystals present in the air voids. The bond to the tile and the underlying mortar bed is excellent. Where the mortar bed is adequately compacted it consists of closely packed sand particles with only a limited amount of cement paste. Portions of the bed are poorly compacted with large fissures. The hardened matrix contains minimal calcium hydroxide and remnant cement grains apart from large crystals of calcium hydroxide growing into the irregular voids.

The petrographic examination of this tile floor was carried out because of 'drummy' (hollow sounding caused by detachment) and cracking tiles. Chemical analysis showed that the cement content of the mortar varied between 1:5 and 1:10 and this, combined with poor workmanship in the laying of the tiles, resulted in them being damaged due to foot traffic and normal wear and tear.

7.6.4 POLYMER FLOOR COATINGS

Polymer floor coatings on concrete and mortar screeds usually consist of filled resins of either epoxy or polyester or polyurethane. Epoxy resins give the most resistant finish. Formulations are usually proprietary and are applied by specialist contractors.

The texture in thin-section of a polymer floor finish is shown in Plate 77. The concrete floor has been roughly screeded and a mortar screed laid on top. Bond between the mortar screed and the concrete is variable with numerous microcracks and fissures present at the interface. In contrast the polymer bond to the mortar screed is excellent. The surface of the mortar screed is reasonably flat and is lightly carbonated to a depth of 1–2 mm. The polymer topping is in two layers. The bottom layer consists of a 1 mm thick layer of isotropic brown polymer which contains a limited amount of aggregate 100–150 μm in size and is full of voids 100–200 μm in diameter. The top layer is a colourless polymer about 0.5 mm thick and is heavily filled with quartz flour.

The polymers were not identified by infrared spectroscopy although they were reputed to be a polyester. In thin-section the resins showed similar moderate negative relief with respect to the epoxy mounting resin which has a refractive index of about 1.58. Polyester resins often have refractive indices of about 1.54. The moderate negative relief observed suggests the polymer floor finish is based on polyester resin.

This polymer floor was cored because the topping was lifting in patches. Coring revealed that no moisture barrier had been used in the floor and moisture movement up through the slab was the cause of the lifting. In spite of the excellent polymer bond to the screed it was insufficient to resist the movement of moisture through the slab.

7.7   Lime materials and products

Both lime and gypsum are among the oldest of construction materials. Various forms of lime were used in mortar and plaster and to make concrete by the early Egyptians and Greeks. It was used extensively by the Romans, in China and other societies (Lea 1970). The problem in adequately discussing lime is the large number of ways this material has been used for construction in the past, many of which have now lapsed in the West although some of these old techniques are still used in some parts of the world (Hill *et al.* 1992).

7.7.1   *LIMESTONE AND LIME*

Both limestone and lime are used extensively in industry, agriculture as well as construction materials. The manufacture and uses of lime are discussed by Searle (1935) and Boynton (1980). Construction uses include the intergrinding of limestone with Portland cement as a diluent (Section 5.2.7) of hydrated lime in mortars and plasters and the use of hydrated lime with a range of siliceous materials to manufacture bricks, blocks and other products. Both hydrated and quick limes are used extensively to stabilise soils especially in road construction.

Hydrated lime is manufactured by crushing either limestone or magnesian or dolomitic limestones and heating them to between 1000 and 1200°C to produce quick lime which is subsequently slaked with water to produce hydrated lime. The slaking process disintegrates the lumps of quick lime to a powder, putty or slurry depending on the amount of water used and grinding is not normally required. In modern production, slaking is carried out under pressure which ensures full hydration and the product may also be ground to control particle size.

$$CaCO_3 + heat \rightarrow CaO + CO_2$$
$$CaCO_3.MgCO_3 + heat \rightarrow CaO.MgO + 2CO_2$$

$$CaO + H_2O \rightarrow Ca(OH)_2$$
$$CaO.MgO + 2H_2O \rightarrow Ca(OH)_2.Mg(OH)_2$$

The mechanism of hardening of a hydrated lime is simply the formation of calcium and magnesium carbonates by absorption and reaction with carbon dioxide from the air as follows.

$$Ca(OH)_2 + CO_2 \rightarrow CaCO_3 + H_2O$$

$$Mg(OH)_2 + CO_2 \rightarrow MgCO_3 + H_2O$$

Atmospheric carbonation is slow and usually partial because the carbonated margins of plaster or mortar impede the access of carbon dioxide. A better idea of conditions controlling carbonation can be derived from artificial carbonation as discussed by Moorhead (1986). Maximum rates of carbonation were attained when the product was exposed to an atmosphere of 100 per cent carbon dioxide and the capillary pores in the lime mortar were half filled with water. Carbonation virtually ceased when the capillaries in the lime mortar were either saturated or dry. Carbonation was also controlled by the thickness of mortar and it was found that rapid artificial carbonation was impractical over depths greater than 25 mm. It is of interest that Moorhead was able to produce within days fully carbonated lime mortars with compressive strengths of 50 MPa.

The reactions given above are applicable to 'fat' limes. These are limes consisting of 90–95 per cent calcium and magnesium hydroxides as specified in BS 890 (1995), ASTM C 206 (1984) and ASTM C 207 (1991). The production of these limes requires a fairly pure limestone as the starting material. However, the majority of limestones are of lesser grade and those traditionally used for building lime production could contain from 5 to 40 per cent of clay, silica and other minor components. When these impure limestones are calcined both calcium and magnesium oxides and phases similar to those found in Portland cement are formed, such as the di-calcium silicate, aluminate and ferrite phases, which give the lime some hydraulic properties and the ability to set under water (Roberts 1956). Hydraulic limes are specified in BS 890 (1995) and ASTM C 141 (1985).

The mechanism of hardening of hydraulic limes is a combination of carbonation and hydration of di-calcium silicate. Where the lime is weakly hydraulic the lime may have good slaking characteristics and still retain some slow hydraulic activity even after slaking. Typically hydraulic limes that may contain up to 26 per cent silica (ASTM C 141 1985) may set within hours but will generally slake poorly and carbonation hardening is limited.

There is another mechanism of setting whereby a 'fat' lime is mixed with a pozzolanic material and the slow formation of a calcium silicate hydrate occurs to harden the mass into a dense and durable product, even under water. This is one of the types of cement used by the Romans and is so durable that it is still occasionally used for marine works. The modern autoclaved lime silica product is merely an extension of the old Roman cement. Lime for use with pozzolanas is specified in ASTM C 821 (1978).

Unlike the production of Portland cement, the techniques of lime burning varied widely. Modern calcining techniques are efficient and use high-grade limestone to produce quality lime, often intended for industrial use as well as building. The least controlled process is the old and simple technique of using a local limestone of indeterminate composition and stacking the crushed limestone and wood in layers in a ventilated pit and burning the wood to provide the heat for calcination. This results in a range of slaking characteristics, 'fatness' and hardening properties of the lime putties used for construction. The wide diversity of burning techniques employed, especially in the past, are beyond the scope of this text and details can be found in Searle (1935) and Boynton (1980).

The main problems of interest to the petrographer that arise from lime burning are caused by variability of raw material, and insufficient control of calcination causing underburning and overburning. In many limes these are interrelated and may be present together. Underburning causes 'core' which is often removed in later processing. This is where the exterior of the limestone particle is calcined and the interior remains as calcium carbonate. Overburning can make the calcium and magnesium oxides resistant to slaking and result in undue later expansion as the calcium and magnesium oxides slowly hydrate in the mortar. Traditionally this problem was avoided by slaking the lime for weeks or even months before use. Another problem arising from overburning is clinkering. This is where fusion occurs in particles containing fluxes so that they form a 'natural cement'. Traditionally, clinker is removed from quick lime as it does not slake well and will not form a powder. Ground up clinker was used as a 'natural cement' and apparently was still being made for this purpose in some countries (Lea 1970).

Underburning and overburning occur not only where there is excessive temperature variation across the calcination zone but also if the raw feed is variable in composition. Each limestone has an optimum burning condition and unless variability in composition can be controlled the calcination cannot be optimised. According to Boynton (1980) variability of composition and properties have been a persistent problem in the production of hydraulic limes. Microscopic methods for determining lime quality and burning characteristics of high-grade pebble lime are given by McKenzie (1994, 1995).

If the hydrated lime has not been adequately calcined or matured it may contain particles of unslaked lime which will gradually absorb water from the atmosphere and will slake and 'blow' after the plaster has been incorporated in the building. Soundness tests for hydrated lime are provided by BS 6463 (1987) and ASTM C110 (1992).

7.7.2   *LIME PLASTER AND MORTAR*

According to Wickens (1992) lime plastering was the most commonly accepted system of plastering in the UK before 1939 but is now virtually obsolete. Some idea of the range of limes used for plaster, renders and mortars is given in Fig. 7.9. Currently BS 890 (1995) specifies only high-calcium, semi-hydraulic and magnesian limes. Methods of test for composition, fineness, soundness, standard consistency and hydraulic strength are now specified in BS 6463. Codes of practice for external renderings are given in BS 5262 (1991) and for internal plastering in BS 5492 (1990) which should be referred to for further details of recommended mixes. Masonry mortars are specified in BS 5224 (1995) and ASTM C 91 (1993).

The varieties of lime possibly used in the UK are shown in Fig. 7.9 The 'fat' limes should primarily be calcium hydroxide or carbonate with magnesium hydroxide and carbonate also being present in magnesian limes. However, the possibility of remnant particles of hardburned calcium and magnesium oxides should always be considered. Other materials likely to be found in lime mortar and plaster are aggregate, Portland cement and fibre reinforcement. The most common aggregate will be sand but both perlite and vermiculite are used in sprayed insulating plaster.

In hydraulic limes silica and alumina in the limestone form dicalcium silicate and calcium aluminates on calcination which give rise to hydrated phases similar to those found in hydrated Portland cement. This poses a problem in differentiating between hydraulic lime and lime intermixed with Portland cement. However, careful microscopic examination of the material in thin-section, or preferably in polished section, should be able to distinguish

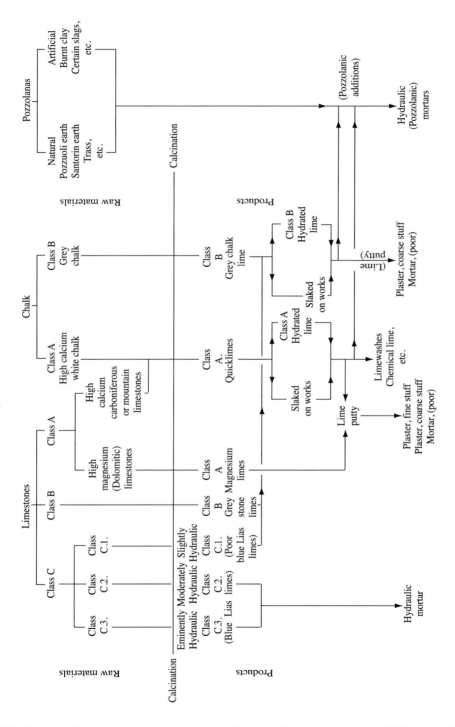

Fig. 7.9 Diagram illustrating the various types and uses of lime produced in the UK in the 1920s. (Cowper 1927).

whether any remnants of alite are present. If remnants of alite are present it indicates the use of Portland cement as alite is not a normal component of hydraulic lime. Since the amount of alite residue will be small identification may be difficult using X-ray diffraction analysis.

As already discussed in Section 7.7.1 the rate of carbonation of lime is dependent both on the concentration of carbon dioxide and the amount of water present in the capillary space inside the plaster or mortar. Because most of the precipitation of the carbonate initially occurs in the capillary space the permeability of the carbonated zone to carbon dioxide is reduced. Thus in most mortars even those that are hundreds of years old there may be a central core of uncarbonated lime remaining.

The mean particle sizes of hydrated limes vary between about 3 and 7 μm with few particles exceeding 30 μm (Boynton 1980). This indicates that a proportion of remnant calcium hydroxide crystals will be large enough to be identifiable under the microscope unless the hydrated lime has been ground. Similarly, particles of hard burnt calcium oxide and magnesium oxide should be readily identifiable as both are isotropic and have distinctive shapes (see Glossary).

The texture of the carbonates formed depends on the conditions of carbonation. At one extreme, the artificially carbonated lime described by Moorhead (1986) consisted of crystals generally below 1 μm and there was evidence to suggest that about half the carbonate crystals were less than 0.1 μm in size and even possibly near amorphous. This is the texture of very rapid carbonation where crystals have had insufficient time to develop. At the other extreme will be plaster and mortar that has been damp but not saturated for very long periods. It would be expected that in these cases large crystals of carbonate would develop in void space.

Ancient lime plasters, mortars and concretes

It has been reported by Malinowski and Garfinkel (1991) that a type of dense lime concrete was used to make polished floors in Jericho dated around 7000 BC. Both the binder and aggregate used consisted of calcareous material and particles of charcoal used to burn the lime were present. The texture of the lime concrete was likened to travertine but as no thin-section examination was carried out further details of the texture they observed are not available.

Sims (1975) found that ancient lime mortars and concretes are quite distinctive in thin-section under the petrological microscope, with the matrix being predominantly finely crystalline calcite (Plate 10) together with residual particles of partially burnt limestone, charcoal contamination from the original lime kiln and patches of more coarsely crystallised calcite (Plate 11) deriving from the slow carbonation of pockets of trapped slaked lime (Sims 1975).

The Greeks and Romans both appear to have understood the differences between hydraulic and non-hydraulic limes and the use of pozzolanas to give added strength and resistance against water (Vitruvius 1962). The pozzolanas they used not only included materials such as the volcanic sands from Pozzuoli, but calcined clays usually in the form of ground-up bricks, tiles and potsherds. Lea (1970) points out that much of the excellence of Roman mortars depended not on any secret in the slaking or composition of the lime but in the thoroughness of mixing and ramming. The Romans also understood the need to use clean, graded sand. According to Davey (1965) Roman mortars and concretes in England contained sand and coarse material which was selected for size and may have been specially screened. The proportions of lime to sand are quite similar to those used in modern lime mortars.

From ancient times until the mid-nineteenth century lime was used in a similar fashion to the Romans in many parts of the East and in both the New and the Old World although in medieval times the use of hydraulic limes seems to have lapsed. While there is a wide variation in materials, petrographically this variation can be assessed on the basis of texture and chemistry. Where a 'fat' lime has been used the only two phases possible are calcium hydroxide and calcium carbonate unless a sand with pozzolanic properties is present. Any fine crystals of calcium hydroxide dispersed among calcium carbonate will be impossible to distinguish unless X-ray diffraction analysis is used. Thus only calcium hydroxide present in discrete areas or pockets will be visible in thin-section.

The presence of pozzolanic sand or calcined clay will alter the hardened matrix which may however still be predominantly calcium hydroxide and calcium carbonate. In the matrix reasonably well-crystallised CSH(I), which can be identified only by careful X-ray diffraction analysis, will form as the result of the pozzolanic reaction. Reaction rims around some pozzolanic grains of sand should be visible as the reaction is slow unless the pozzolana is very finely ground.

Hydraulic mortars contain β-dicalcium silicate and some aluminate and ferrite phases. The coarser portions of both β-dicalcium silicate and the ferrite phases can be surprisingly resistant to hydration and remnants of these phases should be identifiable in thin-section providing sufficient cross-section is examined. X-ray diffraction analysis may be insufficient to identify these phases as residual concentrations of the cementing phases may be below its detection limit. CSH(I) is also formed from the hydration of β-dicalcium silicate and X-ray analysis will not distinguish this from CSH formed by pozzolanic reaction. If both a hydraulic lime and a pozzolana have been used, which apparently was not uncommon, the attributes of both textures described above should be present. Where chemical analysis is required the techniques investigated by Stewart (1982) should be used.

The crystallinity of the calcium carbonate will depend on both the age and the amount of moisture that has been present in the plaster or mortar during its lifetime. Initially, the calcium carbonate crystals will be small but providing the moisture conditions are favourable coarser crystals will grow at the expense of finer crystals. Until the industrial era wood was used to burn lime so that fragments of charcoal are almost invariably present in the ancient lime materials. In some cases hair, reeds or other reinforcing material may have been used which may be identified in thin-section (McCrone and Delly 1973).

7.7.3   CALCIUM SILICATE PRODUCTS

Calcium silicate products consist of a mixture either of siliceous sand or flour and about 10–15 per cent of a well slaked, high calcium lime or pulverised quick lime. The mixture is pressure moulded while still 'green' and the product autoclaved for about 16 hours around 170°C. There is a range of materials and processes used to make calcium silicate products, the commonest products being brick and block (Bessey 1967).

Sand lime and flint lime products use either natural siliceous sand, ranging from 2 mm down to 20 μm in size (Quincke 1967) or crushed flint as the siliceous component and often incorporate pigments. Some European countries add up to 15 per cent of silica flour to accelerate reaction and improve strength and also may add small quantities of clay to aid moulding. These products are specified in BS 187 (1978) and the application of sandlime block in the UK is described by the Building Research Establishment Digest 157 (1981).

Aerated calcium silicate concrete block and building components are used extensively in Scandinavia, where the process was developed, and in eastern Europe. Many details of the patented manufacturing processes used are not available but it is known that the composition essentially consists of high-grade quick lime and silica flour with the possible inclusion of some fine sand and Portland cement. Aeration or foaming is usually achieved by the use of aluminium powder added to the mixture which generates hydrogen by reacting with calcium hydroxide to form a calcium aluminate. By varying the amount of foaming agent and mix proportions lightweight products varying from a density of 0.9 to 0.2 can be made with the range 0.2–0.5 being the most usual.

Silicate concrete was developed in the USSR and is based on milling high grade lime and silica sand to increase reactivity and is used to make bricks, blocks and building units. It differs from sand lime products in that compressive strength ranges from 50 to 80 MPa compared with a strengths for the latter of 10 to 50 MPa. Compressive strengths of aerated products range from 1.5 to 5 MPa dependent on density (Short and Kinniburgh 1978).

Two important physical properties of calcium silicate products that affect durability are drying shrinkage and carbonation. BS 187 (1978) limits the drying shrinkage of sand lime bricks to a maximum of 0.04 per cent. According to Rudnai (1963) the shrinkage of autoclaved aerated products is usually less than 0.05 per cent but from saturated to fully dry condition this may vary up to 0.09 per cent. Precise details of carbonation are not available but it is generally agreed that all lime silica products carbonate with time given the appropriate conditions. Overall the durability of sand lime bricks appears adequate as indicated by the long-term tests carried out by Bessey and Harrison (1970). Short and Kinniburgh (1978) state that damp conditions lead to an increase in strength of autoclaved aerated concrete but in circulating water strength may deteriorate because of leaching of lime.

Petrographic features of calcium silicate products

The siliceous aggregate is the major component in sand lime bricks and since the reaction with quartz or flint grains is only partial they form the dominant constituent when viewed in thin-section. In old bricks the cementing material is partly or wholly carbonated so that at low magnifications the aggregate grains are seen to be embedded in a highly birefringent fine-grained carbonate matrix. Irregular patches of featureless cementing material between quartz grains vary in size with the extent of the carbonation (Plate 78(a)). In some examples small patches of the cementing medium exhibit low birefringence indicating it is more crystalline. Calcium hydroxide is present in these uncarbonated areas of matrix and forms large equant or elongate crystals up to 100 μm in size. The majority of these crystals form a discontinuous fringe around the quartz grains and remain even within areas of partial carbonation as shown in Plate 78(a). Plate 78(b) shows an area of matrix with calcium hydroxide crystals bordering the quartz grains and the development of a small area of more crystalline matrix and Plate 79 shows an unusual crystallisation of dicalcium silicate hydrate and actinolite in a pore.

The principal cementing phase formed in autoclaved calcium silicate products is a poorly crystallised calcium silicate hydrate CSH(I) with a Ca/Si ratio 0.9 to 1.0 (Taylor 1990). In autoclaved aerated lime silica products the pore system, silica flour and relics of the lime used together with associated carbonation and microcracking are visible as shown in Section 5.7.5. Textural details of calcium silicate concrete are not available but it is assumed the texture will be a denser form of that observed in aerated lime silica products. Some textures of foamed blocks containing fly ash are shown in Plates 80 and 81.

7.8   Gypsum, hemihydrate and anhydrite

Hemihydrate is principally used to cast wall boards, building components and for plastering internal surfaces of buildings. It has some unique properties. On setting it undergoes expansion which allows accurate replication of very fine details when used for moulding. Subsequent drying may reverse much of this setting expansion (Blakey 1959). Hemihydrate can be made to set in times that vary from a few minutes to several hours which enable it to be used for a wide range of casting and internal plastering purposes. Once it has set it attains its full strength on drying so that products do not have to be cured like Portland cement. It is non-toxic and has excellent fire resistance. Its main drawback is that gypsum has a solubility of 0.24 per cent in water and thus it is not durable in moist conditions. It is one of the traditional building materials and its use extends back at least 4000 years to the construction of the Egyptian pyramids.

The traditional production of hemihydrate, also known as plaster of Paris or bassanite, is to heat ground gypsum in a large kettle at about 170°C to drive off 1.5 mols of water to form coarse building plaster which is designated β-hemihydrate. Modern production methods now use counter-current rotary kilns and other energy-efficient devices (Daligand 1985). The anhydrous group is produced by calcining gypsum to temperatures in excess of 400°C or greater to produce anhydrite and this group is restricted to plastering.

$$2(CaSO_4.2H_2O) + heat \rightarrow 2(CaSO_4.\tfrac{1}{2}H_2O) + 3H_2O$$

$$CaSO_4.2H_2O + heat \rightarrow CaSO_4 + 2H_2O$$

The setting of hemihydrate and anhydrite with water is the reverse of the above equations to reform gypsum as a set crystalline mass. The natural mineral anhydrite can be used directly by grinding with the addition of a set accelerator to produce an anhydrous plaster without the necessity of heating in its manufacture.

There is considerable complexity in the types of gypsum plasters that can be made which for any one crystalline phase are mainly due to subtle changes in crystallinity. In spite of subdivisions such as α and β hemihydrate based on setting and physical properties there are only four clearly crystalline phases. These are gypsum, hemihydrate, anhydrite and the highly metastable soluble anhydrite (Ridge and Berekta 1969). The range of gypsum plasters used in the early part of this century (Pippard 1938), are shown in Fig. 7.10. All except the hemihydrate types now find little use.

The duration and temperature of calcination affect the setting time and physical characteristics of gypsum plaster. For instance β-hemihydrate can be produced to give setting times between about 10 and 30 minutes without the use of accelerators or retarders. A high grade of hemihydrate can also be made by autoclaving ground gypsum to produce α-hemihydrate commonly known as dental plaster. This plaster requires about half the amount of water as β-hemihydrate to give a fluid mixture and produces much stronger casts as a result.

The manufacture and technology of gypsum plaster is much more complex than is generally realised and much of the industrial information is available only through the patent literature or published in eastern Europe and Japan. There are few texts and the translation of a Russian text by Volzhenskii and Ferronskaya (1974) is difficult to obtain. Kuntze (1976 to 1986) annually reviewed the literature on gypsum technology and technical papers are regularly published in the journals *Zement-Kalk-Gips* and *Ciments, Bétons, Platres, Chaux*. The changes in the types and uses of calcium sulfate plasters in the UK can be seen from comparing the data given by Pippard (1938), Andrews (1948) and Wickens

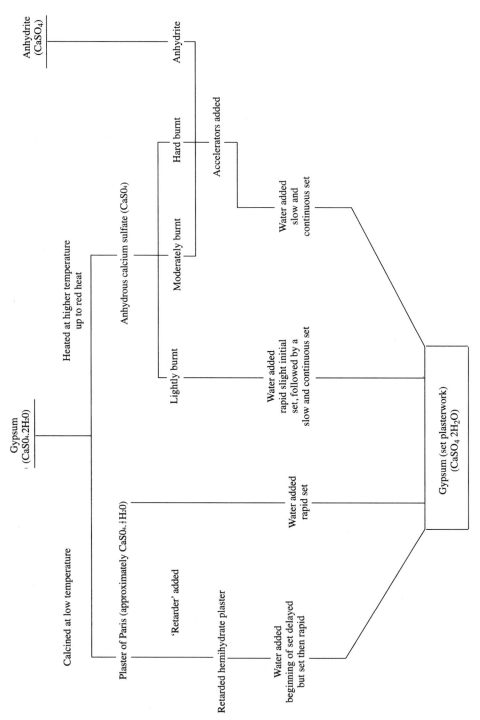

Fig. 7.10 The types of hemihydrate and anhydrite plasters in use in 1938. The accelerated anhydrite plasters were introduced in the early 19th century (after Pippard 1938).

(1992). The properties of cast gypsum as a structural material have been reviewed by Blakey (1959) and the historical use of gypsum plaster is discussed by Davey (1965) and Ashurst and Ashurst (1988).

In the UK gypsum plasters are divided as follows into four classes as specified in BS 1191, Part 1 (1973) as follows with only the first two classes being currently manufactured in the UK.

Class A: Plaster of Paris ($CaSO_4.\frac{1}{2}H_2O$) is used extensively for casting and stopping plasters. It sets too rapidly for most plastering purposes, but is sometimes used for gauging lime.

Class B: Retarded hemihydrate gypsum plaster is available in several grades for under-coats and final coats. It is used neat, or as a 50:50 mix gauged with lime putty.

Class C: Anhydrous gypsum plaster ($CaSO_4$) is used for final coats, sanded coats or undercoats where its gradual and progressive setting allows a high standard of finish.

Class D: Keene's plaster is an anhydrous final coat plaster that is used in specialist applications and can give a very hard smooth surface.

A variety of premixed lightweight gypsum plasters conforming to BS 1191, Part 2 (1973) are mixtures of the retarded hemihydrate gypsum (Class B) and lightweight aggregates such as perlite and vermiculite. ASTM C28 (1986) also specifies a similar range of materials including one containing wood fibre. Gypsum projection plasters are a blend of hemihydrate and anhydrous gypsum plaster formulated to be used with mechanical plastering machines which mix and spray the plasters as a one-coat application on the substrate before being trowelled by conventional methods.

Paper-faced gypsum board is manufactured by automated continuous processes where an accelerated gypsum plaster is rapidly mixed and aerated in a pin mixer and fed onto a continuous belt covered by paper backing. Another layer of paper is rolled onto the top surface to form a sandwich. The plaster sets quickly and the boards are cut to length and kiln dried. Apart from the kiln drying the whole process only takes a few minutes and large production rates are achieved so that this product constitutes a major use of building gypsum. A range of stopping plasters are associated with the installation of paper faced gypsum board. The simplest is based on coarse building plaster with possibly some added methyl cellulose as a trowelling agent. Fine cornice plaster and materials which are based on a fine filler and a polymer addition to provide hardening are used extensively.

There is also a range of hand-made cast gypsum products such as fibrous plaster board, ceiling tiles and ornamental cornices, coves and ceilings. These products were originally reinforced with natural fibres, commonly sisal, but over the last few decades this has been supplanted by 'E glass' fibre. The products containing fibre glass have superior flexural strength and unlike sisal, glass fibre does not interfere with the setting time of plaster. In this respect the use of glass fibre mats in board faces is very effective (King *et al.* 1972). The development of gypsum structural members using low water/plaster ratios achieved by vacuum casting and chopped fibre glass as reinforcement was developed at the Building Research Station (Ali and Grimer 1969). Strengths of 30 MPa were attained enabling the casting of wall floor and ceiling units.

Fig. 7.11 The texture of a gypsum plaster finishing coat. Width of field 1.5 mm. A dark paint layer is to the left. The texture of the gypsum plaster is too fine to show discrete crystals. The crack, highlighted by dye, just below the surface of the plaster is possibly an artefact of sectioning.

7.8.1   THE PETROGRAPHY OF FAULTS IN PRODUCTS MADE FROM HEMIHYDRATE AND ANHYDRITE

The hydration of hemihydrate and anhydrite produces gypsum so that this is the major component of the hardened matrix of products. BS 1191 requires Plaster of Paris to contain not less than 85 per cent hemihydrate. In practice many plasters contain over 95 per cent of hemihydrate. The impurities usually consist of calcium carbonate, clay and more rarely organic matter. Because contaminants are finely divided they may not be detectable by light microscopy and may require chemical and X-ray analysis and possibly scanning electron microscopy for their detection and estimation. Chemical additions such as sugar, starch, alum and potassium and zinc sulfates used as accelerators and proteinaceous retarders are difficult to detect chemically as they are present only at the 0.5 to 1 per cent level.

In thin-section gypsum plasters have a matrix dominated by an interlocking mass of acicular gypsum crystallites which may just be differentiated at the highest magnification. These acicular crystallites are set in anhedral masses of gypsum. Larger gypsum crystals are able to grow into voids (Plate 82) and can be used for optical identification. With Class

C and D gypsum plasters some residual anhydrite remains in the set and hardened matrix and can be used to identify this type of plaster. Anhydrite is easily recognised by its high birefringence and tendency to form rectangular crystals. In many examples it is seen in petrographic thin-section as discrete single grains but euhedral grains also occur within larger, brown-stained impurity particles. (Plate 83).

7.8.2   *GYPSUM PLASTER FINISHES*

Gypsum plastered finishes, especially older finishes, may contain 50–75 per cent of lime in the hardened matrix and be mixed with sand. Some neat finishes were reinforced with hair or natural fibre. The presence of lime, sand and fibre reinforcing are easily identifiable in thin-section and it should be possible to determine the mineralogy of the sand and type of fibre used.

There are a number of problems that can occur with gypsum plaster finishes (Fig. 7.11) which involve both the materials used and the workmanship. Clay and loam in sand can affect both strength and setting, while magnesian limes may cause defects due to delayed hydration of magnesia. Under some conditions magnesian limes may give rise to efflorescence due to the formation of magnesium sulfate. The set of Class A and B plasters can be killed by working the surface after initial set has occurred giving low-strength surfaces prone to dusting. Anhydrous plasters which may take two to three days to set must not be allowed to dry out during this period. Otherwise low-strength and delayed setting expansion can result. Calcium sulfate plasters may with time corrode iron, steel, lead, zinc and aluminium and these metals should be protected from direct contact. Even where lime is added this problem can still occur especially in damp conditions and in damp conditions gypsum loses strength. The absorption of as little as 1 per cent of water can reduce strength by up to 50 per cent. Long-term exposure allows a process known as 'rotting' to occur which involves a breakdown of the crystal interlock. These problems need to be considered by the petrographer when investigating gypsum plaster failures.

Cast gypsum products are invariably made from hemihydrate and do not suffer from many of the problems encountered with plaster finishes. In hand-cast products manufacturers may use excess water for gauging the plaster to obtain increased fluidity for casting and larger output volumes. Coarse casting plaster ideally should be mixed at a powder/water ratio of 0.7 to 0.8 but ratios as high as 1.0 may be used resulting in low strength porous casts. This is difficult to detect and measure (St John 1975). A less common problem can occur where products are overheated during kiln drying causing calcination of the plaster with reduction in strength and cracking. This can usually be detected by X-ray diffraction analysis. Because hand-cast gypsum products use high water/powder ratios they are porous and prone to absorption of smoke and stains which can migrate through the thickness of the product (St John and Markham 1976, Wilson and St John 1979).

The strength of paper-faced gypsum wall board lies in the paper facings and the gypsum is only required to be a low-strength inert filling preferably with a low density. Density is reduced usually by entraining air but other materials such as wood fibre and pumice have been used. Some sections from a Gyproc board are shown in Plate 84.

7.9   Lightweight aggregates and concretes

In concrete construction the dead load of the structure forms a major part of the total loading so there are economic advantages in using lower-density concrete. The advantages are reduction in the size of formwork, foundations and structural members and greater ease of handling and placing the concrete. With blockwork, ease of handling and greater insulating properties are the most important advantages. The disadvantages are lower strength, higher moisture movement, increased carbonation and susceptibility to corrosion of reinforcing steel as the density of the concrete decreases.

Natural lightweight aggregates are pumice, scoria, volcanic tuffs and ashes. These names describe textures of volcanic materials and thus are not rocks of specific composition. Pumice is a white to grey coloured highly cellular glassy lava generally of rhyolitic composition while scoria is usually of basic composition and is characterised by marked vessicularity, dark colour, heaviness and a texture that is partly glassy and partly crystalline. A tuff is a rock formed of compacted volcanic fragments varying from andesite to rhyolite in composition and ash is any loose unconsolidated volcanic material. The fragments making up these two latter materials are usually less than 4 mm in size.

Artificial materials are the most commonly used lightweight aggregates and are specified in BS 3797 (1990), ASTM C 330 (1989), 331 (1989) and 332 (1987). Their use in structural concrete is described by Clarke (1993). Furnace clinker and breeze have been used in the UK for many years and make satisfactory aggregates provided they contain less than 10 per cent of combustible material (Plate 85). Because their sulfur content causes corrosion problems with steel they may not be used in reinforced concrete. A more recent variant of this type of material is furnace bottom ash from power stations. It is somewhat variable in composition but the better grades may be crushed and screened to provide aggregate for block manufacture. Foamed and pelletised slags are manufactured from molten iron and steel slags. Molten slag is foamed with controlled amounts of steam and water and depending on the process aggregate can be produced as rounded pellets with sealed surfaces or material that requires crushing to size. Provided sulfate contents are kept below 1 per cent these materials make reliable lightweight aggregates from 14 mm downwards.

Many shales and clays can be expanded by heating to 1000–1200°C so that on fusion gases generated in the mass expand the material to give a texture that is similar to that formed in a loaf of bread. Dependent on the process, the product is either produced as a sintered cake which requires crushing to produce aggregate or as aggregates which mostly have smooth sealed surfaces. This process is often referred to as bloating. The clay mineral vermiculite has a composition and structure which exfoliates when pellets are heated to 750–1100°C. Perlite is a special glassy form of rhyolite containing 3–5 per cent of water. Graded particles heated to 900–1200°C decrepitate to form rounded aggregate which may be hollow.

Wood particles and sawdust treated with lime are used as lightweight aggregates for manufacture of masonry blocks, wood wool slabs and sawdust concrete (BS 1105 1981). The texture of a sawdust concrete is illustrated in Plate 86.

In Table 7.1 the loose bulk volume densities of the aggregates are listed with data derived mainly from Spratt (1980) and Short and Kinniburgh (1978). These should be considered as only indicative values. Other data is given by Rudnai (1963) but these have not been included. BS 3797 sets limits of 400–1200 kg/m³ for fine lightweight aggregate and 250–1000 kg/m³ for coarse aggregate. For insulating aggregates ASTM C332 specifies dry loose weights for perlite and vermiculite of 90–200 kg/m³ and a maximum density for other combined lightweight aggregates of 1040 kg/m³. Fine aggregates have higher densities because they are often less vesicular than the coarse aggregates. The wide range of bulk densities occurring in any one material are due to both this density difference between fine and coarse aggregates and the fact that most materials can be expanded over a considerable range of densities. However all materials have an optimum range of densities outside which they are not usable.

Table 7.1 Approximate bulk volume densities of air dried aggregates.

Aggregate type	kg/m^3
Normal weight aggregates	1360–1600
Sintered PFA	770–1040
Brick rubble	760–1120
Scoria	720–1300
Volcanic slag	700–1200
Furnace clinker and breeze	720–1040
Expanded slag	700–970
Foamed slag	480–960
Pumice	480–880
Expanded slate	460–800
Sintered diatomite	450–800
Expanded shale & clay	320–960
Wood shavings & sawdust	320–480
Exfoliated vermiculite	60–160
Expanded perlite	50–240

Lightweight aggregates typically are glassy and vesicular in texture and can easily be identified on cut surfaces with the hand lens. In thin-section the walls of the vesicles appear as thin rims of clear glass which often contain scattered crystallites which have failed to melt. In the case of scoria the glass will often be dark and much more crystallisation will be present. Aggregates that have been bloated and not crushed have a continuous dark-brown coloured zone near the amorphous rims. For crushed and most natural lightweight aggregates there are no defined margins. As a general rule lighter density aggregates have thinner glass walls and larger vesicles.

In concrete, lightweight aggregates with sealed rims behave in a similar manner to normal-weight aggregates. For those aggregates with porous rims the interfacial zone is more dense and homogeneous and there is better mechanical locking between cement paste and aggregate (Zhang and Gjorv 1990, Sarkar *et al.* 1992). The penetration of cement paste is limited to the outer porous margin of the aggregates in spite of lightweight aggregates usually having high water absorption (Zhang and Gjorv 1992).

The petrographer is more likely to examine structural lightweight concretes in densities ranging from 1000–2000 kg/m^3 rather than lightweight insulating concretes. The durability of lightweight structural concrete is affected by increasing levels of carbonation and drying shrinkage as the density of the concrete decreases, as illustrated in Section 5.4.5. To reduce these problems it is common practice to replace some of the fines with normal-weight sand. Apart from these attributes the texture of structural lightweight concrete is little different from normal-weight concretes and it appears that, providing cement contents in excess of 400 kg/m^3 are used, it has similar durability (Mays and Barnes 1991). The durability of lightweight concrete in relation to design has been discussed by Vaysburd (1992) and in relation to alkali-aggregate reaction in Section 6.7.8.

Some petrographic textures are illustrated by the following examples. Lytag, which is manufactured by sintering pulverised fuel ash, forms readily identifiable spherical particles

3–5 mm diameter with a rusty brown exterior and a dark almost black inner core. In thin-section the zonal structure can be distinguished and the porous nature of the materials is clearly apparent (Plate 87). In one example of Lytag concrete containing fly ash as a cement replacement, some of the voids close to the carbonation front within the concrete are partially rimmed by a straw coloured mineral with high birefringence which is probably thaumasite (Plate 88). Some voids in the hydrated cement paste are partly filled or wholly filled with acicular ettringite crystals which have grown in from the void margins.

An example of expanded clay aggregate is illustrated by Plate 89(a). The exterior edge of the particle is more dense than the vesicular interior. It is noticeable that relatively few of the pores have been filled with the coloured impregnating resin clearly indicating that only a small percentage of the pores in this Leca expanded clay aggregate interconnect. Expanded clay aggregate varies considerably depending on the type and details of the manufacturing process. An example of an expanded clay where firing is incomplete is illustrated in Plate 89(b), where the opaque core of the particle is surrounded by a more porous outer zone.

Glossary of Minerals

Introduction

This glossary of minerals is intended to bring together optical data on some of the minerals mentioned in the text but excludes data on most aggregates as these are already adequately covered in the literature. The mineral data given has been largely restricted to the type of information that can be gained from thin-section examination. However, in the cases of efflorescences and precipitates the appearance of the mineral in oil immersion has been included. Much of the data on cement and concrete minerals are widely scattered and can be difficult to access and where this is the case references are given. A synoptic listing is given in Highway Research Board (1972). In other cases the general references used are Winchell and Winchell (1951, 1964), Kostov (1968), Troger (1971) and Rigby (1948). For illustrations of textures in colour the reader is referred to McCrone and Delly (1973), Mackenzie *et al.* (1991), Mackenzie and Guilford (1980) and Campbell (1986).

A brief, simplified description of the optical constants and other terminology used in this glossary is given as an aid to the user. The symbols N, ε, ω, α, β, and γ used for the refractive indices have been chosen as they appear to be the ones most commonly used (Bloss 1966). For the reader who is unfamiliar with the practices of optical crystallography, Bloss (1966) is recommended as a simple and clear introduction to the subject.

Isotropic minerals: A limited group of minerals which only have one refractive index (n) and belong to the cubic crystal system. Glasses and amorphous substances are also isotropic and included in this group for convenience.

Biaxial minerals: The bulk of naturally occurring minerals belong to this group which is distinguished by having two optical axes and three refractive indices (α, β, γ). It consists of the triclinic, monoclinic and orthorhombic crystal systems.

α	The smallest refractive index
β	The intermediate refractive index
γ	The largest refractive index
$\gamma-\alpha$	Birefringence
2V	The optic axial angle
Sign	2V is $+$ve when $\beta-\alpha$ is less than $\gamma-\beta$ and $-$ve when $\beta-\alpha$ is greater.

Uniaxial minerals: A more restricted group of minerals in which the crystals only have one optical axis and two refractive indices. The tetragonal and hexagonal (which includes trigonal) crystal systems belong to this group.

ω	Ordinary refractive index
ε	Extraordinary refractive index which may be greater or less than ω and determines sign
Sign	$\varepsilon > \omega$ optically positive, $\varepsilon < \omega$ optically negative
Bir	The positive difference between the two refractive indices

Sign of elongation: Applies to elongate crystals where one of the vibration directions is parallel to the length which is always the case for the hexagonal, tetragonal and orthorhombic systems. In some cases this may apply to the monoclinic system. However even in the monoclinic and triclinic systems the extinction angle of the index in relation to the length of the crystal may often be small and allow a sign of elongation to be assigned.

Length slow:	The vibration direction of the higher refractive index is parallel or sub-parallel to the length and the sign is positive.
Length fast:	The vibration direction of the lower refractive index is parallel or sub-parallel to the length and the sign is negative.

Colour: The colour of the crystal as seen in thin-section and the colour in hand specimen may be different and they should not be confused.

Interference colour: The colour seen under crossed polars in thin-section. The maximum interference colour occurs when the crystal is observed perpendicular to the optic plane. If the birefringence is known the interference colour can be estimated from the Michel-Levy scale or vice versa provided the thickness of the crystal is known. Due to differential absorption in some minerals anomalous interference colours occur which are diagnostic.

Interference figure: Conoscopic examination of a crystal under crossed polars allows determination of optical sign and estimation of 2V without the need to measure the refractive indices. Dispersion of the optical axes may also be observed under favourable circumstances.

Pleochroism: In plane polarised light some crystals vary in colour as they are rotated due to differential absorption of light. This can be an important diagnostic feature.

Relief: When a crystal is immersed in a fluid of identical refractive index its outline and details of relief disappear apart from colour fringes due to refraction at crystal edges. As the difference between the index of the immersion fluid and that of the crystal is increased the outline becomes darker and details of relief are highlighted. The amount of relief can be roughly estimated, that is low, medium or high, and whether the immersion fluid has a higher or lower index than the crystal can be determined. If the fluid is less than the crystal, relief is positive, and if its index is higher, the relief of the crystal is negative. In the mineralogical literature relief is quoted using an immersion index of 1.54 which has been adhered to in the data quoted in this glossary. Thin-sections of concrete are often mounted in epoxy resin of which the index can vary between 1.54 and 1.58 so allowance may need to be made for this change of index when estimating relief in thin-sections of concrete.

Examination of crystals in thin-section: It is possible to determine the optical properties of a crystal provided that it can be manipulated to measure its refractive indices. These properties are sufficient to identify the crystal unless it belongs to the cubic crystal system. However, in thin-section sufficiently precise measurement of refractive indices is rarely possible. In addition, many of the recrystallised products formed in hardened cement paste are too small to give interference figures so that the optic sign and 2V cannot be determined. In many cases observations are restricted to estimates of refractive indices and to relief, sign of elongation, birefringence, pleochroism and anomalous interference colours, crystal habit, cleavage and extinction angle. Identification of a particulate mineral is often possible if at least some of the properties listed above can be observed.

XRD: The first four strongest diffraction peaks are quoted from data published by the JCPDS, International Centre for Diffraction Data, USA, unless otherwise stated.

Cement minerals

Many of the anhydrous cement minerals associated with Portland, high-alumina and other types of hydraulic cements are only of interest in concrete as unhydrated relics. There is a large literature pertaining to the microscopy of these minerals so only brief details are given. For more information on the mineralogy see Campbell (1986), Fundal (1980), Hoffmänner (1973) and Gille *et al.* (1965).

Alite (tricalcium silicate, C_3S)

$3CaO.SiO_2$ Biaxial -ve   Monoclinic
$$\alpha = 1.716\text{--}1.720 \ \gamma = 1.722\text{--}1.724$$

The composition of alite in Portland cement is 96–97 per cent C_3S substituted with a mixture of 3–4 per cent of ferric, magnesium, aluminium, sodium, potassium and phosphorus ions. A number of triclinic, monoclinic and rhombohedral polymorphs exist. In commercial cements the Type 1 monoclinic polymorph predominates (Taylor 1990).

Interference figure: $-2V$ varies from 20 to 60°.

Relief: Relics stand out because of high positive relief.

Birefringence: 0.005 ranging from 0.002 to 0.010 as a function of ionic substitution. Interference colours first-order grey.

Form: Lath, tablet-like or equant. Crystals usually six-sided in cross-section. Normal crystal sizes range from 25 to 65 μm (Plate 1).

Twinning: Rare polysynthetic twinning.

Extinction: Wavy to straight extinction, length slow.

Colour: Colourless to slightly coloured in transmitted light.

Occurrence: Alite is the principal constituent of Portland cement and ranges from 15–65 per cent dependent on type. It occurs in stabilised dolomite bricks and basic steel slags.

Comment: Alite is rare in adequately cured concretes except as crystals embedded in the matrix of large particles of clinker. At very low water/cement ratios alite is quite common because of internal desiccation. Most alite crystals are broken during the milling of the cement so it is difficult to find complete crystals.

XRD: 31-301: 2.789-100, 2.767-70, 2.754-65, 2.613-90. Also see Taylor (1990).

Belite (β-dicalcium silicate, larnite, C_2S)

$2CaO.SiO2$ Uniaxial +ve   Monoclinic
$$\alpha = 1.717 \ \beta = 1.722 \ \gamma = 1.736$$

The composition of belite in Portland cement is 94–96 per cent C_2S substituted with a mixture of 4–6 per cent of aluminium, ferric, magnesium, potassium, sodium, phosphorus, titanium and sulfur ions. A number of trigonal, monoclinic and orthorhombic polymorphs exist. In commercial cements β-dicalcium silicate is the second most common mineral (Taylor 1990).

Interference figure: $+2V$ varies from 64–69°.

Relief: Relics stand out because of high positive relief.

Birefringence: Usually second-order colours as relic grains in concrete are often larger than 40 μm in size.

Form: Rounded grains, often in clusters set in the clinker matrix (Plates 1, 2, 3).

Cleavage: Poor prismatic cleavage.

Twinning: Lamellar structure is distinctive and nearly always present. For the discussion of lamellar structures observed see Campbell (1984).

Colour: Colourless, pale yellow, yellow, amber or shades of green dependent on substitution.

Occurrence: Larnite occurs naturally in high-temperature contact zones but is rare. Belite is the second most common mineral in Portland cement clinker and ranges from 10–60 per cent dependent on cement type.

Comment: In concrete, relic grains often consist of belite crystals with some adherent clinker matrix. The shape, birefringence and lamella twinning of belite are distinctive. Globules of belite are commonly found inside alite crystals.

XRD: 33-302: 2.790-97, 2.783-100, 2.745-83, 2.610-42. Also see Taylor (1990).

Aluminate phases (tricalcium aluminate, C_3A, alkali aluminate)

$3CaO.Al_2O_3$	Isometric	Cubic
	$n = 1.710$	
$Na_2O.8CaO.3Al_2O_3$ (alkali aluminate)	Biaxial -ve	Orthorhombic
	$\alpha = 1.702 \quad \gamma = 1.710$	

Tricalcium aluminate typically contains a mixture of around 13 per cent of ferric, silicon, magnesium, sodium, potassium and titanium ions and in alkali aluminate up to 20 per cent. The alkali content of the cubic form is around 1 per cent and for the orthorhombic form 2–4 per cent which is less than required for NC_8A_3 (Taylor 1990).

Interference figure: $-2V$ varies from 0 to 35°.

Relief: Stands out because of its high positive relief.

Birefringence: Nil to low.

Form: Cubic form typically fills interstices between crystals of belite and ferrite. Alkali aluminate occurs as tablets, laths and stavelike prismatic forms (Plate 1).

Twinning: May be present in alkali aluminate.

Extinction: Alkali aluminate has oblique extinction and is length slow.

Colour: Colourless in white cement. Tan to brown in Portland cements.

Occurrence: In ordinary Portland cement C_3A averages about 12 per cent but in special cements can range from 0–18 per cent. Production clinkers may contain both the cubic lath-shaped forms especially in high-alkali cements.

Comment: Individual particles of the aluminate phases are not observed in concrete. The brown matrix in relic grains commonly seen in concrete probably contains some aluminate phase. In many clinkers the aluminate and ferrite phases are often poorly differentiated and form the matrix in which the silicate phases are embedded. However, X-ray evidence clearly indicates that the matrix in commercial clinker is usually crystalline even when it appears undifferentiated by microscopic examination.

XRD: 38-1429: 2.699-100, 4.080-12, 1.5578-24. Also see Taylor (1990).

Ferrite phases (brownmillerite, C_4AF, ferrite solid solution)

$4CaO.Al_2O_3.Fe_2O_3$	Biaxial -ve	Orthorhombic
	$\alpha = 1.98 \quad \beta = 2.05 \quad \gamma = 2.08$	

The composition of ferrite can vary from C_6A_2F to C_6AF_2. However the typical composition of the ferrite phase in ordinary Portland cement is close to C_4AF containing 10 per cent magnesium, silicon and titanium ions replacing the ferric ion (Taylor 1990).

Interference figure: $-2V$ is moderate.

Relief: Very high positive relief.

Birefringence: White to yellow of the first order.

Form: Either bladed, prismatic, dendritic, fibrous, massive or infilling. Form varies from bladed to dendritic and infilling as the cooling rate of the clinker is increased. In modern cement production with fast efficient clinker coolers the dendritic form is common (Plate 1).

Extinction: Crystals are length slow.

Colour: Brown to yellow. Pleochroic, varying from green to almost opaque.

Occurrence: Brownmillerite is a rare natural mineral. It ranges from 5–15 per cent in cements and is typically about 8 per cent in ordinary Portland cement. A ferrite phase of more variable composition constitutes 20–40 per cent of high-alumina cement.

Comment: Ferrite and belite phases are commonly observed in concrete where fragments of clinker have failed to hydrate either because they are too large or the hardened paste has not fully hydrated.

XRD: 32-226: 7.25-45, 2.673-35, 2.644-100, 1.8149-45.

Periclase (magnesium oxide)

MgO	Isotropic	Cubic
	$n = 1.736 \ (1.732 - 1.738)$	

Relief: High positive relief.

Form: Periclase occurs as small crystals or anhedral crystal aggregates. Synthetic crystals are usually rounded and vary in size and may give a 'sugary' appearance but may also be closely packed hexagons or octagons. In cement (Fig. 4.1) clinker is often dendritic.

Cleavage: Perfect, parallel to cube faces. Not always present but develop at high temperature.

Colour: Colourless when pure. In solid solution with FeO, colour changes from yellow to deep brown with increasing iron content. Also coloured red-brown due to minute crystals of magnesioferrite.

Occurrence: A fairly rare, relatively high-temperature mineral in dolomite contact zones which may be hydrated to brucite. Chief constituent of magnesite brick and present in dolomite and basic refractories. Periclase does not crystallise out in cement clinker until MgO exceeds 2 per cent.

Comment: Magnesia slowly hydrates in concrete and magnesium lime plasters and increases in volume. Thus magnesia is limited to 6 per cent or less in Portland cement to prevent unsoundness due to this volume increase. Where problems of unsoundness occur in Portland cement and hydrated magnesium lime they are monitored for unsoundness by the autoclave expansion test.

XRD: 4-289: 2.106-100, 1.489-52, 0.942-17, 0.860-15.

Monocalcium aluminate, calcium aluminate 12:7, gehlenite and wüstite

$CaO.Al_2O_3$ Biaxial −ve   Monoclinic
$$\alpha = 1.643 \quad \beta = 1.655 \quad \gamma = 1.663$$
$12CaO.7Al_2O_3.0–1H_2O$ (mayenite)   Isometric   Cubic
$$n = 1.61 – 1.62$$
$2CaO.Al_2O_3.SiO_2$ (gehlenite)   Uniaxial −ve Tetragonal
$$\omega = 1.669 \quad \varepsilon = 1.658$$
Fe(Ca)O (substituted wüstite)   Isometric   Cubic
$$n = 2.32 \text{ less with substitution}$$
The typical composition of high-alumina cement is 40–50 per cent monocalcium aluminate, 20–40 per cent ferrite phase, together with smaller proportions of calcium aluminate 12:7, β-dicalcium silicate, gehlenite, pleochroite and wüstite (Taylor 1990). In transmitted light, most of the constituents are either glassy or dark coloured and almost opaque (Lea 1970). Apart from pleochroite, which stands out clearly because of its pleochroism, the relics of high-alumina cement in concrete tend to be undifferentiated as shown in Plates 8, 9 and 28. Monocalcium aluminate is strongly hydraulic.

XRD: CA: 34-440:   2.97-100, 2.52-35, 2.41-18.
        Mayenite: 9-413:   4.89-95, 2.998-45, 2.680-100, 2.189-40.
        Gehlenite: 35-755: 3.063-21, 2.856-100, 1.523-60, 1.299-25.
        Wüstite: 6-615:   2.49-80, 2.153-100, 1.523-60, 1.299-25.

Pleochroite

$6CaO.4Al_2O_3.MgO.SiO_2$ Biaxial +ve   Orthorhombic
$$\alpha = 1.669 \quad \gamma = 1.673–1.680$$

The simplified composition given above is variable and is modified by some substitution of calcium by ferrous ion and aluminium by the ferric ion (Taylor 1990).
  Interference figure: $+2V = 45°$ approximately.
  Relief: High positive relief.
  Birefringence: First-order colours modified by pleochroism.
  Form: Fibrous to acicular.
  Colour: Strongly pleochroic. Distinctively violet.
  Occurrence: Occurs in high-alumina cements made under reducing conditions.
  Comment: The strong violet pleochroism identifies this mineral which can be observed in some concretes made with high-alumina cement.
  XRD: 11-212: 3.70-vs, 2.87-vvs, 2.76-vvs, 1.76-vs.

Melilite (åkermanite–gehlenite solid solution)

$2CaO.Al_2O_3.SiO_2$ (gehlenite)   Uniaxial −ve   Tetragonal
$$\omega = 1.669 \quad \varepsilon = 1.658 \ (n = 1.638 \text{ glass})$$
$2CaO.MgO.2SiO_2$ (åkermanite)   Uniaxial +ve   Tetragonal
$$\omega = 1.632 \quad \varepsilon = 1.641 \ (n = 1.641 \text{ glass})$$

  Birefringence: Akermanite is first-order grey which for åkermanite/gehlenite 60/40 is almost isotropic and then rises to first-order pale yellow for gehlenite. Melilite in slags often has a distinctive anomalous blue interference colour as shown in Plate 30.

Form: Short prisms or tablets often almost square in outline. In slags distinct zonation is often present.

Extinction: Parallel to edges of crystals.

Colour: Colourless.

Occurrence: Found in some contact zones, ores, and basic volcanic rocks. Formed by the action of lime on fire bricks and is constituent of many blastfurnace slags.

Comment: The refractive indices given above are for the synthetic minerals and the natural minerals will vary because they contain other ions. The indices given above for the synthetic glasses should not be used for the glasses found in blastfurnace slags.

XRD: Akermanite: 35-592:   3.087-25, 2.872-100, 2.478-15, 1.761-20
　　　　 Gehlenite: 35-755:   3.0629-21, 2.845-100, 2.431-17, 1.754-25

The calcium silicate hydrates

The calcium silicate hydrates may be conveniently divided into three groups, namely (1) the hydrates that are found in hardened cement paste which has been cured at a temperature of less than about 60°C, (2) a range of hydrates that are either formed hydrothermally in autoclaved products and oil-well cements, geothermal grouts and other high-temperature applications and (3) in non-Portland cement systems such as sand lime bricks and refractories. In addition there is a wide range of calcium silicate hydrates that may be synthesised or are found naturally but these are outside the scope of this book.

Calcium silicate hydrate formed at less than 60°C

Calcium silicate hydrate gel, C-S-H(II). In older texts is often referred to as 'tobermorite' gel. The term CSH is often used without discrimination.

$9CaO.4-6SiO_2.nH_2O$ of variable composition. The CaO/SiO_2 ratio may vary between 1.5 and 2.0 and typically is about 1.8 in mature pastes in concrete. However, the ratio measured depends on the method of sample preparation and may also vary from point to point within the paste on the microscale. The water content in a fully saturated mature paste is 42–44 per cent and in a paste dried at $-79°C$ is in the range 20.4–22.0 per cent (Taylor 1990).

Optical properties: Properties observable are those of the cement paste. Transparent, coloured, isotropic although vague indefinite birefringence may be observed. Mottled and/or wavy appearance to the texture. Where pozzolanic admixtures are present and have removed most of the CH the texture appears to become darker and more featureless. Refractive index of a mature paste is around 1.5.

Appearance: In mortars and concretes the gel appears as an indefinite mass infilling the space and surrounding components such as CH, remnant cement grains, pores and aggregates. It may appear as 'hydration rims' around clinker particles which stand out from the indefinite mass of hydrate material. C-S-H(II) is colourless but because of its intimate mixture with the other hydrates appears coloured. The colour of the paste containing C-S-H(II) is brownish to yellow often with some green tints and may have a mottled appearance.

Physical properties: Density is 1.85–2.45 dependent on state of drying. A fully hydrated cement paste has a density of about 2.13.

Structure: A nearly amorphous gel consisting of chains of polymerised silicate anions. Polymerisation increases with curing temperature. The structural appearance of the gel observed by electron microscopy varies according to the age of the sample and method of sample preparation. Soon after mixing a cement with water the gel formed is fibrous or needle-like (type I) in appearance which soon changes to a honeycomb structure (type II). As curing proceeds the gel takes on a more massive appearance with equant grains (type III) and in the mature paste, that is at 28 days, becomes increasingly featureless (type IV) (Diamond 1986). The gel contains pores with equivalent radii between 0.6 and 1.6 nm and pores between gel particles which range from 1.6 to 100 nm. Although pore volumes are comparatively large, C-S-H(II) remains as a stable xerogel in concrete unless attacked by carbon dioxide.

Occurrence: Calcium silicate hydrate gel is formed from the hydration of the alite and belite phases, present in Portland cement. Both alite and belite react to form similar hydrates only differing in the amount of CH formed during hydrolysis. The gel is the main cementing phase present in all concretes cured below 60°C and is intimately mixed with the aluminate and ferrite hydrates and interspersed with crystallites of CH. The reaction between CH and pozzolanas in concrete produces a similar gel although the CaO/SiO_2 ratio may be reduced towards the lower end of the range.

Comment: The most important distinguishing feature is the association of the C-S-H gel with remnant cement grains as even in a fully hydrated cement paste some grains remain. The appearance of C-S-H gel in mortars and concretes is best examined by electron microscopy but this is not a routine petrographic procedure. Optical examination rarely produces any useful data.

XRD: 29-374: Diffuse peaks at 0.27-0.31 Å and a sharper peak at 0.182 Å. Peaks small and not well defined. Other peaks present usually due to CH. Often in hardened concrete only a broad shallow peak around the 0.3 Å area is observed and many of the peaks listed on Card no. 29–374 are absent. Peaks may become sharper as curing temperatures are increased, especially over 60°C.

Thermal analysis: TGA shows a continuous weight loss from 100°C to 500°C with smaller losses to 800°C. Usually superimposed on these weight losses is the weight loss due to the dehydroxylation of CH between 400 and 500°C. DTA does not give a useful diagnostic curve.

Infrared: Does not give useful diagnostic spectra.

The calcium silicate hydrates formed at elevated temperatures and from non-Portland cement systems

Unlike the calcium silicate hydrates formed from the hydration of Portland cement cured in the normal range of temperatures these hydrates are semi-crystalline to crystalline. However, with only one exception, they remain microcrystalline and require electron microscopy and X-ray analysis for their observation and identification. They may be conveniently grouped into the high-silica and high-calcium types formed under hydrothermal conditions and another group formed by heating concrete in the absence of moisture. Many of these calcium silicate hydrates also occur naturally. In addition there are other ranges of hydrates which can be synthesised or are found naturally (Taylor and Roy 1980, Coleman 1984). These are unlikely to occur in the types of building products encountered by the petrographer and are beyond the scope of this book.

The hydrothermal silicates in Portland cement concretes

	High-silica types	CaO/SiO_2
C-S-H(I)	$C_5S_6H_9$ approx.	0.83
11 Å tobermorite	$C_5S_5H_5$	1.0
Anomalous tobermorite (incl. 0.5Al and 0.1Na)		0.76
Xonotlite	C_6S_6H	1.0
Gyrolite	$C_2S_3H_2$	0.66
Truscotite	$C_7S_{12}H_3$	0.58

	High-calcium types	
α-dicalcium silicate hydrate	$α-C_2SH$	2.0
Tricalcium silicate hydrate, (jaffeite)	$C_6S_2H_3$	3.0

	Types formed on heating	
9 Å tobermorite, (riversideite)	$C_5S_6H_2$	0.83

Occurrence: The high-silica types. These calcium silicate hydrates are formed in hydrothermally cured Portland cement pastes containing ground quartz which is added to concretes used in autoclaved products. In practical terms the minimum temperature of formation for the 11 Å tobermorite starts between 130–150°C and this phase is stable up to 180°C. From 180–250°C anomalous tobermorite may persist by substituting Al and Na into the lattice. Otherwise the stable phase above 180°C is xonotlite which converts to the calcium silicate, wollastonite, above 600°C. In oil-well cements, where the composition is rich in $βC_2S$ and ground quartz is used, truscotite forms in the temperature range 180–350°C. Gyrolite has also been reported in some hydrothermal cementing systems between 140–170°C and forms instead of tobermorite where a more reactive form of silica such as diatomite is used. Below 130°C, even when ground quartz is present, the α-dicalcium silicate hydrate may form together with C-S-H(I) and will only convert to 11 Å tobermorite after extended autoclaving or by increasing the temperature.

High calcium types. These hydrates form above 110°C where the cement does not contain the addition of ground quartz. Up to 180°C α-dicalcium silicate hydrate forms. It has low strength and is the only hydrate that grows large enough to become optically visible. Between 180–400°C tricalcium silicate hydrate is the stable phase. Like $αC_2SH$ it is also a low-strength material.

The degree of crystallinity achieved and the phases formed are dependent both on temperature and the time of heat treatment. For commercial autoclaving, typically carried out at 160–180°C for up to 18 hours, the cement paste may only contain poorly crystalline C-S-H similar to 11 Å tobermorite. Similar phases will be formed for autoclaved products made from sand lime formulations. Where the concrete is steam cured at 60–100°C the phase formed will be C-S-H(II) but increased crystallinity may be detected by better definition of the X-ray diffraction peaks.

The effect of heating concrete, for instance by exposure to fire, is firstly to increase the crystallinity of the phases and then with increasing temperature to contract the layer spacing until above 300°C the 9 Å tobermorite is the major phase which persists to 700°C (Taylor 1957). There appears to be limited information on the composition of the phases in concretes exposed to fire. Xonotlite is also found in boiler scales and lightweight heat-insulating materials.

Mineral descriptions: Mainly from Heller and Taylor (1956).

C-S-H(I)   Probably similar to poorly crystallised natural material from Bally-craigie, Northern Ireland. Seemingly amorphous material with a flaky habit under SEM.
Orthorhombic, $n = 1.49$–1.530
$n = 1.494$ when $d = 2.00$, $n = 1.525$ when $d = 2.20$

11 Å tobermorite
Natural occurrence: Tobermory in Scotland, Northern Ireland and California. Minute radiating aggregates of fibres with distinctive silky lustre.
Orthorhombic, biaxial + ve, low 2V, both length slow and length fast
$\alpha = 1.570$ $\beta = 1.571$ $\gamma = 1.575$ (Ballycraigie)
$n = 1.558$ and 1.545–1.565 (for other samples)
$n = 1.55$–1.565 (synthetic material)
Density $= 2.44$.

Xonotlite
Natural occurrence: Tetela de Xonotla, Mexico. Usually fibrous, colourless to pink.
Monoclinic, biaxial + ve, low 2V, parallel extinction, length slow
$\alpha = 1.578$–1.586 $\beta = 1.583$ $\gamma = 1.590$–1.595
Density $=$ 2.7.

Truscotite
Natural occurrence: Benkulen, Sumatra. White, spheroidal aggregates with fibrous, platy structure and pearly lustre on cleavage.
Hexagonal, uniaxial -ve, parallel extinction, length fast.
Mean $n = 1.560$ $\omega < 1.55$, weak birefringence.
Density $= 2.48$.

Gyrolite
Natural occurrence: Skye, Scotland. Colourless, lamellar crystals, often in radial aggregates.
Hexagonal, uniaxial $-$ve, may appear biaxial with small 2V.
$\omega = 1.549$ $\varepsilon = 1.536$, or $\omega = 1.540$–1.549
Density $=$ 2.39.

α-dicalcium silicate hydrate
Natural occurrence: Colourless, transparent laths or prisms, with parallel growths and twinning.
Orthorhombic, biaxial + ve, 2V $= 68°$, parallel extinction, length fast.
$\alpha = 1.614$ $\beta = 1.620$ $\gamma = 1.633$
Density $= 2.8$.

Tricalcium silicate hydrate
Natural occurrence: Very small, colourless, anisotropic prisms or broad fibres.
Hexagonal, parallel extinction, length slow.
$\alpha = 1.589$ $\gamma = 1.633$
Density $= 2.56$.

9 Å tobermorite (Riversideite)

Natural occurrence: Crestmore, California. Appearance similar to 11 Å hydrate.
Orthorhombic, biaxial +ve, small 2V, both length slow and length fast.

$\alpha = 1.600 \ \beta = 1.601 \ \gamma = 1.605$

Density = 2.6.

Comment: Optically all the hydrates except for $\alpha C_2 SH$ do not form large enough crystals to be identifiable under the microscope (Plate 21). However, Hadley grains (Barnes *et al.* 1978) are more common in the C-S-H matrix compared to concretes cured at normal temperatures. In geothermal samples exposed to downhole temperatures for up to nine months the hardened paste may contain a background of minute, diffuse, birefringent crystals which it is believed are the crystallised hydrates. After three months' exposure down a geothermal well at 160°C, grouts without ground quartz contained laths of $\alpha C_2 SH$ measuring up to 30 μm in length and 10 μm in width which tended to form radial groups with optical properties as listed above (St John and Abbott 1987).

Under the SEM xonotlite is often fibrous, etc.

XRD: C-S-H(I):   12.5-vs, 3.04-vs, 2.80-s, 1.82-s (Taylor 1990).
  11 Å tobermorite 45-1480:   11.3-100, 3.08-92, 2.972-70, 2.806-68.
  Anomalous tobermorite.   No card. Spectrum similar to 11 Å
                                                    tobermorite.
  Xonotlite 29-379:   3.085-100, 2.828-50, 2.697-40, 1.9468-40.
  Truscotite 29-382:   4.21-70, 3.141-100, 2.840-80, 1.839-70.
  Gyrolite 12-217:   22.0-90, 11.1-70, 4.20-70, 3.160-100,
                                                    3.097-90, 2.834-70.
  9 Å tobermorite 29-329:   3.59-17, 3.15-19, 3.014-100, 2.784-18.
  Tricalcium silicate hydrate 29-375: 8.6-95, 2.998-85, 2.898-95, 2.834-100.
  α-dicalcium silicate hydrate 29-373: 4.22-50, 3.302-45, 3.272-100, 2.816-45,
                                                    2.418-60, 2.606-45.

Thermal analysis: Refer to Ramachandran (1969).

The hydrogarnets (grossular, hydrogrossular, katoite, hibschite)

Grossular $3CaO.Al_2O_3.3SiO_2$ Cubic 1.734.
Hydrogrossular $3CaO.Al_2O_3.6H_2O - 3CaO.Al_2O_3.3SiO_2$ Cubic 1.604–1.734.
Katoite $3CaO.Al_2O_3.6H_2O - 3CaO.Al_2O_3.1.5SiO_2.3H_2O$ Cubic.
Hibschite $3CaO.Al_2O_3.1.5SiO_2.3H_2O - 3CaO.Al_2O_3.3SiO_2$ Cubic 1.67–1.68.

Hydrogrossular is a solid solution series which mineralogically is identified as katoite and hibschite as given above. It is not an accepted mineral name. The refractive indices given will vary according to the composition.

Relief: Moderately high.

Birefringence: Hibschite may be weakly birefringent.

Twinning: Present in grossular.

Colour: Grossular ranges from colourless to coloured, similar to other garnets. Hibschite is colourless and it is probable that in cementing systems the other hydrogrossulars are also colourless to white.

Occurrence: Grossular and hibschite occur naturally in contact altered limestones.

Comment: Minor quantities are formed from some composite cements. In Portland cement it is a minor and poorly crystalline hydrated phase. Larger quantities were produced by some older Portland cements and it is a normal hydration product of autoclaved cement-based materials (Taylor 1990). In cementing systems, if sufficient is present, it is detected by X-ray diffraction analysis and scanning electron microscopy.

XRD: Grossular 39-368:   2.962-35, 2.650-100, 2.418-18, 1.583-30.
　　　　Katoite 24-127:   5.13-90, 2.810-80, 2.295-100, 2.039-95.
　　　　Hibschite 45-1447:   3.072-50, 2.748-100, 2.244-60, 1.994-65.

Thermal analysis: Hydrogrossular decomposes at 200–250°C (Taylor 1990).

Merwinite

$3CaO.MgO.2SiO_2$　　　　　　　　　Biaxial +ve　　　　　　　　　Monoclinic
$$\alpha = 1.708 \quad \beta = 1.711 \quad \gamma = 1.724$$

Interference figure: Biaxial positive with 2V = approx. 70°.
Relief: Rather high.
Birefringence: Polarises in greys, whites and yellows. The maximum colour being first-order red.
Form: Crystals are granular or prismatic, tending to lozenge-shaped outline.
Cleavage: Poor.
Twinning: Polysynthetic twinning is common.
Extinction: Generally inclined, the maximum extinction angle is 36°.
Colour: Colourless.
Occurrence: Found in limestone contact zones. Occurs in blastfurnace slags.
Comment: Can be distinguished by its weak birefringence, polysynthetic twinning and biaxial character. Merwinite is etched by nitric acid and stained by hydrofluoric acid.
XRD: 35-591: 2.756-26, 2.687-100, 2.671-65, 2.653-47.

The sulfate minerals

Mirabilite (glauber's salts, sodium sulfate deca-hydrate)
Thenardite (anhydrous sodium sulfate)

Mirabilite $Na_2SO_4.10H_2O$　　　　　Biaxial −ve　　　　　　　　Monoclinic
Thenardite Na_2SO_4　　　　　　　Biaxial +ve　　　　　　　　Orthorhombic
Mirabilite　　　　　　$\alpha = 1.394 \quad \beta = 1.396 \quad \gamma = 1.398$
Thenardite　　　$\alpha = 1.464–1.468 \quad \beta = 1.473–1.474 \quad \gamma = 1.481–1.485$

Interference figure: Mirabilite, −2V = 76°. Thenardite, +2V = 83°.
Relief: Mirabilite has high negative relief compared to thenardite.
Birefringence: Mirabilite gives first-order grey colours. Thenardite gives colours up to red of the second order.
Form: Mirabilite is commonly prismatic to acicular. As massive efflorescent crusts of interlocking fibres. Thenardite is pyramidal, short prismatic or basal plates.
Cleavage: Mirabilite has a perfect 100 cleavage. Thenardite has distinct basal cleavage.
Colour: Both minerals are colourless.

Occurrence: Mirabilite and thenardite are found associated with hot springs and salt lake deposits.

Comment: Mirabilite effloresces and converts to thenardite above 32°C. Occurs as fine silvery needles deposited on surfaces as an evaporation product of groundwater sulfate (St John 1982). Sometimes found as efflorescence on protected surfaces of brick and mortar.

XRD: Mirabilite 11-647:   5.49-100, 3.26-60, 3.21-75, 3.11-60.

           Thenardite 37-1465:   4.657-71, 3.077-55, 2.784-100, 3.181-52, 2.648-52.

Thermal analysis: See Mackenzie (1970).

Ettringite (calcium aluminium sulfate hydroxide, high sulfoaluminate)

$Ca_6Al_2(SO_4)_3(OH)_{12}.26H_2O$ Uniaxial −ve   Pseudo hexagonal, trigonal

                               $\omega = 1.466$ $\varepsilon = 1.462$

Interference figure: Crystals usually too small to obtain an interference figure.

Relief: Moderate negative relief.

Birefringence: Maximum interference colour rarely above first order grey-white due to small dimension of needles.

Form: Always appears as needles which under SEM may prove to be rods. Often present as randomly crossed needles in pores and fissures. In very old weathered concretes may form massive curtain-like material in large fissures in which needles are difficult to observe. Needles rarely more than a few microns in cross-sectional dimension. In hardened cement paste, as opposed to cracks, pores and fissures ettringite is usually submicroscopic (Plates 25, 26).

Extinction: Parallel extinction.

Orientation: Needles are length fast.

Colour: Colourless.

Occurrence: Found in metamorphic limestones. Forms as initial hydration product in Portland cement concretes and as a recrystallisation product in concrete associated with movement of water and alkali-aggregate reaction. Can be formed by delayed reaction in steam-cured products. One of the products formed by sulfate attack on concrete. Important phase in some types of shrinkage compensating cements. Is present as one of the hydration products of super sulfated cement. Optical data from Lerch et al. (1929).

Comment: Ettringite may convert to thaumasite due to prolonged sulfate attack on concrete. Loses crystallinity at 100°C (Buck 1986) and may become amorphous if heated above 60°C (Taylor 1990). Excessive grinding in preparation for XRD analysis can cause anomalous results. Ettringite is prone to atmospheric carbonation (Grounds et al. 1988). The morphology of ettringite is affected by organic additions (Baussant et al. 1989). Ettringite can enter into solid solutions with hydroxyl and carbonate ions (Poellmann and Kuzel 1990).

XRD: 41-1451: 9.72-100, 5.61-76, 3.873-31, 2.569-29.

Thermal analysis: Strong endotherm at 125–130°C (Taylor 1990).

Infrared: see Bensted and Varma (1971).

Tricalcium aluminate monosulfate 12-hydrate (low sulfoaluminate, monosulfate)

$3CaO.Al_2O_3.CaSO_4.12H_2O$ Uniaxial −ve   Hexagonal

                               $\omega = 1.504$ $\varepsilon = 1.488$

Interference figure: Crystals are too small to obtain a figure.

Relief: Moderate negative relief.

Birefringence: Maximum second-order red. In practice, crystals are too small for this colour to be obtained.

Form: Basal plates.

Extinction: Length slow.

Colour: Colourless.

Occurrence: Does not occur naturally. The final product of the reaction between the aluminate and sulfate phases in the hydration of Portland cement.

Comment: Low sulfoaluminate is not visible optically and must be observed by scanning electron microscopy. It is metastable with respect to ettringite which will reform if additional sulfate enters the system. Its composition is variable and complex. See Taylor (1990) for further details.

XRD: 12 hydrate 45-158:   8.9-100, 4.45-70, 2.190-40, 2.070-40.

   14 hydrate 42-62:   9.55-100, 4.78-65, 2.753-30, 2.485-45.

Thermal analysis: See Taylor (1990).

Thaumasite (calcium carbonate silicate sulfate hydroxide hydrate)

$[Ca_3Si(OH)_6.12H_2O](SO_4)(CO_3)$ Uniaxial $-$ve   Hexagonal

$\omega = 1.504$ $\varepsilon = 1.468$

Interference figure: Crystals usually too small to obtain an interference figure.

Relief: Relief varies with rotation of stage.

Birefringence: Maximum interference colour is first-order yellow but rarely above first-order white due to small dimension of needles (Plate 52).

Form: Needles; can be threadlike. May be present as overgrowths on ettringite.

Extinction: Needles length fast with parallel extinction.

Colour: Colourless.

Occurrence: Rare mineral found in altered limestones and basic rocks. Formed from solutions depositing calcite and zeolites in contact with anhydrite. Occurs in concrete and mortars due to sulfate attack (Crammond 1985) where it is favoured by temperatures of 0–5°C and humidities greater than 90 per cent. Close relationship between alkali-silica gel, ettringite and thaumasite claimed. Good literature reviews by Brouxel and Valiere (1992), Berra and Baronio (1987). Solid solution with ettringite and other data (Taylor 1990). Optical data from Erlin and Stark (1966). Occurs as replacement of ettringite in cement mortars in contact with gypsum or lime gypsum mortars (Ludwig and Mehr 1986).

Comment: Appears similar to ettringite and is often misdiagnosed as such. May be present as overgrowths on ettringite. May require a higher optical magnification to distinguish from ettringite or the use of SEM. More thermally stable than ettringite. Usually requires XRD analysis for detection and easily missed in optical examination.

XRD: 25-128: 9.56-100, 5.51-40, 3.41-20, 3.18-16.

Thermal analysis: Strong endotherm at 150°C (Bensted and Varma 1974).

Infrared: Spectral data Bensted and Varma (1974).

Gypsum (selenite, alabaster, satin spar)

$CaSO_4.2H_2O$ Biaxial $+$ve   Monoclinic

$\alpha = 1.520$ $\beta = 1.523$ $\gamma = 1.530$

Interference figure: Biaxial positive with 2V = 58°.

Birefringence: Weak first-order greys up to white.

Form: Anhedral to subhedral aggregates often unevenly grained. May show fibrous structure.

Cleavage: Perfect in {010} direction.

Twinning: Simple twinning common caused by heating the section during preparation.

Orientation: Cleavage traces are parallel to both slow and fast rays.

Colour: Colourless in thin-section.

Occurrence: The chief constituent of gypsum rock, is widespread in many ore deposits and as crystals in sedimentary rocks.

Comment: The differences in the optical appearance of natural and synthetic gypsums have been described by Green (1984). Gypsum often forms in concrete undergoing sulfate attack. Much of the gypsum formed by sulfate attack is submicroscopic but veins of gypsum may be visible in the exfoliation texture of surfaces. These veins have a distinctive texture as shown in Plate 50 and Adams *et al.* (1991). Gypsum is interground with cement clinker to produce Portland cement. Due to the high operating temperatures of many grinding mills most of the gypsum is converted either to hemihydrate or soluble anhydrite.

XRD: 33-311: 7.63-100, 4.283-100, 3.065-100, 2.873-45.

Thermal analysis: On heating, gypsum gives two endothermic peaks which begin at 125°C and 185°C due to the loss of 1.5 and 0.5 molecules of water respectively. However, to produce these peaks it is necessary to use a massive sample holder preferably with a thermocouple protruding into the sample. Modern low-mass, high-sensitivity analysers may fail to give any peak separation. The temperature of the peaks varies with heating rate and sample holder. The detailed thermal curves are described by Ramachandran (1969).

Hemihydrate (bassanite, plaster of Paris, gypsum plaster)

$CaSO_4.0.5H_2O$ Biaxial +ve   Pseudo-hexagonal monoclinic

   α form: α = 1.558–1.5599   β = 1.5595   γ = 1.5836–1.586 (Berg and Sveshnikova 1949).

   β form: α = 1.550   γ = 1.556 (Berg and Sveshnikova 1949).

   β form: n = 1.540 variable (Green 1984).

For β-hemihydrate the indices are variable and dependent on the method by which the hemihydrate is made.

Interference figure: Crystals usually too small to obtain an interference figure. +2V = 14° but reported to vary between 10–15°.

Birefringence: Varies from 0.006 to 0.025. Most commercial gypsum plaster will give grey polarisation colours.

Form: Acicular or fibrous. Particles retain the same shape as the gypsum particles from which they are made. β-hemihydrate particles are porous. α-hemihydrate particles vary from blocky to acicular.

Twinning: Twinning may be present.

Extinction: Fibres have parallel extinction and are length slow.

Colour: Colourless. Some hemihydrates may appear brown.

Occurrence: Very rare mineral found in arid parts of Central Asia and in a few volcanic rocks. It is the chief component of gypsum plaster and forms in Portland cement due to the dehydration of gypsum during grinding. Can occur in thin-sections of rocks containing gypsum due to heat generated by grinding the section.

Comment: The amount of molecular water present in hemihydrate can vary between 0.4 and 0.6 due to over or under dehydration of the gypsum and other factors which make exact characterisation difficult. The more crystalline α form appears to differ from the β form only in degree of crystallinity and crystal size (Taylor 1990). Commercial gypsum plaster consists mainly of the β form together with a little α.

XRD: 41-224: 6.013-80, 3.467-50, 3.006-100, 2.803-90.

Some variation in line intensity occurs between the α and β forms but no other significant differences are found.

Thermal analysis: Refer to gypsum.

Soluble anhydrite (γ anhydrite, anhydrite III)

CaSO$_4$	Uniaxial +ve	Hexagonal

Indices are variable even from crystal to crystal and dependent on method and temperature of preparation.

Indices variable: $\omega = 1.50$ $\varepsilon = 1.56$ (Florke 1952)

Form: Hexagonal basal plates.

Colour: Colourless.

Occurrence: Formed by the low-temperature dehydration of hemihydate. May be present in lightly burned gypsum plaster and Portland cement.

Comment: Appears to be a variable phase which is difficult to characterise under the microscope as in appearance it is similar to hemihydrate. Its presence in Portland cement can lead to premature stiffening and is best determined by a combination of thermal and chemical analysis.

XRD: 45-157: 6.046-100, 3.485-22, 3.016-71, 2.794-22.

Thermal analysis: Refer to gypsum.

Anhydrite (insoluble anhydrite, hard burnt gypsum)

CaSO$_4$	Biaxial +ve	Orthorhombic

$$\alpha = 1.570 \quad \beta = 1.576 \quad \gamma = 1.614$$

Interference figure: $+2V = 42°$.

Birefringence: Strong polarisation colours up to the end of the third order.

Form: Fine to medium grained aggregates sometimes elongated. Well-formed crystals not common.

Cleavage: Three good cleavages intersecting at right-angles parallel to the sides of the crystals.

Twinning: Polysynthetic twinning common.

Extinction: Extinction is parallel to cleavages.

Colour: Colourless.

Occurrence: Usually associated with gypsum and salt deposits and is a relatively widespread mineral. Principal constituent of hard burnt gypsum plasters.

Comment: Green (1984) states that commercial anhydrites almost always consist of block crystals with square corners. The size of crystals and colourful appearance under crossed polars make anhydrite easy to identify (Plate 83).

XRD: 37-1496: 3.499-100, 2.849-29, 2.3282-20, 2.2090-20.

Thermal analysis: Refer to gypsum.

Syngenite

$K_2Ca(SO_4)_2.H_2O$ Biaxial $-$ve   Monoclinic

$\alpha = 1.500\text{–}1.501$ $\beta = 1.515\text{–}1.517$ $\gamma = 1.518\text{–}1.520$

Interference figure: $-2V = 28°$.
Form: Tablets or prisms.
Twinning: Common.
Colour: Colourless.
Occurrence: Found in gypsum and anhydrite beds.
Comment: Forms in Portland cement and is one of the causes of premature stiffening (Smillie *et al.* 1993). Does not survive in concrete. Best detected by thermal analysis and XRD.
  XRD: 28-739: 9.49-40, 5.71-55, 4.624-40, 3.165-75, 2.835-100.
  Thermal analysis: Endotherm at 260°C (Smillie *et al.* 1993).

Ammonium sulfate (mascagnite)

$(NH_4)_2SO_4$ Biaxial $+$ve   Orthorhombic

$\alpha = 1.521$ $\beta = 1.523$ $\gamma = 1.533$

Interference figure: $+2V = 52°$.
Birefringence: 0.01. First-order greys and whites.
Form: Tablets or nearly equant.
Cleavage: Distinct.
Twinning: Twinning gives pseudo hexagonal habit.
Colour: Colourless or stained grey to yellow.
Occurrence: Found in guano and associated with volcanoes. Present in some fertilisers.
Comment: One of the agents which causes sulfate attack on concrete.
XRD: 40-660: 4.392-49, 4.335-100, 3.052-49, 2.096-22.
Thermal analysis: Decomposes at 350°C.

Magnesium sulfate (epsomite, hexahydrate, kieserite)

Epsomite $MgSO_4.7H_2O$	Biaxial $-$ve	Orthorhombic
Hexahydrate $MgSO_4.6H_2O$	Biaxial $-$ve	Monoclinic
Kieserite $MgSO_4.H_2O$	Biaxial $+$ve	Monoclinic
Epsomite	$\alpha = 1.433$ $\beta = 1.455$ $\gamma = 1.461$	
Hexahydrate	$\alpha = 1.426$ $\beta = 1.453$ $\gamma = 1.456$	
Kieserite	$\alpha = 1.523$ $\beta = 1.535$ $\gamma = 1.586$	

Interference figure: Epsomite $-2V = 51°$, hexahydrate $-2V = 38°$, kieserite $+2v = 57°$.
Birefringence: Birefringence of kieserite is strong compared to others.
Form: Epsomite and hexahydrate commonly acicular to fibrous. Kieserite is usually prismatic.
  Cleavage: Cleavages present.
  Colour: Colourless to white or stained.
  Occurrence: Epsomite associated with gypsum and salt deposits and as an efflorescence in some ore deposits. Epsomite dehydrates to hexahydrate in air. Kieserite associated with salt deposits.

Comment: These magnesium sulfate minerals may be associated with sulfate attack on concrete.

XRD: Epsomite 36-149:   5.98-30, 5.34-30, 4.216-100, 4.200-75.
Hexahydrate 33-882:   4.815-75, 3.405-100, 3.351-70, 3.313-70.
Kieserite 24-719:   5.45-50, 5.10-45, 4.39-100, 4.04-45.
Thermal analysis: Refer to Mackenzie (1970).

Arcanite K_2SO_4, *Calcium langbeinite* $K_2SO_4.2CaSO_4$, *Aphthitalite* $(K_2Na)_2SO_4$. These sulfate minerals form in cement clinker but are not observable in concrete. Brief mineralogical details are given by Campbell (1986).

Asbestos and other fibrous materials

A group of hydrous silicates primarily composed of fibrils ranging from 20–200 nm in thickness which form durable, strong fibres that can be either woven or used as reinforcement in materials such as hardened cement and polymeric resins. The optical mineralogy of the asbestiform minerals is well covered in the literature so only a summary of optical data has been given. Crocidolite is a variety of riebeckite and amosite is a commercial name and not a mineral name. The composition of tremolite is $Ca_2Mg_5Si_8O_{22}(OH)_2$ which forms a continuous solid solution with actinolite which is the iron rich member.

	Chrysotile (Antigorite)	Riebeckite (Crocidolite)	Grunerite (Amosite)	Anthophyllite	Tremolite/ Actinolite
Composition	$Mg_3(Si_2O_5)(OH)_4$	$(NaCa)_2(FeMn)_3$ $Fe_2(SiAl)_8O_{22}$ $(OH,F)_2$	$(Fe_{0.9}Mg_{0.1})_7Si_8O_{22}$ $(OH)_2$	$Mg_5Fe_2Si_8O_{22}$ $(OH)_2$	$Ca_2(Mg,Fe)_5$ $Si_8O_{22}(OH)_2$
	Monoclinic	Monoclinic	Monoclinic	Orthorhombic	Monoclinic
α	1.493–1.546	1.693	1.657–1.667	1.598–1.652	1.600–1.628
β	1.504–1.550	1.695	1.684–1.697	1.615–1.662	1.613–1.644
γ	1.517–1.551	1.697	1.699–1.717	1.623–1.676	1.625–1.655
$\gamma - \alpha$	0.011–0.014	0.004	0.042–0.054	0.016–0.025	0.22–0.27
2V	0–50° (+)	5° (+)	10–14° (−)	70–90° (+)	79–85° (−)
Extinction	Parallel	5°	10–15°	Parallel	10–20°
Orientation	Length slow	Length fast	Length slow	Length slow	Length slow
Colour	Colourless	Blue	Colourless	Colourless or pale colour	colourless to pale green
Pleochroism	None	Dark blue to grey blue, strong	Pale brown to yellow, slight	Slightly pleochroic	Trem., none Act., green
Mineral group	Serpentine	Clinoamphibole	Clinoamphibole	Orthoamphibole	Clinoamphibole

Provided that the fibres or fibre bundles are of the order of 1 μm or greater in thickness small numbers of fibres can be detected by the optical method (Plate 72). However, the optical properties of chrysotile are difficult to determine because of the fineness of its fibres. When optical studies are made on fibre bundles, which is often the case for fibre reinforced products, the optical properties are the integrated effects of the individual fibrils. This has the effect that the asbestiform minerals mostly show straight extinction in these cases (Zussman 1979).

X-ray: Antigorite-1M 21-963: 7.29-100, 3.61-80, 2.525-100, 2.458-60.
Riebeckite 19-061: 8.40-100, 3.12-55, 2.801-18, 2.726-40.
Grunerite 31-631: 8.33-100, 3.07-80, 2.766-90, 2.639-70.
Anthophyllite 42-544: 9.37-85, 4.59-80, 4.50-100, 3.23-70.
Tremolite 13-437: 8.38-100, 3.268-75, 3.121-100, 2.705-90.
Actinolite 41-1366: 8.42-75, 3.276-45, 3.117-100, 2.709-55.

X-ray data also given by Zussman (1979) who notes that in powder diffraction of asbestos fibres only the stronger reflections may be detected.

Thermal analysis: *See* Mackenzie (1970).

Infrared: The infrared spectra of the asbestiform minerals provide an unambiguous means of identification on the basis of characteristic absorption bands in the 200 to 1600 cm^{-1} frequency range.

Synthetic and organic fibres

Many fibres are illustrated and discussed in volume 2 of McCrone and Delly (1973) and the Textile Institute (1985). All the fibres have parallel extinction. The data for cellulose are indicative for flax, jute, kraft and sisal fibres.

	$n_\parallel$	$n_\perp$	Bir.	Length
Acrylic	1.511	1.515	0.004	fast
Aramid (kevlar)	> 2.00			
Carbon	opaque black			
Nylon	1.58	1.52	0.06	slow
Polypropylene	1.530	1.496	0.034	slow
Cellulose	1.60	1.53	0.07	slow

Glass

Synthetic glasses

Of the synthetic glasses only a few are of interest to the petrographer. The compositions given below are indicative of the compositional ranges used (Boyd *et al.* 1994, Shand 1958) of the glasses of interest. The relationship of optical properties to composition and density are summarised by Winchell and Winchell (1964).

	SiO_2	Al_2O_3	B_2O_3	Li_2O	Na_2O	MgO	CaO	ZrO_2
Low-expansion borosilicate	81	2	13	–	4	–	–	–
Soda lime	72	1	–	–	15	3	9	–
E-glass	54	14	–	–	–	4.5	17.5	–
Alkali resistant	71	–	–	1	12	–	–	16

	R.I.	Density
Low-expansion borosilicate	1.47	2.23
Soda lime	1.51–1.52	2.46–2.49
E-glass	1.55	2.60
Alkali resistant	1.54	2.62

Remarks: Crushed low-expansion borosilicate glass commonly referred to as 'Pyrex' glass is an aggregate which will give a large expansion in the ASTM C227 mortar bar test and thus can be used as an upper limit. However, its reactivity in the mortar bar test is variable and dependent on its thermal history. Like all glasses, crushed Pyrex exhibits conchoidal fracture. Coloured soda lime glass is occasionally used as an exposed aggregate in concrete or as glass tiles and can result in AAR (Figg 1981).

The common type of fibre-glass used in building products is based on E-glass. Because of the poor alkali resistance of E-glass in concrete an alkali-resistant glass containing zirconia was developed by Pilkingtons under the trade name of 'Cemfil'. To distinguish between this and E-glass, analysis is required as they cannot be differentiated in thin-section. The fibreglass strand used for building products has filaments of about 7 μm in diameter which are combined to form the strand. The strands are treated with sizes or coupling agents which are tailored to the end use of the fibreglass. The specific compositions of these agents are usually confidential to the manufacturer. Where a thick coating of sizing has been used this may be visible in thin-section especially where degradation of the fibre occurs.

Natural glasses

The common textural names used to describe volcanic glasses or rocks containing large amounts of glass are obsidian, perlite, pitchstone, pumice and scoria. Volcanic glasses can vary from acid to basic in composition and these textural terms need to be qualified. For example rhyolitic obsidian, basaltic scoria, etc.

Obsidian, pumice and scoria: Obsidian is usually a dark coloured glassy rock, often banded, with conchoidal fracture. The term is now restricted to glasses of low water content and excludes those of higher water content such as perlite and pitchstone. Obsidians can range from acid to intermediate in composition with rhyolitic obsidian being the most common. In thin-section the glass ranges from clear to brown in colour and usually contains crystallites. However, it is not uncommon to find obsidian fragments in concrete aggregates which do not contain any crystallites and are wholly glassy. Pumice is a light-coloured vesicular rock derived from frothed volcanic glass usually of rhyolitic composition. Pumices of other composition are not common. In thin-section its texture consists of a mass of vesicles separated by thin walls of clear or light coloured glass. Scoria is a dark vesicular rock derived from basalt or andesite and is noticeably heavier than pumice. In thin-section the glass is always dark in colour and may contain considerable quantities of mineral dust and crystallites.

Compositions: The compositions of volcanic glasses will be similar to the bulk composition of the crystalline rock. However George (1924) points out that natural glasses differ in composition from the related crystalline rocks in that they are more acid and contain more alkali with potash dominating over soda.

Refractive index and glass composition: A good correlation between the silica content of natural glass and refractive index has been claimed (George 1924). However, Tilley (1922) points out that this relationship is affected by the water content of the glass as it increase the refractive index. Some comparative curves for silica content versus refractive index are given by Williams *et al.* (1954). An average extrapolation from the curves for glass made by whole rock fusion gives the following refractive indices.

Silica content (%)	Refractive index
75	1.49
70	1.505
65	1.52
60	1.54
55	1.565
50	1.595
47	1.62

The following information is taken from Grout (1932) as it gives some idea of the range of indices to be expected from rocks. However, it is not clear how much of this data is applicable to glass groundmass as opposed to essentially whole rock glass.

Name of rock	Average refractive index	Range recorded
Obsidian or rhyolitic glass	1.492	1.48–1.51
Pitchstone	1.500	1.492–1.506
Perlite	1.497	1.488–1.506
Pumice (Plate 70)	1.497	1.488–1.506
Dacite	1.511	1.504–1.529
Trachyte	1.512	1.488–1.527
Andesite	1.512	1.489–1.529
Leucite tephrite	1.550	1.525–1.580
Tachylite, scoria, basalt glass including palagonite	1.575	1.506–1.562

The glassy groundmass in volcanic rocks

The composition of the glassy groundmass of a volcanic rock is differentiated from the parent rock. It is the portion that remains after other components of the rock have crystallised. Thus an important distinction must be made between natural glasses such as obsidian and the glass which occurs in the groundmass of fresh volcanic rock. Glasses such as obsidian will approximate to the bulk composition of the rock type. However, the glass in the groundmass of a typical andesite may contain between 60 and 70 per cent silica. This increase in silica content occurs because of differentiation as the glass is the residual material remaining after the minerals have crystallised.

The glass in granulated blastfurnace slag

$n = 1.57$ to 1.66 dependent on composition.

The refractive indices of a range of slag glasses have been investigated by Battigan (1986) who found the above results. There appears to be a good relationship between refractive index and the calcium/silica ratio. Addition of alumina and magnesia into the ratio does not significantly affect the correlation indicating that the combined effect of these ions appears to be neutral.

The range in composition of blastfurnace slags is given by Lea (1970) and Taylor (1990). Glass contents vary dependent on method of manufacture. Granulated slags may contain in excess of 95 per cent glass while in pelletised slags it may be as low as 50 per cent. Methods for estimating glass content of slag by transmitted light microscopy have been described by Drissen (1995) and Hooton and Emery (1983). Alternatively, glass content may be estimated by quantitative X-ray diffraction techniques using internal standards and Rietveld analysis (Howard *et al.* 1988) but both microscopy and X-ray diffraction methods will fail if microcrystallinity is present (Taylor 1990).

The silica minerals and mineraloids

Quartz

SiO_2 Uniaxial $+$ve   Trigonal
$\omega = 1.544$ $\varepsilon = 1.553$

Interference figure: Easily obtained if the crystal is large enough. May be pseudo-biaxial with a small 2V angle if the crystal is strained.

Birefringence: In standard 30 μm thickness section maximum birefringence gives an interference colour of white with a slight tint of yellow. For concrete sections which are 27 μm or less in thickness the colour is white to greyish white. Quartz is the main mineral used for estimating the thickness of a thin-section under the microscope.

Form: Quartz occurs in a wide variety of forms which are well-described in literature on optical mineralogy. This literature should be consulted for further details.

Twinning: Although twinning is common in quartz it is rarely observed in thin-section.

Orientation: Euhedral crystals are length slow.

Colour: Colourless

Occurrence: Quartz is a ubiquitous mineral found in many rocks as an essential, accessory, or secondary mineral. It is especially abundant in sandstones, arkoses, quartzites, granites, rhyolites and gneisses. It is one of the most common detrital minerals.

Comment: In loose detrital material quartz can be confused with some of the feldspars. The forms of quartz of particular interest to the concrete petrographer are those exhibiting undulatory extinction and varieties of microcrystalline and disordered quartz of which the latter are difficult to define either in thin-section or by electron microscopy. The method for estimating the undulatory extinction quartz has been described by Smith *et al.* (1992). Measurement of undulatory extinction in quartz is no longer accepted as a method for predicting the reactivity of quartz-bearing rocks (Grattan-Bellew 1992). A method based on the estimation of total grain boundary areas of the quartz crystals appears to provide some correlation with reactivity in some metamorphic rocks (Wigum 1996).

XRD: 33-1161: 4.257-22, 3.342-100, 1.8179-14, 1.5418-9.

Thermal analysis: Gives a sharp exothermic peak at 573°C due to the α to β inversion.

Chalcedony

SiO_2 Av. $n = 1.537$ Cryptocrystalline
                                 (1.530–1.539)

Birefringence: Weak, maximum colour is first-order grey.

Form: Fibrous, often spherulitic, may also show mosaic type structure (Plate 67).

Extinction: Usually parallel to length of fibres and rarely at 30°. Aggregate structures may fail to extinguish. Fibres usually length slow but may also be length fast.

Colour: Colourless in thin-section.

Occurrence: Present as the main constituent of flints, cherts, jaspers and as a cementing agent in quartzites.

Comment: Formerly considered as a distinct mineral but electron microscopy has shown chalcedony to consist of minute quartz crystals. Chalcedony is considered to be a reactive mineral with cement alkalis but work by Rayment (1992) has shown that other disordered forms of quartz are the reactive species in flints and cherts.

XRD: Gives same pattern as quartz.

Cristobalite

SiO_2 Uniaxial $-$ve   Pseudo-isometric (tetragonal)
                                 $\varepsilon = 1.484$ $\omega = 1.487$

Interference figure: Crystals are usually too small to obtain a figure.

Birefringence: Variable. Cristobalite formed below 1470°C is practically isotropic. When formed at higher temperatures both the synthetic and natural cristobalite have grey interference colours giving a characteristic mosaic effect.

Form: When formed below 1470°C cristobalite is minutely crystalline and appears as an isotropic groundmass. Cristobalite exposed for long periods to higher temperatures exhibits a characteristic curved fracture, resembling overlapping roof tiles. Cristobalite crystallising from a glass may be dendritic, fernlike, needles or laths. In volcanic rocks it is found in minute square crystals or aggregates in cavities. It also occurs intergrown with the feldspar fibres in spherulites (Plate 67).

Twinning and extinction: Complex lamellar twinning is common with cristobalite formed at high temperature. Twinning is less common in natural cristobalite. Elongated crystals may show extinction at any angle.

Colour: Colourless in thin-section.

Occurrence: In silica and semi-silica bricks around the periphery of quartz crystals and along cracks. Found in some high-temperature contact rocks and widely found as a late-forming metastable mineral in volcanic rocks. Often forms in cavities but is also present as small pools in the groundmass.

Comment: As little as 2 per cent of cristobalite has been shown to be expansively reactive with cement alkalis (Kennerley and St John 1988) and its presence results in pessimum proportions in alkali-aggregate reaction. Thus its detection is important when examining aggregates. It can be difficult to detect in basalts and requires the field diaphragm to be closed down to detect its negative relief and low refractive index. Chemical detection of the silica minerals such as cristobalite indicates that greater amounts are often present than detected in thin-section (Katayama *et al.* 1989). Synthetic cristobalite has been proposed as a reference aggregate in the ASTM C227 mortar bar test by Lumley (1989).

XRD: 39-1425: 4.040-100, 3.136-8, 2.841-9, 2.487-13.

Thermal analysis: See Mackenzie (1970).

Tridymite

SiO_2 Biaxial +ve   Orthorhombic

synthetic $\alpha = 1.469$ $\beta = 1.470$ $\gamma = 1.473$.

natural $\alpha = 1.477$ $\beta = 1.478$ $\gamma = 1.481$. Indices may vary slightly.

Interference figure: Readily obtained unless crystals are very small which is usually the case in rocks.

Birefringence: Very weak. The maximum polarisation colour is grey.

Form: Wedged or arrowhead crystals are common but may be absent as tridymite may also crystallise as laths (Plate 67).

Twinning: Twinning is very characteristic in both wedge and lath forms.

Colour: Colourless in thin-section.

Axial angle: Minimum $2V = 35°$ with angles of up to $72°$ having been measured.

Occurrence: Tridymite is the principal constituent of silica brick with cristobalite as an associate. Natural tridymite occurs as a metastable mineral in a well-crystallised form in the cavities of acid and intermediate volcanic rocks and less frequently in basalt. In the groundmass it is usually microscopic. Tridymite in general is much more abundant than cristobalite in volcanic rocks and in cases can form up to 25 per cent of the rock (Frondel 1962). May be associated or altered to quartz or cristobalite.

Comment: Tridymite is usually much easier to detect in thin-section than cristobalite because of its distinctive shape and twinning although greater amounts are usually present than observed in thin-section. Hornibrook *et al.* (1943) and Schuman and Hornibrook (1943) tested firebrick composed principally of tridymite and found it gave an expansion of 0.68 per cent comparable to that found with opal. This means that like cristobalite and opaline materials, tridymite contributes to the effect of pessimum proportion and that only small percentages are required to cause expansion.

XRD: Natural orthorhombic 42-1401: 4.28-93, 4.08-100, 3.80-68, 3.242-48.

Thermal analysis: See Mackenzie (1970).

Opal

$SiO_2.nH_2O$ Isotropic

Water content commonly 4–9 per cent but can range up to more than 20 per cent.

Water content (per cent)	3.5	6.33		8.97	28.04
Refractive index			1.459	1.453	1.447	1.409

Birefringence: Usually nil but may show very weak birefringence due to strain.

Form: Often massive without any structure and can be present as a cementing material in sandstone (Plate 67). It is often found as rounded crusts, in veinlets and as a cavity filling or lining. It can replace wood and feldspars.

Colour: Colourless to pale grey or brown in thin-section.

Occurrence: Occurs as the main mineral in opaline shales and opal rock. Opal is a secondary mineraloid in volcanic rocks where it is commonly associated with quartz, chalcedony and tridymite. It is the principal constituent of diatomite and siliceous sinter.

Comment: Opal is a mineraloid consisting of submicroscopic cristobalite with a disordered structure and excess water (Frondel 1962). Its detection in concrete aggregates is important as amounts as small as 1 per cent can cause damaging expansion due to alkali-aggregate reaction.

XRD: Gem quality opal 38-448: 4.08-100, 2.86-10, 2.51-60.

Thermal analysis: Similar to cristobalite (Mackenzie 1970).

Fused silica glass including lechatelierite: $n = 1.458$.

Ignited silica hydrogel: $n = 1.48 - 1.485$ isotropic.

Soluble silicates

Optical details of the anhydrous soluble silicates are given by Vail (1952). Details of some of the crystalline hydrated potassium and sodium silicates are given in Winchell and Winchell (1964). These soluble silicates are either orthorhombic or monoclinic with indices varying from 1.45 to 1.50 and birefringence from 0.007 to 0.009.

Alkali-silica gel

Sodium/potassium silicate gel $n = 1.45–1.51$, mean ~ 1.48.

The composition of the gel is variable and has been reported as: $Na_2O = 1$–25 per cent, $K_2O = 0.5$–19 per cent, $CaO = 0.5$–23 per cent, $SiO_2 = 29$–85 per cent, ignition loss 10–25 per cent (Idorn 1961, Buck and Mather 1978). Older gels, gels remote from the site of reaction and crystallised gels may contain significant concentrations of Ca^{++} ions within the gel/crystal structure.

Relief: Fairly low negative relief.

Birefringence: True gel is isotropic but portions of gel often convert to birefringent material (crystalline gel) with a first-order white colour. May be carbonated near surfaces when it will show higher-order interference colours more typical of carbonates.

Form: Gel often contains conchoidal drying cracks especially where it lines cracks and pores. Has the appearance of a viscous material that has flowed and may be layered. Commonly present as tongues of gel extruding from the mouths of cracks in aggregates. Hardened cement paste saturated with gel at the margins of aggregates and cracks appear darkened under cross polars (Plate 65).

Colour: Can vary from colourless to yellow, brown and almost black. The colour does not appear to be related to crystallisation of the gel.

Occurrence: The product formed by the interaction of pore solution alkalis with siliceous or alumino-siliceous components in the aggregates. Commonly observed where expansion and cracking of concrete has occurred.

Comment: Alkali-silica gel is easy to recognise in cracks and pores where it may be associated with ettringite. It is difficult to observe in the texture of aggregates. Where tongues of gel plug mouths of aggregate cracks, the tongue often sits on an area of finely crystalline birefringent material showing yellow and red colours of possibly the first order. This finely crystalline material, not to be confused with crystallised gel, may be an intermediate product and is claimed to be of the same composition as the adjacent alkali-silica gel (Andersen and Thaulow 1990).

XRD: Indeterminate. Some data given by Buck and Mather (1978).

Carbonates and related calcium minerals

Calcium carbonate (calcite, limestone, marble)

$CaCO_3$ Uniaxial $-$ve   Trigonal (rhombohedral)
$$\varepsilon = 1.486 \quad \omega = 1.658$$

Interference figure: A good uniaxial figure with several rings is obtained from suitable crystals. Occasionally calcite gives a biaxial figure with a small angle.

Relief: Changes markedly on rotation of the stage as relief swings from positive to negative.

Birefringence: The maximum interference colours are pearl grey or white of the higher orders. Coloured fringes can often be seen around crystal edges. In carbonated hardened cement paste the calcium carbonate usually appears golden as it is present as microcrystalline aggregates which lower the birefringence considerably. In some hardened pastes it can be difficult to differentiate carbonation from calcium hydroxide on birefringence alone.

Form: Fine to coarse aggregates, usually anhedral. Euhedral crystals are rare in rock and concrete sections. Fossiliferous limestones often show organic structures. Unless the calcium carbonate is redeposited on surfaces or in cracks it is invariably present in hardened concrete as microcrystalline aggregates (Section 5.4).

Twinning: Polysynthetic twinning is very common and may be generated by pressure on the crystal during grinding of the section. Twin lamellae may show first-order interference colours.

Extinction: Extinction is symmetrical with regard to cleavages.

Colour: Colourless in thin-section but is often cloudy.

Occurrence: Calcite is the principal constituent of limestone and marble. It is a common vein mineral and is found in many rock types as a secondary mineral. Calcite is the principal raw component used to make Portland cement and is ubiquitous in the outer surfaces of hardened cement and concrete where it is produced by the carbonation of calcium hydroxide and the cement hydrates.

XRD: 5-586: 3.035-100, 2.285-18, 2.095-18, 1.913-17, 1.875-18.

Thermal analysis: Endotherm at 800–925°C (Mackenzie 1970, Ramachandran 1969).

Dolomite

$Ca(Mg,Fe,Mn)(CO_3)_2$ Uniaxial $-$ve   Hexagonal
$$\varepsilon = 1.500–1.526 \quad \omega = 1.680–1.716$$

Interference figure: Interference figure has many rings.

Relief: Changes markedly on rotation of stage. High positive relief parallel to the long diagonal of the rhomb.

Birefringence: Extreme and increases with increasing amounts of Fe and Mn.

Form: Fine to coarse grained. Well formed rhombs are common and the crystals may be curved. Zonal structure is common.

Cleavage: Perfect rhombohedral cleavage which usually shows as two intersecting lines at oblique angles.

Twinning: Metamorphic dolomite usually shows polysynthetic twinning.

Extinction: Symmetrical to outlines of crystals and cleavages.

Colour: Colourless to grey in thin-section.

Occurrence: Chief constituent of dolomite rocks and occurs as replacement veins in limestones. Also a common mineral in some ore deposits.

Comment: Raw material used for dolomitic limes. Aggregates based on argillaceous dolomites interbedded with other types of dolomite and limestone can cause expansion in concrete due to the alkali-carbonate reaction (Dolar-Mantuani 1983). The textures of some dolomites in thin-section are given by Adams *et al.* (1991). It is sometimes difficult to differentiate from calcite in thin-section but the stain, Alizarin Red S Table 3.5, can be used for this purpose.

XRD: 34-426: 2.888-100, 2.193-19, 1.7870-13.

Magnesian calcite: 43-697: 3.004-100, 2.263-22, 1.8892-28, 1.8553-21.

Thermal analysis: Endotherms at 800 and 925°C (Mackenzie 1970).

Aragonite

$CaCO_3$ Biaxial −ve   Orthorhombic

$$\alpha = 1.530 \quad \beta = 1.682 \quad \gamma = 1.686$$

Interference figure: $-2V = 18°$

Birefringence: Extreme. Interference colours pearl grey to high-order white.

Form: Columnar or fibrous.

Cleavage: Imperfect parallel to the length of the crystal.

Twinning: Fairly common both as contact and penetration twins.

Extinction: Parallel to fibres or columns.

Colour: Colourless.

Occurrence: Many shells are composed of aragonite. Occurs as a secondary mineral in cavities of basalts and andesites and as seams in limestones and sandstones.

Comment: Above 60°C aragonite forms in cementitious systems in preference to calcite. Common in lower-temperature geothermal cementing systems. May be found as silvery needles on concrete in exposure systems which cycle temperatures up to 60°C. Forms in preference to calcite in seawater attack because of the presence of Mg ions (Taylor 1990).

XRD: 41-1475: 3.397-100, 3.274-50, 2.702-60, 1.9774-55.

Thermal analysis: See Mackenzie (1970).

Vaterite

$CaCO_3$ Uniaxial +ve   Hexagonal

$$\omega = 1.550 \quad \varepsilon = 1.640 – 1.650$$

Birefringence: Extreme colours, pearl grey to high-order white.

Colour: Colourless or white.

Occurrence: Deposition of calcium carbonate in sea basins often follows the transition of calcium carbonate gel → vaterite → aragonite → calcite.

Comment: In the carbonation of hardened cement paste vaterite is formed first and may go through the transition of vaterite → aragonite → calcite (Cole and Kroone 1959–60). Is unlikely to be observable in thin-section.

XRD: 33-268: 3.294-100, 2.730-90, 2.063-60, 1.820-70.

Calcium hydroxide (Portlandite, slaked lime)

$Ca(OH)_2$ Uniaxial −ve   Hexagonal

$$\varepsilon = 1.545–47 \quad \omega = 1.573–75$$

Relief: Indices are close to the standard mountant index of 1.54 or epoxy indices of about 1.57–58. On rotation of stage relief is close to nil in one position.

Birefringence: In standard sections gives first-order red and second-order blue. In thinner sections of concrete more commonly gives first-order yellow, orange and red. Small crystals, such as occur in hydrated lime, appear white. In some hardened pastes the yellow birefringence of small crystals of calcium hydroxide can be confused with the golden birefringence of small aggregates of carbonation.

Form: Crystals occur as minute plates naturally. In hardened cement paste calcium hydroxide when observed by electron microscopy occurs as hexagonal plates, tablets and rods as well as irregular shapes. In thin-sections of concrete calcium hydroxide is observed filling or partially filling void space and is usually irregular in form. Sizes range from microcrystalline to crystals as large as 50 μm in size (Section 5.3).

Cleavage: Perfect basal cleavage which is rarely observed in thin-sections of concrete.

Colour: Colourless

Occurrence: Very rare mineral found in some limestone contacts. Calcium hydroxide is ubiquitous in hardened cement paste and is the principal constituent of slaked and hydraulic limes.

Comment: In hardened cement paste calcium hydroxide is one of the few minerals which is not submicroscopic. The observation of calcium hydroxide is an important diagnostic tool in petrography. In slaked lime, microscopic examination can give information on impurities but the crystals of calcium hydroxide are too small to be readily observed. In hydraulic limes the hydraulic components are best observed in polished section.

XRD: 4-733: 4.9-74, 2.628-100, 1.927-42, 1.796-36.

Thermal analysis: Endotherm 500–600°C (Ramachandran 1969).

Calcium oxide (lime, burnt or quick lime)

CaO	Isotropic	Cubic
	$n = 1.838$	

Form: Usually rounded grains in cement clinker and dolomite bricks. In calcined limestone, calcium oxide essentially retains the shape of the crushed limestone particles and will tend to be polycrystalline. Electron microscopy shows burnt lime to consist of agglomerates 0.5–5 μm in size consisting of minute equant to globular shaped particles.

Cleavage: Perfect cubic cleavage but only apparent if material has been subjected to a high temperature.

Colour: Colourless but often stained yellow or brown by interaction with iron oxide.

Occurrence: Very rare mineral found in some high-temperature limestone contacts. Present in cement clinker, dolomite brick and is the major constituent of burnt lime.

Comment: Calcium oxide is not observable in thin-sections of concrete. In cement clinker it is best observed in polished section (Campbell 1986). The appearance of various lime powders in oil immersion is given by McKenzie (1994). It can be identified in powders and distinguished from periclase and spinel by White's Test (White 1909), which is outlined by Rigby (1948).

XRD: 31-1497: 2.4059-100, 2.777-36, 1.7009-54.

Aluminium minerals

Gibbsite (aluminium hydroxide)

$Al(OH)_3$ Biaxial +ve   Monoclinic

$\alpha \approx \beta = 1.565–1.577 \quad \gamma = 1.58–1.595$ (Winchell and Winchell 1951)
$\alpha = 1.55–157 \quad \beta = 1.55–157 \quad \gamma = 1.57–1.59$ (Rigby 1948)

Interference figure: Crystals usually too small to obtain a figure. 2V very small and may be zero.
Birefringence: Varies between 0.015 and 0.030 so interference colours range from high first order to second order.
Form: Variable. Hexagonal tablets but not common. May be lamellar or fibrous or present mosaic structure similar to chalcedony.
Cleavage: Perfect basal cleavage.
Twinning: Polysynthetic twinning often observed.
Extinction: Fibrous crystals may be length slow or length fast and show inclined extinction.
Colour: Colourless to pearly white. Industrially may be pale brown.
Occurrence: Is a constituent of bauxite, laterites and aluminous clays. Rarely found as a cementing mineral in sandstone.
Comment: Gibbsite is one of the hydration products of high-alumina cement but otherwise is unlikely to be observed in concrete.
XRD: 33-18: 4.849-100, 4.371-70, 4.382-50, 2.452-40. See Brindley and Brown (1980) for further details.
Thermal analysis: See Mackenzie (1970).

Amorphous hydrated alumina (pseudoboehmite, cliachite)

$AlO(OH)_n$ $n = 1.56–1.61$
Form: Fine grained to gelatinous.
Colour: Grey, yellow, white or brown depending on the presence of iron.
Occurrence: Cliachite is the chief aluminium ore. Pseudoboehmite is the name given to the synthetic hydrous alumina used in medicine and industry.
Comment: Amorphous hydrated alumina together with gelatinous silica are the end products of the leaching and corrosion of hardened cement paste. While gelatinous silica is commonly observed amorphous hydrated alumina is rarely identified. The hydrated aluminas are often poorly crystalline and may have some of the optical characteristics of boehmite.
XRD: Broad diffuse bands in the regions of the strongest lines of boehmite.
Boehmite: 21-1307: 6.11-100, 3.164/65, 2.346-55, 1.860-30.
Thermal analysis: See Mackenzie (1970) and Brindley and Brown (1980) for discussion on XRD and thermal data.

Mullite

$3Al_2O_3.2SiO_2$ Biaxial +ve   Orthorhombic
$\alpha = 1.642 \quad \beta = 1.644 \quad \gamma = 1.654$

Indices increase with iron and titanium content.

Interference figure: $+2V = 45$–$50°$.

Relief: Moderately high.

Birefringence: Maximum interference colour is yellow. Usually white to grey.

Form: Most characteristic form is needles or laths.

Extinction: Parallel extinction. Needles are length slow.

Colour: Colourless.

Occurrence: Rare natural mineral found in some high-temperature contact rocks. Mullite is constituent of some porcelains. Crystallises out when slag reacts with fired clay materials. Can often be seen as fine needles crystallising out in the glass spheres in fly ash.

Comment: Mullite needles may be too small to be observed under the optical microscope.

XRD: 15-776: 5.39-50, 3.428-95, 3.390-100, 2.206-60.

Grinding and polishing compounds

Diamond

C   Isometric   Cubic

                                     $n = 2.414$

Relief: High positive relief.

Form: Cubes, octahedrons, etc.

Colour: Colourless, white, yellow, orange, red, green, blue brown, black.

Occurrence: Natural gemstone. Also manufactured.

Comment: Most diamond grinding and lapping compounds are made from synthetic diamond. Diamond particles are observed in polished and thin-sections which have not been adequately cleaned during preparation. They stand out very clearly because of their high relief and colour.

XRD: 6-675: 2.06-100, 1.261-25, 1.075-16, 0.816-16.

Silicon carbide

SiC   Uniaxial $+$ve   Hexagonal

                         $\varepsilon = 2.689$–2.697 $\omega = 2.647$–2.654

Relief: Very high positive relief.

Birefringence: Strong with strong dispersion.

Form: Grinding powder often has splintered appearance with conchoidal fracture showing radiating stress lines.

Colour: Colourless, green, blue or black. Industrial silicon carbide is usually colourless or bluish. Blue crystals are pleochroic.

Occurrence: Rare natural mineral known as moissanite. Manufactured for grinding media which can consist of mixtures of cubic to hexagonal silicon carbide in a number of polytypes.

Comment: Silicon carbide particles are observed in polished and thin-sections which have not been adequately cleaned during preparation. They stand out very clearly because of their high relief and iridescent colours.

XRD: Variable. See 29-1126 through to 29-1131.

Aluminium oxide (corundum, α-alumina, γ-alumina)

| Al_2O_3 Corundum, α-alumina | Uniaxial −ve | Hexagonal |
| | $\varepsilon = 1.760 \quad \omega = 1.768$ | |

| γAl_2O_3 | Isometric | Cubic |
| | $n = 1.696$ | |

Interference figure: Readily obtainable.
Relief: Very high positive relief.
Birefringence: First-order white for corundum.
Form: Natural corundum is often hexagonal in shape. Synthetic corundum consists of angular and irregular shaped aggregates with some signs of conchoidal fracture. γ-alumina appears as irregular aggregations of fine particles.
Twinning: Simple or multiple twinning in corundum.
Colour: Colourless, white or grey.
Occurrence: Found particularly in some contact rocks in low silica environments where it may be called sapphire or ruby. Industrially synthesised for grinding media. Principal constituent of fused alumina brick and other products. γ-alumina is a synthetic precipitated product used as a polishing compound.
Comment: Synthetic corundum is widely used for grinding and polishing of cement clinker and concrete specimens. During grinding the corundum particles fissure and gradually break down to finer particles which makes it an ideal grinding agent. Unless specimens are carefully cleaned residual corundum particles may remain embedded in surfaces. It is not uncommon to find some corundum at the edges of the specimen embedded in the mounting media. γ-alumina is restricted to polishing compounds in the submicron range and is unlikely to interfere with optical examination.
XRD: Corundum: 13-373: 2.41-50, 2.12-50, 1.96-35, 1.395-100.
γ-alumina: 43-1484: 2.551-98, 2.086-100, 1.602-96, 1.374-57.

Cerium dioxide (cerianite, ceria polishing compound)

| CeO_2 | Isometric | Cubic |
| | $n \approx 2.2$ | |

Relief: Very high positive relief.
Form: As irregular shaped aggregates.
Colour: Yellow to brown.
Occurrence: Extracted from bastnasite.
Comment: The polishing compounds are a mixture of cerium dioxide and lanthanum and praseodymium oxides together with other impurities. It is coloured brown by the other rare earth oxides (Kilbourn 1993). Used extensively for polishing glass and may be used to polish cement clinker. Is difficult to remove from surfaces of clinker and is observed as aggregations of orange-brown fine particles.
XRD: Cerianite: 43-1002: 3.124-100, 2.706-27, 1.913-46, 1.632-44.

Pigments and colorants

Red (hematite, red ochres, burnt sienna, burnt umber)

α-Fe$_2$O$_3$ Uniaxial $-$ve   Hexagonal

$$\omega = 3.22\text{--}2.90 \quad \varepsilon = 2.94\text{--}2.69$$

Relief: Extremely high.

Form: Basal plates, rhombohedrons, platy, fibrous or massive.

Colour: Opaque unless crystals are very thin. Thin scales are coloured blood red, red-brown, orange or yellow. May be slightly pleochroic. Steel-grey with metallic lustre in reflected light with a tendency to a marginal red.

Occurrence: A very important iron ore. Rather common as microscopic inclusions and alteration products in many minerals and rocks, colouring them red.

Comment: Hematite is the principal coloured mineral in Spanish and Persian and other red ochres and in burnt sienna and umber. Many of these are impure mixtures especially some of the siennas and umbers which may contain clays and manganese oxide. For industrial use and in concrete red ochre is made synthetically where the quality is uniform and the colour shade is controlled by particle shape and size (Schiek 1982). Both synthetic and natural red iron oxides are classified as CI pigment red 101.

XRD: 33-664: 2.74-100, 2.519-70, 1.841-40, 1.694-45.

Black (magnetite, carbon black)

Fe$_3$O$_4$, (Fe,Mg,Zn,Ni,Mn)Fe$_2$O$_4$ Isometric   Cubic

$$n = 2.42 \text{ varies with substitution}$$

Many natural magnetites are substituted.

Relief: Very high.

Form: Octahedral or granular. Rarely cubes.

Colour: Opaque. Appears black in thin-section. Grey metallic lustre in reflected light.

Occurrence: Widespread in igneous and metamorphic rocks and important in many black ironsands.

Comment: Is also synthesised for use as a pigment. Black colour may be tinged with red unless some manganese is present. Both natural and synthetic iron oxides are classified as CI pigment black 11. Present in slag and fly ash.

XRD: 19-629: 2.967-30, 2.532-100, 1.616-30, 1.485-40.

C (carbon black)   Amorphous
                                     Opaque

Carbon black produced from oil feedstock varies in particle size from 0.5–4 μm in size. Particles less than 0.5 μm are rounded but other particles are sharply angular or sliverlike without apparent agglomerates (McCrone and Delly 1973).

Comment: Carbon black is a superior pigment to iron black and is classified as CI pigment black 6.

Yellow (goethite, lepidocrocite)

FeO(OH)   Uniaxial −ve   Orthorhombic
α-FeO(OH) (goethite) $\alpha = 2.275\text{-}2.15$ $\beta = 2.409\text{-}2.22$ $\gamma = 2.415\text{-}2.23$
γ-FeO(OH) (lepidocrocite) $\alpha = 1.94$ $\beta = 2.20$ $\gamma = 2.51$
   Interference figure: Goethite $-2V = 23°$ with strong dispersion. Lepidocrocite $-2V = 83°$.
   Relief: Very high.
   Birefringence: Goethite crystals are length slow.
   Form: Goethite, prismatic or striated or tabular or fibrous. Lepidocrocite scaly.
   Extinction: Parallel.
   Colour: Goethite has strong variable pleochroism from yellow to orange yellow. Lepidocrocite also pleochroic ranging from colourless to yellow through to red orange.
   Occurrence: Goethite is a common mineral, present in limonite iron ore and is the usual weathering product of siderite, pyrite, magnetite, glauconite and iron silicates. Lepidocrocite is often found associated with goethite. Goethite is the principal constituent of yellow ochre and raw sienna.
   Comment: Limonite is the general term used for undifferentiated iron weathering products and usually consists of goethite and/or lepidocrocite. However, optically it usually appears isotropic with a refractive index of 2.0 to 2.1. The yellow colour of ochres and synthetic iron oxide is due to the presence of one of these hydrated iron oxides. The natural siennas are classified as CI pigment brown 7 and ferrite yellow as CI pigment yellow 42.
   XRD: Goethite: 29-713:   4.183-100, 2.693-35, 2.450-50, 1.792-20.
          Lepidocrocite: 8-98:   6.26-100, 3.29-90, 2.47-80, 1.937-70.
          For further X-ray data see also Brindley and Brown (1980).
   Thermal analysis: See Mackenzie (1970).

Green (eskolaite, green cinnabar, phthalocyanine green)

Cr_2O_3 (eskolaite)   Uniaxial −ve   Hexagonal
                                            $n = $ about 2.5

   Relief: Very high.
   Birefringence: Thin plates are anisotropic but polarisation colours are masked by natural green colour.
   Form: Equant, tabular or scales.
   Cleavage: Parallel to the faces of the rhomb.
   Colour: In thick sections is opaque. Thin-sections exhibit a bright emerald green colour.
   Occurrence: Natural mineral known as green cinnabar. Occurs in refractory mixes containing chromic oxide.
   Comment: Synthetic chromic oxide is the principal green colorant used with cement.
   XRD: Eskolaite: 38-1479: 2.665-100, 2.480-93, 1.6274-87, 1.4316-39.

$C_{31}H_xCu.15\text{-}16Cl$ (phthalocyanine green) n about 1.40, laths (Gettens and Stout 1966). A chlorinated derivative of phthalocyanine blue classified as C.I. pigment green 7. Variations also listed as CI pigment green 42 and 43 are now listed as CI pigment green 7 and pigment green 36. CI pigment green 36 is a polybromochloro derivative of different shade. The systematic name is polychloro-tetrabenztetraazaporphyrin.
   XRD: CI pigment green 7: 36-1870: 14.8-83, 13.2-57, 5.21-40, 3.34-100.

Blue (cobalt blue, phthalocyanine blue)

Co.Al$_2$O$_3$ (cobalt blue)   Isometric   Cubic

$n > 1.78$ (red) $n = 1.74$ (blue)

This is CI pigment blue 28.
  XRD: 38-816: 2.439-100, 2.021-21, 1.557-36, 1.429-42.

C$_{32}$H$_{16}$Cu (copper phthalocyanine blue) Biaxial   Monoclinic

n about 1.38, laths (Gettens and Stout 1966).

This is classified as CI pigment blue 15 and forms α and β modifications which are slightly different in shade and stability. There are a number of variations of this pigment classified as 15.1, 15.2 and 15.3 which are phthalocyanine substituted with other ions such as nickel and cobalt. The systematic name is copper tetrabenztetraazaporphyrin.
  XRD: 22-1686: 13.2-100, 21.1-100, 8.93-40, 5.71-40.

White (Anatase, rutile)

α-TiO$_2$ (rutile)   Uniaxial +ve   Tetragonal

$\omega = 2.609$–2.616 $\varepsilon = 2.895$–2.903

β-TiO$_2$ (anatase)   Uniaxial −ve   Tetragonal

$\omega = 2.561$–2.562 $\varepsilon = 2.488$–2.489

  Interference figure: Both rutile and anatase may give anomalous biaxial figures with a small 2V.
  Relief: Highest relief of any rock-forming mineral.
  Birefringence: Extreme birefringence which does not show well on account of total reflection.
  Form: Small prismatic to acicular crystals.
  Cleavage: Parallel to length of crystals.
  Twinning: Common.
  Extinction: Parallel.
  Colour: Yellowish to reddish brown in thin-section. In reflected light shows adamantine lustre.
  Occurrence: Widely distributed accessory minerals in metamorphic rocks and occur as detrital minerals. Rutile is much more widespread than anatase.
  Comment: The synthetic pigment grades of rutile and anatase appear as dark agglomerates of very fine particles which under crossed polars appear white with some reddish brown coloration in the larger agglomerates. Individual fine particles exhibit brownian motion and may show some visible birefringence.
  XRD: Rutile: 21-1276: 3.247-100, 2.487-50, 2.188-25, 1.687-60.
Anatase: 21-1272: 3.52-100, 2.378-20, 1.892-55, 1.700-20.

Magnesium oxychloride cement hydrates (Sorel cement)

Formed by mixing calcined magnesite and calcium chloride solution. A range of hydrates are possible with 3Mg(OH)$_2$.MgCl$_2$.nH$_2$O and 5Mg(OH)$_2$.MgCl$_2$.nH$_2$O usual hydration products at ambient temperatures. The 5-form is metastable and converts to the 3-form. Also these hydrates are unstable in air and slowly convert to the magnesium chlorocarbonate, $Mg(OH)_2.MgCl_2.2MgCO_3.6H_2O$. Formation of the calcium oxychloride, $3Ca(OH)_2.CaCl_2.11H_2O$ also occurs in this system. At higher temperatures more crystalline 9-form and 2-form hydrates occur. Data given is from Demediuk *et al.* (1955), Cole and Demediuk (1955) and Newman (1955).

Optical data: The 3-form occurs as a gel-like aggregate of possibly monoclinic, minute crystals which are irregular in shape. The mean refractive index lies between 1.505 and 1.510 and the birefringence is very low. The 5-form tends to be more crystalline forming well-shaped needles about 10 μm long. The mean refractive index is about 1.525 and the birefringence is weak. The needles have parallel extinction and are length fast. The calcium oxychloride is possibly the $Ca_4O_3Cl_2.15H_2O$ listed by Winchell and Winchell (1964). Colourless, orthorhombic, acicular crystals with basal cleavage. $\alpha = 1.481$, $\beta = 1.536$, $\gamma = 1.543$, $-2V = 44°$.

XRD: 3-form: 8.03-s, 3.83-s, 2.431-ms, 1.985-ms.

5-form: 7.57-s, 4.14-ms, 2.64-m, 2.420-s.

From Cole and Demediuk (1955). Data variable with hydration. JPCDS data difficult to correlate.

$Ca_4Cl_2O_3.15H_2O$: 2-280: 4.08-100, 2.75-80, 2.61-80, 2.50-70.

Thermal analysis: See Cole and Demediuk (1955).

Appendix 1 Recent Descriptions of Attack by Chemicals on Concrete

It is over twenty years since Biczoc's monograph on concrete corrosion was published and there does not appear to be any more recent publication in English. A selection of references on attack by industrial chemicals are listed in an attempt to give the reader an update on the literature. (Abbreviations: CA – Chemical Abstracts; CRA – Cements Research Progress.)

Akman, M.S. and Gulseren, H. 1994: The influence of heat treatment on the durability of concrete. *Durability of Concrete. Third International Conference, Nice.* ACI SP-145. The effect of atmospheric steam curing on the properties of concretes containing Portland/pozzolana cement, high-strength concrete and high-performance concrete containing a superplasticiser was investigated. Exposure to 5 per cent ammonium nitrate solution indicated that steam curing was less detrimental to high-performance concrete and most detrimental to the Portland/pozzolana concrete.

Akman, M.S. and Guner, A. 1984: Applicability of Sonreb method on damaged concrete. *Materials and Structures, Research and Testing,* **17(99)**, 195–200.

Akman, M.S. and Yildirim, M. 1987: Loss of durability of concrete made from Portland cement blended with natural pozzolanas due to ammonium nitrate. *Durability of Building. Materials,* **4(4)**, 357–69 (CRP 1987: 174). Corrosion by ammonium nitrate based fertiliser increased with nitrate concentration and resistance to corrosion decreased with pozzolana addition greater than 30 per cent.

Al-Amoudi, O.S.B., Rasheeduzzafar, Maslehuddin, M. and Abduljauwad, S.N. 1994: Influence of chloride ions on sulfate deterioration in plain and blended cements. *Magazine of Concrete Research,* **46(167)**, 113–23. Chloride ions mitigate the sulfate attack on plain cements but have only a marginal effect on cements blended with BFS and silica fume.

Alexander, M., Addis, B. and Basson, J. 1994: Case studies using a novel method to assess aggressiveness of waters to concrete. *ACI Materials Journal,* **91(2)**, 188–96. Practical application of a previously published method, by Basson and Addis (1992), to assess the aggressiveness of water toward concrete is discussed together with comments received.

Attiogbe, E.K. and Rizkalla, S.H. 1992: Response of concrete to sulfuric acid attack. *ACI Materials Journal,* **85(6)**, 481–8. Changes in weight, dimension and sulfate content of mortars exposed to 1 per cent sulfuric acid solution were measured. It was concluded that all three parameters could be used to measure attack. The rate of attack was increased by alternate wetting with the acid solution and drying.

Babachev, G. and Petrova, M. 1976: Corrosion of cement stone and concrete by sulfuric acid. *Stroit. Mater. Silik. Prom-st.*, **17(11–12)**, 8–10. (CA 86: 160090). The effect of sulfuric acid on hardened cement and concrete is reported.

Babushkin, V.L. and Zinov, V.G. 1978: Study of the effect of salt physical corrosion on the moisture state of cement stone. (In Russian.) *Gidratatsiya Tverd. Tsem.*, **3: 42–56**, (CA 93:136785). The equilibrium water content of OPC pastes were determined at $293\,^{\circ}K$ after a varying number of cycles of exposure to 5 per cent potassium chloride solutions and drying at $100\,^{\circ}C$. As a result of the salt corrosion the body of the cement paste changed from a capillary-porous state to a colloidal capillary-porous state. The increase in water content during the salt corrosion was attributed to the formation of a hygroscopic colloidal crystalline material in the pores of the cement paste resulting in an increase in specific area and loosening of the structure. The data indicated the possibility of predicting the increase in water content in the cement paste from the compressive strength or vice versa.

Bajza, A. 1992: Corrosion of hardened cement pastes by formic acid solutions. *9th International Congress on the Chemistry of Cement, New Delhi*, **V**, 402–8. Attack by formic acid leads to deterioration of concrete forming a layer structure.

Bajza, A., Rousekova, I. and Vrana, O. 1986: Corrosion of hardened cement pastes by NH_4NO_3 solutions. *8th International Congress on the Chemistry of Cement, Rio de Janeiro*, **V**, 99–103. The corrosion of OPC, 20 per cent BFS and 40 per cent BFS blends by 0.5–5 per cent ammonium nitrate solutions was investigated. It was found that the nitrate solutions severely corroded the pastes with the BFS cement blends demonstrating better resistance in the short term. Corrosion was related to concentration and time of exposure and resulted in decreased paste density and strength and increases in non-evaporable water content, porosity and defects in the paste structure. The main corrosion product was identified as $C_3A.Ca(NO_3)_2.10H_2O$.

Basson, J.J. and Addis, B.J. 1992: An holistic approach to the corrosion of concrete in aqueous environments using indices of aggressiveness. *Durability of Concrete*, ACI SP-131: 33–65. By considering all relevant influences as part of the total corrosive environment it is claimed that it is possible to quantify aggressivenes as indices and to use these to select the appropriate technology. Codes of practice in South Africa and other countries are discussed.

Bermejo-Munoz, M., Sagrera-Moreno, J. and Gaspar-Tebar, D. 1987: Chemical resistance of concrete. XXVII. Study of hydrated cement P-550-ARI-de-ionised water system. *Mater. Constr. (Madrid)*, **37(205)**, 37–47 (CA 107: 160341). The leaching of hydrated cement by de-ionised water is discussed.

Bock, E. 1987: Method and apparatus for the determination of the resistance of materials, e.g., building materials to biogenic acid attack. *German patent DE 3,604,912* (CA 107: 182312).

Bock, E. and Sand, W. 1986: Applied electron microscopy on the biogenic destruction of concrete and blocks. Use of the transmission electron microscope for identification of mineral acid producing bacteria. *Proceedings of the International Conference on Cement Microscopy* 285–302. The effects of substrate material on biogenic corrosion and the use of the TEM to differentiate bacteria is described.

Bulatova, Z.I. and Abyzov, A.N. 1976: Concrete with binders made from treated synthetic slags in steel making. *Stroit. Mater. Izdeliya Osn. Otkhodov Prom-st. Vermikulita*, 70–5 (CA 88: 26822). Concrete containing slag cements can give improved acid resistance.

Calleja, J. 1980: Durability. *7th International Congress on the Chemistry of Cement, Paris, 1,* Theme VII-2: 1–48. The durability of concrete with respect to chemical attack, carbonation, carbonic acid attack, alkali-aggregate reactions, attack by chlorides and deicing salts, and sulfate attack are reviewed together with a discussion of standards, methods of tests and types of cements applicable.

Chandra, S. 1988: Hydrochloric acid attack on cement mortar – an analytical study. *Cement and Concrete Research,* **18,** 193–203. The zones of attack by hydrochloric acid on mortars and concrete are described.

Chernov, A.V. 1983: Classification and evaluation of corrosivity of organic media. (In Russian) *Beton Zhelezobeton,* **(8),** 13–14 (CA 99:163069). Corrosion of concrete by organic media was classified into three groups. Saturated hydrocarbons, water-insoluble alcohols, aromatic hydrocarbons, chlorbutadiene and styrene were the least aggressive. Water-soluble alcohols, polyhydric alcohols, water soluble ketones, urea, dicyanodiamide and melamine were classified as being of medium aggressiveness. Carboxylic acids, dimethyl formamide, phenols, formaldehyde and dichlorbutene were classified as being highly aggressive.

Chernov, A.V. and Kuznetsova, O.Y. 1978: Durability of concrete in powdered mineral fertilizers during freezing. (In Russian) *Zashchita Stroit. Konstruktsii ot Korrozzi, Rostov N/D,* 108–15 (CA 92:219876).

Collepardi, M. and Monosi, S. 1992: Effect of the carbonation process on concrete deterioration by $CaCl_2$ aggression. *9th International Conference on the Chemistry of Cement, New Delhi,* V, 389–93. Carbonation of the cement paste makes mortars durable against 30 per cent calcium chloride attack.

Danilov, I.N. and Ishmaeva, A.M. 1983: Corrosion resistance of non-metallic materials in the treatment and purification of petroleum refinery waste water. (In Russian) *Plast. Massy,* **(5),** 47–49 (CA 99:23682). The corrosion resistance of plastics, enamels, sealing compounds, asbestos-cement and concrete was studied. The corrosion resistance decreased with increasing temperature and pressure and by the action of oxygen and ozone. Fluoropolymers, concrete, Pentaplast and polypropylene had the highest resistance.

Duber, A. and Mirski, J. 1990: Resistance of cement fly ash materials to the action of agricultural silage acids. (In Polish.) *Ochr. Koroz.,* **33(11),** 257–9 (CA 115:77390). Concrete containing fly ash increased the corrosion resistance to 7:3 lactic/acetic acid mixtures ranging from 2.2 to 10 per cent in concentration.

El Hakim, F.A. 1987: Rate of chemical degradation of normal weight concrete subject to HCl solution. *KFAS Proc. Ser.* 1984, 2(Corrosion): 567–78 (CA 107:120071). Two groups of prisms mixed at water/cement ratios of 0.60 and 0.75 were used to predict the rate of reduction in dynamic and static moduli of elasticity and compressive strength in hydrochloric acid solutions. The prisms were cured for varying periods and then subjected to varying periods of exposure. Analysis of the results showed it is possible to predict reductions in dimension and strength for severe chemical attack.

Engvall, A. 1986: Mineral materials. *Biotechnology,* **8,** 607–26 (CA 106: 181557). Bio-corrosion of structural materials is reviewed.

Fattuhi, N.I. and Hughes, B.P. 1988: The performance of cement paste and concrete subjected to sulfuric acid attack. *Cement and Concrete Research,* **18(4),** 545–53 (CRP 1988: 129). Changes in weight, dimension and sulfate content were used to indicate deterioration in 1 per cent sulfuric acid. Wetting and drying increased the deterioration.

Fattuhi, N.I. and Hughes, B.P. 1988: Ordinary Portland cement mixes with selected admixtures subjected to sulfuric acid attack. *ACI Mater. J.*, **85(6)**, 512–18 (CRP1988: 129). Lower water/cement ratio, polyvinyl alcohol and lower acid concentrations but not cement content gave less deterioration to mortars exposed to 13 per cent sulfuric acid.

Fiertak, M. 1987: Selection of a new index of acid corrosiveness for cementitious materials. (In Polish.) *Cem. Wapno-Gips*, **(10)**, 205–7 (CA 109:97901). It was found that pH did not predict the corrosiveness of hydrochloric, acetic, lactic and citric acids and ammonium chloride and phenol solutions to concrete. Studies showed that acid concentration expressed in terms of acidity gave a good index of corrosivity.

Fonlupt, J. 1979: Behaviour of various materials in contact with ozone. (In French) *Journ. Eur. Appl. Ozone Trait. Eaux*, 193–213 (CA 93:210022). Concrete, metal alloys, some plastics and rubbers used in water purification installations are corroded by ozone. In general the action of ozone is similar to that of hydrogen peroxide and the corrosion rate is increased by intermittent wetting. A study of ozone in dry air, humid air and water at ambient temperatures for up to 20×10^3 hours showed that cast iron, stainless steel, rigid PVC, vinyl ester resin, Teflon, Rilsan, chlorsulfonated polyethylene-propene rubber, concrete, glass and ceramics have high resistance. Fritted stainless steel and silicone rubber were rapidly attacked.

Fritzke, H. 1986: Experimental and theoretical study on the reaction of hot liquid sodium with concrete. Report GKSS, 86/E/14: 145 (CA 106: 107023). The reaction of liquid sodium with concrete was found to depend on the composition of the concrete and the temperature of the sodium.

Fuji, T. and Nakamura, A. 1994: Behaviour of composite hardened cement pastes exposed to acid and deicer solutions. *JCA Proceedings of Cement & Concrete*, **48**, 542–7 (English abstract). Hardened cement pastes containing ground granulated blastfurnace slag reinforced with carbon or aramid fibres were exposed to either sulfuric acid at pH3 or 3 per cent sodium chloride or 3 per cent calcium magnesium acetate solution or mixtures of these solutions. It was concluded that the combined action of the acid and deicer solution is more deleterious to the cement-based composite materials than the sulfuric acid solution alone.

Garibov, S.M., *et al.* 1978: Corrosion of concrete in mineral salt solutions. (In Russian.) *Uch. Zap. Azerb. Inzh.-stroit. Int*, **10(2)**, 63–9 (CA 92:219873).

Gerdes, A. and Wittmann, F.H. 1992: Electrochemical degradation of cementitious materials. *9th International Congress on the Chemistry of Cement, V:* 409–415. Dissolution of solid phases in the concrete can be caused by long lasting electric fields with the reinforcement acting as an electrode.

Gilbert, P.D., Steele, A.D., Morgan, T.D.B. and Herbert, B.N. 1984: Concrete Corrosion. *Microb. Probl. Corros. Oil Prod. Storage, Pap. Microbiol. Group Symposium. 1983.* (Editor E.C. Hill), London, Institute of Petroleum: 71–80 (CA 102:83407). A review, with 23 references is given on the contribution of microbial activity in concrete oil-water storage systems to environments aggressive to concrete.

Goncalves, A. and Rodrigues, X. 1991: The resistance of cements to ammonium nitrate attack. *Durability of Concrete, ACI SP-126*: 1093-1118. High alumina cement gave the best resistance followed by 60 per cent blastfurnace slag cement to attack by ammonium nitrate.

Greschuchna, R. 1974: Cement paste crystals in sulfate and acidic solutions. *The VI International Congress on the Chemistry of Cement, Moscow*, Section II-6, Paper 96:

1–10. The reactions of sulfates and hydrochloric acid on mortars is reported.

Gudev, N., Todorov, R., Chuleva, D. and Angelov, B. 1982: Corrosion of concrete in some organic compounds I. Corrosion in monocarboxylic acids and their esters. (In Bulgarian.) Exposure to 1–12 per cent acetic, 1–10 per cent maleic and 1–10 per cent phthalic acids and saturated butyl acetate for 10–500 days resulted in yellowish surfaces and reductions in strength. Attack increased with increasing concentration of solution. Addition of butyl alcohol to the butyl acetate and ethyl acetate to the acetic acid solutions retarded attack.

Gunnar, A. and Akman, M.S. 1984: The influence of addition of blastfurnace slag on the corrosion resistance of concrete. *Proceedings of the 3rd International Conference on the Durability of Building Materials and Components, Finland,* 14–25. Concretes containing ordinary or pozzolanic cements are damaged by fertilisers and the use of blastfurnace slag was investigated to try to improve corrosion resistance. Partial replacement by BFS increased resistance to sulfate attack but not against ammonium salts such as ammonium sulfate and nitrate and diammonium phosphate.

Gutt, W.H. and Harrison, W.H. 1977: Chemical resistance of concrete. *Concrete,* **11(5)**, 35–7 (CRP 1977: 211). The chemical resistance of concrete is reviewed.

Guzeev, A.E. and Pimenov, A.N. 1986: Acid-resistant high strength concrete. *Sposoby Povysh. Korroz. Stoikosti Betona i Zhelezobetona, M.* 37–42 (CA 108: 61388). The properties of acid resistant high strength concrete are reported.

Hanus, Z., Tomiska, J. and Burget, C. 1976: Acid corrosion of lightweight cement slurries. II. Effect of inorganic acids on hardened lightweight plugging cements. *Pr. Ustavu Geol. Inz.,* **34**, 229–49 (CA 86: 94929).

Harner, A.L., Copeland, C.H. and Crim, B.G. 1992: Destruction of concrete by fertilisers urea ammonium nitrate vs. concrete. *Proceedings of the Fourteenth International Conference on Cement Microscopy* 53–63. Urea ammonium nitrate fertiliser attacks concrete by reaction with calcium hydroxide and weakening of the matrix. Rinsing and air drying of concrete surfaces did not prevent further deterioration.

Hollander, J. and Lanting, R. 1986: Effects of acid deposition on construction materials. *Stud. Environ. Sci.* 233–49 (CA 106: 89354). The effects of acidic air pollutants on construction materials are reviewed.

Hudec, P.P., MacInnis, C. and McCann, S.P. 1994: Investigation of alternate concrete de-icers. *Durability of Concrete. Third International Conference, Nice.* ACI SP-145, 65–84. Phosphate/chloride mixtures were investigated as deicing agents. The most promising were mono-phosphates of sodium, potassium and calcium mixed in specific proportions with chlorides. A deicer mixture of sodium chloride with a small amount of added phosphate may prove to be an effective and benign deicer.

Ilgner, R. 1978: Corrosive effect of urea on concrete and concrete protection systems. Part I. Effect of urea on hydrated cement and uncovered concrete. (In German.) *Baustoffindustrie,* **21(1)**, 6–9 (CA 89:151483). Urea dust, from prills in humid atmospheres and in saturated solution causes surface deterioration, decreases strength and makes the concrete more sensitive to freeze-thaw cycles. The strength continues to decrease with time of exposure.

Izumu, I., Kita, T. and Maeda, T. 1986: *Durability of Concrete Structures Series: Neutralization.* (In Japanese) Tokyo, Gihodo Shuppan, 112 (CA 107:182256).

Jambor, J. 1976: Possibilities of more precise evaluation of aggressivity of medium and resistance of concrete. *Slavebnicky Casopis,* **24(4)**, 273–83 (CRP 1977:211). The aggressivity of media to concrete is used to calculate corrosion rates and degree of corrosion to predict life spans of concrete.

Jaspers, M.J. 1977: Contribution to the experimental study of the chemical resistance of cement by means of immersion of microprisms (MIM) in aggressive solution. *Ciments, Betons, Platres, Chaux,* **(704)**, 51–8 (CRP 1977: 211). The chemical resistance of cements using Le Chatelier-Anstett, ASTM and MIM tests are compared. The MIM test is critically reviewed and the chemical resistance of OPC and BFS cements are analysed in terms of their compositions.

Kaltwasser, H. 1976: Destruction of concrete by nitrification. *Eur. J. Appl. Microbiol.,* **3(3)**, 185–92 (CRP 1977: 215). Asbestos cement was attacked by nitric and hydrochloric acids formed from ammonium chloride by bacterial action.

Karavaiko, G.L. and Zherebyat'eva T.V. 1989: Bacterial corrosion of concrete. (In Russian.) *Dokl. Akad. Nauk SSSR,* **306(2)**, 477–81 (CA 111:54086). Corrosion of concrete cooling towers was associated with heterotrophic sulfate reducing, sulfur oxidising, nitrifying and photosynthetic bacteria. The most common bacteria identified were *Synechoccucus, Nitrosomonas, Nitrobacter, Nitrosocyclys* and *Thiobacillus thioparus.* The number of bacteria were positively correlated with the degree and extent of corrosion.

Karlina, I. and Kondratskaya, S.I. 1986: Concrete corrosion in sodium citrate and hydrogen citrate solutions. (In Russian.) *Beton Zhelezobeton,* **(11)**, 30–1 (CA 106:107028). Concrete stored in sodium hydrogen citrate solutions for two months suffered a 40–45 per cent decrease in strength compared with 3.5 months in sodium citrate solution for the same decrease.

Karlina, I.N. and Chernoc, A.V. 1977: Corrosion of concrete in dichlorbutene. *Beton Zhelezobeton* (CA 86: 126041). Dichlorobutene, which has acid properties, penetrates concrete rapidly and reacts with calcium hydroxide to form calcium chloride which leads to corrosion of steel.

Kawagashi, T., Inoue, S., Miyagawa, T., Fujii, M., Nagaoka, S. and Kobayashi, T. 1993: Study of concrete corrosion in a sewage treatment facility. *JCA Proceedings of Cement & Concrete,* **47**, 492–7 (English abstract.) Corrosion of concrete in a sewage-treatment facility using oxygen-activated sludges is reported. The cause was found to be due to the fact that the oxygen-activated processes produced water with a lower pH than conventionally treated sludges.

Keifer, O. 1987: Major damage to concrete and reinforcing steel by N_2O_4 and concentrated nitric acid. *Concrete Durability, ACI SP-100,* 1549–74. The effects of attack by N_2O_4 forming 18 per cent nitric acid in a missile silo is reported.

Kengerli, A.D. 1979: Study of the effect of acidified and untreated recirculating sea water on the corrosion of concrete samples. (In Russian.) *Tr. – Bakinskii Fil.Vses. Nauchno-Issled. Inst. Vodosnabzh., Kanaliz., Gidrotech. Sooruzh. Inzh. Gidrogeol.* **15**, 163–7 (CA 95:49127). The corrosion rate of concrete in acidified and untreated sea water studied in a model return water supply system. The corrosion in acidified sea water was considerably lower than in untreated water because a gel crystalline film of corrosion products form (on surfaces) which have a high resistance to diffusion.

Kerger, B.D., Nichols, P.D., Sand, W., Bock, E. and White, D.C. 1987: Association of acid-producing thiobacilli with degradation of concrete: Analysis by 'Signature' fatty acids from the polar lipids and lipopolysaccharide. *J. Ind. Microbiol.,* **2(2)**, 63–9 (CRP 1987: 174).

Kim, I.P., Kuznetsov, A.P., Anichkhina, N.P., Dukhov, V.G., Latunin, V.V., Sobolev, A.V. and Zhirov, I.D. 1987: Mortar containing cement, inorganic aggregate and aqueous polymer dispersion. *USSR patent SU 1,337,362* (CA 108: 10473). A concrete using an aqueous dispersion of the sodium salt of a condensed formaldehyde naphthalenesulfonic acid to increase resistance to oil is described.

Kireeva, Y.I. 1978: Study of the corrosion resistance of concrete with new antifrost additions by a resonance method. (In Russian.) *Zashchita Stroit. Konstruktsii ot Korrozzi, Rostov N/D*, 84–95 (CA 92:219874).

Kirtania, K.R. and Maiti, S. 1986: Polymer impregnated concrete. Part 3. Effect of corrosive environment on strength and durability properties. *Journal of Materials Science*, **21**, 341–5. The durability properties of some polymer-impregnated concretes to mineral acids, chlorides, alkalis, sea water and crude oil media were investigated. The reasons for the improved resistance of polymer-impregnated concrete of the polymer used were established.

Knofel, D. and Rottger, K.G. 1985: Behaviour of cement-bonded building materials in sulfur dioxide enriched atmosphere. (In English and German.) *Betonwerk Fertigteil-Tech.* **51(2)**, 107–14 (CA 103:41553). The effects of moist, sulfur dioxide enriched atmospheres on a range of concretes and mortars were studied at concentrations typically found in chimneys. Moist sulfur dioxide attack leads to the formation of ettringite and gypsum, has a leaching effect and decreases strength. The formation of gypsum has a sealing effect on the surface leading to less penetration than occurs in carbonation. Initially attack is rapid and then decreases to a constant rate. Only the outer millimetre of the surface is penetrated by the sulfate attack and results in a decrease of pore volume. Low quality mortars and concretes reacted more readily and BFS cements were less resistant than OPC.

Kobayashi, W. and Okabayashi, S. 1975: Resistance against chlorides of various types of cements. *Sem. Gijutsu Nempo*, **29**, 66–70 (CA 86:140201). The resistance of cement mortars containing OPC and BFS cements to magnesium chloride solutions was studied. Brucite formed on surfaces especially in the case of the OPC mortars and the OPC mortars suffered losses in strength but no cracking or warping was observed. The chloride, which formed chloroaluminates, penetrated to the centre of the OPC specimens but only a small distance into the BFS specimens.

Koelliker, E., 1985: On the hydrolytic decomposition of cement paste and the behaviour of calcareous aggregate during the corrosion of concrete by water. *Betonwerke Fertigteil Tech.*, **52(4)**, 234–9 (CA 105: 28821). The decomposition of mortar between pH3 and pH11 is described. It was found that the rate of decomposition was dependent on pH. Calcite was found to decompose more rapidly than cement paste at $< pH4.8$.

Kong, H.L. and Orbison, J.G. 1987: Concrete deterioration due to acid precipitation. *ACI Materials Journal*, **84(2)**, 110–16. Corrosion of concrete in mixtures of sulfuric acid increased with decreasing pH levels of the solution and increasing strength of the concrete.

Krowicka, I., Gorayski, S. and Marianowska, L. 1977: Effect of mineral fertilisers on structural elements and protective coatings in storage facilities. (In Russian.) *Wybrane Zagadnienia Ochr. Koroz. Pr. Zbiorowa, Konf Nauk-Tech. Technol. Rob. Antykoroz, 4th*, 124–30 (CA 92:219853).

Kurochka, P.N. and Chernov, A.V. 1985: Role of capillary-mechanical phenomena in fracture of porous materials. (In Russian.) *Izv. VUZOV, Stroit. Arkhit*, **(7)**, 59–62 (CA 104:38805). Concrete made from OPC, quartz sand and granite aggregate was immersed in a water-fatty acid medium so that $\frac{1}{4}$ of the specimen was in water, $\frac{1}{2}$ in fatty acid and $\frac{1}{4}$ in air. Valeric, caprylic, lauric, margaric and arachic acids were used. After 30 days the portions immersed in acid, at the water-acid interface and acid-air interface had strengths of 3.13, 1.72 and 4.79 MPa respectively. Cement paste specimens were moved through the acid-water interface from acid to water. The change of pressure exerted on the cement paste after a displacement of 0.5 mm increased with increasing molecular weight of the acid. Thus the capillary-mechanical phenomena, caused by the change in the sign of the curvature of the meniscus of capillary entrapped water resulting from hydrophobisation of capillary surfaces by the acids, led to significant stresses in and weakening of the cement paste.

Kuwahara, T., Yoneyama, Y., Ohnishi, S. and Kashiwaya, M. 1987: Corrosion and its counter-measure of concrete construction by hydrogen sulfide at the Gakunan sewerage system. I. Elucidation of cause. *Gesuido Kyokaishi*, **24(282)**, 77–86 (CA 108: 42915). Corrosion of concrete in a sewerage system and laboratory simulation using hydrogen sulfide and sulfur bacteria are described.

Lewry, A.J., Asiedu-Dompreh, J., Bigland, D.J. and Butlin, R.N. 1994: The effect of humidity on the dry deposition of sulfur dioxide onto calcareous stones. *Construction and Building Materials*, **8(2)**, 97–100. Tests to simulate the deposition of sulfate from a polluted atmosphere with controlled sulfur dioxide and relative humidity on calcareous material were performed. It was found sulfate deposition was dependent on the moisture history of the surface with deposition being an order of magnitude higher for wet surfaces than for dry surfaces.

Ludwig, U., Vakil, M. and Wolter, A. 1980: Frost resistance of concrete aggregates with regard to chemical antifreeze agents. (In German.) *TIZ*, **104(9)**, 625–9 (CA 94: 144539). Damage due to the use of either 10 per cent urea or 10 per cent ethylene glycol antifreeze agents on concrete airport runways was studied. The use of these agents over 10–20 freeze thaw cycles showed significant attack on concrete. Only the addition of natural quartz gravel or crushed limestone aggregates showed promise for counteracting the effects of freeze thaw although impurities in these aggregates had some effect.

MacDonald, C.N. 1992: Durability comparisons of fiber reinforced concrete in Chemical plant applications. *Durability of Concrete, Second International Conference, Montreal*, **2**, 773–82. The durability of non-fibre, polypropylene and steel fibre reinforced concretes are compared in chemical plant environments since 1980. Steel fibre reinforced concrete was the most durable, polypropylene fibre second and non-fibre reinforced concrete third.

McVay, M., Rish, J., Sakezles, C., Mohseen, S. and Beatty, C. 1995: Cements resistant to synthetic oil, hydraulic fluid and elevated temperature. *ACI Materials Journal*, **92(2)**, 155–63. Scaling of concrete by aircraft lubricants and heat was investigated in the laboratory by refluxing coated and uncoated specimens with water, lubricating and hydraulic oils. OPC lost 55 per cent of its strength after seven days while neutral pH cements showed no reaction. Only polyvinyl alcohol and polyacrylic coatings showed a significant reduction in attack.

McVay, M.C., Smithson, L.D. and Manzione, C. 1993: Chemical damage to airfield concrete aprons from heat and oils. *ACI Materials Journal*, May–June, 253–8. Damage to airfield concrete aprons by heat and oils was investigated and it was found that both calcium hydroxide and the hydrated phases were being attacked.

Marusinova, S., Augusta, I., Rieger, Z. and Vlach, K. 1977: Lime-sand bricks with increased resistance to acids. *Czech Patent 168,858, April 15, 1977* (CA 88: 26910). Acid-resistant sand-lime bricks have been developed.

Matvienko, V.A., Drozd, G.Ya., Kovalenko, L.I. and Zatolokin, N.E. 1984: Corrosion of cement stone in sewage treatment plants. (In Russian.) *Beton Zhelzobeton*, **(8)**, 40–1 (CA 102:30908) Cellulose and proteins in sewage-treatment tanks evolve 70 per cent methane, 20 per cent carbon dioxide, 8 per cent nitrogen and 8 per cent hydrogen with traces of hydrogen sulfide and ammonia by microbial action. The stability of concrete in model solutions was related to bicarbonate, ammonium and sulfide ion contents as well as to pH. Corrosion of concrete in methane tanks occurs by the action of carbon dioxide released during the fermentation of the sediments in the sewage.

May, D. 1982: The corrosion and protection of concrete in an ammonium nitrate environment. *Proc. -Fert. Soc*, **(207)**, 21–35 (CA 97: 77545). A discussion (no references) on why ammonium nitrate corrodes and deteriorates concrete. Methods for the recognition of contaminated concrete, asbestos-cement, coatings and sealants, linings and overlays and mixtures to improve impermeability are given.

Medgyesi, I., Bozsidar, M. and Marton, M. 1975: Derivatographic results in the research of concrete corrosion. *Proceedings 4th International Conference on Thermal Analysis*, **3**, 533–42 Heyden, London. The corrosion resistance of mortar to carboxylic acids and esters, ethanol, ethylene glycol and glycerol was studied.

Medgyesi, I. 1976: Concrete corrosion by organic compounds. *Epitoanyag*, **28(11)**, 410–18 (CA 86: 144692). The corrosion resistance of mortar to carboxylic acids and esters, ethanol, ethylene glycol and glycerol was studied.

Mehta, P.K. 1977: Properties of blended cements made from rice husk ash. *Journal of the American Concrete Institute*, **74(9)**, 440–2 (CRP 1977: 215). The use of rice husk ash as a pozzolana can improve acid resistance.

Mehta, P.K. 1985: Studies on chemical resistance of low water/cement ratio concretes. *Cement and Concrete Research*, **15(6)**, 969–78. Low water/cement ratio concrete containing silica fume has better corrosion resistance to acids than that containing a styrene-butadiene latex.

Midgley, H.G. 1980: The chemical resistance of high-alumina cement. *7th International Congress on the Chemistry of Cement, Paris, III*, Theme V, 85–7.

Mizukami, K. 1986: *Durability of Concrete Structures Series: Chemical Corrosion.* (In Japanese.) Tokyo, Gihodo Shuppan, 144 (CA 107:182257).

Mohan, D. and Rai, M. 1980: Deterioration of concrete in fertilizer factories. *Durability of Building Materials and Components, Ottawa, ASTM STP 691*, 288–396. The attack of ammonium salts was found to be highly aggressive to concrete with the ammonia reducing the pH and assisting deterioration. Intense reactions resulted in loss of CH and formation of sulfates, chlorides, nitrates, sulfoaluminates, chloroaluminates and nitroaluminate hydrates. Crystallisation of ammonium sulfate caused cracking but chloride and nitrate led to serious deterioration without cracking.

Morgan, T.D.B. and Steele, A.D. 1987: Factors affecting the durability of reinforced concrete under semi-stagnant offshore conditions. *Proc. Inst. Pet., London*, **1**, 39–47 (CRP 1987: 174). Various aspects of microbial corrosion of concrete and reinforcing steel in sea water are described.

Mullick, A.K., Rajkumar, C. and Jain, N.K. 1992: Performance of concrete structures in industrial environments. *Durability of Concrete, Second International Conference, Montreal*, **1**, 577–97. A condition survey of industrial structures, mainly associated with fertiliser production, built over the last thirty years in India is reported. It was found that in the absence of protective coating the onset of distress usually occurred within five years. Subsequent repairs were not successful because of problems in neutralising the effects of salts and ions which had penetrated the concretes.

Nazurski, D. 1981: Corrosion of stressed concrete in liquid corrosive media during periodic wetting and drying. (In Bulgarian.) *Stroit. Mater. Silik. Prom-st*, **22(7)**, 7–10 (CA 96:23972). Cyclic wetting of concrete loaded in flexure with 1 per cent ammonium sulfate solution, followed by drying, caused a rapid 40 per cent decrease in strength after 30 wetting and drying cycles. The decrease was attributed to the destruction of structural bonds and to the growth of new crystal phases during the drying cycle.

Okulova, L.I. 1978: Corrosion resistance of lightweight concretes based on porous sands in wall constructions of livestock buildings. (In Russian.) *Zashchita Stroit. Konstruktsii ot Korrozzi, Rostov N/D*, 125–31 (CA 92:219877).

Pareides y Gaibrois, G. de. 1976: Durability of concrete. I. Chemical resistance of cements. *Mater. Constr.*, **163**, 65–82 (CRP 1977: 211). The chemical resistance of cements and concretes is reviewed.

Pavlik, V. 1994: Corrosion of hardened cement paste by acetic and nitric acids. Part 1: Calculation of corrosion depth. *Cement and Concrete Research*, **24(3)**, 551–62. The rates of corrosion of hardened cement paste in solutions of nitric, hydrochloric and formic acids were compared. The relationships of the concentration and time of action of the acid solutions on the depth of corrosion were determined and to a large extent shown to depend on the acid's dissociation constant. Part II: Formation and chemical composition of the corrosion products layer. *Cement and Concrete Research*, **24(8)**, 1495–508. Layers of corrosion products formed by the attack of nitric and acetic acids were chemically analysed and found to consist of an outer white coloured zone composed of mostly silica, and an intermediate brown coloured zone containing additional hydrated oxides of aluminium and iron. At lower concentrations an inner zone of cement paste formed which is depleted in calcium hydroxide and contains an increased sulfate content.

Pereima, A.A., Pertakov, Y.I. and Trusov, S.B. 1986: Corrosion resistance of cement stone in hydrogen sulfide-containing media. *Neft. Khoz.*, **(3)**, 29–32 (CA 104:212103). It was found that unless a water soluble polyelectrolyte corrosion inhibitor was used concrete exposed to hydrogen sulfide deteriorated due to the formation of ettringite.

Perkins, P.H. 1977: The protection of Portland cement concrete against sulfuric acid formed by bacterial action. *Inst. Pet. Tech. Pap. IP77-001*, 27–36 (CA 87:156174). The protection of OPC concrete from sulfuric acid produced by bacterial action is reviewed.

Peter, G., Yang, Q.W. and Brunold, B. 1986: Durability of concrete in contact with water and wastewater. *J. Mater. Sci.*, **66(7)**, 453–59 (CA 105: 138787). The corrosion of concrete in contact with wastewaters is reviewed.

Peterson, O. 1995: Chemical effects on cement mortar of calcium magnesium acetate as a deicing salt. *Cement and Concrete Research*, **25(3)**, 617–26. Two different Ca/Mg ratios of calcium magnesium acetate were tested in concentrated and dilute solution. Loss of mass was more rapid in the concentrated solution and at 20°C as opposed to 5°C. Greater decreases of flexural and compressive strength occurred with the dilute solutions.

Phol, M. and Bock, E. 1986: Bacterial damage to structural materials. *Goldschmidt Inf.*, **64**, 19–25 (CA 106: 142931). Biocorrosion of structural materials is reviewed.

Plum, D.R. and Hammersley, G.P. 1984: Concrete attack in an industrial environment. *Concrete*, **18(5)**, 8–11. The types of aggressive substances and their effects on concrete are discussed together with methods of examination and diagnosis.

Popov A, *et al.* 1978: Improving the corrosion resistance of pile foundations in service in corrosive liquid media. (In Russian) *Tr. NII Prom. Str-va*, **(21)**, 108–11 (CA 92:219878).

Prudil, S. 1977: Model of concrete behaviour in aggressive environments. *Cement and Concrete Research*, **7(1)**, 77–83 (CRP 1977:211). Various mathematical models for corrosion by leaching of lime, acidic and salt exchange and sulfate attack are examined using published data.

Rajagopalan, P.R., Jain, V.K. and Kai, M. 1976: Ultrasonic testing of cement mortar subjected to corrosive chemicals. *Indian Concrete Journal*, **50(7)**, 213–16 (CRP 1977: 211). Ultrasonic measurements and flexural strength tests of OPC and BFS cement mortars exposed to corrosive solutions were used. It was concluded that ultrasonic tests could be used to predict strength and give an estimate of the extent of deterioration.

Reinhardt, H.W. and Aufrecht, M. 1995: Simultaneous transport of an organic liquid and gas in concrete. *Materials and Structures*, **28(175)**, 43–51. The transport of organic fluids in concrete is governed by capillary absorption and diffusion. A test method and model has been devised which shows that the gas front always precedes the liquid front. The results of tests show that concrete can be made so that it is impervious to gas for a stipulated period of time.

Robertson, K.R. and Rashid, M.A. 1976: Effects of solutions of humic compounds on concrete. *Journal of the American Concrete Institute*, **73(10)**, 577–80 (CRP1976:176). The corrosive effect on mortars of fresh and salt water with and without humic acid was studied. Calcium was found to increase more rapidly in the solutions containing humic acid. An organic coating formed on the mortars which limited further corrosion.

Rubetskaya, T.V., Shneiderova, V.V. and Lyubarskaya, G.V. 1975: Corrosion of concrete in acid corrosive media and method of evaluating protective properties of coatings (In Russian.) *Tr. Nauchno-Issled. Inst. Betona Zhelezobetona*, **(19)**, 17–23 (CA 88:171906). A 30 μm thick coat of chlorinated PVC protects concrete against corrosion in 0.1N hydrochloric acid for up to four years. The corroded layer consists mostly of silica gel from which calcium chloride is absent. The corrosion proceeds by diffusion of the hydrochloric acid through the coating and is inhibited by the formation of a gel layer under the coating.

Sakai, T., Kawai, T. and Karibe, K. 1976: [Stability] of hardened concrete in marine environments. (In Japanese.) *Nippon Daigaku Seisan Kogakubu Hokoku, A*, **9(1)**, 35–9 (CA 86:33496). The attack of brines on reinforced concrete was tested in solutions at 20°C containing 1 per cent sodium chloride and 500 ppm of hydrogen sulfide, or adjusted to pH2 with sulfuric acid. The compressive strength increased in the early stages but then decreased with exposure. Thus sulfide, sulfate and chloride ions attacked the concrete causing disintegration or the formation of expansive sulfoaluminates.

Sand, W. 1987: Importance of hydrogen sulfide, thiosulfate and methylmercaptan for growth of thiobacilli during simulation of concrete corrosion. *Appl. Environ. Microbiol.*, **53(7)**, 1645–8 (CRP 1987: 174). It was found that in the corrosive action of thiobacilli, hydrogen sulfide causes severe corrosion, thiosulfate moderate corrosion and methylmercaptan negligible corrosion.

Sand, W., Bock, E. and White, D. 1987: Biotest system for rapid evaluation of concrete resistance to sulfur–oxidising bacteria. *Mater. Perform.*, **26(3)**, 14–17 (CRP 1987:174).

Saricimen, H., Maslehuddin, M., Shamim, M. and Allam, I.M. 1987: Case study of deterioration of concrete in sewerage environment in an Arabian Gulf country. *Durability Build. Mater.*, **5(2)**, 145–54 (CRP 1986: 174). It is shown that hydrogen sulfide is produced by anaerobic bacteria in sewage and that aerobic bacteria converts the hydrogen sulfide to sulfuric acid which attacks concrete.

Sayward, J.M. 1984: Salt action on concrete. *Special Report 84–25, U.S. Army Cold Regions Research and Engineering Laboratory*, 69. The salt weathering of concretes was investigated to show that deterioration of bridge decks could also occur by this mechanism.

Scholl, E. and Knoffel, D. 1991: On the effect of SO_2 and CO_2 on cement paste. *Cement and Concrete Research*, **21(1)**, 127–36. To simulate conditions in smoke stacks the effects of 10 times normal CO_2 and 1000 times normal SO_2 concentrations on cement mortars were investigated.

Scholz, E. 1978: Contribution to the technical classification of aggressive media in agriculture with regard to corrosion protection. (In German.) *Baustoffindustrie, Ausg. B*, **21(2)**, 13–14 (CA 89: 151485). The corrosion of concrete by various agricultural materials was classified by the depth of attack on one face of a concrete cube immersed in solution. The materials tested were urea, superphosphate, potassium fertilisers, ammoniumsulfate, potassium ammonium nitrate, phosphorus potassium fertiliser, silage and acetic acid.

Shirayama, K. and Kasai, Y. *et al.* 1984: [Testing] method for chemical resistance of concrete in aggressive solution. II. (In Japanese.) *Semento Konkuriito*, **(444)**, 29–39 (CA 100: 179282). A comprehensive study of the resistance of mortar and concrete to hydrochloric and sulfuric acids and sodium sulfate and related solutions. Strength, mercury intrusion porosity and pH profiles were measured. Corrosion of mortars and concrete containing OPC, BFS, fly ash and sulfate resisting cement was greater in the acids and was noticeable in ammonium and aluminium sulfate solutions.

Shpynova, L.G. *et al.* 1976: Corrosion of concrete in structures of cryolite plants. (In Russian.) *Izv. Vyssh. Uchebn. Zaved., Stroit, Arkhit.*, **(10)**, 83–8 (CA 86:144675). The corrosion of plain and reinforced concrete in the cryolite industry was studied. Hydrofluoric acid decomposes the calcium hydroxide and calcium silicate hydrates with the formation of CaF_2, $CaSiF_6$ and Na_2SiF_6. CaF_2 causes a decrease in volume and $CaSiF_6$ and Na_2SiF_6 an increase in volume. These increases and decreases in volume cause internal stresses resulting in cracking and rupture of the concrete.

Shpynova, L.G., Tuzyak, V.E., Nikonets, I.I. and Feklin, V.I. 1980: Corrosion of concrete in alumina production. (In Russian.) *Ukr. Khim. Zh.*, **46(1)**, 58–60 (CA 93:136832). XRD, IR and electron microscopy were used to study the corrosion of concrete in alumina plants from the corrosive action of aluminate solutions containing free sodium hydroxide and carbonate. Complete disintegration of the cement paste was observed in sodium carbonate aluminate solutions containing up to 330g/l of sodium oxide and 170g/l of alumina. The corrosion products formed were $Na_2SiO_3.5H_2O$, $Ca(OH)_2$, $3CaO.Al_2O_3.6H_2O$, $4CaO.Al_2O_3.19H_2O$ and $2CaO.Al_2O_3.8H_2O$.

Shvidko, Ya I., Epshtein, V.S., Sobolev, E.V., Degtyarev, E.V., Tomofeeva, L.K., Ukolov, V.S., Chaika, E.A. 1987: Concrete mix. U.S.S.R. Patent SU 1,321,713 (CA 108: 10464). The addition of an aniline condensation product with acetaldehyde and furyl acid for increased resistance of concrete to acids and alkalis is described.

Skenderovic, B. and Lomic, G. 1992: Effects of air-pollution on chemistry and dynamics of concrete carbonation. *9th International Congress on the Chemistry of Cement, New Delhi*, **V**, 383–8. It was established that outdoor climatic conditions and higher concentrations of air pollution containing sulfates increases the speed of carbonation in cement mortars.

Spicker, G. 1992: Chemical resistance of AAC. (ed. Wittmann, F.H.) *Advances in Autoclaved Aerated Concrete*: 165–70. The chemical resistance of aerated, autoclaved concrete was determined by compressive strength measurements. AAC shows good resistance to acyclic and aromatic hydrocarbons and carbohydrates and good resistance to alkaline solutions but is attacked by acids.

Spirin, P., Shcherbina, G., Alferova, F. and Pavlenko, L. 1985: Optimisation of the composition of a paste for producing high strength acid-resistant concrete. *Silik. Keram. Stenovye Mater. Izdeliya Primem. Poputynykh Prod.* 93–7 (CA 106: 200807). Equations are presented for calculating the strength of acid-resistant concrete.

Subbotkin, M.I., Bessonov, V.S. and Shmakov, S.A. 1984: Corrosion of concrete in sugar refineries. (In Russian.) *Korrozion. Stoilost Betona i Zhelezobetona v Aggresiv. Sredakh, M.* 71–6 (CA 103:75380).

Szlezak, P.R. and Pisarski, M. 1976 Non-destructive studies on the ammonia corrosion of concrete. *Zes. Nauk. Szk. Gl. Gospod. Wiejsk. – Akad Roln. Warszawie, Melior. Rolne,* **15**, 107–13 (CA 88:26826). Ultrasonic testing was found useful in estimating ammonia attack. Several methods of destructive and non-destructive testing for corrosion were compared.

Tashiro, C., Shintanin, T. and Mise, A. 1977: Sulfuric acid resistance of hardened barium silicate cement. *Zairyo*, **26(283)**, 367–71 (CA 87: 10387). In strong acid the formation of dense barium sulfate and amorphous silica slows deterioration. In 0.1 and 0.3 per cent sulfuric acid almost complete loss of strength occurs.

Tashiro, C. and Kubota, W. 1976: Sulfuric acid resistance of cement mortar impregnated with pulp waste liquid. *Cement and Concrete Research*, **6(6)**, 727–32. A 26 to 52 per cent decrease in the rate of attack by sulfuric acid on cement mortar impregnated with waste wood pulp liquid was observed.

Tazaki, K., Nonaka, T., Mori, T. and Noda, S. 1990: Mineralogical investigation of microbial corroded concrete. II. Microbial corrosion of mortar bar. (In Japanese.) *Nendo Kagaku*, **30(3)**, 178–86 (CA 115:77391). Mortar specimens exposed to 25–300 ppm of hydrogen sulfide for 10 months at 12–30°C contained gypsum, calcite, ettringite and traces of barite in the deteriorated regions of the specimens. Deterioration proceeded through the generation of microcracks, expansion and softening. The deterioration was explained by the production of ettringite at pH5–6 and gypsum at pH2.

Tazawa, E., Moringa, T. and Kawai, K. 1993: Effect of metabolites of microorganisms on concrete. *JCA Proceedings of Cement & Concrete*, **47**, 522–7 (English abstract.) It was found that the metabolites of aerobic hydrogen sulfide producing bacteria were composed of acetic acid, propionic acid, sulfate ion, nitrate ion, hydrogen sulfide and carbonate ion, and that organic acids were present in considerable quantities.

Tazawa, E., Moringa, T. and Kawai, K. 1994: Effect of metabolites of microorganisms on concrete. *JCA Proceedings of Cement & Concrete*, **48**, 642–7 (English abstract.) It has been suggested that concrete deterioration could be caused by common aerobic microorganisms which would metabolise organic acids. A study of Bacillus subtillus. Esherichia coli, Penicillium expansum and Aureobasidium pullulans showed that carbonic and organic acids were produced which attacked concrete in a similar manner to that when sulfur bacteria was present.

Tazawa, E.I., Morinaga, T. and Kawai, K. 1994: Deterioration of concrete derived from metabolites of microorganisms. *Durability of Concrete, Third International Conference, Nice*, ACI SP-145, 1087–95. Possible attack on an underground structure by soil microorganisms was investigated. It was found that calcium was dissolved from mortar soaked in a culture medium containing aerobic bacteria derived from the soil around the structure. The main metabolites found were acetic, propionic and carbonic acids which could be the cause of the attack.

Tenenbaum, G.V. and Yakovlev, V.V. 1989: Evaluation of the corrosiveness of a chlorine-containing atmosphere to concrete. (In Russian.) *Vopr. Povysh. Dolgovechnosti Stroit. Konstruktsii v Agressiv. Sredakh, Ufa*, 22–30 (CA 113:196659). The use of residual concrete strength as measured by a combination of the ultrasonic pulse velocity and the Schmidt index to assess the damage to concrete from fire and chemical attack is recommended.

Tomasevic, M. and Skenderovic, B. 1981: Corrosion of concrete by meat industry waste waters. (In Serbo-Croat) *Technol. Mesa,* **22(11)**, 316–19 (CA 97: 77588). Extraction of calcium hydroxide is considerably more intense in waste waters than in normal water due to reaction with fat and gaseous protein decomposition products. Differences in corrosion resistance were found to depend on the type of cement used.

Tomiska, J., Hanus, Z. and Burget, C. 1976: Acid corrosion of lightweight cement slurries. I. Methods for studying inorganic acid corrosion of hardened plugging cement. *Pr. Ustavu Geol. Inz.,* **34**, 213–27 (CA 86: 94928).

Torii, K. and Kawamura, M. 1994: Effects of fly ash and silica fume on the resistance of mortar to sulfuric acid and sulfate attack. *Cement and Concrete Research,* **24(2)**, 361-70. Attack on mortars over three years by 2 per cent sulfuric acid and 10 per cent solutions of sodium and magnesium sulfate was investigated. The amount of fly ash and silica fume necessary to prevent attack varied significantly with the concentration of the attacking solutions.

Ukraincik, V., Bjegovic, D. and Djurekovic, A. 1980: Concrete corrosion in a nitrogen fertiliser plant. *Durability of Building Materials and Components, Ottawa, ASTM STP 691*, 397–409. The corrosion of concrete by calcium ammonium nitrate in a fertiliser factory is reported.

Velica, P. and Ilca, A. 1983: Some aspects of concrete behaviour under the action of propionic and treated cereals (in Romanian, English abstract only). *Materiale de Constructii,* **13(2)**, 83–6. It was found that if more than 1.2 per cent of propionic acid was used to treat cereals a strong corrosive effect occurred in concrete storage facilities.

Wafa, F.F. 1994: Accelerated sulfate attack on concrete in a hot climate. *Cement, Concrete, and Aggregates, CCAGPD,* **16(1)**, 31–5. Cubes made from Type V cement using a high water/cement ratio were immersed in 10 per cent magnesium sulfate solution exposed to humid ambient temperatures ranging from 40 to 50 °C. A 50 per cent loss of strength occurred after 15 months' immersion but by increasing the cement content and reducing the water/cement ratio to produce a dense impermeable concrete these strength losses were eliminated.

Wakeley, L.D., Wong, G.S. and Burkes, J.P. 1992: Petrographic techniques applied to cement-solidified hazardous wastes. *Proceedings of the Fourteenth International Conference on Cement Microscopy*, 274–89. The stability of cement-based composites in magnesium brines was investigated and it was found that the first evidence of deterioration was depletion of calcium from the paste.

Wakeley, L.D., Poole, T.S., Ernzen, J.J. and Neeley, B.D. 1993: Salt saturated mass concrete under chemical attack, (ed. Zia, P.). *High Performance Concrete in Severe Environments,* ACI SP-140: 251–57. Deterioration of concrete in brines containing high concentrations of Mg, Cl and SO_4 is due principally to the reaction of Mg with Ca. The removal of Ca is accompanied by softening of the hardened cement paste. Accumulation of Mg does not directly indicate deterioration as Mg tends to deposit on surfaces. Calculation of Ca loss from the hardened paste should be a useful tool to investigate systems for the containment of radioactive wastes.

Webster, R.P. and Kukacka, L.E. 1986: Effects of acid deposition on Portland cement concrete. *ACS Symp. Ser. 318, Mater. Degrad. Caused by Acid Rain*, 239–49 (CRP 1986: 312). A review of the acid deposition on concrete is given.

Wozybun, I. 1977: Quantitative method for the determination of the rate of corrosion of concrete in acidic media. (In Polish.) *Wybrane Zagadnienia Ochr. Koroz. Pr. Zbiorowa,*

Konf Nauk-Tech. Technol. Rob. Antykoroz, 4th, 79–86 (CA 92: 219879). Cement pastes were immersed in glass cells containing 0.08, 0.1, 0.2 and 0.4N hydrochloric acid which was continuously circulated and monitored for concentration. The quantity of cement paste dissolved was measured every 1–2 days and calculated as calcium oxide and the depth of corrosion measured. It was found that the quantity of calcium oxide produced was proportional to the time of reaction and more cement paste was attacked in the stronger acid solutions.

Xie, P. and Tang, M. 1988: Effect of Portland cement paste aggregate interface on electrical conductivity and chemical corrosion resistance of mortar. *Cemento*, **85(1)**, 33–42 (CRP 1988: 129). It was found that clinker aggregate conferred more protection on mortar than the use of sand aggregates.

Ya Lavrega, L. *et al.* 1989: Increase in concrete durability under the action of organic acid environments (In Russian, English abstract only). *Beton i Zhelezobeton*, **(3)**, 20–2. It was found that all concretes require corrosion protection from weak acetic, formic and lactic acids and that service was improved by the use of polymers.

Yakovlev, V.V. 1986: Prediction of corrosion resistance of concrete in liquid acid media. *Beton Zhelezobeton* **(7)**, 15–16 (CA 105: 157860).

Yakovlev, V., Golovacheva, T. and Shurkova, T. 1986: Development of rapid methods for the study of corrosion resistance of concrete in acid solutions. *Stroit. Konstruktsii i Mater. Ufa* 95–106 (CA 107: 27506). Rapid methods for studying the corrosion resistance of concrete against acid are reported.

Yoda, A. and Yokomuro, T. 1986: Chemical resistance of concrete. II. Dependence on different types and concentrations of dipping solutions. (In Japanese.) *Kenkyu Shuroku Ashikaga Kogyo Daigaku*, **(12)**, 203–8 (CA 106:54882). Concrete made from OPC, SR and BFS cements were immersed in 0.5–10 per cent sulfuric acid, 0.2–4 per cent hydrochloric acid, 1.5–30 per cent ammonia and 10 per cent sodium sulfate solution for 90–340 days and resistance to attack measured by strength, elastic modulus, expansion and visual inspection. Deterioration was most serious for the acids and the least for ammonia. BFS and SR cements were more resistant than OPC and resistance was enhanced by lower water/cement ratios and entrained air contents.

Yoda, A. and Yokomura, T. 1983: Concrete immersed in Kusatsu Hot Spring (Japan) and in solutions of hydrochloric acid, sulfuric acid and sodium sulfate. (In Japanese.) *Seménto Gijutsu Nenpo*, **(37)**, 322–5 (CA 102:66420). Concretes containing OPC, sulfate resisting and BFS cements were immersed in 2 per cent hydrochloric acid, 5 per cent sulfuric acid and 10 per cent sodium sulfate solutions at 20°C and in the hot spring at pH1.0 at 93°C and pH1.1 at 63°C for 28 and 91 days. Deterioration was measured by changes in compressive and bending strengths and neutralisation by phenolphthalein. There was a marked deterioration in strength in all the solutions except for the sodium sulfate solution. Sulfate resisting and BFS cements had greater resistance than OPC.

Yoda, A., Yokomuro, T., Noumi, T. and Onuki, M. 1985: Chemical resistance of concrete. I. Soaking [of concrete] in Kusatsu hot spring. *Kenkyu Shuroku – Ashikaga Kogyo Daigaku*, **(11)**, 5–10 (CA 104:55313). Concrete made with OPC, sulfate resisting cement and BFS was immersed in the hot spring for 28–91 days at pH1.0–1.1 and at temperatures of 63 and 93°C and the strength, porosity and depth of neutralisation measured. The BFS cement and sulfate-resisting cements were more resistant than the OPC. The porosity of the samples was reduced due to the formation of gypsum.

Zivica, V. 1992: Aggressivity of phenol homologues solutions and caused corrosion of cement composites. *9th International Congress on the Chemistry of Cement, New Delhi*,

V, 396–401. Among the phenol analogues it was found that resorcinol was the chemical which was most aggressive to concrete.

Appendix 2 Micrometric Determination of Mix Proportions[1]

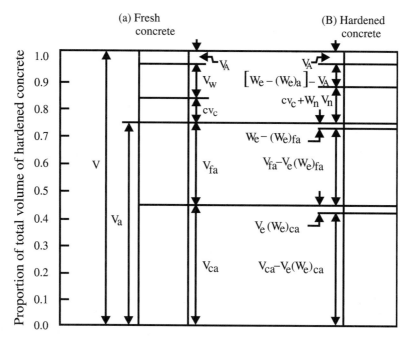

Fig. A2.1 Physical composition of fresh and hardened concrete (Axon 1962).

The symbols in Fig. A2.1 are defined as

V = total volume of concrete
V_A = volume of air voids
cv_c = volume of cement
V_a = total volume of aggregate
V_{ca} = volume of coarse aggregate
V_{fa} = volume of fine aggregate

[1] From Figg (1989), reproduced with the kind permission of The Concrete Society, Slough, UK.

V_w = volume of mixing water (after bleeding)
W_n = weight of non-evaporable water
W_e = weight of total evaporable water
$(W_e)_a = (W_e)_{ca} + (W_e)_{fa}$ = weight of evaporable water in aggregate
V_c = specific volume of cement
V_n = specific volume of non-evaporable water
V_e = specific volume of total evaporable water

Estimation of original mix proportions

If it is assumed that all voids are water filled, then the following relationship applies (Figg and Bowden 1971):

$$P + V_A = \frac{cD}{3.12} + \frac{h_cD}{1.22} + \frac{p_cD}{1.00}$$

where:

D = bulk relative density of concrete

p_c = capillary porosity, expressed in grams per 100g concrete (equivalent to W_e-$(W_e)_a$ in Axon's terminology as above).

h_c = combined water content, expressed in grams per 100 g concrete (equivalent to W_n in Axon's terminology as above).

c = cement content as a percentage by mass.

1.22 and 1.00 are the relative densities of combined and free water respectively. The relative density of cement is 3.12.

The cement content can then be calculated from:

$$c = \frac{3.12}{D}(P + V_A - p_cD - 0.82\,h_cD)$$

Where P = volume percentage of cement paste

V_A = volume percentage of air voids

If an assumption is made about the degree of hydration of the concrete (maturity) and the amount of water required for full hydration, then a determination of h_c can be avoided. Full hydration of 1g of cement requires 0.25 g of water, but in practice a figure of 0.23 is considered average (Axon 1962, Figg and Bowden 1971.)

Axon (1962) used a maturity factor (m) which was estimated on the basis of age and environment (Table A2.1). The values for m given in the table are probably rather low for modern Portland cements.

Substituting 0.23 mc for h_c in the expression for c above,

$$c = \frac{3.12}{D}\frac{(P + V_A - p_cD)}{1 + 0.59\ m}$$

Various modifications to the above formulae exist and some of these are discussed by Figg and Bowden (1971), others are currently being evaluated.

From a determination of the cement content, the water/cement ratio can be determined from the following relationship (Figg and Bowden 1971):

$$P = \frac{cD}{3.12}(1 + 3.12 \text{ w/c})$$

where 3.12 is the relative density for cement.

$$\text{w/c} = \frac{P}{cD} - 0.321$$

Table A2.1 Estimated values for maturity factor, m, for different environments and ages (Axon 1962).

Age of concrete	Value of m for environment			
	A	B	C	D
7 days	0.54	0.50	0.50	0.50
14 days	0.63	0.58	0.58	0.58
28 days	0.71	0.67	0.63	0.63
2 months	0.79	0.75	0.67	0.67
3 months	0.83	0.79	0.71	0.71
6 months	0.92	0.83	0.75	–
1 year	0.96	0.88	0.79	–
2 years	1.00	0.92	0.83	–
3 years	–	0.94	0.88	–
4 years	–	0.96	0.90	–
5 years	–	0.98	0.92	–
6 years	–	1.00	0.94	–
7 years	–	–	0.96	–
8 years	–	–	0.98	–
9 years	–	–	0.99	–
10 years	–	–	1.00	–

Key to environments:
A   continuous moist curing in laboratory
B   in structures which can obtain moisture from soils, as roads and retaining walls
C   in structures which dry out but are wetted by rain, as bridge decks and hand rails
D   in structures which dry out and are not re-wetted, as beams and columns in a building

Table A2.2 Comparison of actual (A) and determined (D) cement contents and water/cement ratios (Axon 1962).

	Cement/content (kg/m^3)	Water/cement ratio
Mix 1(1)		
A	315	0.48
D	304	0.49
Mix 1(2)		
A	317	0.49
D	313	0.49
Mix 2b		
A	215	0.74
D	219	0.72
Mix 3b		
A	349	0.49
D	349	0.47
Mix 3d		
A	289	0.49
D	273	0.52
Mix 4a		
A	280	0.59
D	251	0.64
Mix 4b		
A	240	0.72
D	242	0.68

References

Aarre, T. 1995: Influence of measurement technique on the air void structure of hardened concrete. *ACI Materials Journal* **92(6)**, 599-604.

Adams, A.E., Mackenzie, W.S. and Guildford, C. 1991: *Atlas of sedimentary rocks under the microscope.* Longman Scientific, UK.

Addis, B.J. and Alexander, M.G. 1994: Cement-saturation and its effects on the compressive strength and stiffness of concrete. *Cement and Concrete Research* **24(5)**, 975–86.

Ahmed, W.U. 1991: Advances in sample preparation for clinker and concrete microscopy. *Proceedings of the 13th International Conference on Cement Microscopy*, 17–29.

Ahmed, W.U. 1994: Petrographic methods for analysis of cement clinker and concrete microstructure. In: DeHayes, S.M. and Stark, D. (eds) *Petrography of cementitious materials*, ASTM STP 1215. American Society for Testing and Materials, Philadelphia, USA.

AIA Recommended Test Method No. 1 (RTM1). 1982: *Reference for the determination of airborne asbestos dust fibre concentrations at workplaces by light microscopy (membrane filter method).*

Aïtcin, P. (ed.) 1983: *Silica Fume* 52. Les Editions de L'universite de Sherbrooke, Canada.

Al-Adeeb, A.M. and Matti, M.A. 1984: Leaching and corrosion of asbestos cement pipes. *The International Journal of Cement Composites and Lightweight Concrete* **6(4)**, 233–40.

Al-Amoudi, O.S.B., Maslehuddin, M. and Saadi, M.M. 1995: Effect of magnesium sulfate and sodium sulfate on the durability performance of plain and blended cements. *ACI Materials Journal* **92(1)**, 15–24.

Alexander, M., Addis, B. and Basson, J. 1994: Case studies using a novel method to assess aggressivement of waters to concrete. *ACI Materials Journal* **91(2)**, 188–96.

Ali, M.A. and Grimer, F.J. 1969: Mechanical properties of glass fibre-reinforced gypsum. *Journal of Materials Science* **4(5)**, 389–95.

Al-Kadhimi, T.K.H. and Banfill, P.G.F. 1996: The effect of electrochemical re-alkalisation on alkali-silica expansion in concrete. *Proceedings of the 10th International Conference on Alkali-aggregate Reaction in Concrete*, 637–44, Melbourne, Australia.

Allen, R.T.L. 1993: Investigation and diagnosis (eds R.T.L. Allen and J.D.N. Shaw) *The Repair of Concrete Structures*, 15–38. Glasgow, J.D.N. Blackie.

Allen, T. 1990: *Particle size measurement*, 4th edn. Chapman & Hall, London.

Allman, M. and Lawrence, D.F. 1972: *Geological laboratory techniques.* London, Blandford Press.

American Concrete Institute. Annual: *ACI manual of concrete practice*, Detroit, USA.

American Concrete Institute. 1980: *Performance of Concrete in Marine Environment*, ACI

SP-65 Symposium, New Brunswick, Canada. Detroit, USA.

American Concrete Institute. 1984: Causes, evaluation, and repair of cracks in concrete. ACI Committee 224. *ACI Materials Journal* **81(3)**, 211–30, Detroit, USA.

American Concrete Institute. 1985: *Guide to the use of waterproofing, damp-proofing, protective and decorative barrier systems for concrete*, ACI Committee 515, Report 515.1R-79(85), Detroit, USA.

American Concrete Institute. 1987: Silica fume in concrete. Preliminary committee report. ACI Commitee 226. *ACI Materials Journal* **84(2)**, 158–66, Detroit, USA.

American Concrete Institute. 1988: Measurement of properties of fiber reinforced concrete. ACI Committee 544. *ACI Materials Journal* **85(6)**, 583–93, Detroit, USA.

American Concrete Institute. 1989: Chemical admixtures for concrete. ACI Committee 212. *ACI Materials Journal* **86(3)**, 297–327, Detroit, USA.

American Concrete Institute. 1991: Proposed revision of: Guide to durable concrete. ACI Committee 201. *ACI Materials Journal* **88(5)**, 544–82, Detroit, USA.

American Concrete Institute. 1991: *Standard practice for selecting proportions of normal heavyweight and mass concrete.* ACI Committee 211, Report 211.1, Detroit, USA.

American Concrete Institute. 1991: *Guide for concrete highway bridge deck construction.* ACI Committee 345, Report 345R, Detroit, USA.

American Concrete Institute. 1992: Guide for making a condition survey of concrete in service. ACI Committee 201. *ACI Manual of Concrete Practice*, 1994 edn, Detroit, USA.

American Concrete Institute. 1992: *Guide to durable concrete.* ACI Committee 201, Report 201.2R, Detroit, USA.

American Concrete Institute. 1992: *Requirements for reinforced concrete.* ACI Committee 318, Report 318M/318RM. *Requirements for structural plain concrete.* ACI Report 318.1M/318.1RM, Detroit, USA.

American Concrete Institute. 1993: Guide for evaluation of concrete structures prior to rehabilitation. ACI Committee 364. *ACI Materials Journal* **90(5)**, 479–98, Detroit, USA.

American Concrete Institute. 1993: Guide to Portland cement plastering. ACI Committee 524. *ACI Materials Journal* **90(1)**, 69–93, Detroit, USA.

American Concrete Institute. 1994: Proposed report: Use of natural pozzolanas in concrete. ACI Committee 232. *ACI Materials Journal* **91(4)**, 410–26, Detroit, USA.

American Concrete Institute. 1995: *Guide for the use of silica fume in concrete.* ACI Committee 234, Report 234R, Detroit, USA. (Also abstract in *ACI Materials Journal* **92(4)**, 437–40.)

American Concrete Institute. 1995: Specification for materials, proportioning and application of shotcrete. ACI Committee 506. *ACI Manual of Concrete Practice 1995, Part 5*, ACI 506R-90, Detroit, USA.

American Concrete Institute. 1995: Guide for the evaluation of shotcrete. ACI Committee 506. *ACI Manual of Concrete Practice 1995, Part 5*, ACI 506R.90, Detroit, USA.

American Concrete Institute. 1995: State-of-the-art report on fiber-reinforced shotcrete. ACI Committee 506. *ACI Manual of Concrete Practice 1995, Part 5*, ACI 506.1R-84, Detroit, USA.

American Concrete Institute. 1996: *Specifications for structural concrete for buildings.* ACI Committee 301, Report 301, Detroit, USA.

American Public Health Association. 1992: *Standard Methods for the Examination of Water and Waste Water*, 18th edn, Washington.

American Society of Civil Engineers. 1995: *Standard practice for shotcrete, Vol. 11: Technical Engineering & Design Guides as Adapted from the US Army Corps of Engineers.*

Andersen, K.T. and Thaulow, N. 1990: The study of alkali-silica reactions in concrete by the use of fluorescent thin sections *Petrography Applied to Concrete and Concrete*

Aggregates, ASTM STP 1061, 71–89.

Andersen, P.E. and Petersen, B.H. 1961: Drilling of concrete cores and preparation of thin sections. *Rilem Bulletin* **(11)**, 94–106.

Andrade, C., Alonso, C., Santos, P. and Macia, A. 1986: Corrosion of steel during accelerated carbonation of solutions which simulate the pore concrete solution. *8th International Congress Chemistry of Cements, Rio de Janeiro* **V**, 256–62.

Andrews, H. 1948: Gypsum and anhydrite plasters. *National Building Studies Bulletin No. 6, DISR*, London, HMSO.

Anon. 1976: Air entrainment and concrete. *Concrete Construction*, March, 105–11.

Arnold, D.R. (Chairman) 1980: Pigments for integrally colored concrete (Task Group of ASTM Subcommittee Section C09.03.08.05 on methods of testing and specifications for admixtures). *Cement, Concrete, and Aggregates, CCAGDP* **2(2)**, 74–7.

Asakura, E., Yoshida, H., Nakato, T. and Namamura, T. 1993: Effect of characters of silica fume on physical properties of cement paste. *JCA Proceedings of Cement and Concrete* **47**, 178–83, (in Jaapanese, English abstract).

Asgeirsson, H. 1986: Silica fume in cement and silane for counteracting of alkali-silica reactions in Iceland. *Cement and Concrete Research* **16(3)**, 423–8.

Ash, J.E. 1972: Bleeding in concrete – a microscopy study. *ACI Journal* **69(4)**, 209–11. Title No. 69–19.

Ashurst, J. and Ashurst, N. 1988: *Practical building conservation. English Heritage technical handbook Volume 3. Mortars, plasters and renders.* Gower Technical Press, UK.

ASTM C28-86, 1986: *Standard specification for gypsum plasters.* Philadelphia, USA.

ASTM C28-92, 1992: *Standard specification for gypsum plasters.* Philadelphia, USA.

ASTM C33-93, 1993: *Standard specification for concrete aggregates.* Philadelphia, USA.

ASTM C73-94, 1994: *Standard specification for calcium silicate face brick (sand-lime brick). Philadelphia*, USA.

ASTM C73-96, 1996: *Standard specification for calcium silicate face brick (sand-lime brick).* Philadelphia, USA.

ASTM C76-90, 1990: *Standard specification for reinforced concrete culvert, storm drain, and sewer pipe.* Philadelphia, USA.

ASTM C76-95, 1995: *Standard specification for reinforced concrete culvert, storm drain, and sewer pipes.* Philadelphia, USA.

ASTM C88-90, 1990: *Standard test method for soundness of aggregates by use of sodium sulfate or magnesium sulfate.* Philadelphia, USA.

ASTM C90-94, 1994: *Standard specification for load-bearing concrete masonry units.* Philadelphia, USA.

ASTM C90-96, 1996: *Standard specification for load-bearing concrete masonry units.* Philadelphia, USA.

ASTM C91-93, 1993: *Standard specification for masonry cement.* Philadelphia, USA.

ASTM C91-95, 1995: *Standard specification for masonry cement.* Philadelphia, USA.

ASTM C94-96, 1996: *Standard specification for ready-mixed concrete.* Philadelphia, USA.

ASTM C110-92, 1992: *Standard test methods for physical testing of quicklime, hydrated lime, and limestone.* Philadelphia, USA.

ASTM C110-95, 1995: *Standard test methods for physical testing of quicklime, hydrated lime, and limestone.* Philadelphia, USA.

ASTM C125-96, 1996: *Standard terminology relating to concrete and concrete aggregates.* Philadelphia, USA.

ASTM C129-94, 1994: *Standard specification for non-load-bearing concrete masonry units.* Philadelphia, USA.

ASTM C129-96, 1996: *Standard specification for non-load-bearing concrete masonry units.*

Philadelphia, USA.

ASTM C141-85(94), 1985: *Standard specification for hydraulic lime for structural purposes.* Philadelphia, USA.

ASTM C145-85, 1985: *Standard specification for solid load-bearing concrete masonry units.* Philadelphia, USA, (discontinued 1992 and incorporated in ASTM C90).

ASTM C150-96, 1996: *Standard specification for Portland cement.* Philadelphia, USA.

ASTM C151-93, 1993: *Standard test method for autoclave expansion of Portland cement.* Philadelphia, USA.

ASTM C183-95, 1995: *Standard practice for sampling and the amount of testing of hydraulic cement.* Philadelphia, USA.

ASTM C184-94, 1994: *Standard test method for fineness of hydraulic cement by the 150 μm (no. 100) and 75 μm (no. 200) sieves.* Philadelphia, USA.

ASTM C204-96, 1996: *Standard test method for fineness of hydraulic cement by air permeability apparatus.* Philadelphia, USA.

ASTM C206-84(92), 1984: *Standard specification for finishing hydrated lime.* Philadelphia, USA.

ASTM C207-91, 1991: *Standard specification for hydrated lime for masonry purposes.* Philadelphia, USA (reapproved 1992).

ASTM C227-90, 1990: *Standard test method for potential alkali reactivity of cement-aggregate combinations (mortar-bar method).* Philadelphia, USA.

ASTM C231-91b, 1991: *Standard test method for air content of freshly mixed concrete by the pressure method.* Philadelphia, USA.

ASTM C289-94, 1994: *Standard test method for potential alkali-silica reactivity of aggregates (chemical method).* Philadelphia, USA.

ASTM C294-86, 1986: *Standard descriptive nomenclature for constituents of natural mineral aggregates.* (Reapproved 1991).

ASTM C295-90, 1990: *Standard guide for petrographic examination of aggregates for concrete.* Philadelphia, USA.

ASTM C330-89, 1989: *Standard specification for lightweight aggregates for structural concrete.* Philadelphia, USA.

ASTM C331-89, 1989: *Standard specification for lightweight aggregates for concrete masonry units.* Philadelphia, USA.

ASTM C332-87, 1987: *Standard practice for lightweight aggregates for insulating concrete.* Philadelphia, USA (reapproved 1991).

ASTM C331-94, 1994: *Standard specification for lightweight aggregates for concrete masonry units.* Philadelphia, USA.

ASTM C441-89, 1989: *Standard test method for effectiveness of mineral admixtures or ground blast-furnace slag in preventing excessive of concrete due to the alkali-silica reaction.* Philadelphia, USA.

ASTM C457-90, 1990: *Standard test method for microscopical determination of parameters of the air void system in hardened concrete.* Philadelphia, USA.

ASTM C476-91, 1991: *Standard specification for grout masonry.* Philadelphia, USA.

ASTM C476-95, 1995: *Standard specification for grout masonry.* Philadelphia, USA.

ASTM C497M-94, 1994: *Test methods for concrete pipe, manhole sections, or tile.* Philadelphia, USA.

ASTM C497-96, 1996: *Test methods for concrete pipe, manhole sections, or tile.* Philadelphia, USA.

ASTM C500-91, 1991: *Standard test methods for asbestos-cement pipe.* Philadelphia, USA (reapproved 1995).

ASTM C586-92, 1992: *Standard test method for potential alkali reactivity of carbonated rocks for concrete aggregates (rock cylinder method).* Philadelphia, USA.

ASTM C595-95, 1995: *Standard specification for blended hydraulic cements.* Philadelphia, USA.

ASTM C618-96, 1996: *Standard specification for fly ash and raw or calcined natural pozzolan for use as a mineral admixture in Portland cement concrete.* Philadelphia, USA.

ASTM C786-94, 1994: *Standard test method for fineness of hydraulic cement and raw materials by the 300 μm (no. 100), and 75 μm (no. 200) sieves by wet methods.* Philadelphia, USA.

ASTM C821-78(95), 1978: *Standard specification for lime for use with pozzolanas.* Philadelphia, USA.

ASTM C823-95, 1995: *Standard practice for examination and sampling of hardened concrete in constructions.* Philadelphia, USA.

ASTM C856-95, 1995: *Standard practice for petrographic examination of hardened concrete.* Philadelphia, USA.

ASTM C897-88, 1988: *Standard specification for aggregate for job-mixed Portland cement-based plasters.* Philadelphia, USA.

ASTM C897-95, 1995: *Standard specification for aggregate for job-mixed Portland cement-based plasters.* Philadelphia, USA.

ASTM C926-90, 1990: *Standard specification for application of Portland cement-based plaster.* Philadelphia, USA.

ASTM C926-95, 1995: *Standard specification for application of Portland cement-based plaster.* Philadelphia, USA.

ASTM C938-80, 1980: *Standard practice for proportioning grout mixtures for pre-placed-aggregate concrete.* Philadelphia, USA (reapproved 1991).

ASTM C938-85, 1985: *Standard practice for proportioning grout mixtures for pre-placed-aggregate concrete.* Philadelphia, USA (reapproved 1991).

ASTM C946, 1991: *Practice for construction of dry-stacked surface bonded walls 04.05.* Philadelphia, USA.

ASTM C979-82, 1986: *Standard specification for pigments for integrally colored concrete.* Philadelphia, USA (reapproved 1986).

ASTM C989-95, 1995: *Standard specification for ground granulated blast-furnace slag for use in concrete and mortars.* Philadelphia, USA.

ASTM C1071-91, 1991: *Standard test method for hydraulic activity of ground slag by reaction with alkali.* Philadelphia, USA.

ASTM C1073-91: *Test method for hydraulic activity of ground clay by reaction with alkali.* Philadelphia, USA.

ASTM C1089-88, 1988: *Standard specification for spun prestressed concrete poles.* Philadelphia, USA.

ASTM C1105-95, 1995: *Test method for length change of concrete due to alkali carbonate rock reaction.* Philadelphia, USA.

ASTM C1107-91, 1991: *Standard specification for packaged dry, hydraulic-cement grout (nonshrink).* Philadelphia, USA.

ASTM C1116-91, 1991: *Standard specification for fiber-reinforced concrete and shotcrete.* Philadelphia, USA.

ASTM C1116-95, 1995: *Standard specification for fiber-reinforced concrete and shotcrete.* Philadelphia, USA.

ASTM C1141-89, 1989: *Standard specification for admixtures for shotcrete.* Philadelphia, USA.

ASTM C1141-95, 1995: *Standard specification for admixtures for shotcrete.* Philadelphia, USA.

ASTM C1260-94, 1994: *Standard test method for potential alkali reactivity of aggregates (mortar-bar method)*. Philadelphia, USA.

ASTM D75-87, 1987: *Standard practice for sampling aggregates.* Philadelphia, USA (reapproved 1992).

ASTM E20-85(94), 1985: *Standard practice for particle analysis of particulate substances in the range of 0.2 to 75 micrometers by optical microscopy.* Philadelphia, USA.

ASTM E20-85, 1985: *Standard specification for standard load-bearing concrete masonry units.* Philadelphia, USA (discontinued 1992 and incorporated in ASTM C90-946).

ASTM E105-58, 1958: *Standard practice for probability sampling of materials.* Philadelphia, USA (reapproved 1996).

Attiogbe, E.K. 1993: Mean spacing of air voids in hardened concrete. *ACI Materials Journal* **90(2)**, 174–81, Title No. 90-M19 (discussion in **91, (1)**, 1994, 121–3).

Axon, E.O. 1962: Method of estimating the original mix composition of hardened concrete using physical tests. *Proc. Am. Soc. Test Mater* **62**, 1068–80.

Bache, H.H., Idorn, G.M., Nepper-Christensen, P. and Nielsen, J. 1966: Morphology of calcium hydroxide in cement paste. *Highway Research Board Special Report No.* **90**, 154–74.

Bager, D.H. and Sellevold, E.J. 1979. How to prepare polished cement product surfaces for optical microscopy without introducing visible cracks. *Cement and Concrete Research* **9(5)**, 653–4.

Bager, D.H. and Sellevold, E.J. 1986: Ice formation in hardened cement paste. Part I – Room temperature cured pastes with variable moisture contents. Part II – Drying and resaturation of room temperature cured pastes. *Cement and Concrete Research* **16(5)**, 709–20; **(6)**, 835–44.

Bager, D.H. and Sellevold, E.J. 1987: Ice formation in hardened cement paste. Part III – Slow resaturation of room temperature cured pastes. *Cement and Concrete Research* **17(1)**, 1–11.

Baker, A.F. 1992: Structural investigations. *Durability of Concrete* (ed. G. Mays) 37–76, London, E. & F.N. Spon.

Ballim, Y. and Alexander, M.G. 1991: Carbonic acid attack of Portland cement-based concretes. *Proceedings of the 5th International Conference on the Durability of Materials and Components*, 179-84. E. & F. Spon, Brighton.

Barker, A. 1984: Structural and mechanical characterisation of calcium hydroxide in set cement and the influence of additives. *World Cement*, **15(1)**, 25–31.

Barker, A. 1990: The use of electron-optical techniques to study the hydration characteristics of blended cements. *Proceedings of the 12th Ineternational Conference on Cement Microscopy*, 107–20.

Barnard, J. L. 1967: Corrosion of sewers. *CSIR Research Report*, 1–16, Pretoria.

Barnes, B.D., Diamond, S. and Dolch, W. 1978: Hollow shell hydration particles in bulk cement paste. *Cement and Concrete Research*, **8**, 263–71.

Barnes, B.D., Diamond, S. and Dolch, W. 1978: The contact zone between Portland cement paste and glass aggregate surfaces. *Cement and Concrete Research* **8(2)**, 233–43.

Barnes, P., Clark, S.M., Fentiman, C.H. *et al.* 1992: The rapid conversion of high-alumina cement hydrates as revealed by synchrotron energy-dispersive diffraction. *Advances in Cement Research* **4(14)**, 61–7.

Baron, J. and Douvre, C. 1987: Technical and economical aspects of the use of limestone filler additions in cement. *World Cement*, April, 100–14.

Barta, R. 1972: Cements. In Mackenzie, R. C. (ed.) *Differential Thermal Analysis 2*, 207–28, Academic Press, London.

Bassett, H. 1934: Notes on the system lime-works, and on the determination of calcium. *Journal of the Chemical Society, Part II*, **276**, 1270–5.

Bate, S.C.C. 1974: Report on the failure of roof beams at Sir John Cass Foundation and Red Coat Church of England School, Stepney. *BRE Current Paper CP58/74*. Building Research Establishment.

Bates, R.C. 1964: *Determination of pH – Theory and Practice*. New York, Wiley.

Battigan, A.F. 1986: The use of the microscope for estimating the basicity of slags in slag-cements. *8th International Congress Chemistry of Cement*, Rio de Janeiro, Theme 3, **4**, 17–21.

Baussant, J.B., Vernet, C. and Defosse, C. 1989: Growth of ettringite in diffusion controlled conditions – influence of additives on the crystal morphology. *Proceedings 11th International Conference on Cement Microscopy*, 186–97, USA.

Ben-Dor, L. 1983: *Advances in Cement Technology* (ed. A. Ghosh), UK, Pergamon Press.

Bensted, J. and Varma, S.P. 1971: Studies of ettringite and its derivatives. *Cement Technology*, 73–6.

Bensted, J. and Varma, S.P. 1974: Studies of thaumasite. *Silicate Industriels* **39**, 11–19.

Bensted, J. 1993: High alumina cement – present state of knowledge. *Zement-Kalk-Gips* **(9)**, 560–6.

Bentur, A. and Mindess, S. 1990: *Fibre reinforced cementitious composites*, 22–32, London, Elsevier Applied Science.

Berg, L.G. and Sveshnikova, V.N. 1949: On modifications of hemihydrate calcium sulfate. *Mineralogical abstracts* **10**, 464–5.

Berger, R.L. and McGregor, J.D. 1972: Influence of admixtures on the morphology of calcium hydroxide formed during tricalcium silicate hydration. *Cement and Concrete Research* **2**, 43–55.

Berger, R.L. and McGregor, J.D. 1973: Effect of temperature and water-solid ratio on growth of $Ca(OH)_2$ crystals formed during the hydration of Ca_3SiO_5. *Journal of the American Ceramic Society* **56(2)**, 73–9.

Berra, M. and Baronio, G. 1987: Thaumasite in deteriorated concretes in the presence of sulfates. In: Scanlon, J.M. (ed.), *Concrete durability: Katharine and Bryant Mather international conference*, ACI SP-100, 2073-2089. Paper SP100-106. American Concrete Institute, Detroit, USA.

Berra, M. Di Maggio, R., Mangialardi, T. and Paolini, A.E. 1995: Fused quartz as a reference reactive aggregate for alkali-silica reaction studies. *Advances in Cement Research* **7(25)**, 21–32.

Berry, E.E. and Malhotra, V.M. 1987: Fly ash in concrete. In: Malhotra, V.M. (ed.), *Supplementary cementing materials for concrete*, Chapter 2, 35–163. Canada Centre for Mineral and Energy Technology (CANMET), Canadian Government Publishing Centre, Ottawa, Canada.

Bérubé, M.A., Chouinard, D., Boisvert, L., Frenette, J. and Pigeon, M. 1996: Influence of wetting-drying and freezing-thawing cycles, and effectiveness of sealers on ASR. *Proceedings of the 19th International Conference on Alkali-aggregate Reaction in Concrete*, 1056–63. Melbourne, Australia.

Bérubé, M.A., Duchesne, J. and Rivest, M. 1996: Alkali contribution by aggregates to concrete. *Proceedings of the 10th International Conference on Alkali-aggregate Reaction in Concrete*. 899–906. Melbourne, Australia.

Bessey, G.E. 1950: *Investigations into building fires. Part 2: The visible changes in concrete or mortar exposed to high temperatures*, 6–18, National Building Studies Technical Paper No. 18, London, HMSO, 6–18.

Bessey, G.E. 1967: The history and present day development of the autoclaved calcium silicate building products industries. *Symposium Autoclaved Calcium Silicate Building Products, London 1965*, 3–6. Society of Chemical Industry.

Bessey, G.E. and Harrison, W.H. 1970: Some results of exposure tests on durability of calcium silicate bricks. *Building Research Current Paper 24/70.*

Bhatty, J.I. 1991: A review of the application of thermal analysis to cement-admixture system. *Thermochimica Acta* **189**, 313–50.

Bickley, J.A., Hemmings, R.T., Hooton, R.D. and Balinksi, J. 1994: Thaumasite related deterioration of concrete structures. *Proceedings of V.M. Malhotra Symposium, Concrete Technology Past, Present and Future.* ACI SP144, 159–75.

Biczoc, I. 1967: *Concrete corrosion and concrete protection.* New York, Chemical Publishing Co.

Bied, J. 1926: *Recherches industrielles sur les chaux, ciments en mortiers.* Dunod, Paris, France. (Completed by Chaix, M. after the death of Bied, J.)

Bier, T.A., Kropp, J. and Hilsdof, H.K. 1989: The formation of silica gel during carbonation of cementious systems containing slag cements. *Proceedings of Third International Conference on Fly Ash, Silica Fume, Slag and Natural Pozzolans in Concrete*, Norway, ACI SP-114, **2**, 1413–28.

Birk, G. 1984: *Instrumentation and techniques for fluorescence microscopy.* Sydney, Wild Leitz (Australia).

Blackwell, B.Q. and Pettifer, K. 1992: Alkali reactivity of greywacke aggregates in Maentwrog Dam (North Wales) *Magazine of concrete research* **44(161)**, 255–64.

Blakey, F.A. 1959: Cast gypsum as a structural material. *CSIR Division of Building Research Report Z.7*, Melbourne.

Blaschke, R. 1987: Natural building stone damaged by slime and acid producing microbes. *Proceedings of the Ninth International Conference on Cement Microscopy*, 70–81.

Bloom, M.C. and Goldberg, L. 1965: Fe_2O_3 and the passivity of iron. *Corrosion Science* **8(9)**, 623–30.

Bloss, F.D. 1966: *An introduction to the methods of optical crystallography.* New York, Holt, Rinehart and Winston.

Bloss, F.D. 1981: *The spindle stage: principles and practice.* Cambridge, Cambridge University Press.

Bock, E. and Sand, W. 1986: Applied electron microscopy on the biogenic destruction of concrete blocks–use of the transmission electron microscope for identification of mineral acid producing bacteria. *Proceedings of the Eighth International Conference on Cement Microscopy*, 285–302.

Bogue, R.H. 1955: *The chemistry of Portland cement.* 2nd edn, Reinhold, New York, USA.

Bonen, D. and Cohen, M.D. 1992: Magnesium sulfate attack on Portland cement paste – II. Chemical and mineralogical analyses. *Cement and Concrete Research* **22(4)**, 707–18.

Bonen, D. and Diamond, S. 1992: Investigation on the coarse fraction of a commercial silica fume. *Proceedings of the 14th International Conference on Cement Microscopy*, 103–13. California.

Boyd, D.C., Danielson, P.S. and Thompson, D.A. 1994: Glass. *Kirk-Othmer Encyclopedia of Chemical Technology* **12**, 551–627. New York, Wiley Interscience.

Boynton, R.S. 1980: *Chemistry and technology of limestone.* New York, John Wiley.

Bradbury, S. 1991: *Basic Measurement Techniques for Light Microscopy*, Oxford University Press – Royal Microscopical Society.

Brandt, A.M. 1995: *Cement-based composites–materials, mechanical properties and perform-ance.* E. & F.N. Spon, London, UK.

Brandt, I. 1993: Connection between chemical admixtures and the structure of $Ca(OH)_2$ in thin sections of hardened concrete. (Lindqvist, J.E. and Nitz, B.) *Proceedings of the Fourth Euroseminar on Microscopy Applied to Building Materials*, Swedish National Testing and Research Institute, Building Technology SP Report 1993, 15.

Brindley, G.W. and Brown, G. 1980: *Crystal structures of clay minerals and their x-ray identification.* Mineralogical Society Monograph no. 5, London.

British Cement Association. 1992: *Alkali-aggregate reaction 1977–1992. A Bibliography.* British Cement Association. 1992: *The Diagnosis of Alkali-Silica Reaction, Report of a working party*, Wexham Springs, British Cement Association.

Brouxel, M. and Valiere, A. 1992: Thaumasite as the final product of alkali-aggregate reaction: A case study. *9th International Conference on Alkali-Aggregate Reaction, London* 136–44.

Brown, B.V. 1982: Air entrainment (Part 1) (Concrete Society Current Practice Sheet no. 80), *Concrete* (Journal of the Concrete Society) **16(12)**, 59–60.

Brown, B.V. 1983: Air entrainment (Part 2) (Concrete Society Current Practice Sheet no. 81), *Concrete* (Journal of the Concrete Society) **17(1)**, 45–6.

Brown, J.H. 1991: The effect of exposure and concrete quality: field survey results from some 400 structures. *Proceedings of the 5th International Conference on the Durability of Materials and Components, Brighton, 1990*, 249–59. E. & F.N. Spon.

Brown, L.S. and Carlson, R.W. 1936: Petrographic studies of hydrated cements. *Proceedings of the American Society of Testing Materials Part II*, **36**, 332–50.

Brown, P.W. and Bothe, J.V. 1993: The stability of ettringite. *Advances in Cement Research* **5(18)**, 47–63.

Bruere, G.M. 1960: The effect of type of surface-active agent on the spacing factors and surface areas of entrained bubbles in cement pastes. *Australian Journal Applied Chemistry* **11**, 289–94.

Brunauer, S., Kantro, D.L. and Copeland, L.E. 1958: The stoichiometry of the hydration of β-dicalcium silicate and tricalcium silicate at room temperature. *Journal of the American Chemical Society* **80(4)**, 761–7.

BSI 12, 1991: *Specification for Portland cement.*

BSI 146, 1991: *Specification for Portland blastfurnace cement.*

BSI 187, 1978: *Specification for calcium silicate (sandlime and flintlime) bricks.*

BSI 486, 1981: *Specification for asbestos-cement pressure pipes and joints* (withdrawn). (Note: replaced by BS EN 512, 1995).

BSI 776, 1972: *Specification for materials for magnesium oxychloride (magnesite) flooring (metric units) Part 2* (withdrawn 1994).

BSI 812, 1975: *Methods of sampling and testing of mineral aggregates, sands and fillers.*

BSI 812, 1985: *Testing aggregates. Part 103, Method of determination of particle size distribution, Section 103:1, Sieve tests.*

BSI 812, 1989: *Testing aggregates. Part 102, Methods for sampling.*

BSI 812, 1989: *Testing aggregates. Part 121, Methods for determination of soundness.*

BSI 812, 1989: *Testing aggregates. Part 105, Method for determination of particle shape, Section 105:1, Flakiness index for coarse aggregates.*

BSI 812, 1990: *Testing aggregates. Part 105, Method for determination of particle shape, Section 105:2 Elongation index for coarse aggregates.*

BSI 812, 1994: *Method for qualitative and quantitative petrographic examination of aggregates. Part 104.*

BSI 812, 1995: *Testing aggregates. Part 120.*

BSI 882 and 120, 1954: *Specification for aggregates from natural sources for concrete (including granolithic).*

BSI 882, 1201 Part 2, 1973: *Specification for aggregates from natural sources for concrete (including granolithic).*

BSI 882, 1983: *Specification for aggregates from natural sources for concrete.*

BSI 882, 1992: *Specification for aggregates from natural sources for concrete*

BSI 890, 1972 (1995): *Specification for building limes.*

BSI 1014, 1992: *Specification for pigments for Portland cement and Portland cement products.*

BSI 1105, 1981 (1994): *Specification for wood wool cement slabs up to 125mm thick.*

BSI 1191, 1973 (1994): *Specification for gypsum building plasters. Part 1: Excluding premixed lightweight plasters. Part 2: Premixed lightweight plasters.*

BSI 1198, 1199 and 1200, 1976: *Specification for building sands from natural sources.*

SBSI 1377, 1990: *Methods of test for soils for civil engineering purposes. Part 3, Chemical and electrochemical tests.*

BSI 1881, 1983: *Testing concrete Part 106, Methods for determination of air content of fresh concrete.*

BSI 1881, 1988: *Testing concrete Part 124, Methods for analysis of hardened concrete.*

BSI 3797, 1990: *Specification for lightweight aggregates for concrete.*

BSI 4027, 1996: *Specification for sulfate-resisting Portland cement.*

BSI 4131, 1973: *Specification for terrazzo tiles.*

BSI 4246, 1991: *Specification for high slag blastfurnace cement.*

BSI 4248, 1974: *Specification for supersulphated cement.*

BSI 4550, 1978: *Methods of testing cement. Part 3, Physical tests, Section 3:7, Soundness test.* (Note: replaced by BS EN 196-3:1995).

BSI 4551, 1980: *Methods of testing mortars, screeds and plasters.*

BSI 4624, 1981 (1992): *Methods of test for asbestos-cement building products.*

BSI 4625, 1970: *Specification for prestressed concrete pipes (including fittings).*

BSI 5178, 1975: *Specification for prestressed concrete pipes for drainage and sewerage* (withdrawn). (Note: replaced by BS 5911, Part 103, 1994).

BSI 5224, 1976 (1995): *Specification for masonry cement.*

BSI 5262, 1991: *Code of practice for external renderings.*

BSI 5328, 1990: *Concrete Part 1: Guide to specifying concrete.*

BSI 5337, 1976: *Code of practice for the structural use of concrete for retaining aqueous liquids.* (Note: replaced by BS 8007: 1987).

BSI 5492, 1990: *Code of practice for internal plastering.*

BSI 5991, 1984: *Precast concrete pipes and fittings for drainage and sewerage. Part 1.*

BSI 6073, 1981 (1988): *Precast concrete masonry units. Part 1-2.*

BSI 6089, 1981: *Guide to assessment of concrete strength in existing structures.*

BSI 6110, 1985: *Pozzolanic cement with pulverized-fuel ash as pozzolana.*

BSI 6432, 1984 (1990): *Methods for determining properties of glass fibre reinforced concrete material.* (Reissued 1990).

BSI 6463, 1984-7: *Quicklime, hydrated lime and natural calcium carbonate. Part 1-4.*

BSI 6588, 1985: *Portland pulverized-fuel ash cement.*

BSI 6610, 1991: *Specification for pozzolanic pulverized-fuel ash cement.*

BSI 6699, 1986: *Specification for ground granulated blastfurnace slag for use with Portland cement.* (Note: replaced by BS 6699: 1992).

BSI 6699, 1992: *Specification for ground granulated blastfurnace slag for use with Portland cement.*

BSI 7583, 1992: *Specification for Portland limestone cement.*

BSI 8007, 1987: *Code of practice for design of concrete structures for retaining aqueous liquids.*

BSI 8110, 1985: *Structural use of concrete. Part 1, Code of practice for design and construction.*

BSI 8204, 1987-94: *Screeds, bases and in-situ flooring. Part 1-5.*

BSI CP 204, 1970: *In-situ floor finishes.* (Note: replaced by BS 8204: 1987–94)

BSI DD ENV 197-1, 1995: *Cement-composition, specifications and conformity criteria: Part 1 – common cements* (Note: this specification will be issued as BS EN 197-1 in due course.)

BSI EN 196-6, 1992: *Methods of testing cement. Determination of fineness.*

BSI EN 196-7, 1992: *Methods of testing cement. Methods of taking and preparing samples of cement.*

BSI EN 572, 1995: *Fibre cement products, pressure pipes and joints.*

Buck, A.D. 1986: Relationship between ettringite and chloroaluminate, strength and expansion in paste mixtures. *8th International Congress Chemistry of Cement, III*, Theme 2, 417–24.

Buck, A.D., Mather, K. 1978: Alkali-silica reaction products from several concretes: Optical, chemical and X-ray diffraction data. *Proceedings of the 4th International Conference on the Effects of Alkalies in Cement and Concrete, Purdue University*, 73–85.

Buckingham, W.F. and Spaw, J.M. 1988: Direct measure of spacing factor in air entrained concrete. *Proceedings 10th International Conference on Cement Microscopy, USA*: 82–92.

Buenfeld, N.R. and Newman, J.B. 1984: The permeability of concrete in a marine environment. *Magazine of Concrete Research* **36**, (127), 67–80.

Buenfeld, N.R. and Newman, J.B. 1986: The development and stability of surface layers of concrete exposed to sea-water. *Cement and Concrete Research*, **16**(5), 721–32.

Building Research Establishment. 1973: *Report on the collapse of the roof of the assembly hall of the Camden School for Girls.* Dept of Education & Science, HMSO, London, UK.

Building Research Establishment. 1981: *Assessment of chemical attack of high alumina cement concrete.* BRE Information Paper IP22/81 Watford, UK.

Building Research Establishment. 1981: Calcium silicate (sandlime, flintlime) brick work. *BRE Digest 157:* 8, Watford, UK.

Building Research Establishment. 1991: Shrinkage of natural aggregates in concrete. *BRE Digest 357*, Watford, UK.

Building Research Establishment. 1991: Sulfate and acid resistance of concrete in ground. *BRE Digiest 363*, Watford, UK.

Building Research Establishment. 1994: Assessment of existing high alumina cement concrete construction in the UK. *BRE Digest 392*, Watford, UK.

Bullock, P., Federoff, N., Jongerius, A., Stoops, G. and Tursina, T. 1985: *Handbook for soil thin section description.* Waine Research Publications, England.

Bungey, J.H. 1982: *The testing of concrete in structures.* Surrey University Press.

Burrells, W. 1961: *Industrial microscopy in practice.* Fountain Press, London.

Bye, G.C. 1983: *Portland cement composition, production and properties.* Institute of Ceramics/Pergamon Press, Oxford, UK.

Byfors, K., Klingstedt, G., Lehtonen, V., Pyy, H. and Romben, L. 1989: Durability of concrete made with alkali activated slags. *Proceedings 3rd International Conference on Fly Ash, Silica Fume, Slag and Natural Pozzolans in Concrete*, Norway, ACI SP-114, **2**, 1429–66.

Caberra, J.G. and Woolley, G.R. 1985: A study of twenty-five year old pulverized fuel ash concrete used in foundation structures. *Proc. Instn Civ. Engrs*, Part 2, **79**, 149–65, Paper 8885.

Cahill, J., Dolan, J.C. and Inward, P.W. 1994: The identification and measurement of entrained air in concrete using image analysis. *Petrography of Cementitious Materials, ASTM STP* **1215**, 111–24.

Caldarone, M.A., Gruber, K.A. and Burg, R.G. 1994: High-reactivity meta-kaolin: A new generation mineral admixture. *Concrete International* **16(11)**, 37–40.

Calleja, J. 1980: Durability. Sub-theme VII-2, *7th International Congress Chemistry of Cement, Paris* **1**, Sub-theme VII-2, 2/1–2/48.

Campbell, D.H. 1986: *Microscopical examination of interpretation of Portland cement and clinker*. Construction Technology Laboratories, Portland Cement Association, Skokie, Ilinois, USA.

Campbell-Allen, D. 1979: *The reduction of cracking in concrete*. 165 Sydney, University of Sydney and Cement and Concrete Association of Australia.

Candlot, E. 1890: *Bulletin de la Societie d'Encouragement d'Industrie Nationale*, **5**, 685–716.

Carles-Gibergues, A., Saucier, F., Grandet, J. and Pigeon, M. 1993: New-to-old concrete bonding: Influence of sulfates type of new concrete on interface structure. *Cement and Concrete Research* **23(2)**, 431–41.

Castillo, C. and Durrani, A.J. 1990: Effect of transient high temperature on high-strength concrete. *ACI Materials Journal* **87(1)**, 47–53.

Caveny, B. 1992: Cement hydration study using the environmental scanning electron microscope. *Proceedings 14th International Conference on Cement Microscopy*, 29–52.

Cement and Concrete Association of New Zealand. 1991: *Alkali-aggregate reaction. Minimising the risk of damage to concrete. Guidance notes and model specification clauses 63*. Cement and Concrete Association Report no. TR3.

Cement Association of Japan. 1971: Qualitative analysis of quartz (phosphoric acid method). *Standard Testing Method of the Association (CAJS 1–31)* 103–19. (In Japanese.)

Chandra, S. 1988: Hydrochloric acid attack on cement mortar – an analytical study. *Cement and Concrete Research* **18(2)**, 193–203.

Chatterji, S. 1982: Probable mechanisms of crack formation at early ages of concretes: A literature survey. *Cement and Concrete Research* **12**, 271–376.

Chatterji, S. 1984: The spacing factor in entrained air-bubbles. Has it any significance? *Cem Concr. Res.* **14(5)**, 757–8.

Chatterji, S. and Gudmundsson, H. 1977: Characterisation of entrained air bubble systems in concretes by means of an image analyzing microscope. *Cement and Concrete Research* **7**, 423–8.

Chayes, F. 1956: *Petrographic modal analysis*. J. Wiley & Sons, New York.

Chemical Society, 1975: *Recommendations for the testing of high alumina cement concrete samples by thermoanalytical techniques*. Thermal Methods Group, Analytical Division of the Chemical Society, Lyme Regis, UK.

Chew, Y.L. 1988: Assessing heated concrete and masonry with thermoluminescence. *ACI Materials Journal* **85(6)**, 537–43.

Childe, H.G. 1961: *Concrete Products and Cast Stone*. (9th edn) 263. London, Concrete Publications.

Chinchon, S., Guirado, F., Gali, S. and Vazquez. 1994: Cement content in concrete made with aluminous cement. *Materials and Structures* **27(169)**, 285–7.

Chrisp, T.M., Waldron, P. and Wood, J.G.M. 1994: Development of a non-destructive test to quantify damage in deteriorated concrete. *Magazine of Concrete Research* **45(165)**, 247–56.

Christensen, P., Gudmundsson, H., Thaulow, N., Damgard-Jensen, A.D. and Chatterji, S. 1979: Structural and ingredient analysis of concrete – methods, results and experience. *Nordisk Betong* **3**, 4–9. (C & CA translation from Swedish, T292, 1984).

Clear, C.A. 1985: The effects of autogenous healing upon the leakage of water through cracks in concrete. *Cement and Concrete Association Technical Report no. 559*, 31.

Cocke, D.L. 1990: The binding chemistry and leaching mechanisms of hazardous substances in cementitious solidification/stabilisation systems. *Journal of Hazardous Materials* **24**, 231–53.

Cohen, M.D. and Bentur, A. 1988: Durability of Portland cement-silica fume pastes in magnesium sulfate and sodium sulfate solutions. *ACI Materials Journal* **85(3)**, 148–57.

Cohen, M.D. and Mather, B. 1991: Sulfate attack on concrete – research needs. *ACI*

Materials Journal **88(1)**, 62–9.

Cole, W.F. and Beresford, F.D. 1980: Influence of basalt aggregate on concrete durability. In: Sereda, P.J. and Litvan, G.G. (eds) *Durability of Building Materials and Components*, Proceedings of the First International Conference, Ottawa, 1978. ASTM STP 691, 617–28. American Society for Testing and Materials, Philadelphia, USA.

Cole, W.F. and Demediuk, T. 1955: X-ray, thermal, and dehydration studies of magnesium oxychlorides. *Australian Journal of Chemistry* **8(2)**, 234–51.

Cole, W.F. and Kroone, B. 1959–60: Carbon dioxide in hydrated Portland cement. *Proceedings of the American Concrete Institute* **56**, 1275–95.

Coleman, S.E. 1984: Calcium silicate hydrates formed under hydrothermal conditions. *Proceedings 18th National ACS Meeting, Advances in Oil Field Chemicals and Chemistry*. St Louis.

Coles, J.A. 1978: Colour and pigments. *Concrete (Journal of the Concrete Society)* **12(3)**, 16–18.

Collepardi, M. and Monosi, S. 1992: Effect of the carbonation process on the concrete deterioration by $CaCl_2$ aggression. *Proceedings of the 9th International Congress on the Chemistry of Cement*, New Delhi V, 389–95.

Collins, A.R. 1944: The destruction of concrete by frost. *Journal of the Institution of Civil Engineers* **23(1)**, 29–41.

Collins, R.J. and Bareham, P.D. 1987: Alkali-silica reaction: suppression of expansion using porous aggregate. *Cement and Concrete Research* **17(1)**, 89–96.

Collins, R.J. and Gutt, W. 1988: Research on long-term properties of high alumina cement concrete. *Magazine of Concrete Research* **40(145)**, 195–208 (and Discussion, **41(149)**, 1989, 243–4.

Comité Euro-International du Béton. 1992: *Durable Concrete Structures, Design Guide*. London, Thomas Telford.

Commission of the European Communities Council. 1993: *Directive Draft of the Landfill of Wastes*. SYN 335, Brussels.

Commission Internationale des Grands Barrages. Exposure of dam concrete to special aggressive waters. Guidelines. *Bulletin 71*.

Concrete Society. 1973: Fibre-reinforced cement composites. *Technical Report 51.067*.

Concrete Society. 1979: Specification for sprayed concrete. *Publication 53.029*.

Concrete Society. 1980: Code of practice for sprayed concrete. *Publication 53.030*.

Concrete Society. 1982: Non-structural cracks in concrete. *Report of a Concrete Society Working Party, Concrete Society Technical Report No. 22*.

Concrete Society. 1987: Changes in the properties of ordinary Portland cement and their effects on concrete. *Report of a Concrete Society Working Party, Concrete Society Technical Report No. 29*.

Concrete Society. 1987: Concrete core testing for strength. *Concrete Society Technical Report No. 11*.

Concrete Society. 1988: Analysis of hardened concrete. *Report of a Joint Working of the Concrete Society and Society of the Chemical Industry. Concrete Society Technical Report No. 32*.

Concrete Society. 1989: Analysis of hardened concrete. A guide to tests, procedures and interpretation of results. Report of joint working party of the Concrete Society and Society of the Chemical Industry. *Concrete Society Technical Report No. 32*.

Concrete Society. 1989: Analysis of hardened concrete. *Report of a Joint Working Party of the Concrete Society and Society of Chemical Industry. Concrete Society Technical Report No. 32, 117*.

Concrete Society. 1990: Assessment and repair of fire-damaged structures. *Concrete Society*

Technical Report No. 33.

Concrete Society. 1991: *The use of ggbs and pfa in concrete. Concrete Society Technical Report No. 40.*

Concrete Society. 1992: Non-structural cracks in concrete. *Report of a Concrete Society Working Party. Concrete Society Technical Report No. 22.*

Concrete Society. 1993: Microsilica in concrete. *Report of a Concrete Society Working Party. Technical Report No. 41, 54.*

Concrete Society. 1994: Polymers in Concrete. *Cement and Concrete Association, Technical Report No. 39.*

Concrete Society. 1995 Alkali-silica reaction. Minimising the risk of damage to concrete. Guidance notes and model specification clauses. *Report of a Concrete Society Working Party. Concrete Society Technical Report No. 30.*

Conjeaud, M.L. 1980: Mechanism of sea water attack on cement mortar. *Performance of Concrete in Marine Environment*, ACI SP-65, 39–61.

Connor, J.R. 1990: *Chemical fixation and solidification of hazardous waste forms*, New York, Vand Nostrand, Reinhold.

Cooke, P.M. 1994: Chemical Microscopy. *Analytical Chemistry* **66(12)**, 558R–594R.

Copeland, L.E. and Kantro, D.L. 1964: Chemistry of hydration of Portland cement at ordinary temperature. In: Taylor, H.F.W. (ed.), *The chemistry of cements*, Vol. 1, Chapter 8, 313–70. Academic Press, London, UK.

Corish, A. 1989: European cement standards – a manufacturer's view. *Concrete Journal of the Concrete Society* **23(7)**, 19–20.

Coutinho, A. de S. 1979: Aspects of sulfate attack on concrete. *Cement, Concrete and Aggregates* **1(1)**, 10–12.

Coutts, R.S.P. 1983: Autoclaved beaten wood fibre-reinforced cement composites. *Composites* **18(2)**, 139–43.

Coutts, R.S.P. 1988: Wood fibre reinforced cement composites. (ed. R.N. Swamy) *Natural fibre reinforced cement and composites*, 1–62 Glasgow, Blackie.

Cowie, J. and Glasser, F.P. 1991/92: The reaction between cement and natural waters containing dissolved carbon dioxide. *Advances in Cement Research* **4(15)**, 119–34.

Cowper, A.D. 1927: Lime and lime mortars. *DSIR Building Research Special Report no 9*, 91. HMSO, London.

Cowper, A.D. 1950: Sands for plasters, mortars and external renderings. *National Building Studies, Bulletin no. 7*, HMSO, London.

Crammond, N.J. 1984: Examination of mortar bars containing varying percentages of coarsely crystalline gypsum as aggregate. *Cement and Concrete Research* **14(2)**, 225–30. (and personal communication 1990).

Crammond, N.J. 1985a: Quantitative X-ray diffraction analysis of ettringite, thaumasite and gypsum in concretes and mortars. *Cement and Concrete Research* **15(3)**, 431–41.

Crammond, N.J. 1985b: Thaumasite in failed cement mortars and renders from exposed brickwork. *Cement and Concrete Research* **15(6)**, 1039–50.

Crammond, N.J. 1990: Personal communication. See Crammond (1984).

Crammond, N.J. and Currie, R.J. 1993: Survey of condition of precast high-alumina cement concrete components in internal locations in 14 existing buildings. *Magazine of Concrete Resarch* **45(165)**, 275–9.

Crammond, N.J. and Halliwell, M.A. 1995: *The thaumasite form of sulfate attack in concretes containing a source of carbonate ions: A microstructural overview*. In Malhotra, V.M. (ed.) *Advances in concrete technology*, 2nd Symposium. ACI SP 154, Paper SP 154–19, 357–80. American Concrete Institute, Detroit, USA.

Cripwell, B. 1992: What is pafa? *Concrete* (Journal of the Concrete Society) **26(3)**, 11–13.

Crumpton, C.F. 1988: Lightweight aggregate concrete sometimes grows. Blame AAR if its the most likely cause. *Concrete Construction* **33(6)**, 618–19, **(12)**, 1103–5.

CSA. 1983: *Cementious hydraulic slag*, A363-M1983.

CSIR, South Africa. 1967: Corrosion of concrete sewers. *CSIR Research Report 163*, Pretoria.

Currie, R.J. and Crammond, N.J. 1994: Assessment of existing high alumina cement construction in the UK. *Proc. Instn Civ. Engrs Structs & Bldgs* **104**, 83–92.

Cusino, L. and Negro, A. 1980: *7th International Conference Chemistry of Cements, III*, V-62.

Czernin, W. 1980: *Cement chemistry and physics for civil engineers,* translated by Amerongen, C. van. 2nd English edn. George Godwin Ltd, London, UK for Bauverlag Gmbh, Wiesbaden, Germany.

Daligand, D. 1985: Le platre et ses techniques de production. *Ciments, Betons, Platres, Chaux* **(753)**, 83–8 **(755)**, 222–30.

Danilatos, G.D. 1991: Review and outline of environmental SEM at present. *Journal of Microscopy*, 162, (Pt 3), 391–402.

Davey, N. 1961: *A history of building materials.* Phoenix House, London, UK.

Davey, N. 1965: *A history of building materials.* Chapter 12, Limes and cements. Phoenix House, London, UK. 1961, reprinted 1965.

Davies, G. and Oberholster, R.E. 1986: The alkali-silica reaction product. A mineralogical and an electron microscope study. *Proceedings 8th International Conference on Cement Microscopy*, 303–26.

Deane, Y. 1989: *Finishes.* Mitchell's Building Series. London. Mitchell.

De Ceukelaire, L. 1991: Concrete surface deterioration due to expansion by the formation of jarosite. *Cement and Concrete Research* **21(4)**, 563–74.

De Ceukelaire, L. 1992: Alkali-silica reaction in a lightweight concrete bridge. *Proceedings of the 9th International Conference on Alkali-Aggregate Reaction in Concrete*, 231–9, London.

De Ceukelaire, L. 1992: The effects of hydrochloric acid on mortar. *Cement and Concrete Research* **22(5)**, 903–14.

De Ceukelaire, L. and Van Niewenberg, D. 1993: Accelerated carbonation of a blast-furnace cement concrete. *Cement and Concrete Research* **23(2)**, 442–52.

Deelman, J.C. 1984: Textural analysis of concrete by means of surface roughness measurements. *Materials and Structures* **17(101)**, 359–67.

DeHoff, R.T. and Rhines, F.N. 1968: Quantitative microscopy. New York, McGraw-Hill Books.

Delesse, A. 1848: Procédé mechanique pour determiner la composition des roches. *Annal. des Mines*, fourth series **1**, 379–88.

Delly, J.G. 1986: A basic microscope library. *Microscope* **34(1)**, 11–25.

Demediuk, T., Cole, W.F. and Hueber, H.V. 1955: Studies of magnesium and calcium oxychlorides. *Australian Journal of Chemistry* **8(2)**, 215–33.

Demoulian, E., Gourdin, P., Hawthorn, F. and Vernet, C. 1980: Influence of slags chemical composition and texture on their hydraulicity. *Proceedings of the 7th International Congress on the Chemistry of Cement, Paris, Proc.* Vol. III, 89–94.

Demoulian, E., Vernet, C., Hawthorn, F. and Gourdin, P. 1980: Slag content determination in cements by selective dissolutions. *Proceedings of the 7th International Congress Chemistry of Cements, Paris* **2(III)**, 151–6.

Deng Min, Hong Dongwen, Lan Xianghui and Tang Minshu. 1995: Mechanism of expansion in hardened cement pastes with hardburnt free lime. *Cement and Concrete Research* **25(2)**, 440–8.

Deng Min and Tang Minshu. 1993: Mechanism of dedolomitization and expansion in dolomitic rocks. *Cement and Concrete Research* **23(6)**, 1397–408.

Deng Min, Tang Mingshu. 1994: Formation and expansion of ettringite crystals. *Cement and Concrete Research* **24(1)**, 119–26.

Dengler, T.R. and Montgomery, J.R. 1984: The estimation of activity and glass content of blast furnace slag by microscopic observation. *Proceedings of the 6th International Conference on Cement Microscopy*, 265–75.

Department of Transport. 1991: *Specification for highway works*. DoT, London, UK.

Deruelle, S. 1991: Role du support dans le croissance des microorganismes. *Materials and Structures* **24**, 163–8.

Determann, H. and Lepusch, F. Undated: *The Microscope and its Application*, Ernst Leitz, Germany.

Dewar, J.D. (Convenor). 1976: *Concrete core testing for strength – report of a Concrete Society working party. Concrete Society Technical Report no. 11* including Addendum (1987). Reprinted with addendum 1987. The Concrete Society. London, UK.

Dewar, J.D. and Anderson, R. 1992: *Manual of ready-mixed concrete*, 2nd edn. Blackie Academic & Professional (Chapman & Hall), Glasgow, UK.

Dhir, R.K., Jones, M.R. and Amned, H.E.H. 1990: Determination of total and soluble chlorides in concrete. *Cement and Concrete Research* **20**, 579–90.

Dhir, R.K., Jones, M.R., Byars, E.A. and Shaaban, I.G. 1994: Predicting concrete durability from its absorption. *Durability of Concrete. Third International Conference, Nice*, ACI-145, 1177–94.

Diamond, S. 1975: A review of alkali-silica reaction and expansion mechanisms. I: Alkalies in cements and in concrete pore solutions. *Cement and Concrete Research* **5**, 329–45.

Diamond, S. 1976: Cement paste microstructure. In: *Hydraulic cement pastes: their structure and properties – an overview at several levels.* Proceedings of a conference, Sheffield, UK, 2-30. Cement & Concrete Association, Slough, UK.

Diamond, S. 1986: The microstructure of cement paste in concrete. *8th International Congress on the Chemistry of Cement, Rio de Janeiro* **1**, 123–47.

Diamond, S. 1989: Another look at mechanisms. *Proceedings 8th International Conference on Alkali Aggregate Reaction,* 83–94. Kyoto.

Diamond, S. 1992: Alkali aggregate reactions in concrete: An annotated bibliography 1939–1991. *SHRP-C/UPW-92-601.*

Dickson, J.A.D. 1966: Carbonate identification and genesis as revealed by staining. *Journal of Sedimentary Petrology* **361(2)**, 491–505.

Dolar-Mantuani, L. 1983: *Handbook of concrete aggregates – a petrographic and technological evaluation.* Noyes Publications, New Jersey, USA.

Dolch, W.L. 1984: Air entraining admixtures. *Concrete Admixtures Handbook* (ed. V.S. Ramachandran), 269–302. Noyes Publications, USA.

Douglas, E., Malhotra, V.M. and Emery, J.J. 1985: Cementitious properties of non-ferrous slags from Canadian sources. *Cement, Concrete, and Aggregates CCAGDP* **7(1)**, 3–14.

Douglas, E. and Malhotra, V.M. 1987: A review of the properties and strengths of non-ferrous slags-Portland cements binders. *Supplementary Cementing Materials for Concrete* (ed. V.M. Malhotra) Canmet.

Dressler, L. 1980: Polarization-optical examinations of birefringent effects using a full-wave plate in sub-parallel position. *Jena Review* **(3)**, 123–6.

Drissen, R. 1995: Determination of the glass content in granulated blastfurnace slag. *Zement-Kalk-Gips* **48(1)**, 59–62.

Dron, R. and Brivot, F. 1996: Solid-liquid equilibria in $K-C-S-H/H_2O$ systems. *Proceedings of the 10th International Conference on Alkali-aggregate Reaction in Concrete* 927–33. Melbourne, Australia.

Dubrolubov, G. and Romer, B. 1985: Guidelines for determining and testing the frost as well as frost-salt resistance of cement-concrete. *Bulletin of Betonstrassen AG. Research and Consulting in Concrete Road Construction (Concrete Roads Ltd.)* Special Number, June. Wildeg/Switzerland.

Duchesne, J. and Bérubé, M.A. 1994: The effectiveness of supplementary cementing materials in suppressing expansion due to ASR: Another look at the reaction mechanisms Part 1: Concrete expansion and portlandite depletion. Part 2: Pore solution chemistry. Cement and Concrete Research **24(1)**, 73–82, **24(2)**, 221–30.

Dunstan, E.R. 1982: A spec odyssey – Sulfate resistant concrete for the 1980s. *George Verbeck Symposium on Sulfate Resistant Concrete*, ACI-SP77: 41–62.

Dunster, A. and Weir, I. 1996: Appraisal of a building containing precast HAC concrete roof beams. *Construction Repair*, May/June, 2–6.

Durning, T.A. and Hicks, M.C. 1991: Using microsilica to increase concrete's resistance to aggressive chemicals. *Concrete International* **13(3)**, 42–8.

Dutruel, F. and Guyader, R. 1977: Considerations sur les efflorescences des betons apparents. *Ciments Betons Platres Chaux*, (6/77), 340–9.

Dutton, A.R. 1977: Manufacture of concrete pipe in South Africa (ed. D.S. Fulton) *Concrete Technology. A South African Handbook*, 337–43. The Portland Cement Institute, Johannesburg.

Dziezak, J.D. 1988: Microscopy and image analysis for R & D. *Food Technology* **42(7)**, 110–24.

Edwards, A.G. 1970: *Scottish aggregates: rock constituents and suitability for concrete.* Current Paper CP 28/70. Building Research Station, Ministry of Public Buildings and Works, HMSO, London, UK.

Efes, Y. 1988: Determination of the composition of hardened concrete by image analysis. *Betonwerk + Fertigteil-Technik*, Heft **11**, 86–91; **12**, 62–8.

Eglinton, M.S. 1987: *Concrete and its chemical behaviour*, Thomas Telford Ltd., London, UK.

Elias, H. 1971: Three-dimensional structure identified from single sections. *Science* **174**, 993–1000.

Elsen, J., Lens, N., Aarre, T., Quenard, D. and Smolej, V. 1995: Determination of the w/c ratio of hardened cement paste and concrete samples on thin sections using automated image analysis techniques. *Cement and Concrete Research* **25(4)**, 827–34.

ENDS. 1992: *Environmental News Data Services.* Report 205, February. 11–33.

Erlin, B. 1962: Air content of hardened concrete by a high pressure method. *J. Portland Cement Association* **5(3)**, 240–9.

Erlin, B. and Stark, D.C. 1965: Identification and occurrence of thaumasite in concrete. *Highway Research Record No. 113*, 108–13.

Erntroy, H.C. 1992: The effect on compressive strength of the glass content of blastfurnace slag when used as a cementitious constituent. *Zement-Kalk-Gips* **45(10)**, 533–5.

Everett, A. 1994: *Materials.* Mitchell's Building Series. Mitchell, London.

Fairbanks, E.E. 1925: Modification of Lemberg's staining method. *American Mineralogist* **10**, 126–7.

Fairhall, G.A. 1989: Effect of process operational variables on the product properties of encapsulated intermediate level wastes. *Radioactive Waste Management 2, Proceedings of the International Conference Organised by the British Nuclear Energy Society.* 79–84. London, Telford.

Fairhall, G.A. 1994: The treatment and encapsulation of intermediate level wastes at Sellafield. *Proceedings of the 4th International Conference on Nuclear Fuel Reprocessing and Waste Management Reconditioning, British Nuclear Industrial Forum, April 1994* **1**.

Farmer, V.C. 1974: *The Infrared Spectra of Minerals.* Monograph 4, 539. Mineralogical

Society, London.

Fattuhi, N.I. and Hughes, B.P. 1986: Resistance to acid attack of concrete with different admixtures or coatings. *The International Journal of Cement Composites and Lightweight Concrete* **8(4)**, 223–30.

Feigl, F. 1958: Spot tests in inorganic analysis. Elsevier Publishing Co, New York.

Figg, J. 1983: *Chloride and sulfate attack on concrete.* Chemistry and Industry, 17 October, 770–5.

Figg, J.W. (Chairman) 1989: *Analysis of hardened concrete – a guide to tests, procedures and interpretation of results – report of a joint working party of the Concrete Society and Society of Chemical Industry.* Technical Report no. 32. The Concrete Society, London, UK.

Figg, J.W., Moore, A.E. and Gutteridge, W.A. 1976: On the occurrence of the mineral trona ($Na_2CO_3.NaHCO_3.H_2O$) in concrete deterioration products. *Cement and Concrete Research* **6(5)**, 691–6.

Figg, J.W. 1977: *Alkali-aggregate (alkali-silica and alkali-silicate) reactivity in concrete. Bibliography.* Cembureau.

Figg, J.W. 1981: Reaction between cement and artificial glass in concrete. *Proceedings of the Fifth International Conference on Alkali-Aggregate Reaction*, Cape Town, paper no. S252/7.

Figg, J.W. and Bowden, S.R. 1971: *The analysis of concretes.* Building Research Establishment HMSO, London, UK.

Fiskaa, O., Hansen, H. and Moum, J. 1971: Betong i alunskifer. *Norwegian Geotechnical Institute, Publication No. 86*, 1–32. (In Norwegian with summary.)

Fleischer, M., Wilcox, R.E. and Matzko, J.J. 1984: *Microscopic determination of nonopaque minerals.* 3rd edn Bulletin 1627, USGS, Washington.

Florke, O.W. 1952: Kristallographische und rontgenometrische Untersuchungen in System $CaSO_4$-$CaSO_4.2H_2O$. *Neues Jahrb. Min. Abhandl.* **84**, 189–240; *Mineralogical Abstracts* **12**, 94, 1953–55.

Fookes, P.G. 1977: A plain man's guide to cracking in the Middle East. *Concrete in the Middle East.* A Viewpoint Publication. Cement and Concrete Association, Slough, UK, 22–4. (Originally published in *Concrete*, **10(9)**, 20–2 (1976).

Fookes, P.G. 1993: Concrete in the Middle East – past, present and future: a brief review. *Concrete* **27(4)**, 14–20.

Fookes, P.G. and Collis, L. 1975: Aggregates and the Middle East. *Concrete (Journal of the Concrete Society)* **9(11)**, 14–19.

Fookes, P.G. and Collis, L. 1975: Problems in the Middle East. *Concrete* **9(7)**, 12–19.

Fookes, P.G. and Collis, L. 1977: Problems in the Middle East. *Viewpont Publication,* Cement and Concrete Association, Slough, UK, 1–7.

Fraay, A.L.A., Bijen, J.M. and de Haan, Y.M. 1989: The reaction of fly ash in concrete – a critical examination. *Cement and Concrete Research* **19(2)**, 235–46.

Frearson, J.P.H. and Sims, I. 1991: Sandberg on slag. *Concrete (Journal of the Concrete Society)* **25(6)**, 37–40.

Freitag, S.A. 1990: Alkali-aggregate reactivity of concrete sands from the Rangitikei and Waikato rivers. *N.Z. Works Consultancy Services, Central Laboratories Report*, 90-B4207:27.

French, W.J. 1991a: Concrete petrography: a review. *Quarterly Journal of Engineering Geology* **24(1)**, 17–48.

French, W.J. 1991b: Comments on the determination of the ratio of ggbs to Portland cement in hardened concrete. *Concrete (Journal of the Concrete Society)* **25(6)**, 33–6.

French, W.J. 1992: Determination of the ratio of pfa to Portland cement in hardened concrete. *Concrete (Journal of the Concrete Society)* **26(3)**, 43–5.

French, W.J. 1993: Quantitative concrete petrography. Science and Technology Conference, The Institute of Materials. Unpublished.

Frondel, C. 1962: *Dana's System of Mineralogy. Volume III. Silica Minerals.* John Wiley and Sons, New York.

Fu, Y., Gu, P., Xie, P. and Beaudoin, J.J. 1995: A kinetic study of delayed ettringite formation in hydrated Portland cement paste. *Cement and Concrete Research* **25(1)**, 63–70.

Fulton, F.S. 1977: *Concrete technology. South African Handbook.* Portland Cement Institute, Johannesburg.

Fundal, E. 1980: Microscopy of cement raw mix and clinker. *FLS-Review 25*, F.L. Smidth Laboratories, Copenhagen.

Gabrisova, A., Havlica, J. and Sahu, S. 1991: Stability of calcium sulphoaluminate hydrates in water solutions with various pH values. *Cement and Concrete Research* **21(6)**, 1023–7.

Galopin, R. and Henry, N.F.M. 1972: *Microscopic study of opaque minerals.* W. Heffer, Cambridge.

Garrett, H.L. and Beaman, D.R. 1985: A method for preparing steel reinforced mortar or concrete for examination by transmitted light microscopy. *Cement and Concrete Research* **15**, 917–20.

George, C.M. 1983: Industrial aluminous cements. *Structure and Performance of Cements* (ed. P. Barnes). Applied Science, 415–70. London.

George, W.O. 1924: The relation of the physical properties of natural glasses to their chemical composition. *The Journal of Geology* **32**, 353–72.

Gerdes, A. and Wittmann, F.H. 1992: Electrochemical degradation of cementitious materials. *9th International Congress on the Chemistry of Cement* **V**, 409–15.

Gettens, R.J. and Stout, G.L. 1966: *Painting Materials.* Dover Publications, New York.

Ghorab, H.Y., Heinz, D., Ludwig, U., Meskendahl, Y. and Wolter, A. 1980: On the stability of calcium aluminate sulphate hydrates in pure systems and in cements. *7th International Congress on the Chemistry of the Cement, Paris* **IV**, 496–503.

Gille, F., Dreizler, I., Grade, K., Kramer, H. and Woermann, E. 1965: *Microscopy of cement clinker – Picture atlas.* Benton-Verlag GmbH, English translation by P. Schmid 1980 Dusseldorf, Germany.

Gillott, J.E. and Rogers, C.A. 1994: Alkali-aggregate reaction and internal release of alkalis *Cement and Concrete Research* **24(5)**, 99–112.

Gillott, J.E. and Swenson, E.G. 1969: Mechanism of the alkali-carbonate rock reaction. *Quarterly Journal of Engineering Geology*, 7–24.

Glanville, W.H. (ed.) 1950: *Modern Concrete Construction* **1**, 168–320. The Caxton Publishing Co. London.

Glasser, F.P., Damidot, D. and Atkins, M. 1995: Phase development in cement in relation to the secondary ettringite problem. *Advances in Cement Research* **7(26)**. 57–68.

Goguel, R.L. 1996: Selective dissolution techniques in AAR investigation: application to an example of failed concrete. *Proceedings of the 10th International Conference on Alkali-Aggregate Reaction in Concrete*, 783–90. Melbourne, Australia.

Goguel, R.L. 1995: A new consecutive dissolution method for the analysis of slag cements. *Cement, Concrete and Aggregates CCAGDP* **17(1)**, 84–91.

Goguel, R.L. and St John, D.A. 1993a: Chemical identification of Portland cements in New Zealand concretes – I. Characteristic differences among New Zealand cements in minor and trace element chemistry. *Cement and Concrete Research* **23(1)**, 59–68.

Goguel, R.L. and St John, D.A. 1993b: Chemical identification of Portland cements in New Zealand concretes. II. The Ca-Sr-Mn plot in cement identification and the effect of aggregates. *Cement and Concrete Research* **23(2)**, 283–93.

Goltermann, P. 1995: Mechanical predictions of concrete deterioration–Part 2: Classification of crack patterns. *ACI Materials Journal* **92(1)**, 58–63.

Gomà, F. 1989: The chemical analysis of hardened concrete containing fly ashes, slags, natural pozzolans, etc. In: Alasali, M. (ed.), Supplementary Papers, 3rd CANMET/ACI International Conference: *Fly ash silica fume, slag and natural pozzolans in concrete*, 828–45. Trondheim, Norway.

Gooding, P. and Halstead, P.E. 1952: The early history of cement in England. *Proceedings of the 3rd Symposium on the International Chemistry of Cement*, 1–29 London, Cement and Concrete Association.

Gouda, G.R., Roy, D.M. and Sarker, A. 1975: Thaumasite in deteriorated soil-cements. *Cement and Concrete Research* **5(5)**, 519–22.

Grade, K. 1968: Determination of the fluorescence of granulated slags. In *Proceedings of the 5th International Symposium on the Chemistry of Cement* **IV** 168–72. Tokyo.

Grandet, J. and Ollivier, J.P. 1980: Etude de la formation du monocarboaluminate de calcium hydrate au contact d'un granulat calcair dans une pate de ciment Portland. *Cement and Concrete Research* **10(6)**, 759–70.

Granger, F. (ed. and transl.) 1962: *Vitruvius on architecture*. Vols I and II. William Heinemann Ltd., London, UK. 1st edn 1931, reprinted 1962.

Grattan-Bellew, P.E. 1989: Test methods and criteria for evaluating the potential reactivity of aggregates. *Proceedings of the 8th International Conference on Alkali–Aggregate Reaction*, 279–94. Kyoto, Japan.

Grattan-Belew, P.E. 1992: Microcrystalline quartz, undulatory extinction and the alkali-silica reaction. *Proceedings of the 9th International Conference on Alkali-Aggregate Reaction in Concrete*, 383–94. London.

Grattan-Bellew, P.E. 1996: A critical review of accelerated ASR tests. *Proceedings of the 10th International Conference on Alkali-Aggregate Reaction in Concrete*, 27–38. Melbourne, Australia.

Green, G.W. 1984: Gypsum analysis with the polarizing microscope (ed. R.A. Kuntze). *The chemistry and technology of gypsum, ASTM STP 861*, 22–47.

Grehn, J. 1977: *Leitz Microscopes for 125 Years*. Ernst Leitz, Germany.

Grounds, T., Midgeley, H.G. and Nowell, D.V. 1988: The carbonation of ettringite by atmospheric carbon dioxide. *Thermochimica Acta* **135**, 347–52.

Grout, F.F. 1932: *Petrography and petrology*. McGraw-Hill Book Co. New York.

Grove, R.M. 1968: The identification of ordinary Portland cement and sulphate resisting cement in hardened concrete samples. *Silicates Industriels* **10**, 317–20.

Groves, G.W. 1981: Microcrystalline calcium hydroxide in Portland cement pastes at low water/cement ratio. *Cement and Concrete Research* **11**, 713–18.

Groves, G.W., Hodges, D.J. and Zhang, Xiaozhong. 1987: TEM studies of reactive aggregates and their reaction with cement. *Proceedings of the International Conference on Cement Microscopy*, 458–65.

Groves, G.W. and Richardson, I.G. 1994: Microcrystalline calcium hydroxide in pozzolanic cement pastes. *Cement and Concrete Research* **24(6)**, 1191–6.

Grube, H. and Rechenberg, W. 1989: Durability of concrete structures in acidic water. *Cement and Concrete Research* **19(5)**, 783–92.

Gudmundsson, H. and Chatterji, S. 1979: The measurement of paste content in hardened concrete using automatic image analyzing technique. *Cement and Concrete Research* **9**, 607–12.

Gunnar, A. and Akman, M.S. 1984: The influence of addition of blastfurnace slag or the corrosion resistance of concrete. *Proceedings of the 3rd International Conference on the*

Durability of Building Materials and Components, Finland, 14–25.

Gustaferro, A.H. and Scott, N.L. 1990: Reading structural concrete cracks. *Concrete Construction* December, 994–1003.

Gutmann, P.F. 1988: Bubble characteristics as they pertain to compressive strength and freeze-thaw durability. *ACI Materials Journal* **85(5)**, 361–6. (Title No. 85–M40.)

Gutt, W. 1992: BS 6699: 1992 Specification for ground granulated blastfurnace slag for use with Portland cement. *Concrete (Journal of the Concrete Society)* **26(1)**, 37–8.

Gutt, W., Nixon, P.J., Smith, M.A., Harrison, W.H. and Russell, A.D. 1974: *A survey of the locations, disposal and prospective uses of the major industrial by-products and waste materials.* Current Paper CP 19/74, Building Research Establishment, DOE, Garston, Watford, UK.

Gutteridge, W.A. 1979: On the dissolution of the interstitial phases in Portland cement. *Cement and Concrete Research* **9(3)**, 319–24.

Gy, P.M. 1982: *Sampling of particulate materials. Theory and practice, 431.* Elsevier, Oxford.

Haavik, D.J. and Mielenz, R.C. 1991: Alkali-silica reaction causes concrete pipe to collapse. *Concrete International*, May, 54–7.

Hadley, D.W. 1968: Field and laboratory studies in the reactivity of sand-gravel aggregates. *Journal of the PCA Research and Development Laboratories*, January, 17–23.

Hall, C. 1994: Barrier performance of concrete: A review of fluid transport theory. *Materials and Structures* **27(169)**, 291–306.

Hallimond, A.F. 1953: *Manual of the polarizing microscope.* York, Cooke, Troughton & Simms.

Halse, Y., Pratt, P.L., Dalziel, J.A. and Gutteridge, W.A. 1984: Development of microstructure and other properties in fly ash OPC systems. *Cement and Concrete Research* **14(4)**, 491–8.

Hamilton, J.J. and Handegord, G.O. 1968: The performance of ordinary Portland cement concrete in Prairie soils of high sulfate content. *Performance of Concrete, Symposium in Honour of Thorbergur Thorvaldson*, 135–58.

Hammersley, G.P. 1980: The identification of the primary constituents of hardened grouts, mortars and concretes. MSc Dissertation, Queen Mary College, University of London (unpublished).

Hannawayya, F. 1984: Microstructural study of accelerated vacuum curing of cement mortar with carbon dioxide. *World Cement* **15(9)**, 326–34; **(10)**, 378–84 and *Proceedings of the 9th International Conference on Cement Microscopy, 1987*, 337–54.

Hansen, T.C. 1989: Marine concrete in hot climatesss – designed to fail. *Materials and Structures* **22**, 344–6.

Harboe, E.M. 1982: Longtime studies and field experience with sulfate attack. *George Verbeck Symposium on Sulfate Resistant Concrete*, ACI-SP77, 1–20.

Harmathy, T.Z. 1968: Determining the temperature history of concrete constructions following fire exposure. *ACI Journal, Proceedings* **65(11)**, 959–64.

Harris, P.M. and Sym, R. 1990: Sampling of aggregates and precision test standards. *Standards for Aggregates*, 19–63. Ellis Horwood, New York.

Harrison, W.H. 1992: Assessing the risk of sulphate attack on concrete in the ground. *BRE Information Paper, P 15/92*.

Harrison, W.H. 1992: Sulphate resistance of buried concrete. The third report on long-term investigations at Northwick Park and on similar concretes in sulphate solutions at BRE. *Building Establishment Report.* BRE, Garston.

Harrison, W.H. and Munday, R.S. 1975: *An investigation into the production of sintered pfa aggregate.* BRE Current Paper CP2/75, HMSO, Watford, UK.

Harrison, W.H. and Teychenné, D.C. 1981: Sulphate resistance of buried concrete; second interim report on long-term investigation at Northwick Park. *Building Research Establishment Report*, HMSO.

Hartshorne, N.H. and Stuart, A. 1969: *Practical optical crystallography* 326. American Elsevier, New York.

Hartshorne, N.H. and Stuart, A. 1970: *Crystals and the polarising microscope*. Edward Arnold, London.

Hyat, M.A. 1989: *Principles and techniques of electron microscopy*. CRC Press, USA.

Heinz, D. and Ludwig, U. 1986: Mechanisms of subsequent ettringite formation in mortars and concretes after heat treatment. *8th International Congress on the Chemistry of Cement, Rio de Janeiro*, 189–95.

Heinz, D. and Ludwig, U. 1987: Mechanisms of secondary ettringite formation in mortars and concretes subjected to heat treatment. In Scanlon, J.M. (ed.) *Concrete durability: Katharine and Bryant Mather international conference*, ACI SP-100, 2059–2071. Paper No. SP100-105. American Concrete Institute, Detroit, USA.

Heller, L. and Taylor, H.F.W. 1956: *Crystallographic data for the calcium silicates*. HMSO, London.

Highway Research Board, 1972: *Guide to compounds of interest in cement and concrete research*. Highway Research Board Special Report 127, Washington, USA.

Hill, N., Holmes, S. and Mather, D. (eds.) 1992: *Lime and other alternative cements*. Intermediate Technology Publications, London.

Hills, C.D. 1993: *The hydration of ordinary Portland cement during cement based solidification of hazardous wastes*. PhD. Thesis, University of London (unpublished).

Hime, W.G. 1974: Multitechnique approach solves construction materials failure problems. *Analytical Chemistry* **46**, 1230A.

Hinsch, J. 1979: Critical focusing in low-power photomicrography. *American Laboratory*, April, 35–7.

Hjorth, L. 1965: Large rotating stage for polarising microscope. *RILEM Bulletin* **(26)**, 111–16.

Hjorth, L. 1982: Microsilica in concrete. *Nordic Concr. Res.* **1**, 9.1–9.18.

Hobbs, D.W. 1988: *Alkali-silica reaction in concrete*. Thomas Telford, London.

Hodgkinson, L. and Rostam, O. 1991: Admixtures in air entrained concrete. *Concrete (Journal of the Concrete Society)* **25(2)**, 11–13.

Hoffmänner, F. 1973: *Microstructure of Portland cement clinker* 33–7. Holderbank, Switzerland.

Hooton, R.D. 1987: The reactivity and hydration products of blast-furnace slag. In: Malhotra, V.M. (ed.) *Supplementary cementing materials for concrete*, Canadian Centre for Mineral and Energy Technology (CANMET), SP 86-8E, Chapter 4, 245–88. Canadian Government Publishing Centre, Ottawa, Canada.

Hooton, R.D. and Emery, J.J. 1983: Glass content determination and strength development predictions for vitrified blast furnace slag. Proc. of the CANMET/ACDI 1st Int. Conf. on *The use of fly ash, silica fume, slag and other mineral by-products in concrete*, Montebello, Quebec, Canada. ACI Special Publication SP-79, 943–62.

Hornain, H., Marchand, J., Ammouche, A., Commene, J.P. and Moranville, M. 1995: Microscopic observations of cracks in concrete – A new sample preparation technique using dye impregnation. *Proceedings of the 17th International Conference on Cement Microscopy*, 271–82.

Hornibrook, F.B., Insley, H. and Schuman, L. 1943: Effects of alkalis in Portland cement on the durability of concrete. An appendix to the Report of Committee C1. *Proceedings American Society of Testing Materials* **43**, 218.

Hoshino, M. 1988: Difference of the w/c ratio, porosity and microscopical aspect between the upper boundary paste and the lower boundary paste of the aggregate in concrete. *Materials and Structures* **21**, 336–40.

Houst, Y.F. and Wittmann, F.H. 1994: Influence of porosity and water content on the diffusivity of CO_2 and O_2 through hydrated cement paste. *Cement and Concrete Research* **24(6)**, 1165–76.

Hover, K. 1993a: Why is there air in concrete? (Part 1 of 4). *Concrete Construction* **38(1)**, 11–15.

Hover, K. 1993b: Air bubbles in fresh concrete. (Part 2 of 4). *Concrete Construction* **38(2)**, 148–52.

Hover, K. 1993c: Measuring air in fresh and hardened concrete (Part 3 of 4). *Concrete Construction* **38(4)**, 275–8.

Hover, K. 1993d: Specifying air-entrained concrete (Part 4 of 4). *Concrete Construction* **38(5)** 361–7.

Howard, C.J., Hill, R.J. and Sufi, M.A.M. 1988: Quantitative phase analysis by Rietveld analysis of X-ray and neutron diffraction patterns. *Chemistry in Australia* October: 367–9.

Hubbard, F.H., Dhir, R.K. and Ellis, M.S. 1985: Pulverized-fuel ash for concrete: compositional characterisation of United Kingdom pfa. *Cement and Concrete Research* **15(1)**, 185–98.

Hubbard, F.H., McGill, R.J., Dhir, R.K. and Ellis, M.S. 1984: Clay and pyrite transformations during ignition of pulverised coal. *Mineralogical Magazine* **48(347)**, 251–6.

Hudec, P.P. 1991: Common factors affecting alkali reactivity and frost durability of aggregates. *Proceedings of the 5th International Conference on the Durability of Materials and Components* 77–86. 1990, E. & F.N. Spon, Brighton, UK.

Hutchinson, C.S. 1974: *Laboratory handbook of petrographic techniques.* John Wiley, New York.

Ichikawa, M. and Komukai, Y. 1993: Effect of burning conditions and minor components on the colour of Portland cement. *Cement and Concrete Research* **23**, 933–8.

Idorn, G.M. 1961: Studies of disintegrated concrete. Part I. *Progress report, Series N – No. 2, Danish National Institute of Building Research, Committee on Alkali Reactions in Concrete.*

Idorn, G.M. 1961: Studies of disintegrated concrete – Parts I–V. *Danish National Institute of Building Research and the Academy of Technical Sciences. Committee on Alkali Aggregate Reactions in Concrete,* Progress Report N2–N6.

Idorn, G.M. 1965: Thin section photography. *Rilem Bulletin* **(26)**, 111–16.

Idorn, G.M. 1967: *Durability of Concrete Structures in Denmark – a study of field behaviour and microscopic features.* Technical University of Denmark, Copenhagen, Denmark.

Idorn, G.M. 1969: *The durability of concrete.* Technical Paper PCS 46. The Concrete Society, London, UK.

Idorn, G.M. and Thaulow, N. 1983: Examination of 136-years-old Portland cement concrete. *Cement and Concrete Research* **13**, 739–43.

Ignatiev, N. and Chatterji, S. 1992: On the mutual compatibility of mortar and concrete in composite members. *Cement & Concrete Composites* **14**, 179–83.

Insley, H. 1936: Structural characteristics of some constituents of Portland cement clinker. *Journal of Research, National Bureau of Standards* **17**, 353–61.

Insley, H. 1940: Nature of the glass in Portland cement clinker. *Journal of Research, National Bureau of Standards* **25**, 295–300.

Insley, H. and McMurdie, H.F. 1938: Minor constituents in Portland cement clinkers. *Journal of Research, National Bureau of Standards,* **20**, 173.

Insley, H. and Fréchette, van D. 1955: *Microscopy of ceramics and cements – including glasses, slags, and foundry sands.* Academic Press, New York, USA.

Institution of Structural Engineers. 1992: *Structural effects of alkali-silica reaction. Technical guidance on the appraisal of existing structures*, London.

Ivanov, F.M. 1981: Attack of aggressive fluids. *Cement, Concrete, and Aggregates* **3(2)**, 105–7.

Japan Society of Civil Engineers. 1988: Recommendation for design and construction of concrete containing ground granulated blast-furnace slag as an admixture. *Translation from the Concrete Library No. 63.*

Jennings, H.M. and Pratt, P.L. 1980: Use of high voltage electron microscope and gas reaction cell for the microstructural investigation of wet Portland cement. *Journal of Materials* **15**, 250–3.

Jennings, H.M. and Sujata, K. 1992: New experimental techniques for characterizing cement-based materials. *Advanced Cementitious Systems: Mechanisms and Properties, Boston, 1991* **245**, 243–52. Pittsburgh, Materials Research Society.

Jensen, A.D., Chatterji, S., Christensen, P. and Thaulow, N. 1984: Studies in alkali-silica reaction – Part II. Effect of air entrainment on expansion. *Cement and Concrete Research* **14(3)**, 311–14.

Jensen, A.D., Eriksen, K., Chatterji, S., Thaulow, N. and Brandt, I. 1985: Petrographic analysis of concrete. *Danish Building Export Council* 12.

Jensen, V. 1995: Personal Communication.

Jepsen, B.B. and Christensen, P. 1989: Petrographic examination of hardened concrete. *Bulletin of the International Association of Engineering Geology* **39**, 99–103.

Johansen, V., Thaulow, N. and Idorn, G.M. 1994: Expansion reactions in mortar and concrete. *Zement-Kalk-Gips* **47(5)**, E150–55.

John, P.H., Lock, M.A. and Gibbs, M.M. 1977: Two new methods for obtaining water samples from shallow aquifers and littoral sediments. *Journal Environmental Quality* **6(3)**, 322–4.

Johnson, N.C. 1915: The microstructures of concretes. *Proceedings of the American Society of Testing Materials* 15, Part II, 171–213.

Jones, A.E.K. and Clark, L.A. 1996: A review of the Institution of Structural Engineers report 'Structural effects of alkali-silica reaction (1992).' *Proceedings of the 10th International Conference on Alkali-aggregate Reaction in Concrete*, 394–401. Melbourne, Australia.

Jones, F.E. and Tarleton, R.D. 1958: Reactions between aggregates and cement. Part VI. Alkali aggregate interaction. Experience with some forms of rapid and accelerated tests for alkali aggregate reactivity. *Recommended test procedures, National Building Studies Research Paper 25*. London, HMSO.

Jones, J.C., Hawes, R.W.M. and Dyson, J.R. 1966: Semi-automatic preparation of ultra-thin and large-area thin sections. *Transactions of the British Ceramic Society* **65**, 603–12.

Jones, L.W. 1988: *Interference Mechanisms in Waste Stabilisation/Solidification Processes. Literature Review*. Hazardous Waste Engineering Research Laboratory, Office of Research and Development, US, E.P.A., Cincinnati.

Junior, E.M., Munhoz, F.A.C., Junior, J.S. and Placido, W.F. 1990: Characterization and quantitative determination of calcium aluminate clinker phases through reflected light microscopy. *Proceedings of the 12th International Conference on Cement Microscopy*, 1–15.

Jun-yuan, H., Scheetz, B.E. and Roy, D.M. 1984: Hydration of fly ash – Portland cements. *Cement and Concrete Research*, **14(4)**, 505–12.

Kaneda, K., Irie, N., Fujii, H., Yamashita, H., Tazawa, E. and Kawai, K. 1995: A system for observing three dimensional microscopic structures using multiple optical-microscopic images. *Proceedings of the 7th Conference on Cement Microscopy*, 94–101.

Kaplan, M.F. 1960: Effects of incomplete consolidation on compressive and flexural strength, ultrasonic pulse velocity, and dynamic modulus of elasticity of concrete. *Journal American Concrete Institute* **31**, 853–67.

Kapralik, I. and Hanic, F. 1980: Studies of the system CaO-Al$_2$O$_3$-MgO-SiO$_2$ *Transactions and Journal of the British Ceramic Society* **79**, 128–33.

Karol, R.H. 1990: *Chemical Grouting.* 2nd edn Vol. 8, Civil Engineering Series, Marcel Dekker.

Katayama, T. 1992: A critical review of carbonate rock reactions – is their reactivity useful or harmful? *Proceedings of the 9th International Conference on Alkali-Aggregate Reaction in Concrete* 508–18. London.

Katayama, T. and Bragg, D.J. 1996: Alkali-aggregate reaction combined with freeze/thaw in New Foundland, Canada – Petrography using EPMA. *Proceedings of the 10th International Conference on Alkali-aggregate Reaction in Concrete.* Melbourne, Australia.

Katayama, T., St John, D.A. and Futugawa, T. 1989: The petrographic comparison of some volcanic rocks from Japan and New Zealand – potential reactivity related to interstitial glass and silica minerals. *Proceedings of the 8th International Conference on Alkali-Aggregate Reaction, Kyoto,* 537–42.

Kay, K.A., Fookes, P.G. and Pollock, D.J. 1982: Deterioration related to chloride ingress. Concrete in the Middle East. *Viewpont Publication* 23–32. Eyre & Spottiswoode, London.

Kaye, G.W.C. and Laby, T.H. 1986: Tables of Physical and Chemical Constants 15th edn, Longman, London and New York.

Kelly, J.W. 1981: Cracks in concrete. *Concrete Construction* Sept., 725–34.

Kennerley, R.A. 1965: Ettringite formation in dam gallery. *Journal American Concrete Institute* December, 559–76.

Kennerley, R.A. and St John, D.A. 1988: Part I – Reactivity of New Zealand aggregates with cement alkalis. (ed. St John, D.A.) *Alkali-Aggregate Studies in New Zealand, DSIR Chemistry Division, Report No. CD 2390,* 60–1.

Keriené, J. and Kaminskas, A. 1993: Investigation of mineral wool fibre in alkaline solution and humidity using scanning electron microscopy. (eds Lindqvist, J.E. and Nitz, B.) *Proceedings of the 4th Euroseminar on Microscopy applied to Building Materials.* Swedish National Testing and Research Institute, Building Technology SP Report 1993, 15.

Kerrick, D.M. and Hooton, R.D. 1992: ASR of concrete aggregate quarried from a fault zone: results and petrographic interpretation of accelerated mortar bar tests. *Cement and Concrete Research* **22(5)**, 949–60.

Kessler, D.W. 1948: Terrazzo as affected by cleaning materials. *Journal of the American Concrete Institute* **20(1)**, 33–40.

Kilbourn, B.T. 1993: Ceria and cerium products. *Kirk-Othmer Encyclopedia of Chemical Technology, 4th edn* **5**, 728–49. John Wiley, New York.

King, G.A., Walker, G.S. and Ridge, M.J. 1972: Cast gypsum reinforced with glass fibres. *Building Materials and Equipment* **15** (Aug/Sept), 41–3.

Klein, C. and Hurlburt, C.S. 1993: *Manual of Mineralogy* (after James D. Dana). 21st edn. John Wiley & Sons Inc.

Klieger, P. 1980: Something for nothing – almost (1979 Raymond E. Davis Lecture). *Concrete International Design & Construction* **2(1)**, 15–23.

Klug, H.P. and Alexander, L.E. 1974: *X-Ray Diffraction Procedures for Polycrystalline and Amorphous Material.* Wiley Interscience, New York.

Knofel, D.F.E. and Rottger, K.G. 1985: Contribution on the influence of SO$_2$ on cement mortar and concrete *Proceedings of the 7th International Conference on Cement Microscopy,* 333–58.

Knudsen, T. 1992: The chemical shrinkage test; some corrolaries. *Proceedings of the 9th International Conference on Alkali-Aggregate Reaction in Concrete,* 543–9. London.

Kodak, 1970: *Kodak filters for scientific and technical uses.* Eastman Kodak.

Koelliker, E. 1985: Method for microscopic observation of corrosion layers on concrete and mortar. *Cement and Concrete Research* **15(5)**, 909–13.

Kostov, I. 1968: *Mineralogy.* Oliver and Boyd, Edinburgh.

Kostuch, J.A., Walters, G.V. and Jones, T.R. 1993: High performance concretes in corporating metakaolin – A review. *Concrete 2000.* Dundee, Scotland.

Koxholt, P.M. 1985: Iron oxide pigments – the largest colour. *Ind. Miner. London, Suppl.,* May 1985, 22–5.

Kramf, L. and Haksever, A. 1986: Possibilities of assessing the temperatures reached in concrete building elements during fire (ed. T.Z. Harmarthy) *ACI, SP-92,* 115–42.

Kuenning, W.H. 1966: Resistance of Portland cement mortar to chemical attack – a progress report. *Symposium on Effects of Aggressive Fluids on Concrete, Highway Research Record Number 113, (Publication – National Research Council No. 113),* 43–71.

Kuntze, R.A. 1976–1986: Gypsum and plasters. *Cements Research Progress,* Cements Division, American Ceramic Society.

Kupper, D. and Schmid-Meil, W. 1988: Process engineering and raw materials factors affecting the lightness and hue of 'white' clinker and cement. *Zement-Kalks-Gips* **(2)**, 30–2.

Kuzel, H.J. 1990: Reactions of CO_2 with calcium aluminate hydrates in heat treated concrete. *Proceedings of the 12th International Conference Microscopy,* 219–27. Vancouver.

Lange, D.A., Jennings, H.M. and Shah, S.P. 1994: Image analysis techniques for characterization of pore structure of cement-based materials. *Cement and Concrete Research* **24(5)**, 841–53.

Larsen, G. 1961: Microscopic point measuring: A quantitative petrographic method of determining the $Ca(OH)_2$ content of the cement paste in concrete. *Magazine of Concrete Research,* **13(38)**, 71–6.

Larsen, E.S. and Berman, H. 1934: *Microscopic determination of the monopaque minerals.* 2nd edn USGS, Bulletin 848. Washington.

Lauer, K.R. and Slate, F. 1956: Autogeneous healing of cement paste. *Journal American Concrete Institute,* June, 1083–98.

Laurencot, J.L., Pleau, R. and Pigeon, M. 1992: The microscopical determination of air voids' characteristics in hardened concrete: development of an automatic system using image analysis techniques applied to micro-computers. *Proceedings of the 14th International Conference on Cement Microscopy, USA* 259–73.

Lawrence, C.D. 1990: Sulfate attack on concrete. *Magazine of Concrete Research* **42(153)**, 249–64.

Lawrence, C.D. 1993: Laboratory studies of concrete expansion arising from delayed ettringite expansion. *British Cement Association Publication C/16.*

Lawrence, C.D. 1995: Mortar expansions due to delayed ettringite formation – effects of curing period and temperature. *Cement and Concrete Research* **25(4)**, 903–14.

Lawrence, C.D., Dalziel, J.A. and Hobbs, D.W. 1990: *Sulphate attack arising from delayed ettringite formation.* BCA Ref. ITN 12, British Cement Association, Slough, UK.

Laxmi, S., Ahluwalia, S.C. and Chopra, S.K. 1984: Microstructure of coloured clinkers. *Proceedings of the 6th International Conference on Cement Microscopy* 61–77. Albuquerque, USA.

Le Chatelier, H. 1882: *Comptes Rendues hebdomandaires des seances de l'academie des Sciences* **94**, 13.

Le Chatelier, H. 1905: *Experimental researches on constitution of hydraulic mortars* (translated by J.L. Mack), McGraw-Hill, New York, USA.

Lea, F.M. 1968: Some studies on the performance of concrete structures in sulphate-bearing environments. *Performance of Concrete, Symposium in Honour of Thorbergur Thorvaldson* 56–65.

Lea, F.M. 1970: *The chemistry of cement and concrete* 3rd edn. Edward Arnold Lmited, London, UK.

Lee, A.R. 1974: *Blastfurnace and steel slag – production, properties and uses.* Edward Arnold Limited, London, UK.

Leek, D.S. 1997: A study of the effects of chloride and sulphate on the hydration of Portland cement and the corrosion of carbon steel reinforcement using electron-optical techniques and energy dispersive x-ray analysis. PhD thesis, London University (unpublished).

Lees, G. 1964: The measurement of particle shape and its influence in engineering materials. *Journal of British Granite and Whinstone Federation* **4**, 1–22.

Leifeld, G., Munchberg, W. and Stegmaier, W. 1970: Ettringit und Thaumasit als Treibursache in Kalk-Gips-Puzen. *Zement-Kalks-Gips* **(4)**, 174–7.

Lerch, W.L., Ashton, F.W. and Bogue, R.H. 1929: The sulphoaluminates of calcium. *Journal of Research, National Bureau of Standards* **2**, 715–31.

Levitt, M. 1982: *Precast Concrete. Materials, Manufacture, Properties and Usage* 227. Applied Science Publishers, London.

Levitt, M. 1985: Pigments for concrete and mortar (Concrete Society Current Practice Sheet No. 99). *Concrete (Journal of the Concrete Society)* **19(3)**, 21–2.

Lewry, A.J., Asiedu-Dompreh, J., Bigland, D.J. and Butlin, R.N. 1994: The effect of humidity on the dry composition of sulfur dioxide onto calcareous stones. *Construction and Building Materials* **8(2)**, 97–100.

Lord, G.W. and Willis, T.F. 1951: Calculation of air bubble size distribution from results of a Rosiwal traverse of aerated concrete. *ASTM Bulletin* October, 56–61.

Loveland, R.P. 1970: *Photomicrography. A comprehensive treatise. 1 & 2.* Wiley, New York.

Ludwig, U. 1980: Durability of cement mortars and concretes. *Durability of Building Materials and Components,* ASTM STP 691: 269–81.

Ludwig, U. and Mehr, S. 1986: Destruction of historical buildings by the formation of ettringite or thaumasite. *8th International Congress on the Chemistry of Cement, Rio de Janeiro, V* 181–8.

Ludwig, U. and Rudiger, I. 1993: Quantitative determination of ettringite in cement pastes, mortars and concretes. *Zement-Kalk-Gips* **46(5)**, E153–156.

Lukas, W. 1975: Betonzerstörung durch SO_3 – angriff unter bildung von thaumasite und woodfordite. *Cement and Concrete Research* **5(5)**, 503–18.

Luke, K. and Glasser, F.P. 1987: Selective dissolution of hydrated blast furnace slag cements. *Cement and Concrete Research* **17**, 273–82.

Lumley, J.S. 1989: Synthetic cristobalite as a reference reactive aggregate. *Proceedings of the 8th International Conference on Alkali-Aggregate Reaction* 561–6. Kyoto, Japan.

Lynsdale, C.J. and Cabrera, J.G. 1989: Coloured Concrete – a State of the Art Review. *Concrete* **23(7)**, 29–34.

McCarthy, G.J., Swanson, K.D., Keller, L.P. and Blatter, W.C. 1984: Mineralogy of western fly ash. *Cement and Concrete Research* **14(4)**, 471–8.

McClaren, F.R. 1984: Sulfide and corrosion prediction and control. *American Concrete Pipe Association, Vienna, USA* 90.

McCrone, W.C. 1983: Trick photomicrography. *Microscope* **31**, 245–61.

McCrone, W.C. and Delly, J.G. 1973: *The particle atlas, 1 & 2.* Ann Arbor Science Publishers, USA.

MacInnis, C. and Racic, D. 1986: The effect of superplasticizers on the entrained air void system in concrete. *Cement and Concrete Research* **16(3)**, 345–52.

McIver, J.R. and Davis, D.E. 1985: A rapid method for the detection and semi-quantitative assessment of milled granulated blastfurnace slag in hardened concrete. *Cement and Concrete Research* **15(3)**, 545–8.

McKenzie, L. 1994: The Microscope: a method and tool to study lime burning and lime quality. *World Cement* **25(11)**, 56–9.

McKenzie, L. 1995: Lime particle size: Its effects on feed and products of a hydrator circuit. A plant case study. *Proceedings of the 17th Conference on Cement Microscopy*, 402–23.

Mackenzie, R.C. 1970: (ed.) *Differential Thermal Analysis, Vol. 1, Fundamental Aspects.* 1972: *Vol. 2, Applications*, Academic Press, London.

Mackenzie, W.S., Donaldson, C.H. and Guildford, C. 1991: *Atlas of igneous rocks and their textures*. Longman Scientific, UK.

Mackenzie, W.S. and Guilford, C. 1980: *Atlas of rock-forming minerals in thin-section.* Longman, London.

Mailvaganam, N.P. 1984: Miscellaneous admixtures. (ed. V.S. Ramachandran) *Concrete Admixtures Handbook* 480–557. Noyes Publications, USA.

Majumdar, A.J. and Singh, B. 1992: Properties of some blended high-alumina cements. *Cement and Concrete Research* **22(6)**, 1101–14.

Majumdar, A.J., Singh, B. and Edmonds, R.N. 1990a: Hydration of mixtures of 'Ciment Fondu' aluminous cement and granulated blast furnace slag. *Cement and Concrete Research* **20(2)**, 197–208.

Majumdar, A.J., Singh, B. and Edmonds, R.N. 1990b: Blended high-alumina cements. *Conferences on Advances in Cementitious Materials, Ceramic Transactions* **16**, 661–78. The American Ceramic Society.

Male, P. 1989: Properties of microsilica concrete – An overview of microsilica concrete in the UK. *Concrete (Journal of the Concrete Society)* **23(8)**, 31–4.

Malhotra, H.L. 1956: The effect of temperature on the compressive strength of concrete. *Cement and Concrete Research* **8(23)**, 85–94.

Malhotra, V.M. 1976: Testing of Hardened Concrete: Non-Destructive Methods. Monograph no. 2. American Concrete Institute.

Malhotra, V.M. 1976: No-fines concrete – its properties and applications. *ACI Journal* November, 628–44.

Malhotra, V.M. (ed.) 1987: *Supplementary cementing materials for concrete.* CANMET (Canada Centre for Mineral and Energy Technology) SP 86–8E, Canadian Government Publishing Centre, Ottawa, Canada.

Malhotra, V.M. and Carette, G.G. 1983: Silica fume concrete – properties, applications and limitations *Concrete International Design & Construction* **5(5)**, 40–6.

Malhotra, V.M. and Hemmings, R.T. 1995: Blended cements in North America – A review. *Cement and Concrete Composites* **17**, 25–35.

Malinowski, R. 1982: Durable ancient mortars and concretes. *Nordic Concrete Research*, The Nordic Concrete Federation, **19(1)**, 1–22.

Malinowski, R. and Garfinkel, Y. 1991: Prehistory of concrete. *Concrete International* **13(3)**, 62–8.

Malisch, W.R. 1990: A contractors' guide to air-entraining and chemical admixtures – their effect on concrete properties that affect job progress. *Concrete Construction* **35(3)**, 279–86.

Manns, W. and Wesche, K. 1968: Variation in strength of mortars made of different cements due to carbonation. *Proceedings of the 5th International Symposium on the Chemistry of Cements* **III**, 385–93. Tokyo, Cement Association of Japan.

Marchand, J., Sellevoid, E.J. and Pigeon, M. 1994: The deicer salt scaling deterioration of concrete – An overview (ed. V.M. Malhotra) *Durability of Concrete. Third International Conference, Nice, France, ACI SP-145*, 1–46.

Marfil, S.A. and Maiza, P.J. 1993: Zeolite crystallization in Portland cement concretes.

Cement and Concrete Research **23(5)**, 1283–8.

Marten, A., Knofel, D. and Strunge, J. 1994: Quantitative structure analysis via image analysis. A suitable tool for quality control of Portland cement clinker. *World Cement* **25(8)**, 49–52.

Martineau, F., Guedon-Dubied, J.S. and Larive, C. 1996: Evaluation of the relationships between swelling, cracking, development of gels. *Proceedings of the 10th International Conference on Alkali-Aggregate Reaction in Concrete*, 798–805. Melbourne, Australia.

Marusin, S.L. 1994: A simple treatment to distinguish alkali-silica gel from delayed ettringite formations in concrete. *Magazine of Concrete Research* **46(168)**, 163–6.

Marusin, S.L. 1995: SEM case studies of discoloured surfaces of various building materials exposed to the weather. *Proceedings of the 17th International Conference on Cement Microscopy*, 18–28.

Massazza, F. 1985: Concrete resistance to seawater and marine environment. *Il Cemento* **82**, 3–26.

Massazza, F. 1993: Pozzolanic cements. *Cements and Concrete Composites* **15**, 185–214.

Massazza, F. and Costa, U. 1986: Bond, paste-aggregate, paste-reinforcement and paste-fibres. *8th International Congress Chemistry of Cements, Rio de Janeiro* **1**, 159–80.

Materials Research Society Symposium Proceedings. 1991: *Advanced Cementitious Systems: Mechanisms and Properties* **245**, 105–49.

Mather, K. 1952: Applications of light microscopy in concrete research. *Symposium on Light Microscopy*, ASTM STP 143, 51–70.

Mather, K. 1966: Petrographic examination. Significance of Tests and Properties of Concrete and Concrete-Making Materials, ASTM STP No. 169-A, 125–43.

Mather, K. 1982: Current research in sulfate resistance at the Waterways Experiment Station. *George Verbeck Symposium on Sulfate Resistant Concrete* ACI-SP77, 63–74.

Mather, K., Buck, A.D. and Luke, W.I. 1964: Alkali-silica and alkali-carbonate reactivity of some aggregates from South Dakota, Kansas, and Missouri. *Highway Research Board* **4**, 72–109.

Mathieu, C. 1996: Principles and applications of the variable pressure scanning electron microscope. *Microscopy and Analysis* **(55)**.

Matthews, J.D. 1992: The resistance of PFA concretes to acid groundwaters. *9th International Conference on the Chemistry of Cement, New Delhi* **V**, 355–62.

Mayfield, B. 1990: The quantitative evaluation of the water/cement ratio using fluorescence microscopy. *Magazine of Concrete Research* **42(150)**, 45–9.

Mays, G. (ed.) 1992: *Durability of concrete structures*. Spon, London.

Mays, G.C. and Barnes, R.A. 1991: The performance of lightweight aggregate concrete structures in service. *The Structural Engineer* **69(20)**, 351–61.

Mehlmann, M. 1989: Current state of research on renderings and mortars. *Zement Kalks Gips* **(3)**, 70–2, translation (1/89).

Mehta, P.K. 1985: Influence of fly ash characteristics on the strength of Portland-fly ash mixtures. *Cement and Concrete Research* **15(4)**, 669–74.

Mehta, P.K. 1987: Natural pozzolans. In: Malhotra, V.M. (ed.), *Supplementary cementing materials for concrete*, Chapter 1, 1–33, CANMET (Canada Centre for Mineral and Energy Technology), Canadian Government Publishing Centre, Ottawa, Canada.

Mehta, P.K. 1991: Durability of concrete – fifty years' progress? *Durabiity of Concrete, Second International Conference*, Montreal, Canada, ACI SP-126, 1–31.

Mehta, P.K. 1993: Sulfate attack on concrete – A critical review. J. Skalny (ed.) *Materials Science of Concrete III* 105–30. American Ceramic Society.

Mehta, P.K., Schiessi, P. and Raupach, M. 1992: Performance and durability of concrete systems. *9th International Congress on the Chemistry of Cement, New Delhi* **1**, 571–659.

Meyer, A. 1968: Investigations on the carbonation of concrete. *Proceedings 5th International Symposium Chemistry of Cement* 3, 394–401.

Meyer, A. 1987: The importance of the surface layer for the durability of concrete structures. *Concrete Durability, Katherine and Bryant Mather International Conference, ACI SP-100* 1, 49–61.

Middleton, A.P. 1979: The use of asbestos and asbestos-free substitutes in buildings. (eds L. Michaels and S.S. Chissick). *Asbestos, Volume 1, Properties, Applications, and Hazards* 306–34. Wiley Interscience, Chichester.

Midgley, H.G. 1957: A compilation of X-ray powder diffraction data of cement minerals. *Magazine of Concrete Research* **9(25)**, 17–24.

Midgley, H.G. 1958: The staining of concrete by pyrite. *Magazine of Concrete Research* **10(29)**, 75–8.

Midgley, H.G. 1964: The formation and phase composition of Portland cement clinker. In: Taylor, H.F.W. (ed.), *The chemistry of cements*, Vol. 1, Chapter 3, 89–130. Academic Press, London, UK.

Midgley, H.G. 1967: The mineralogy of set high-alumina cement. *Transactions of the British Ceramic Society* **66(4)**, 161–87. (Also Building Research Station, Current Paper 19/68, March 1968).

Midgley, H.G. 1979: The determination of calcium hydroxide in set Portland cements. *Cement and Concrete Research* **9(1)**, 77–82.

Midgley, H.G. and Midgley, A. 1975: The conversion of high alumina cement. *Magazine of Concrete Research* **27(91)**, 59–77.

Midgley, H.G. and Pettifer, K. 1971: The micro-structure of hydrated super sulfated cement. *Cement and Concrete Research* **1**, 101–4.

Midgley, H.G. and Rosaman, D. 1960: The composition of ettringite in set Portland cement. In: *Chemistry of Cement, Proceedings of the 4th International Symposium, Washington, 205*, 259–62. National Bureau of Standards Monograph 43, US Department of Commerce.

Mielenz, R.C. 1962: Petography applied to Portland-cement concrete. In: Flurh, T., Legget, R.F. (eds), *Reviews in Engineering Geology*, **1**, 1–38. The Geological Society of America, USA.

Mielenz, R.C. 1983: Mineral admixtures – history and background. *Concrete International* **5(8)**, 34–42.

Milestone, N.B. 1984: Identification of concrete admixtures by differential thermal analysis in oxygen. *Cement and Concrete Research* **14(2)**.

Milestone, N.B., St John, D.A., Abbott, J.H. and Aldridge, L.P. 1986: CO_2 corrosion of geothermal cement grouts. *Proceedings of the 8th International Congress on the Chemistry of Cement, Rio de Janeiro* **V**, 141–4.

Minato, H. 1968: Mineral composition of blastfurnace slag. Supplementary Paper IV–110. *Proceedings of the 5th International Symposium Chemistry of Cement* 263–9. Tokyo.

Mindess, S. and Gilley, J.C. 1973: The staining of concrete by an alkali-aggregate reaction. *Cement and Concrete Research* **3(6)**, 821–8.

Mindess, S. and Young, F.J. 1981: *Concrete* 83. Prentice-Hall, USA.

Moir, G.K. and Kelham, S. 1989: Performance of limestone-filled cements: *Report of Joint BRE/BCA/Cement Industry Working Party*. Paper No. 7. Durability. Seminar held at the Building Research Establishment, 28 Novewmber.

Mollring, F.K. 1976: *Microscopy from the very beginning*. Carl Zeiss, Germany.

Monteiro, P.J.M. and Mehta, P.K. 1986: Interaction between carbonate rock and cement paste. *Cement and Concrete Research* **16(2)**, 127–34.

Montgomery, D.G., Hughes, D.C. and Williams, R.I.T. 1981: Fly ash in concrete – a microstructural study. *Cement and Concrete Research* **11(4)**, 591–603.

Moorhead, D.R. 1986: Cementation by the carbonation of hydrated lime. *Cement and Concrete Research* **16(5)**, 700–8.

Moorhead, D.R. and Morand, P. 1975: Some observations of the carbonation of portlandite $(Ca(OH)_2)$ and its relationship to the carbonation hardening of lime products. In: *Science and research of silicate chemistry*, Proceedings of the 3rd International Congress.

Moreland, G.C. 1968: Preparation of polished thin sections. *American Minerologist* **53**, 2070–4.

Moriconi, G., Castellano, M.G. and Collepardi, M. 1994: Mortar deterioration of the masonry walls in historic buildings. A case history: Vanvitelli's Mole in Ancona. *Materials and Structures* **27(171)**, 408–14.

Morrey, C. (ed.) 1991: *Department of Environment Digest of Environmental Protection and Waste Statistics, No. 14*. HMSO, London.

Muhamaed, M.N., Barnes, P., Fentiman, C.H., Hausermann, D., Pollmann, H. and Rashid, S. 1993: A time resolved synchrotron energy dispersive diffraction study of the dynamic aspects of the synthesis of ettringite during minepacking. *Cement and Concrete Research* **23**, 267–72.

Natesaiyer, K.C. and Hover, K.C. 1988: In situ identification of ASR products in concrete. *Cement and Concrete Research* **18**, 455–63.

Natesaiyer, K.C. and Hover, K.C. 1989: Further study of an in-situ identification method for alkali-silica reaction products in concrete. *Cement and Concrete Research* **19**, 770–8.

Neck, U. 1988: Auswirkungen de Warmebehandlung auf Festigkeit und Dauerhaftigkeit von Beton. *Beton*, **V(38)**, 488–93.

Negro, A. 1985: Modele mathematique de l'expansion de l'ettringite. *Ciments, Betons, Platres, Chaux* **(756)**, 319–27.

Nepper-Christensen, P., Kristensen, B.W. and Rasmussen, T.H. 1994: Long-term durability of special high strength concretes. *Durability of Concrete. Third International Conference, Nice*, ACI SP-145: 173–90.

Neville, A.M. 1975: *High alumina cement*, Lancaster, The Construction Press.

Neville, A.M. 1988: *Properties of Concrete*, 3rd edn. UK Longman Scientific and Technical.

Neville, A.M. and Brooks, J.J. 1987: *Concrete Technology*. Longman Group Limited, Harlow, UK.

Newman, E.S. 1955: A study of the system magnesium oxide-magnesium chloride-water and the heat of formation of magnesium oxychloride. *Journal of Research of the National Bureau of Standards* **54(6)**, 347–55.

New Zealand Portland Cement Association. 1962: Design of concrete floors for commercial and industrial use. *N.Z. Portland Cement Association Technical Bulletin ST. 26.*

Nielson, A. 1994: Development of alkali silica reactions on concrete structures with time. *Cement and Concrete Research* **24(1)**, 83–5.

Nielsen, J. 1966: Investigation of resistance of cement paste to sulphate attack. *Symposium on Effects of Aggressive Fluids on Concrete, Highway Research Record No. 113, (Publication – National Research Council No. 1335)* 114–17.

Nishikawa, T., Suzuki, K. and Ito, S. 1992: Decomposition of synthesized ettringite by carbonation. *Cement and Concrete Research* **22(1)**, 6–14.

Nixon, P.J., Collins, R.J. and Rayment, P.L. 1979: The concentration of alkalies by moisture migration in concrete – a factor influencing alkali aggregate reaction. *Cement and Concrete Research* **9(4)**, 417–23.

Nixon, P.J. and Sims, I. 1992: RILEM TC106 alkali aggregate reaction – accelerated tests. Interim report and summary of survey of national specifications. *Proceedings of the 9th International Conference on Alkali-Aggregate Reaction in Concrete*, 731–8. London.

Norrish, K. and Chappell, B. 1977: X-ray fluorescence spectrometry In: (ed. J. Zussman) *Physical Methods in Determinative Mineralogy* 2nd edn, 201–72. Academic Press, London.

Norrish, K. and Hutton, J.T. 1969: An accurate X-ray spectrographic method for the analysis of a wide range of geological samples. *Geochemica et Cosmochimica Acta* **33**, 431–53.

Novak, G.A. and Colville, A.A. 1989: Efflorescent mineral assemblages associated with cracked and degraded residential concrete foundations in Southern California. *Cement and Concrete Research* **19(1)**, 1–6.

Novokshchenov, V. 1987: Investigation of concrete deterioration due to sulfate attack – a case history. *Durability of Concrete*, ACI SP-100: 1979–2006.

Nurse, R.W. and Midgley, H.G. 1951: The mineralogy of blastfurnace slag. *Silicates Industries* **16(7)**, 211–17.

Oberholster, R.E. and Davies, G. 1986: An accelerated method for testing the potential alkali reactivity of siliceous aggregates. *Cement and Concrete Research* **16(2)**, 181–9.

Oberholster, R.E., Du Toit, P. and Pretorius, J.L. 1984: Deterioration of concrete containing a carbonaceous sulphide-bearing aggregate. *Proceedings of the 6th International Conference on Cement Microscopy*, 360–73.

Oberholster, R.E., Maree, H. and Brand, J.H.B. 1992: Cracked prestressed concrete railway sleepers: alkali-silica reaction or delayed ettringite formation. *Proceedings of the 9th International Conference on Alkali-aggregate Reaction in Concrete*, London, 739–49.

Oberste-Padtberg, R. 1985: Degradation of cements by magnesium brines. *Proceedings of the 7th International Conference on Cement Microscopy*, 24–36.

Odler, I. and Chen, Y. 1995: Effect of cement composition on the expansion of heat-cured cement pastes. *Cement and Concrete Research* **25(4)**, 853–62.

Odler, I. and Gasser, M. 1988: Mechanism of sulfate expansion in hydrated Portland cement. *Journal of the American Ceramic Society* **71(11)**, 1015–20.

Ohgishi, S. and Ono, H. 1983: Study to estimate the depth of neutralization on concrete members. *CAJ Review*, 168–70.

O'Neill, R.C. 1992: Identification and effects of sulfate materials in hardened concrete. *Proceedings of the 14th International Conference on Cement Microscopy*, 198–208.

Opoczky, L. and Péntek, L. 1975: Investigation of the 'corrosion' of asbestos cement fibres in asbestos cement sheets weathered for long times (ed. A. Neville) *Rilem Symposium 1975, Fibre Reinforced Cement and Concrete*, 269–76. Construction Press, UK.

Osborne, G.J. 1991: The sulphate resistance of Portland blastfurnace slag cement concretes. *Durability of Concrete*, ACI SP-126, 1047–71.

Owens, P.L. 1980: Pulverised-fuel ash, Parts 1 & 2 (Concrete Society Current Practice Sheets 54 & 57), *Concrete (Journal of the Concrete Society)* **14**, **(7 & 10)**, 33–4, 35–6.

Owens, P.L. 1982: Pulverised-fuel ash, Parts 3 & 4 (Concrete Society Current Practice Sheets 75 & 76), *Concrete (Journal of the Concrete Society)*, **16**, **(6 & 7)**, 39–41.

Ozyildirim, C. and Halstead, W.J. 1994: Improved concrete quality with combinations of fly ash and silica fume. *ACI Materials Journal* **91(6)**, 587–94. Title No. 91–M59.

Palmer, D. 1992: The diagnosis of Alkali-Silica Reaction. Report of a Working Party. British Cement Association, Wexham Springs, Slough, UK.

Papadoulos, G. and Suprenant, B. 1988: Identifying fly ash, slag and silica fume in blended cements and hardened concrete. *Proceedings of the 10th International Conference on Cement Microscopy*, 344–56.

Parker, D.G. 1985: Microsilica concrete, Part 1: The material (Concrete Society Current Practice Sheet No. 104). *Concrete (Journal of the Concrete Society)* **19(10)**, 21–2.

Parker, D.G. 1986: Microsilica concrete, Part 2: In use (Concrete Society Current Practice Sheet No. 110). *Concrete (Journal of the Concrete Society)*, **20(3)**, 19–21.

Parker, T.W. 1952: The constitution of aluminous cement. *Proceedings of the 3rd International Symposium Chemistry of Cements*, 485–529. London.

Parker, T.W. and Hirst, P. 1935: Preparation and examination of thin sections of set cement. *Cement and Cement Manufacture* **8**(10), 235–41.

Parrott, L.J. 1987: *A review of carbonation in reinforced concrete*. UK, Cement and Concrete Association, 42 and appendix.

Parrott, L.J. 1990: Damage caused by carbonation of reinforced concrete. *Materials and Structures* **23**(135), 230–4.

Parrott, L.J. 1991: Assessing carbonation in concrete structures. *Proceedings of the 5th International Conference on the Durability of Building Materials and Components, Brighton, 1990*, 575–85. E. & F.N. Spon.

Parrott, 1991: Factors influencing relative humidity in concrete. *Magazine of Concrete Research* **43**(154), 45–52.

Parrott, L.J. 1992: Water absorption in cover concrete. *Materials and Structures*, **25**(149), 284–92.

Parrott, L.J. 1994: Moisture conditioning and transport properties of concrete test specimens. *Materials and Structures* **27**, 460–8.

Patel, H.H., Bland, C.H. and Poole, A.B. 1995. The microstructure of concrete cured at elevated temperatures. *Cement and Concrete Research* **25**(3), 485–90.

Patzelt, W.J. 1986: *Polarized light microscopy*, Ernst Leitz, Germany.

Pavlik, V. 1994: Corrosion of hardened cement paste by acetic and nitric acids. Part II: Formation of chemical composition of the corrosion products layer. *Cement and Concrete Research* **24**(8), 1495–508.

Petrov, I. and Schlegel, E. 1994: Application of automatic image analysis for the investigation of autoclaved aerated concrete. *Cement and Concrete Research* **24**(5), 830–40.

Petruk, W. (ed.) 1989: A short course on image analysis applied to mineral and earth sciences. *Mineralogical Association of Canada* **16**, May.

Philleo, R.E. 1983: A method for analyzing void distribution in air-entrained concrete. *Cement, Concrete, and Aggregates CCAGDP* **5**(2), 128–30.

Piasta, J., Piasta, W. and Sawicz, Z. 1987: Sulfate resistance of mortars and concretes with calciferous aggregates. *Durability of Concrete*, ACI SP-100: 2153–2170.

Piasta, J., Sawicz, Z. and Piasta, W.G. 1989: Durability of high alumina cement pastes with mineral additions in water sulfate environment. *Cement and Concrete Research* **19**(1), 103–13.

Pigeon, M. and Pleau, R. 1995: *Durability of Concrete in Cold Climates*. E. & F.N. Spon, London.

Pippard, W.R. 1938: Clacium sulphate plasters. *DSIR Building Research Bulletin No. 13*. HMSO, London.

Pistilli, M.F. 1983: Air-void parameters developed by air-entraining admixtures, as influenced by soluble alkalies from fly ash and Portland cement. *ACI Journal*, May–June, 217–22. (Title No. 80–22.)

Pitard, F.F. 1989: *Pierre Gy's Sampling Theory and Sampling Practice*. 2 volumes. CRC Press, USA.

Placido, F. 1980: Thermoluminescence test for fire-damaged concrete. *Magazine of Concrete Research* **32**(11), 112–16.

Poellmann, H., Kuzel, H.J. and Wenda, R. 1990: Solid solution of ettringites. *Cement and Concrete Research*, **20**, 941–7.

Pomeroy, R.D. 1977: *The Problem of Hydrogen Sulphide in Sewers*. Clay Pipe Development Association.

Poole, A.B. 1981: Alkali carbonate reactions in concrete. *Proceedings of the 6th International*

Conference on Alkali-Aggregate Reaction. Cape Town: S252/13, 1–13.

Poole, A.B. 1992: Introduction to alkali-aggregate reaction in concrete. (ed. R.N. Swamy.) *The alkali-silica reaction in concrete*. Blackie, Glasgow.

Poole, A.B., McLachlan, A. and Ellis, D.J. 1988: A simple staining technique for the identification of alkali-silica gel in concrete and aggregate. *Cement and Concrete Research* **18**, 116–20.

Poole, A.B., Patel, H.H. and Shiekh, V. 1996: Alkali silica and ettringite expansions in 'steam-cured' concretes. In: Shayan, A. (ed.), *Alkali-aggregate reaction in concrete, Proceedings of the 10th International Conference*. Melbourne, Australia, 943–8.

Poole, A.B. and Thomas, A. 1975: A staining technique for the identification of sulphates in aggregates and concrete. *Mineralogical Magazine* **40**, 315–16.

Popovics, S. 1987: A classification of the deterioration of concrete based on mechanism. *Concrete Durability – Katherine and Bryant Mather International Conference*, ACI SP-100, **1**, 131–42.

Portilla, M. 1989: Enthalpy of dehydration of Portland and pozzolanic cements. *Cement and Concrete Research* **19**, 319–26.

Portland Cement Association (PCA). 1981: Effects of substances on concrete and protective treatments. *Portland Cement Association – Concrete Information*, Skokie, USA.

Poulsen, E. 1958: Preparation of samples for microscopic investigation. *The Danish National Institute of Building Research and the Academy of Technical Sciences. Committee on Alkali Reactions in Concrete. Progress Report M1.*

Pourbaix, M. 1974: Applications of electrochemistry in corrosion science and practice. *Corrosion Science* **14(1)**, 25–82.

Powers, T.C. 1949a: The non-evaporable water content of hardened Portland cement paste; its significance for concrete research and its method of determination, *ASTM Bulletin* **158**, 68–76.

Powers, T.C. 1949b: Air requirement of frost-resistant concrete. *Proceedings of the Highway Research Board* **29**, 184–211.

Powers, T.C. 1956: Resistance to weathering-freezing and thawing. *Significance of Tests and Properties of Concrete and Concrete Aggregates*, ASTM STP No. 169, 182–7.

Powers, T.C. and Brownyard, T.L. 1948: *Studies of the physical properties of hardened Portland cement paste*. Bulletin 22, Research Laboratories of the Portland Cement Association, Chicago, USA. (reprinted from *Journal of the American Concrete Institute*, 1947, **43**, 101–992).

Powers, T.C., Copeland, L.E. and Mann, H.M. 1958: Capillary continuity and discontinuity in cement pastes. *Journal of the Portland Cement Association Research and Development Laboratories* **1(2)**, 38–48.

Powers-Couche, L.J. 1992: Microscopical examination of a slag cement concrete. *Proceedings of the 14th International Conference on Cement Microscopy*, 256–8.

Price, R. and Caveny, B. 1992: FT-IR microscope spectroscopy study of cement crystal phases. *Proceedings of the 14th International Conference on Cement Microscopy, USA*, 115–33.

Prince, W., Perami, R. and Espagne, M. 1994: Une noevelle approche du mecanisme de la reaction alcali-carbonate. *Cement and Concrete Research* **24(1)**, 62–72.

Pullar-Strecker, P. 1987: *Corrosion Damaged Concrete: assessment and repair* 97. Butterworths, London.

Quincke, J.E. 1967: Broadening the particle size range of natural sands for calcium silicate manufacture. *Symposium Autoclaved Calcium Silicate Building Products, London 1965* 11–17. Society of Chemical Industry.

Raask, E. 1968: Cenospheres in pulverized-fuel ash, *Journal of the Institute of Fuel* **41**,

339–44.

Rajhans, G.S. and Sullivan, J.L. 1981: *Asbestos sampling and analysis*, Ann Arbor Science.

Ramachandran, V.S. 1969: *Applications of differential thermal analysis in cement chemistry.* Chemical Publishing Co. New York.

Ramachandran, V.S. (ed.) 1984: *Concrete admixtures handbook: properties, science and technology.* Noyes Publications, New Jersey, USA.

Rasmachandran, V.S. 1988: Thermal analyses of cement compounds hydrated in the presence of calcium carbonate. *Thermochimica Acta* **(127)**, 385–94.

Ramachandran, V.S. (ed.) 1995: *Concrete admixtures handbook: properties, science and technology*, 2nd edn. Noyes Publications, New Jersey, USA.

Ramachandran, V.S., Feldman, R.F. and Beaudoin, J.J. 1981: *Concrete science. Treatise on current research.* 380–7. Heyden, London.

Randolph, J.L. 1991: Thin section identification of various admixtures and their effects on the cement paste morphology in hardened concrete. *Proceedings of the 13th International Conference on Cement Microscopy* 281–93. Florida, USA.

Rankin, G.A. and Wright, F.E. 1915: The ternary system $CaO-Al_2O_3-SiO_2$ (optical study by F.E. Wright). *American Journal of Science* **39(4)**, 1–79.

Rashid, S., Barnes, P., Bensted, J. and Turrillas, X. 1994: Conversion of calcium aluminate cement hydrates re-examined with synchrotron energy-dispersive diffraction. *J. Mat. Sci. Lett.* **13**, 1232–4.

Rashid, S., Barnes, P. and Turrillas, X. 1992: The rapid conversion of high alumina cement hydrates, as revealed by synchroton energy-dispersive X-ray diffraction. *Adv. Cement Res.* **4(14)**, 61–7.

Ravenscroft, P. 1982: Determining the degree of hydration of hardened concrete, *John Laing Forum*, 4.

Rayment, D.L. and Majumdar, A.J. 1982: The composition of C-S-H phases in Portland cement pastes. *Cement and Concrete Research* **12(6)**, 753–64.

Rayment, P.L. 1992: The relationship between flint microstructure and alkali-silica reaction. *Proceedings of the 9th International Conference on Alkali-Aggregate Reaction in Concrete*, London, 843–50.

Rayment, P.L. and Haynes, C.A. 1996: The alkali-silica reactivity of flint aggregates. *Proceedings of the 10th International Conference on Alkali-Aggregate Reaction in Concrete* 750–57. Melbourne, Australia.

Rayment, P.L., Pettifer, K. and Hardcastle, J. 1990: The alkali-silica reactivity of British concreting sands, gravels and volcanic rocks. Department: Transport Contractor Report 218.

Reardon, E.J. 1990: An ion interaction model for the determination of chemical equilibria in cement/water systems. *Cement and Concrete Research* **20(2)**, 175–92.

Redler, L. 1991: Quantitative X-ray diffraction analysis of high alumina cements. *Cement and Concrete Research* **21(5)**, 873–84.

Reed, F.S. and Mergner, J.L. 1953: Preparation of rock thin sections. *American Mineralogist* **53**, 1184–203.

Reffner, J.A., Coates, J.P. and Messerschmidt, R.G. 1987: Chemical microscopy with FTIR microspectrometry. *International Laboratory* July/August, 18–25.

Regourd-Moranville, M. 1989: Products of reaction and petrographic examination. *Proceedings of the 8th International Conference on Alkali-Aggreate Reaction* 445–56. Kyoto, Japan.

Reid, W.P. 1969: Mineral Staining Tests. *Colorado School of Mines Mineral Industries Buletin* **12(3)**, 1–20.

Richardson, I.G. and Groves, G.W. 1990: The microstructure of blastfurnace slag/high alumina cement pastes. In: *Calcium Aluminate Cements, Proceedings of a Symposium,* Queen Mary & Westfield College, University of London, 282–93. QMW, London, UK.

Richardson, J.G. 1991: *Quality in Precast Concrete: Design – Production – Supervision.* Longman Scientific & Technical, England.

Ridge, M.J. and Berekta, J. 1969: Calcium sulphate hemihydrate and its hydration. *Reviews of Pure and Applied Chemistry* **19(17)**, 17–44.

Rigby, G.R. 1948: *The Thin-Section Mineralogy of Ceramic Materials.* British Refractories Research Association.

RILEM. 1988: Siliceous by-products for use in concrete – final report (RILEM technical report – RILEM Committee 73-SBC), *Materials and Structures* **21(121)**, 69–80.

RILEM Committee 104-DC. 1994: Draft recommendations for damage classification on concrete structures. *Materials and Structures* **27(170)**, 362–9.

Riley, M.A. 1991: Possible new method for the assessment of fire-damaged concrete. *Magazine of Concrete Research* **43(155)**, 87–92.

Rixom, M.R. (ed.) 1977: *Concrete admixtures: uses and applications.* Construction Press, Longman, New York, USA.

Rixom, M.R. 1978: *Chemical Admixtures for Concrete* 234. E. & F.N. Spon, London.

Rixom, M.R. and Mailvaganam, N.P. 1986: Chemical Admixtures for Concrete. 2nd edn. E. & F.N. Spon, UK.

Roberts, L.R. and Scali, M.J. 1984: Factors affecting image analysis for measurement of air content in hardened concrete. *Proceedings of the 6th International Conference on Cement Microscopy, USA,* 402–19.

Roberts, M.H. 1956: The constitution of hydraulic lime. *Cement and Lime Manufacture* **29(3)**, 27–36.

Roberts, M.H. 1986: Determination of the chloride and cement contents of hardened concrete. *BRE Information Paper,* IP 21/86, 4.

Roberts, M.H. and Jaffrey, S.A.M.T. 1974: *Rapid chemical test for the detection of high-alumina cement concrete.* BRE Information Sheet IS 15/74, Building Research Establishment, Watford, UK.

Robinson, G., Vaughan, F. and Hordell, G. 1992: Tiling. *Specification 92,* London, MBC Architectural Press and Building Publications.

Robson, T.D. 1962: *High-alumina cements and concretes.* Contractors Record Limited, London, UK.

Robson, T.D. 1964: Aluminous cement and refractory castables. In: Taylor, H.F.W. (ed.), *The chemistry of cements,* Vol. 2, Ch. 12, 3–35. Academic Press, London, UK.

Rogers, C.A. and Hooton, R.D. 1989: Leaching of alkalies in alkali-aggregate reaction testing. *Proceedings of the 8th International Conference on Alkali-Aggregate Reaction, Kyoto* 327–32.

Rogers, D.E. 1973: Determination of carbon dioxide and aggressive carbon dioxide in waters. *New Zealand Journal of Science* **16**, 875–983.

Roper, H., Cox, J.E. and Erlin, B. 1964: Petrographic studies on concrete containing shrinking aggregates. *Journal of the PCA Research and Development Laboratories* **6(3)**, 2–18.

Rosiwal, A. 1898: On geometric rock analysis. A simple surface measurement to determine the quantitative content of mineral constituents of a stony aggregate. *Verhandl. K.K. Geol. Recih. Wien* 5–6, 143.

Roy, D.M. 1988: Cementitious materials in nuclear waste management. *Cements Research Progress,* 261–91.

Rudnai, G. 1963: *Lightweight Concretes.* Publishing House of the Hungarian Academy of Sciences.

Russ, J.C. 1990: *Computer-Assisted Microscopy. The Measurement and Analysis of Images.* Plenum Press, New York.

Russell, P. 1983: *Efflorescence and the discolouration of concrete.* Viewpoint Publications, UK.

Ryder, J.F. 1975: Applications of fibre cement (ed. A. Neville), *Rilem Symposium 1975, Fibre Reinforced Cement and Concrete*, 23–35. Construction Press, UK.

Sadtler Research Laboratories. *Sadtler Standard Spectra, Sadtler Commercial Spectra.* USA

Samarai, M.A. 1976: The disintegration of concrete containing sulfate-contaminated aggregates. *Magazine of Concrete Research* **28(96)**, 130–42.

Sandberg, A. and Collis, L. 1982: Toil and trouble on concrete bubbles. *Consulting Engineer* November, 32–5.

Sandberg, Messrs. 1987: *The estimation of equivalent water/cement ratio of hardened Portland cement concrete by fluorescence microscopy.* Messrs Sandberg Test Procedure TP/G8/2, Messrs Sandberg, London, UK.

Sarkar, S.L. and Aïtcin, P-C. 1990: The importance of petrological, petrographical and mineralogical characteristics of aggregates in very high strength concrete. In: Erlin, B. and Stark, D. (eds), *Petrography applied to concrete and concrete aggregates*, ASTM STP-1061, 129–44.

Sarkar, S.L., Chandra, S. and Berntsson, L. 1992: Interdependence of microstructure and strength of structural lightweight aggregate concrete. *Cement and Concrete Composites* **14**, 239–48.

Sarkar, S.L., Jolicoeur, C. and Khorami, J. 1987: Microchemical and microstructural investigations of degradation in asbestos-cement sheet. *Cement and Concrete Research* **17(6)**, 864–74.

Saucier, F., Pigeon, M. and Cameron, G. 1991: Air-void stability, Part V: temperature, general analysis, and performance index. *ACI Materials Journal* **88(1)**, 25–36 (Title No. 88-M4).

Saylor, C.P. 1966: Accurate microscopical determination of optical properties on one small crystal (eds R. Baker and V.E. Cosslett) *Advances in Optical and Electron Microscopy*, 42–76. Acdemic Press, London.

Schiek, R.C. 1982: Pigments (Inorganic). *Kirk-Othmer Encyclopedia of Chemical Technology 3rd edn* **17**, 788–838. John Wiley, New York.

Schmidt, M., Harr, K. and Boeing, R. 1993: Blended cement according to ENV 197 and experiences in Germany. *Cement, Concrete and Aggregates. CCAGDP* **15(2)**, 156–64.

Schimmel, G. 1970: Basic calcium carbonates. *Naturwissenschaften* **57**, 38.

Schrader, E. and Kaden, R. 1987: Durability of Shotcrete (ed. J.M. Scanlon) *Concrete Durability, Katherine and Bryant Mather International Conference, ACI SP-100* **2**, 1071–101.

Schuman, L. and Hornibrook, F.B. 1943: Discussion to paper by T.E. Stanton: Studies to develop an accelerated test procedure for the detection of adversely reactive cement-aggregate combinations. *Proceedings of the American Society of Testing Materials* **43**, 875–904.

Scott, P.W., Critchley, S.R. and Wilkinson, F.C.F. 1986: The chemistry and mineralogy of some granulated and pelletised blastfurnace slags. *Mineral Magazine* **50**, 141–7.

Scrivener, K.L. and Taylor, H.F.W. 1990: Microstructural development in pastes of a calcium aluminate cement. In: Mangabhai, R.J. (ed.), *Calcium Aluminate Cements*, Proceedings of the International Symposium (dedicated to the late Dr H.G. Midgley), Queen Mary and Westfield College, University of London, 41–51. E. & F.N. Spon (Chapman & Hall), London, UK.

Scrivener, K.L. and Taylor, H.F.W. 1993: Delayed ettringite formation: a microstructural and microanalytical study. *Advances in Cement Research* **5(20)**, 139–46.

Searle, A.B. 1935: *Limestone and its Products.* Ernest Benn, London.

Sellevold, E.J., Bager, D.H., Klitgaard Jensen, K. and Knudsen, T. 1982: *Silica fume – cement pastes: hydration and pore structure.* Report BML 82.610, 19–50. The Norwegian Institute of Technology, Trondheim, Norway.

Sellevold, E.J. and Nilsen, T. 1987: Condensed silica fume in concrete: a world review. In: Malhotra, V.M. (ed.), *Suplementary Cementing Materials for Concrete*, Chapter 3, 167–243. The Canada Centre for Mineral and Energy Technology (CANMET), Canadian Government Publishing Centre, Ottawa, Canada.

Senior, S.A. and Franklin, J.A. 1987: *Thermal characteristics of rock aggregate materials.* Report RR 241, Ontario Ministry of Transportation and Communications, Downsview (Toronto), Ontario, Canada.

Shand, E.B. 1958: *Glass engineering handbook.* McGraw-Hill Book Co., New York.

Sharman, W.R. and Vautier, B.P. 1986: Accelerated durability testing of autoclaved wood fibre reinforced cement sheet composites. *Durability of Building Materials* **3**, 255–75.

Shaw, J.D.N. 1985: Cementitious grouts – an update. *Civil Engineering* (October) 50–9.

Shayan, A. 1988: Deterioration of a concrete surface due to the oxidation of pyrite contained in pyritic aggregates. *Cement and Concrete Research* **18(5)**, 723–30.

Shayan, A. and Quick, G.W. 1991/92: Relative importance of deleterious reactions in concrete: formation of AAR products and secondary ettringite. *Advances in Cement Research* **4(16)**, 149–57.

Shayan, A. and Quick, G.W. 1992: Microscopic features of cracked and uncracked railway sleepers. *ACI Materials Journal* **89(4)**, 348–61.

Shayan, A., Quick, G.W. and Lancucki, C.J. 1993: Morphological, mineralogical and chemical features of steam-cured concrete containing densified silica fume and various alkali levels *Advances in Cement Research* **5(20)**, 151–62.

Sheppared, W.L. 1982: *Chemically Resistant Masonry.* Marcel Dekker, New York.

Shi, D. 1988: An automatic quantitative image analysis system for cement and concrete research. *Computer Applications in Concrete Technology, ACI AP* 106, 139–57. American Concrete Institute.

Short, A. and Kinniburgh, W. 1978: *Lightweight Concrete*, 3rd edn. Applied Science Publishers, London.

Sibilia, J.P.A. 1987: *A Guide to Materials Characterization and Chemical Analysis.* VCH. Publishers.

Sims, I. 1975: Analysis of concrete and mortar samples from Sparsholt Roman Villa, near Winchester, Hampshire. Report prepared for the Department of the Environment (not published).

Sims, I. 1977: High-alumina cement concrete: general discussion and testing procedures and case study examples. In PhD thesis, Investigations into some chemical instabilities of concrete, Chapter 7, 244–63, Queen Mary College, London University, UK (unpublished).

Sims, I. 1992: Alkali-silica reaction – UK experience in the alkali-silica reaction in concrete (ed. R.N. Swamy) Chap. 5, 122–87. Blackie & Son Ltd., Glasgow.

Sims, I. 1996: Phantom, opportunistic, historical and real AAR. Getting diagnosis right. *Proceedings of the 10th International Conference on Alkali-Reaction in Concrete*, 175–82. Melbourne, Australia.

Sims, I. and Brown, B.V. 1997: Concrete aggregates. In: Hewlett, P.C. (ed.), *Lea's The Chemistry of Cement and Concrete*, 4th edn. Edward Arnold (Publishers) Limited, London, UK.

Sims, I., Hunt, B.J. and Miglio, B.F. 1992: Quantifying microscopical examinations of concrete for alkali aggregate reactions (AAR) and other durability aspects. In: Holm, J. (ed.), *Durability of Concrete*, ACI SP-131, 267–87.

Sims, I. and Poole, A.B. 1980: Potentially alkali-reactive aggregates from the Middle East. *Concrete* **14(5)**, 27–30.

Skalny, J. 1993: Simultaneous presence of alkali-silica gel and ettringite in concrete. *Advances in Cement Research* **5(17)**, 23–9.

Skalny, J., Clark, B.A. and Lee, R.J. 1992: Alkali-silica reaction revisited. *Proceedings of the 14th International Conference on Cement Microscopy*, 309–24.

Smillie, S., Moulin, E., Macphee, D.E. and Glasser, F.P. 1993: Freshness of cement: conditions for syngenite $CaK_2(SO_4)_2.H_2O$ formation. *Advances in Cement Research* **5(19)**, 93096.

Smith, A.S., Dunham, A.C. and West, G. 1992: Undulatory extinction of quartz in British hard rocks. *Proceedings of the 9th International Conference on Alkali-Aggregate Reaction in Concrete* 1001–8. London.

Smith, M.R. and Collis, L. (eds) 1993: Aggregates: sand, gravel and crushed rock aggregates for construction purposes. *Geological Society Engineering Geology Special Publication No. 9*. 2nd Edition, The Geological Society, London, UK.

Smolczyk, H.G. 1980: Slag structure and identification of slags. *7th International Congress on the Chemistry of Cement, Paris*, Volume I, Principal Reports, Sub-Theme III, 1/3 to 1/17.

Society of Dyers and Colourists. Revised annually: *Colour Index*. Bradford, Society of Dyers and Colourists and the American Association of Textile Chemists and Colourists.

Sommer, H. 1979: The precision of the microscopical determination of the air-void system in hardened concrete. *Cement, Concrete, and Aggregates, CCAGDP* **1(2)**, 49–55.

Sorrentino, D.M., Clement, J.Y. and Goldberg, J.M. 1992: A new approach to characterise the chemical reactivity of the aggregates. *Proceedings of the 9th International Conference on Alkali-Aggregate Reaction in Concrete* 1009–16. London.

Sourie, A. and Glasser, F.P. 1991: Studies of the mineralogy of high alumina cement clinkers. *Transactions and Journal of the British Ceramic Society* **90(3)**, 71–6.

Spence, F. 1988: Coloured concrete for architectural use. *Concrete Forum* **1**, 15–18.

Spence, R.D. (ed.) 1993: *Chemistry and Microstructure of Solidified Waste Forms*. Oak Ridge National Laboratory, Lewis Publishers.

Spratt, B.H. 1980: *An introduction to lightweight concrete*. 6th edn, Cement and Concrete Association, UK.

Stanton, T.E., Porter, O.J., Meder, L.C. and Nicol, A. 1942: California experience with the expansion of concrete through reaction between cement and aggregate. *Journal American Concrete Institute* **13(3)**, 209–36 and discussion in Supplement to November 1942.

Stark, D. 1982: Longtime study of concrete durability in sulfate soils. *George Verbeck Symposium on Sulfate Resistant Concrete*, ACI-SP77, 21–40.

Stark, D. 1985: Alkali-silica reactivity in five dams in the southwestern United States. *US Department of the Interior, Bureau of Reclamation*, REC-ERC-85-10.

Stark, D. 1987: Deterioration due to sulphate reactions in Portland cement-stabilized slag aggregate concrete. *Durability of Concrete*, ACI SP-100: 2019–2102.

Stark, D., Bollman, K. and Seyfarth, K. 1992: Investigation into delayed ettringite formation in concrete. *9th International Conference on the Chemistry of Cement, New Delhi, V*, 248–354.

Stimson, C.C. (Chairman) 1997: *The 'Mundic' Problem – a Guidance Note – Recommended sampling, examination and classification procedure for suspect concrete building materials in Cornwall and parts of Devon*. 2nd Edition, The Royal Institution of Chartered Surveyors, London, UK.

St John, D.A. 1975: Some physical properties of fibrous plaster sheet manufactured in New Zealand. *New Zealand Journal of Science* **18**, 417–31.

St John, D.A. 1982: An unusual case of ground water sulfate attack on concrete. *Cement and Concrete Research*, **12(5)**, 633–9.

St John, D.A. 1983: The petrographic examination of an unusual texture in some failed concrete piles. *Proceedings of the 5th International Conference Cement Microscopy*, 204–15. USA.

St John, D.A. 1985: The application of microscopy in the determination of alkali-aggregate reaction. *Proceedings of the 7th International Conference on Cement Microscopy*, 395–406.

St John, D.A. 1990: The use of large-area thin sectioning in the petrographic examination of concrete. In: Erlin, B. and Stark, D. (eds), *Petrography applied to concrete and concrete aggregates*, ASTM STP-1061, 55–70. American Society for Testing and Materials, Philadelphia, USA.

St John, D.A. 1991: Petrographic examination of corroded stormwater pipe samples from the business district, Rotorua (BM290). *New Zealand DSIR Chemistry Division Internal Report*.

St John, D.A. 1992: AAR investigations in New Zealand – work carried out since 1989. In: *Proceedings of the 9th International Conference on Alkali-Aggregate Reaction in Concrete*, **2**, 885–93. The Concrete Society, Slough, UK.

St John, D.A. 1994a: The use of fluorescent dyes and stains in the petrographic examination of concrete. *Industrial Research Limited Report No. 243*, The New Zealand Institute for Industrial Research and Development, Lower Hutt, New Zealand.

St John, D.A. 1994b: The dispersion of silica fume. *Industrial Research Limited Report No. 244*, The New Zealand Institute for Industrial Research and Development, Lower Hutt, New Zealand.

St John, D.A. and Abbott, J.H. 1983: Semi-automatic production of concrete thin-sections. *Industrial Diamond Review* **43(494)**, 13–16.

St John, D.A. and Abbott, J.H. 1987: The microscopic textures of geothermal cement grouts exposed to steam temperatures of 150°C and 260°C. *Proceedings of the 9th International Conference on Cement Microscopy*, 280–300, USA.

St John, D.A. and Goguel, R.L. 1992: Pore solution/aggregate enhancement of alkalies in hardened concrete. *Proceedings of the 9th International Conference on Alkali-Aggregate Reaction* 894–901. The Concrete Society, London.

St John, D.A., McLeod, L.C. and Milestone, N.B. 1993: An investigation of the mixing and properties of DSP mortars made from New Zealand cements and aggregates. *Industrial Research Limited Report No. 41*, The New Zealand Institute for Industrial Research and Development, Lower Hutt, New Zealand.

St John, D.A., McLeod, L.C. and Milestone, N.B. 1994: Durability of DSP mortars exposed to conditions of wetting and drying. *Proceedings of the International Workshop on High Performance Concrete*, Bangkok.

St John, D.A., McLeod, L.C. and Milestone, N.B. 1996: The durability of DSP mortars exposed to conditions of wetting and drying. *International Workshop, High Performance Concrete*, Bangkok, 1994. ACI SP-159, 45–56. American Concrete Institute, Detroit, USA.

St. John, D.A. and Markham, K.R. 1976: Paint-disfiguration by migration of antioxidant from wall-board adhesive. *Journal of the Oil Colour Chemist's Association*, **59**, 331–3.

St John, D.A. and Penhale, H.R. 1980: Estimation of ground water corrosion of concrete sewer pipes in the Hutt Valley, New Zealand. *Durability of Building Materials and Components*, ASTM STP-691, 377–87.

St John, D.A. and Smith, L.M. 1976: (ed. A.B. Poole) Expansion of concrete containing

New Zealand argillite aggregate. *The Effect of Alkalis on the Properties of Concrete* 319–52. Cement and Concrete Association, London.

St John, D.A. and Smith, LM. 1976: Expansion of concrete containing New Zealand argillite aggregate. *The Effect of Alkalis on the Properties of Concrete. Proceedings of Symposium, London,* 319–54.

Stewart, J. 1982: Chemical techniques of historic mortar analysis. *Bulletin-Association for Preservation Technology* **14(1)**, 11–16.

Strunge, H. 1988: Carbonation measured by thin section microscopy and by phenolph-thalein. *Carbonation of concrete, Nordic Miniseminar, SINTEF, Proc.,* 15–25.

Strunge, H. and Chatterji, S. 1989: Microscopic observations on the mechanism of concrete neutralization. (eds Berntsson *et al.*) *Durability of Concrete: aspects of admixtures and industrial by-products. 2nd International Seminar,* 229–35. Swedish Council for Building Research, Stockholm.

Stumm, W. and Morgan, J.J. 1970: *Aquatic Chemistry.* Wiley Interscience, New York.

Stutterheim, N. 1954: Excessive shrinkage of aggregates as a cause of deterioration of concrete structures in South Africa. *Trans. S. African Instn. Civ. Engre.* **4(12)**, 351–67.

Stutterheim, N., Laurie, J.A.P. and Shand, N. 1967: Deterioration of a multiple arch dam as a result of excessive shrinkage of aggregate. *Proceedings of the 9th International Conference on Large Dams,* 227–4. Istanbul.

Sulcek, Z. and Povondra, P. 1989: *Methods of decomposition in inorganic analysis* 325. Roca Baton, Florida.

Summer, M.S., Hepher, N.M. and Moir, G.K. 1989: The influence of a narrow particle size distribution on cement paste and concrete water demand. *Ciments, Betons, Platres, Chaux,* **(778)**, 164–8.

Sutherland, A. 1974: *Air-entrained concrete.* Cement and Concrete Association, Ref. 45.022.

Sveinsdottir, E.L. and Gudmundsson, G. 1993: Condition of hardened Icelandic concrete. A microscopic investigation. (eds J.E. Linqvist and B. Nitz) *Proceedings of the 4th Euroseminar on Microscopy Applied to Building Materials.* Swedish National Testing and Research Institute, Building Technology SP Report 1993: 15.

Swamy, R.N. 1988: *Natural fibre reinforced cement and concrete* 288. Blackie, Glasgow.

Swamy, R.N. (ed.) 1992: *The alkali-silica reaction in concrete,* 333. Blackie, Glasgow.

Swamy, R.N. 1992: Role and effectiveness of mineral admixtures in relation to alkali silica reaction. *Durability of Concrete,* ACI SP-131: 219–54.

Swamy, R.N. 1993: Fly ash and slag: standards and specifications – help or hindrance. *Materials and Structures* **26(164)**, 600–13.

Swamy, R.N. and Jiang, E.D. (1992): Pore structure and carbonation of lightweight concrete after 10 years exposure. *Structural Lightweight Aggregate Concrete Performance, ACI SP-136,* 377–95.

Sylla, H.M. 1988: Reaktionen im Zementstein durch Warmebehandlung. *Beton,* **V(38)**, 449–54.

Tabor, L.J. 1992: Repair materials and techniques (ed. G. Mays) *Durability of Concrete Structures.* E. & F.N. Spon, London.

Takashima, S. 1958: Systematic dissolution of calcium silicate in commercial Portland cement by organic acid solution. *Review 12th General Meeting, Tokyo, Japan Cement Engineering Association,* 12–13.

Takashima, S. 1972: Studies on belite in Portland cement. *Review 26th General Meeting, Tokyo, Technical Session, Tokyo, Japan Cement Engineering Association,* 27–8.

Talling, B. and Brandstetr, J. 1989: Present state and future of alkali-activated slag concretes. *Proceedings of the 3rd International Conference on Fly Ash, Silica Fume, Slag and Natural Pozzolans in Concrete,* ACI SP-114, **2,** 1519–45. Norway.

Talvite, N.A. 1964: Determination of free silica: Gravimetric and spectrophotometric procedures applicable to air-borne and settled dust. *Industrial Hygiene Journal*, March–April, 169–78.

Tang Mingshu. 1992: Classification of alkali-aggregate reaction. *Proceedings of the 9th International Conference on Alkali-Aggregate Reaction in Concrete*, 6438–53. London.

Tang Mingshu, Deng Min, Lan Xianghui and Han Sufen. 1994: Studies on alkali-carbonate reaction. *ACI Materials Journal*, **91(1)**, 26–9.

Tang Mingshu, Deng Min, Xu Zhongzi, Lan Xianghui and Han Sufen. 1996: Alkali-aggregate reactions in China. *Proceedings of the 10th International Conference on Alkali-Aggregate Reactions in Concrete*, 195–201. Melbourne, Australia.

Tang, Mingshu, Lan Xiangshu and Han Sufen. 1994: Autoclave methods for the identification of alkali-reaction carbonate rocks. *Cement and Concrete Composites* **16**, 163–7.

Taylor, H.F.W. 1957: The dehydration of tobermorite. *Proceedings of the 6th National Conference on Clays and Clay Minerals*, 101–9. Pergamon Press, California.

Taylor, H.F.W. 1964: The calcium silicate hydrates. In: Taylor, H.F.W. (ed.), *The chemistry of cements*, Vol. 1, Ch. 5, 167–232. Academic Press, London, UK.

Taylor, H.F.W. 1984: Studies on the chemistry and microstructure of cement pastes. *British Ceramic Proceedings* **(35)**, September, 65–82.

Taylor, H.W.F. 1990: *Cement Chemistry*. Academic Press, London.

Taylor, H.F.W., Mohan, K. and Mori, G.K. 1985: Analytical study of pure and extended Portland cement paste: fly ash and slag-cement pastes. *Journal American Ceramic Society* **68(12)**, 685–90.

Taylor, W.H. 1974: The production, properties and uses of foamed concrete. *Precast Concrete* February, 83–96.

Tenoutasse, N. and Moulin, E. 1990: Influence of water-cement ratio on concrete microstructure. *Proceedings of the 12th International Conference on Cement Microscopy*, 323–38.

Tepponen, D. and Ericksson, B.E. 1987: Damage in concrete railway sleepers in Finland. *Nordic Concrete Research*, **(6)**, 199–209.

Terry, R.D. and Chillingar, G.V. 1955: Summary of 'concerning some additional aids in studying sedimentary formations' by M.S. Shvetsov. *Journal of Sedimentary Petrology* **25(3)**, 229–34.

Terzaghi, R.D. 1949: Concrete deterioration due to carbonic acid. *Journal of the Boston Society of Civil Engineers* **36**, 136–60.

Textile Institute. 1985: *Identification of Textile Materials*. 7th edn. The Textile Institute, Manchester, UK.

Teychenné, D.C., Franklin, R.E. and Erntroy, H.C. 1975: Design of Normal Concrete Mixes, 1–31, HMSO.

Thaulow, N. 1984: *Thin section investigation of old slag cement concrete*. Report 60-251-4420. Danish Technical Institute, Building Technology, Tåstrup, Denmark.

Thaulow, N., Damgaard-Jensen, A., Chatterji, S., Christensen, P. and Gudmundsson, H. 1982: Estimation of the compressive strength of concrete samples by means of fluorescence microscopy. *Nordisk Betong*, 2–4.

Thaulow, N. and Geiker, M.R. 1992: Determination of the residual reactivity of alkali silica reaction in concrete. *Proceedings of the 9th International Conference on Alkali-Aggregate Reaction in Concrete*, 1050-57. London.

Thistlethwayte, D.K.B. 1972: *The control of sulphides in sewerage systems*. Butterworths, London.

Thomas, M.D.A. and Matthews, J.D. 1991: *Durability studies of pfa concrete structures*. BRE Information Paper IP 11/91. Building Research Establishment, Watford, UK.

Thomas, M.D.A. and Matthews, J.D. 1993: Performance of fly ash concrete in UK structures. *ACI Materials Journal* **90(6)**, 586-93, Title No. 90-M59.

Thompson, M.L., Grattan-Bellew, P.E. and White, J.C. 1994: Application of microscopic and XRD techniques to investigate alkali-silica reactivity potential of rocks and minerals. *Proceedings of the 16th International Conference on Cement Microscopy*, 174-92.

Thornton, H.T. 1978: Acid attack on concrete caused by sulfur bacteria action. *ACI Journal* 577-84.

Tilley, C.E. 1992: Density, refractivity, and composition relations of some natural glasses. *The Mineralogical Magazine* **19(96)**, 275-94.

Toennies, H.T. 1960: Artificial carbonation on concrete masonry units. *Journal American Concrete Institure* **31**, 737-55.

Tong Liang and Tang Mingshu, 1996: Concrete of alkali-silica and alkali-dolomite reaction. *Proceedings of 10th International Conference on Alkali-Aggregate Reaction in Concrete*, 742-9. Melbourne, Australia.

Torii, K., Kawamura, M., Matsumoto, K. and Ishii, K. 1996: Influence of cathodic protection on cracking and expansion of the beams due to alkali-silica reaction, *Proceedings of the 10th International Conference on Alkali-Aggregate Reaction in Concrete*, 653-60. Melbourne, Australia.

Törnebohm, A.E. 1897: The petrography of Portland cement. *Tornindustrie Zeitung* **21**, 1148-50, 1157-59.

Törnebohm, A.E. 1897: *Tornindustrie-Zeitung* **21**, 1148 (also 1910-11) *Baumaterielenkunde* **6**, 142.

Tremper, B. 1932: The effect of acid waters on concrete: *Proceedings American Concrete Institute* **28(1)**, 1-32.

Tröger, W.E. 1971: *Optische bestimmung der gesteinsbilderenden minerale. Pt I Identification tables.* Newly revised edition by Bambauer, H.F., Taborsky, F. and Trochim, H.D.E. Schweizerbart'sche Verlagsbuchhandlung, Stuttgart.

Turriziani, R. 1964: The calcium aluminate hydrates and related compounds. In: Taylor, H.F.W. (ed.), *The chemistry of cements*, Vol. 1, Chapter, 6, 233-86. Academic Press, London, UK.

Urabe, K. 1990: Phase changes of calcium silicate hydrate on heating. *CAJ Proceedings of Cement and Concrete, Cement Association of Japan* **(44)**, 36-41. (In Japanese with English abstract.)

US Bureau of Reclamation. 1955: p. 120. (referred to in Mielenz 1962).

US Department of the Interior. 1981: *Concrete Manual.* Water Resources Technical Publication, 8th edn revised. Denver, US Department of the Interior.

Usher, P.M. (Convenor). 1980: *Guide to chemical admixtures for concrete - report of a joint working party.* Concrete Society Technical Report No. 18, The Concrete Society, London, UK.

Vail, G.V. 1952: *Soluble Silicates. Their Properties and Uses. Volume 1: Chemistry.* Reinhold Publishing Corp. New York.

Valenta, O. and Modry, S. 1969: A study of the determination of surface layers of concrete. *International Symposium on the Durability on Concrete, Prague*, Final Report, **1**, A55-A64.

Van Aardt, J.H.P. and Visser, S. 1975: Thaumasite formation: A cause of deterioration of Portland cement and related substances in the presence of sulfates. *Cement and Concrete Research* **5(3)**, 225-32.

Van der Plas, L. and Tobi, A. 1965: A chart for judging the reliability of point counting results. *American Journal of Science* **263(1)**, 87-90.

Varma, S.P. and Bensted, J. 1973: Studies of thaumasite. *Silicates Industriels* **38**, 29-32.

Vaysburd, A.M. 1992: Durability of lightweight concrete and its connections with the composition of concrete, design and construction methods (eds T.A. Holme and A.M. Vaysburd). *Structural Lightweight Aggregate Concrete Performance*, ACI SP-136, 295-317.

Venkateswaran, D., Narayan, S., Chakraborty, I.N., Biswas, S.K. 1991: Microstructure of high purity alumina cement clinkers and its effect on cement properties. *Proceedings of the 13th International Conference on Cement Microscopy*, 71-85.

Vennesland, O. 1995: Electrochemical chloride removal and realkalisation - maintenance methods for concrete structures. *Improving Civil Engineering Structures Old and New* (ed. W.J. French), 113-24. Geotechnical Publishing.

Verbeck, G.J. 1958: Carbonation of hydrated Portland cement. *Cement and Concrete*, ASTM STP 205: 17-36.

Vicat, L.J. 1837: *Treatise on Calcareous Mortars and Cements*. Translated with additions by J.T. Smith, London.

Vitruvius. 1962: *Vitruvius on Architecture*. 2 vols. Translated by F. Granger, William Heinemann, London.

Vivian, H.E. 1947: The effect of void space on mortar expansion. *Commonwealth of Australia, CSIR Bulletin No. 229*, Part III, 47–54.

Vivian, H.E. 1987: The importance of Portland cement quality on the durability of concrete. In: Scanlon, J.M. (ed.), *Concrete durability: Katherine and Bryant Mather International Conference*, ACI SP-100, 1691–1701. Paper No. SP 100-86. American Concrete Institute, Detroit, USA.

Volzhenskii, A.V. and Ferronskaya, A.V. 1974: *Gypsum Binders and Articles*, Stroiizdat, (Chem. Abst. 126219, **81**, 1974) Moscow.

Wahlstrom, E.E. 1955: *Petrographic mineralogy*. New York, Wiley, 44–68.

Wakely, L.D., Poole, T.S. Weiss, C.A. and Burkes, J.P. 1992: Petrographic techniques applied to cement-solidified hazardous wastes. *Proceedings of the 14th International Conference on Cement Microscopy*, 333–50.

Walker, H.N. 1979: Evaluation and adaptation of the Dubrolubov and Romer method of microscope examination of hardened concrete. *Virginia Highway and Transportation Reserch Council* VHTRC 79-R42. April.

Walker, H.N. 1980: Formula for calculating spacing factor for entrained air voids. *Cement, Concrete, and Aggregates CCAGDP* **2(2)**, 63–6.

Walker, H.N. 1981: Examination of Portland cement concrete by fluorescent light microscopy. *Proceedings of the 3rd International Conference on Cement Microscopy, USA* 257–78.

Walton, P.L. 1993: Effects of alkali-silica reaction on concrete foundations. *BRE Information Paper, IP 16/93*, 1–4.

Webb, J. 1993: The battle of Britain's nuclear dustbin. *New Scientist*, 6 November 14–15.

Weigand, W.A. 1994: Progress toward a standard procedure for point-counting the phases in cement clinker with reflected light microscopy. In: DeHayes, S.M., Stark, D., *Petrography of Cementitious Materials*, ASTM STP 1215. American Society for Testing and Materials, Philadelphia, USA.

Werner, D. and Giertz-Hedstrom, S. 1934: Physical and chemical properties of cement and concrete. *Engineer* **135**, 235–8.

Werner, O.R. (Chairman). 1987: Ground granulated blast-furnace slag as a cementitious constituent in concrete (ACI Committee 226 Report). *ACI Materials Journal* **8(4)**, 327–42, Title No. 84–M34.

Weyers, R.E., Brown, M., Al-Quadi, I.L. and Henry, M. 1993: A rapid method for measuring the acid-soluble chloride content of powdered concrete samples. *Cement, Concrete and Aggregates* **15(1)**, 3–13.

White, A.H. 1909: Free lime in Portland cement. *The Journal of Industrial and Engineering Chemistry* **1(1)**, 5–11.

Wickens, D. 1992: Plasterwork and rendering. *Specification 92*. MBC Architectural Press and Building Publications, London.

Wigum, B.J. 1995: Examination of microstructural features of Norwegian cataclastic rocks and their use for predicting alkali-reactivity in concrete. *Engineering Geology* **40**, 195–214.

Wigum, B.J. 1996: A classification of Norwegian cataclastic rocks for alkali-reactivity. *Proceedings of the 10th International Conference on Alkali-Aggregate Reaction in Concrete*, 758–66. Melbourne, Australia.

Wilk, W. and Dobrolubov, G. 1984: Microscopic quality control of concrete during construction. *Proceedings of the 6th International Conference on Cement Microscopy*, 330–43, USA.

Wilk, W., Dobrolubov, G. and Romer, B. 1984: The development of quantitative and qualitative microscopic control of concrete quality and durability and of a frost-salt resistance test with rapid cycles. *Proceedings of the 6th International Conference on Cement Microscopy*, 309–29, USA.

Williams, A. (ed.) 1994: *Specification 1994* (published in three volumes: Technical; Products; Clauses) Emap Business Publishing (Emap Architecture), London.

Williams, H., Turner, F.J. and Gilbert, C.M. 1954: *Petrography. An Introduction to the Study of Rocks in Thin Section*. W.H. Freehman and Co., San Francisco.

Wilson, R.B. 1973: *Preparation of Microscope Slides of Rocks, Minerals and Other Research Materials*, 1–12. Logitech Ltd., Glasgow.

Wilson, R.B. 1978: The preparation of microscope slides of rocks, minerals and other research materials. *3rd Australian Conference Science Technology*, Canberra University.

Wilson, R.B. and Milburn, G.T. 1978: The preparation of microscope slides of rocks, minerals and other research materials. *3rd Australian Conference on Science Technology*, Australian National University, Canberra.

Wilson, R.D. and St John, D.A. 1979: The avoidance of staining in cast Australian gypsum. *Zement-Kalk-Gips* **68(1)**, 41–3.

Winchell, A.N. 1954: *Optical properties of organic compounds*. Academic Press, New York.

Winchell, A.N. and Winchell, H. 1951: *Elements of Optical Mineralogy, Part II – Descriptions of Minerals*. John Wiley and Sons, New York.

Winchell, A.N. and Winchell, H. 1964: *The Microscopical Characters of Artificial Inorganic Solid Substances: Optical Properties of Artificial Minerals*. Academic Press, New York.

Wirgot, S. and Cauwelaert, F. Van 1991: The measurement of impregnated cement fluorescence by means of image analysis. *Proceedings of the 3rd Euroseminar on Microscopy applied to Building Materials, Barcelona.*

Wirgot, S. and Cauwelaert, F. Van 1994: The influence of cement type and degree of hydration on the measurement of w/c ratio on concrete fluorescent thin sections. In: DeHayes, S.M. and Stark, D. (eds) *Petrography of Cementitious Materials*, ASTM STP 1215, 91–110. American Society for Testing and Materials, Philadelphia, USA.

Wood, J.G.M., Nixon, P.J. and Livesey, P. 1996: Relating ASR structural damage to concrete composition and environment. *Proceedings of the 10th International Conference on Alkali-Aggregate Reaction in Concrete*, 450–547. Melbourne, Australia.

Woolf, D.O. 1952: Reaction of aggregate with low-alkali cement. *Public Roads* **27(3)**, 49–56.

Yilmaz, V.T. and Glasser, F.P. 1991: Crystallization of calcium hydroxide in the presence of sulphonated melamine formaldehyde superplasticizer. *Journal of Material Science Letters* **10**, 712–15.

Yilmaz, V.T. and Glasser, F.P. 1991: Refraction of alkali-resistant glass fibres with cement. Part 1. Review, assessment, and microscopy. Part 2. Durability in cement matrices conditioned with silica fume. *Glass Technology* **32(4)**, 91–8, 138–47.

Yingling, J., Mullings, G.M. and Gaynor, R.D. 1992: Loss of air content in pumped concrete. *Concrete International* **14(10)**, 57–61.

Zappia, G., Sabbioni, C., Pauri, M.G. and Gobbi, G. 1994: Mortar damage due to airborne sulfur compounds in a simulation chamber. *Materials and Structures* **27**, 469–73.

Zhang, M. and Gjorv, O.E. 1990: Microstructure of the interfacial zone between lightweight aggregate and cement paste. *Cement and Concrete Research* **20(4)**, 610–18.

Zhang, M. and Gjorv, O.E. 1992: The penetration of cement paste into lightweight aggregate. *Cement and Concrete Research* **22(1)**, 47–55.

Zivica, V. 1992: Aggressivity of phenol homologues solutions and caused corrosion of cement composites. *9th International Congress on the Chemistry of Cement*, New Delhi, **V**, 396–401.

Zoldners, N.G. 1968: Thermal properties of concrete under substantial elevated temperatures *Temperature and Concrete, ACI SP-25*, 1–31.

Zussman, J. 1979: The mineralogy of asbestos. *Asbestos* (eds L. Michaels and S.S. Chisick), 45–56. John Wiley and Sons, Chichester.

Index

The starting page of main subject entries is shown in **bold**. Referenced author entries are given in *italics*; references with more than one author are only included under the first author. Glossary entries are shown in parentheses. Colour plate entries are prefixed **P**.